BORATE PHOSPHORS

BORATE PHOSPHORS

Processing to Applications

Edited by
S. K. Omanwar, R. P. Sonekar, and N. S. Bajaj

CRC Press
Taylor & Francis Group
Boca Raton London New York

CRC Press is an imprint of the
Taylor & Francis Group, an **informa** business

First edition published 2022
by CRC Press
6000 Broken Sound Parkway NW, Suite 300, Boca Raton, FL 33487-2742

and by CRC Press
2 Park Square, Milton Park, Abingdon, Oxon, OX14 4RN

© 2022 selection and editorial matter, S. K. Omanwar, R. P. Sonekar, N. S. Bajaj; individual chapters, the contributors

CRC Press is an imprint of Taylor & Francis Group, LLC

Library of Congress Cataloguing-in-Publication Data
A catalog record has been requested for this book

ISBN: 978-1-032-07574-7 (hbk)
ISBN: 978-1-032-07575-4 (pbk)
ISBN: 978-1-003-20775-7 (ebk)

DOI: 10.1201/9781003207757

Typeset in Times
by MPS Limited, Dehradun

Contents

Preface

Borates also play important roles in the family of material science due to its simple structural and chemical properties. The borate-based phosphors have attracted much attention, due to their high optical stability, can be synthesized by using low-cost synthesis through conventional and non-conventional methods. The technology based on these materials are also environmentally friendly. Moreover, it allows to select the number of borates compounds with different structures. Hence, depending on their selection, they can be used in a variety of applications based on phosphor technology that can start from your daily utilities and end up at medical and radiological applications. However, the synthesis of borate phosphors with desired structures is at best tricky and hence the selection of proper technique is plays a vital role.

In this book, we have discussed the structural and chemical parameters of borates as phosphors, the suitable synthesis methods and way of proper characterization of materials. Also, we have made a tremendous literature survey on borate materials based on their applications. This book covers the entire electromagnetic spectra utilized for fetching luminescence from the prepared borate material through the globe. On the basis of that, the chapters were distributed and formulated; likewise the book starts with an introduction to borates, their properties and synthesis technique and ends on their modification in properties of borate functional groups when mixed up or substituted with other metallic functional groups.

Editors' Biographies

Dr. S. K. Omanwar recently retired as a Sr. Professor (HAG) and Former Head, Department of Physics, on June 30, 2019, and since July 1, 2019, he has been working as a distinguished UGC-BSR Faculty Fellow at Sant Gadge Baba Amravati University, Amravati – 444602 (M.S.), India.

Dr. Omanwar has more than 32 years of teaching and administrative experience and a distinctive research career in developing inorganic luminescent materials for various applications such as mercury free lamps, SSL, display devices, LED-based photo-therapy devices, spectral matching phosphors for solar PV panels, PL-LCD panels, PDP panels, CFL bulbs and TLD and OSL materials for personnel radiation monitoring, as well as biomaterials. He has developed the commercially viable competitive products such as TLD, OSL and SSL devices. He has also developed a cost-effective method such as simple combustion method with little modification for the easy synthesis of these materials. Besides having sponsored research projects, he has received good citations with SCOPUS (163) and Thomson & Reuters (122). He has delivered several invited talks at many international events and is a life member of 11 national organizations. He has been a mentor for many Ph.D. scholars and PG students from 1986 till date. He has been the recipient of 26 awards and 178 recognitions.

Dr. R. P. Sonekar, Ph.D., Professor, Department of Physics, G. S. College Khamgaon (MH), India

Dr. R. P. Sonekar was awarded a master's degree from the RTM Nagpur Univesity, Nagpur, in 1989 and his Ph.D. (physics) on borates from SGB Amravati University in 2008. Looking at his excellence soon after his master's degree, he was appointed as assistant professor at G.S. Arts, Commerce and Science College, Khamgaon, affiliated with the SGB Amravati University, Amravati. During his career, he has achieved many milestones in the field of research. Ge has completed two minor and one major project on borates. He also obtained a FIP fellowship from UGC to complete his Ph.D. work in borates. Dr. Sonekar has expertise in borate phosphors till date and he has guided 5+ students strictly on synthesis and application of borates. He has excellent publications in indexed journals.

Dr. N. S. Bajaj, Ph.D., UGC- Fellow, Assistant Professor, Department of Physics, Toshniwal Arts, Commerce and Science College, Sengaon, Dist: Hingoli (MH), India.

Dr. N. S. Bajaj obtained his master's degree from the Vidhyabharti College, Amravati, affiliated with the SGB Amravati University, Amravati, in 2008 and his Ph.D. (physics) from the same university in 2014. He served from 2010–2015 as a UGC-Project Fellow at SGB Nagpur University, in a project sanctioned under the supervision of Dr. S. K. Omanwar. He also served various colleges under the same university as a visiting lecturer. In 2015, he became an assistant professor at the Toshniwal Arts, Commerce and Science College, Sengaon affiliated to SRTM University, Nanded. During his short career, he has achieved many milestones in the field of research and received a couple of international awards and recognitions. Recently, he was awarded the most precious award: Marathwad Bhushan Award. Dr. Bajaj has expertise in the synthesis of phosphors and phosphor application radiation dosimeter. He has excellent publications in indexed journals.

Contributors

A. B. Chauhan
Department of Physics
Vidyabharti Mahavidyalay
Maharashtra, India

Devayani Chikte (Awade)
Department of Physics
G. N. Khalsa College
Maharashtra, India

A. B. Gawande
Department of Physics
MVPS G.M.D. Arts
B.W. Commerce & Science College
Maharashtra, India

K. A. Koparkar
M. S. P. Art & Science and K. P. T. Commerce
    College Manora
Maharashtra, India

Rupali Korpe
Department of Physics
Shri Shivaji Science College
Maharashtra, India

Yogesh K. More
Department of Physics
Bhiwapur Mahavidyalaya
Maharashtra, India

P. A. Nagpure
Department of Physics
Shri Shivaji Science College,
Maharashtra, India

C. B. Palan
Department of Physics
B S Patel ACS College Pimpalgaon
Maharashtra, India

Vaishali Raikwar
Ramniranjan Jhunjhunwala College
Maharashtra, India

N. S. Sawala
Government Polytechnic
Maharashtra, India

Pritee K. Tawalare
Department of Physics
Jagadamba Mahavidyalaya
Maharashtra, India

1 Introduction to Borate Phosphors

P. K. Tawalare and A. B. Gawande

CONTENTS

1.1 INTRODUCTION

The varied aspects of luminescence and the complex processes involved in the origin of light emission, offer interesting challenges for researchers in this field. This is one of the research fields, wherein diverse application area exists, which range from radiation monitoring for health and safety, phosphors for lamps and display purposes to X-ray imaging and other means of medical diagnostics.

Luminescence is a well-established field of scientific research. In 1652, Zechi made an important contribution to the understanding of photoluminescence. It is the emission of light, which persists after the excitation agency, is removed (luminescence). Moreover; he proved experimentally that the color of the phosphorescence light in a material is independent of the color of the exciting light and also clearly distinguished the phenomenon from scattering. About 200 years later, Stoke showed that the incident and emitted light differed in color and enunciated his well-known Stoke's law regarding the increase in wavelength, which accompanies photoluminescence. In 1867, E. Bequerel distinguished two types of phosphorescence or after-glow, which were attributed respectively to monomolecular and to bimolecular decay mechanism.

The last few decades have witnessed dramatic changes in research on luminescence. There has been a phenomenal growth in the subject, and a significant progress has been made in the field of luminescence research. Recent research is characterized by strong interaction among other branches of solid-state physics and between different areas of luminescence using inorganic and organic materials. Both experimental as well as theoretical approaches have been made.

DOI: 10.1201/9781003207757-1

Luminescent materials are called phosphors. The first systematic study of luminescent crystals was made by Lenard [1] and his school, at the beginning of the 20th century. The phosphors they studied are called 'Lenard Phosphors'. The practical interest in luminescent materials for use in efficient cathode ray screens, and eventually for luminescent lamps, which were then developed in 1930s, stimulated the study of crystal luminescence in a very substantial way.

1.2 PHOSPHORS

Phosphors are solid, luminescent materials that emit photons when excited by an external energy source, such as an electron beam (cathodoluminescence) or ultraviolet light (photoluminescence). Phosphors are composed of an inert host lattice, which is transparent to the excitation radiation and an activator, typically a 3d or 4f electron metal, which is excited under energy bombardment. The process of luminescence occurs by adsorption of energy at the activator site, relaxation, and subsequent emission of a photon and a return to the ground state. The efficiency of a phosphor depends on the amount of relaxation that occurs during the activation and emission. Relaxation is the process in which energy is lost to the lattice as heat; it needs to be minimized in order to extract the highest luminous efficiency. The luminous efficiency is defined as the ratio of the energy emitted to the energy absorbed.

1.3 APPLICATIONS OF PHOSPHORS

The substantial advances in understanding luminescent phenomena and the discoveries of unusual luminescent processes, for example, up-conversion and quantum splitting, present unusual opportunities for the applications of luminescence. In some instances, these possible applications depend on improvements in efficiencies and stabilities of inorganic luminescent materials; in other instances, on the problems of adapting the available scientific understanding of luminescent phenomena to established techniques. Luminescent materials find applications ranging from as commonplace as lighting to very sophisticated such as lasers. Some of the applications of luminescence are given in Table 1.1.

1.3.1 LUMINESCENCE PHENOMENON

1.3.1.1 Principle

Luminescence is defined as the emission of light by bodies, which is in excess of that attributable to black body radiation, and persists considerably longer than the periods of electromagnetic radiations in the visible range after the excitation stops.

Important characteristics which distinguish luminescence from other light emitting phenomena are time lag between the excitation and emission, spectral distribution of emission and its temperature dependence. Thus, for black body radiation, the emission maximum shifts to shorter wavelengths with increasing temperature, while the reverse is true for photoluminescence (PL). Again, intensity of emission increases with temperature for black body radiation, while the photoluminescence intensity decreases with temperature due to thermal quenching.

The luminescent system generally consists of a host lattice and a luminescent centre, often called an 'activator'. In general, the host needs to be transparent to the radiation source used for the excitation process. The activator absorbs the exciting radiation and is raised to an excited state. The excited state returns to the ground state by emission of radiation or by non-radiative decay. It is necessary to suppress this non-radiative process. In some materials, the excitation radiation is not absorbed by the activator but the other ion may absorb the exciting radiation and subsequently transfer it to the activator. In this case, the absorbing ion is called a 'sensitizer'. In many cases, the host lattice transfers its excitation energy to the activator so that the host lattice acts as the sensitizer. High-energy excitation always excites the host lattice. Direct excitation of an activator is only possible with ultraviolet and visible radiation [2].

TABLE 1.1

Classification of luminescence with excitation source and applications

Luminescence type	Excitation source	Applications
Photoluminescence	Photons	Fluorescent lamps, PL-LCD, plasma display, LASERs, LSCs, paints, inks, upconversion material
Cathodoluminescence	Electrons	TV set, FED, oscilloscope, monitors, storage tubes, flying spot scanners, radars
Electroluminescence	Electric field	LEDs, EL displays, diode lasers
Radioluminescence	Ionizing radiations such as X-rays or gamma rays	X-ray imaging, scintilators, dosimetry
Optically stimulated luminescence	Visible Photons	X-ray radiography, dosimetry
Lyoluminescence	Chemical reaction	Detectors, analytical devices, lyoluminescence dosimetry
Chemiluminescence	Chemical reaction	Analytical chemistry
Bioluminescence	Biochemical reaction	Analytical chemistry
Thermoluminescence	Ionizing radiations	Radiation dosimetry, archeological and geological dating, forensic science
Triboluminescence	Mechanical energy	Mechanical engineering, energy, biological monitoring, and sensors as well as lighting, imaging, and displaying
Sonoluminescence	Ultrasound	Estimating the extreme temperatures generated in the bubbles during implosion

1.3.1.2 Excitation

In general, luminescence may be excited by a number of agents such as light, cathode rays or positive ion bombardment or X-rays, by contact with flame, or by friction. The region of wavelengths for which a given material can be excited by optical means with high efficiency usually consists of one or more broadbands, which are characteristic of both the host material and the activator. The position of the bands evidently is of much importance for practical purposes. For example, it is important that the given material possess a prominent excitation band at 253.7 nm, if it is to be excited by radiation from a low-pressure mercury discharge. It should be emphasized that not all crystals possessing a high efficiency for optical excitation, also posses a high efficiency for excitation by cathode rays. It has generally been observed that the efficiency of excitation of luminescence decreases, reversibly, as the temperature is raised sufficiently.

1.3.1.3 Emission

The emission spectra of luminescent material usually consists of one or more bands whose position is related to the activator. In general, peaks become narrower and narrower in the crystalline materials as the temperature of the specimen is lowered, and they approximate sharp lines near the absolute zero of the temperature. It is clear that if the electrons, which absorb energy and radiate light, are in the well-shielded inner shells of the atom (for example, rare earth ions), excitation may have relatively little effect upon the chemical binding, in which case the degradation of energy will be smaller than it would be if the electrons that are to be excited are in the outermost shell.

1.3.1.4 Decay Characteristics

An examination of the decay properties of the luminescent materials indicate that they fall in to two broad classifications [3]:

In the first type, the decay equation is given by:

$$I(t) = I_o \exp(-\alpha t)$$

where I_o is the initial intensity, $I(t)$ is the intensity at time t and α is a constant.

This resembles closely the process governing the progress of monomolecular reaction. This behavior suggests that in these cases the luminescence takes place by simple excitation with subsequent optical emission in the active centre, the excitation energy remaining closely localized in the centre between excitation and emission. The decay constant is independent of temperature and is small.

Most of the luminescent materials, which are valued for their long decay characteristics, obey a decay equation of second type:

$$I(t) = I_0/(\beta t + 1)^n$$

where I_o is the initial intensity, $I(t)$ is the intensity at time t and β and n are constants.

This equation is similar to the rate equation for the bimolecular reaction. The constant β is dependent on temperature. The atoms or clusters of atoms become ionized during the excitation and the luminescent radiation is emitted during recombination of the free electrons and the ionized centres. Johnson [4,5] has suggested that essentially all centres become ionized during excitation and that a majority of free electrons are recaptured into a state, which has a very long lifetime (of the order of milliseconds), because the optical transition to the ground state is forbidden. These electrons contribute an exponential component to the decay curve. The remaining electrons are captured at the trapping centres and are released over a period of time that is long compared to the lifetimes of the excited state of the fluorescing centre. The second class of electrons is responsible for the bimolecular component of the decay curve.

1.3.1.5 Mechanism

The configurational coordinate model describes the electronic transitions of absorption and emission [6]. It shows the potential energy curves of the absorbing centre as a function of configurational coordinates. In optical absorption, the centre is promoted from its ground state to the high vibrational level of the excited state. The centre returns first to the lowest vibrational level of the excited state, giving up excess energy to the surroundings. This is schematically shown in Figure 1.1.

The activator ions possess energy levels that can be populated by direct activation or indirectly by energy transfer, and are responsible for the luminescence. Generally, two types of activator ions can be distinguished. In the first type the energy levels of the activator ions involved in the emission process show only weak interactions with the host lattice (e.g. many of the lanthanide ions Ln^{3+}). The characteristic line emission spectra can be observed in this case. The second types of activator ions strongly interact with the

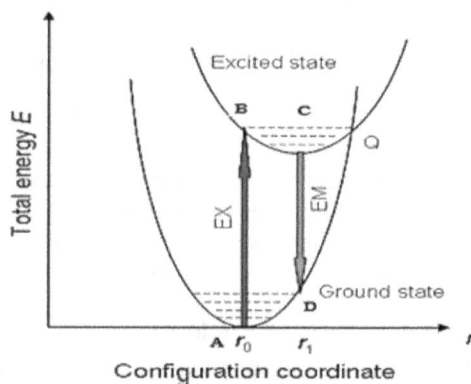

FIGURE 1.1 A schematic illustration of a configurational coordinate model.

host lattice (e.g. Mn^{2+}, Eu^{2+}, Ce^{3+}, Pb^{2+}, etc.). The strong couplings of the electronic states with vibrational modes of the lattice mainly lead to more or less broadbands in the spectrum. The half width (FWHM) is related to the Stokes shift S, which is the energy difference between absorption and emission maximum.

$$FWHM = \sqrt{8} \ln 2 \ \sqrt{2kT} \ \sqrt{S} [eV]$$

$$S = S_e h\omega_e + S_g h\omega_g$$

S_e and S_g are Huang-Rhys parameters for the excited and ground state, respectively.

Phosphors that show an emission with a large Stokes shift usually exhibit a low quenching temperature, which is disadvantageous for many applications.

In general, the luminescent process can be divided into the steps of energy absorption, energy transfer and emission. Energy absorption need not take place at the activator ion itself but can occur at random places in the lattice. This implies that energy transfer of the absorbed energy to the luminescent centre takes place before emission can occur. The migration of energy absorbed by the lattice can take place through one of the following processes:

- Migration of electric charge (electrons, holes),
- Migration of excitons,
- Resonance between atoms with sufficient overlap integrals,
- Re-absorption of photons emitted by another activator ion or sensitizer.

The occurrence of energy transfer within a luminescent material has far-reaching consequences for its properties as a phosphor. On the other hand, the absorbed energy can migrate to the crystal surface or to the lattice defects, where it is lost by radiation-less deactivation. As a consequence, the quantum efficiency of the phosphor declines.

Besides the quantum yield, the quality of a phosphor material is further characterized by its colour points, the lumen equivalent, the reflection spectrum and the emission lifetime under given excitation conditions. The colour point is derived from the spectral energy distribution of the emission and is defined according to the convention of the *Commission Internationale d'Eclairage* (CIE) in a normalized two-dimensional coordinate system shown in Figure 1.2.

For lighting applications, the colour saturation of the phosphors is of less importance. In contrast, the luminescent materials should emit a spectrum with a high lumen equivalent. This value is calculated by multiplication of the spectral power distribution $P(\lambda)$ of the phosphor

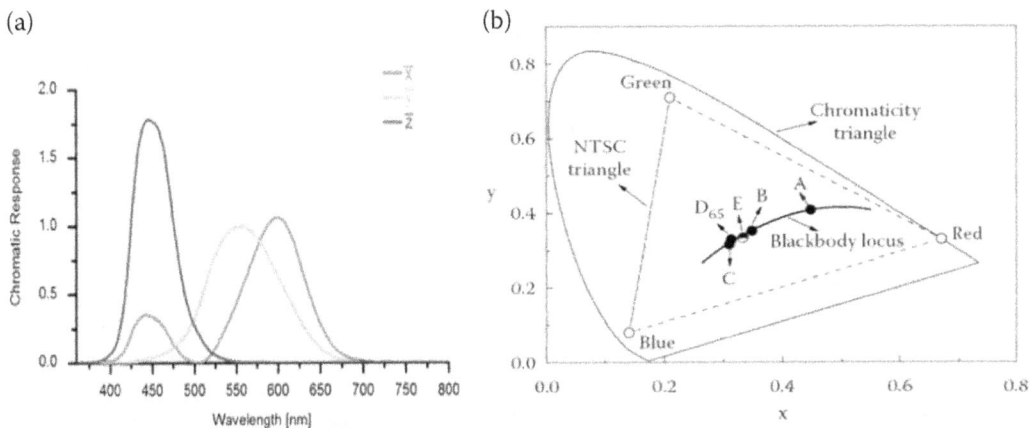

FIGURE 1.2 (a) 1931 CIE XYZ colour-matching. (b) 1931 CIE chromaticity diagram.

emission, which integral has been normalized with the spectral luminous efficiency (LE) for the human eye V(λ):

$$LE = \int_{380}^{780} V(\lambda)\ P(\lambda)\ d\lambda \quad [lm/W]$$

To obtain white light with a high lumen equivalent, it is important that the phosphors applied in a mixture display very sharp emission spectra or even better emission lines, rather than broad emission spectra; otherwise a light is generated in spectral areas where the eye sensitivity is too low. Naturally, also the emission maximum of line emitting phosphors should not be too far off the eye sensitivity curve maximum.

To determine just how efficient a phosphor is, the term 'quantum efficiency' (QE) is defined. It is the ratio between the number of photons emitted and the number of photons absorbed by the phosphor. Generally, the phosphors that have QEs of 80% or greater are considered to be efficient phosphors.

Besides the quantity, there is also a quality performance of light sources with one factor usually being traded off against the other. Light quality is mainly determined by the colour rendering of the light source, which is the ability to display the colours of an irradiated object in a natural way. A qualitative measure is the colour-rendering index (CRI), which by definition can adopt values between 0 and 100. This value is calculated by comparing the reflection spectra of selected test colours obtained by irradiation with the light source under investigation with the reflection spectra when irradiated in the same way with a black body radiator. By definition, a black body radiator has a colour-rendering index of 100. In contrast, a line emitter with a single emission line at any part of the visible spectral region has a CRI of zero, because colours cannot be displayed under such an irradiation source. While a combination of line emitters yields high light output but moderate CRI, broadband emitters enable higher CRI values to be obtained. In addition, colour rendering is dependant on the spectral position of the emission maximum and thus on the spectral power distribution of the phosphor. A compromise has to be found between colour rendering and high light output.

The luminescence efficiency of a phosphor falls at higher temperatures. This is called 'temperature quenching'. The activator concentration in excess may also cause reduction in the intensity of the phosphor, which is termed 'concentration quenching'. There are some impurities whose presence, even in very small amount, will reduce the intensity of the phosphor. Such impurities are called as 'killer impurities'.

Classification of Luminescence

'Luminescence' is the general term that covers both fluorescence and phosphorescence. The concept of fluorescence originated from the mineral fluorite because the phenomenon of light emission under ultraviolet radiation was first recognized in that mineral. Phosphorescence is named after the well-known optical property of the element phosphorous. In Greek, the term 'phosphor' means 'light bearer'. But the chemical element phosphorous is not phosphor. The difference between fluorescence and phosphorescence is to some extent arbitrary. Historically, the temperature independent decay is known as fluorescence while the temperature-dependent decay is phosphorescence. According to modern conventions, fluorescence refers to the emission, which persists for 10^{-3} s to 10^{-8} s while the phosphorescence persists for a considerably longer duration from 10^{-3} s to several seconds. The various types of luminescence are distinguished by a prefix, which denotes the mode of excitation, e.g. photoluminescence is caused by the absorption of photons of visible or near-visible radiation, electroluminescence is caused due to electric field etc. Thermoluminescence is rather a misnomer. Thermoluminescence is the thermally stimulated luminescence excited by other means and the thermal energy is not the source of excitation.

There are a variety of luminescence phenomena observed in nature or in man-made materials. The nomenclature given to these is invariably related to the exciting agent, which produce the luminescence. Following description summarizes the main types of the luminescence emission phenomena.

Photoluminescence

This is the emission produced by excitation with the light photons. The fluorescent lamp used in household and general lighting is the principal example of this. A 254 nm UV radiation from the mercury vapor discharge is absorbed by one of activator impurities in the phosphor coated on the inner side of the glass tube. Some of this energy is transferred by resonance to a second impurity. By adjusting the relative concentrations of these activator impurities, one can produce desired modification in the colour of the light. There are a large variety of organic and inorganic phosphors, which are used in the consumer items like in road and traffic signals, displays, laundry whiteners, etc. in addition to those used in industrial and scientific applications. An example of high technology is called LASER, a kind of photoluminescence in which emission is coherent.

Cathodoluminescence

When electron beams generated at the electrical cathodes do excite, the emission produced is called cathodoluminescent. The screens of cathode ray tubes and television tubes glow by this kind of emission. In cathode ray tubes, zinc and cadmium sulfide phosphors are used.

Radioluminescence

When X-rays or nuclear radiations provide the excitation energy, the resulting luminescence is called radioluminescence.

Scintillation

This phenomenon is the same as radio-luminescence. It is given this name because it is used as a technique to detect individual light pulses generated by the incidence of each X-ray, or gamma ray photon or a nuclear particle. Such light pulses are called scintillations, since like a spark they are very short lived. Thallium-activated sodium iodide is a well-known scintillation detector used for gamma ray spectrometry. The intensity of the scintillation (light pulse) is directly proportional to the energy of the incident gamma ray photon (when it is totally absorbed). The measurement of this pulse intensity, therefore, provided the means for knowing the gamma ray energy.

Electroluminescence

Application of electric fields can produce luminescence in many phosphors. There is another type of electroluminescence, known as injection luminescence. In this, electrons are injected from an external supply across a semiconductor p-n junction. On applying a DC voltage across the junction, such that the electrons flow to the p-region, luminescence is produced by the electron-hole recombination in that region. The light emitting diodes (LEDs), which are now commonly used as display devices in many scientific instruments, are based on this principle.

Chemiluminescence

Some chemical reactions are the source of luminescence. The oxidation of white phosphorous in air is the best-known example of this. All chemical molecules are not capable of luminescence. Lyo-luminescence, which is caused during the dissolution of certain compounds, which have been bombarded by X-rays beforehand, is a kind of chemiluminescence. A well-known example is the case of X-irradiated NaCl, which emits a flash of light when quickly dissolved in water.

Bioluminescence

Biochemical reactions inside the cell of living organisms can produce electronic excited states of the biomolecules. Fire flies, glowworms, some bacteria and fungi and many sea creatures, both near the surface and at great depths, are the striking examples of luminescence in living beings.

Triboluminescence

A large number of inorganic and organic materials subjected to mechanical stress emit light, which is called triboluminescence. It has also been named mechanoluminescence by some authors. The spectra of triboluminescence light are similar to those of photoluminescence in many substances.

Thermoluminescence

Unlike the various types of luminescence phenomena listed above, the prefix 'thermo' here does not mean the form of excitation energy, but to stimulation of luminescence, which was excited in a different way. The primary agent for the induction of TL in a material is the ionizing radiation (X-rays or gamma radiation) or sometimes even UV rays to which the material is exposed. The light produced by subsequent heating of material is called TL.

The most broadly investigated and utilized of all thermally stimulated phenomena is the emission of light during the heating of a solid sample, previously excited. The initial excitation (typically by irradiation) is the source of energy, whereas the heating serves only as a trigger, which helps in releasing this accumulated energy. The term thermally stimulated luminescence (TSL) is more descriptive; however, 'thermoluminescence' is traditionally more often utilized and popularly accepted.

As noted except for very unusual cases, the occurrence of the TL curve following a given irradiation is a 'one-shot' effect. Cooling the sample and re-heating it normally does not result in a second TL emission.

Luminescent Materials

The general materials used as phosphors may be classified according to their use in the application or according to the chemical forms. Considering the synthesis viewpoint they are classified here on the basis of their chemical forms. The main categories are as follows:

a. Inorganic materials: These are mainly solid-state compounds that consist of crystals 1–10 µm in size. The inorganic phosphor usually consists of a host lattice with activator ions doped into it in small concentrations, typically a few mole percent or less. The activator ions posses energy levels that can be populated by direct excitation or indirectly by energy transfer, and are responsible for the luminescence. Generally, two types of activator ions can be distinguished. In the first type, the energy levels of the activator ion involved in the emission process show only weak interaction with the host lattice. Typical examples are many of the lanthanide ions Ln^{3+}, where the optical transitions take place solely between 4f terms that are well shielded from their chemical environment by outer electrons. As a consequence, characteristic line emission spectra can be observed. The second type of activator ions strongly interact with the host lattice. This is the case when d electrons are involved. For example. in Mn^{2+}, Eu^{2+}, and Ce^{3+}, as well as for s2 ions like Pb^{2+} or Sb^{3+}, or for complex anions such as $MoO4^{2-}$ or $NbO4^{2-}$. The strong coupling of the electronic states with vibrational modes of lattice mainly lead to more or less broadbands in the spectrum.

Some of the host materials used for inorganic phosphors are aluminates, arsenates, borates, bromides, carbonates, chlorides, chromates, cuprates, fluorides, ferrites, germanates, gallates, indates, iodides, mangnates, nitrides, oxides, phosphates, selenides, silicates, sulphates, sulphides, tantalates, titanates, tungstates, uranates, vanadates, zirconates, halophosphates and apatite structures.

b. Organic materials: These are either polymers or low molecular weight materials applied as thin films or solid solutions. The most recent class of materials comprises main chain polymers with isolated chromophores and side chain polymers with linked chromophores. They are mainly used in organic light emitting diodes (OLEDs) [7].

c. Hybrid compounds: This class generally includes the organically modified silicates (ORMOSIL) [8] and hybrid organic-inorganic complexes such as Eu^{3+}, Tb^{3+}, or their complexes with β-diketones, aromatic carboxylic acids and heterocyclic ligands such as 2,2′-bipyridine and 1,10-phenanthroline, into various matrices [9].

1.4 BORATE-BASED PHOSPHORS

Borates are naturally occurring minerals containing boron, the fifth element on the periodic table. The element boron does not exist by itself in nature. Rather, boron combines with oxygen and other elements to form inorganic salts called borates. Boron has an ionic radius 0.11 Å and hence can occur in both triangular (BO_3) and tetrahedral (BO_4) coordination where bonded to oxygen [10]. BO_3 groups have an average B-O bond-valance approximately equal to 1 valence unit and BO_4 groups have an average B-O bond-valance approximately equal to 3/4 valence unit. Hence, both (BO_3) and (BO_4) groups can polymerize by sharing corners without violating the valance sum rule. Such polymerization is very common in both minerals and synthetic inorganic compounds. In general, a borate structure contains clusters of corner sharing (BO_3) and (BO_4) polyhedra, which occur as discrete polyanions to form larger clusters, chains, sheets or frameworks [11]. Since the boron atom is capable of coordination in either trigonal or tetragonal mode [12–14], borate anions exist in numerous structural types. There are hundreds of different structures with various borate anionic groups as basic structural units in the known borate crystals [15]. However, there are only a few types of basic structural units of borates of practical interest [12,16]:

i. $(BO_3)^{3-}$
ii. $(BO_4)^{5-}$
iii. $(B_2O_5)^{4-}$
iv. (iv) $(B_2O_7)^{8-}$
v. $(B_3O_6)^{3-}$
vi. $(B_3O_7)^{5-}$
vii. $(B_3O_8)^{7-}$
viii. $(B_3O_9)^{9-}$
ix. $(B_5O_{10})^{5-}$
x. $(B_4O_9)^{6-}$

A variety of pre-decided inorganic borate host materials attempted for successful synthesis and verifying the potential photoluminescence characteristics are listed in Table 1.2. In addition to the pre-decided list, a few new materials (hosts) are successfully synthesized.

1.4.1 CLASSIFICATION OF BORATES

Inorganic borate host compounds are classified into various groups on the basis of chemical composition and crystal structure.

1.4.1.1 Classification of Borates Based on Chemical Formula
1.4.1.2 Classification of Borates Based on Crystal Structure

Borate host inorganic compounds exist in numerous crystal structures. On the basis of crystal structure, borate compounds were classified as follows:

i. Aragonite-Type Borate

The aragonite-type borates crystallize in the structure of the mineral aragonite ACO_3, A = Ca, Ba, Pb, Sr. The structure of the aragonite borate is composed of the triangular borate ion group (BO_3), with a boron at the centre of the triangle and three oxygens at each corner. It has an orthorhombic symmetry.

Examples: $LaBO_3$ [10], $NdBO_3$ [11], $CeBO_3$ [12].

ii. Calcite-Type Borate

The calcite-type borates are the polymorphs of aragonite with trigonal symmetry.

Figure 1.3 represents the $InBO_3$ unit cell with (110) orientation.

Examples: $InBO_3$, $LuBO_3$, $ScBO_3$ [13].

TABLE 1.2

Inorganic borates are classified on the basis of a chemical formula

Sr no.	Borate type	Metal oxide	Examples
1	Metaborate	1 B_2O_3	$NaBO_2$ ($Na_2O{:}B_2O_3$), $LiBO_2$ ($Li_2O{:}\ B_2O_3$), CaB_2O_4 ($CaO{:}\ B_2O_3$), BaB_2O_4 ($BaO{:}\ B_2O_3$), SrB_2O_4 ($SrO{:}\ B_2O_3$), $CsBO_2$ ($Cs_2O{:}\ B_2O_3$)
2	Diborate	2 B_2O_3	$Li_2B_4O_7$ ($Li_2O{:}\ 2\ B_2O_3$), $Na_2B_4O_7$ ($Na_2O{:}\ 2\ B_2O_3$), ZnB_4O_7 ($ZnO{:}\ 2\ B_2O_3$)
3	Triborate	3 B_2O_3	LiB_3O_5 ($Li_2O{:}\ 3\ B_2O_3$), EuB_3O_6 ($Eu_2O_3{:}\ 3\ B_2O_3$), CeB_3O_6 ($Ce_2O_3{:}3\ B_2O_3$), BiB_3O_6 ($Bi_2O_3{:}\ 3\ B_2O_3$)
4	Tetraborate	4 B_2O_3	AgB_4O_7 ($Ag_2O{:}\ 4\ B_2O_3$), $Na_2B_8O_{13}$ ($Na_2O{:}\ 4\ B_2O_3$)
5	Pentaborate	5 B_2O_3	CsB_5O_8 ($Cs_2O{:}\ 5\ B_2O_3$), $Na_2B_{10}O_{16}$ ($Na_2O{:}\ 5\ B_2O_3$)
6	Hexaborate	6 B_2O_3	SrB_6O_{10} (2 $SrO{:}\ 6\ B_2O_3$)
7	Heptaborate	7 B_2O_3	$Li_3B_7O_{12}$ (3 $Li_2O{:}\ B_2O_3$)
8	Octaborate	8 B_2O_3	$Bi_2B_8O_{15}$ (2 $Bi_2O_3{:}\ 8\ B_2O_3$), BaB_8O_{13} (2 $BaO{:}\ 8\ B_2O_3$)
9	Nonaborate	9 B_2O_3	$NaBaB_9O_{15}$ ($Na_2O{:}2BaO{:}9B_2O_3$), $LiBaB_9O_{15}$ ($Li_2O{:}2BaO{:}9B_2O_3$), $LiSrB_9O_{15}$ ($Li_2O{:}2SrO{:}9B_2O_3$), $GdBaB_9O_{16}$ ($Gd_2O_3{:}2BaO{:}9B_2O_3$)
10	Deaborate	10 B_2O_3	$K_2B_{10}O_{16}$ ($K_2O{:}\ 10\ B_2O_3$)

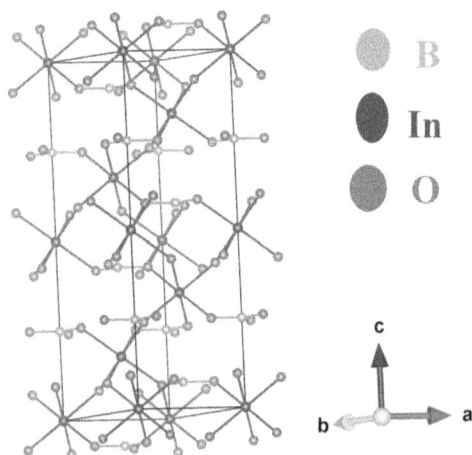

FIGURE 1.3 Ball-and-stick representation of $InBO_3$ unit cell with (110) orientation.

iii. Vaterite-Type Borate

The vaterite-type borates are the polymorphs of aragonite with hexagonal symmetry.

Examples: $SmBO_3$, $EuBO_3$ [14–16].

iv. Stillwelite-Type Borate

Mineral stillwelite is a family of trigonal borates with the general formula X [BO(SiO$_4$)] with X = Ce, La, Pr, Nd, Th. Stillwelite contains helical chains of (BO$_4$) tetrahedra polymerized by sharing corners [17]. These chains are decorated by SiO$_4$ tetrahedra that share two corners with adjacent (BO$_4$) tetrahedra. Boron is coordinated by four oxygen atoms in a tetrahedral arrangement. The (BO$_4$) tetrahedra polymerize by sharing corners to form a helical chain extending along the z-axis. The periphery of the chain is decorated by (SiO$_4$) tetrahedra that share two corners with an adjacent (BO$_4$) tetrahedral. The resultant [BSiO$_5$] chain is the fundamental building block of the structure that is repeated by the lattice translations to form a hexagonal columnar array.

The lanthanide borate-silicate materials LnBSiO$_5$ [18,19] (Ln = La, Gd, Y, Pr, Nd) exhibits stillwelite structure of LaBSiO$_5$, shown in Figure 1.4.

v. Melitite-Type Borates

Melitite represents a family of natural and synthetic compounds A$_2$XZ$_2$O$_7$, where A = Na, Ca, Sr, Ba, Cd, Pb, Y, Ln; X = Be, Mg, Co, Fe, Mn, Cu, Zn, Cd, Al, Ga; Z = Be, B, Al, Si, Ge. Bi$_2$ZnB$_2$O$_7$, CaBiGaB$_2$O$_7$ and CdBiGaB$_2$O$_7$ are the only examples of synthetic diborate members of the melitite family [20].

vi. Leucite-Type Borates

The leucite-type structure (AlSi$_2$O$_6$) exists over a wide range of alkali cations from Na to Cs [21]. Al atoms can be replaced by B [22–25]. The boron-substituted leucite is known as boro leucite [26]. The examples of boroleucite are KBSiO$_6$ [27] and RbBSiO$_6$ [28]. The structure of leucite [29] represents a continuous three- dimensional skeleton, formed by (Si,Al)O$_4$ tetrahedra, each of which shares all its oxygen with its neighbours [30,31].

vii. Warwickite-Type Borate

The warwickite is an orthorhombic oxyborate minerals [31] with the general formula M$_2$OBO$_3$ (M = Mg^{2+}, Mn^{2+}, Fe^{3+}, Ti^{4+}, Al^{3+}). Most transition metal oxyborates with the general formula M^{2+}M^{3+}BO$_4$ crystallizes in an orthorhombic structure of warwickite (Mg$_{1.5}$Ti$_{0.5}$BO$_4$), representing a system of linear, weakly interacting ribbons comprising two internal and external chains in which octahedrally coordinated divalent and trivalent transition metal atoms are randomly distributed over nonequivalent crystallographic positions of two types [32]. MgGaBO$_4$ is isostructural to the mineral warwickite (Fe,Mg)Ti(BO$_4$)$_2$ [33]. The structure of MgGaBO$_4$ along [001] contains ribbons of four edge-sharing octahedral formed from two pseudo-packed layers of oxygen atoms between which magnesium atoms occupy the octahedral interstices shown in Figure 1.5.

viii. Mineral Huntite Borate

These borate compounds crystallizes in the structure of the mineral huntite. These materials generally adopt a noncentrosymmetric structure, trigonal space group R32. The general formula of huntite borates is LnM$_3$(BO$_3$)$_4$ where Ln = Lanthanide and M = Al, Sc, Cr, Fe, Ga.

Examples: CeSc$_3$(BO$_3$)$_4$ [34], TbAl$_3$(BO$_3$)$_4$ [35].

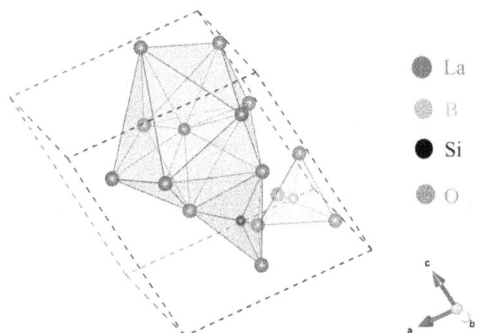

● La

● B

● Si

● O

FIGURE 1.4 LaBSiO$_5$ with stillwellite structure.

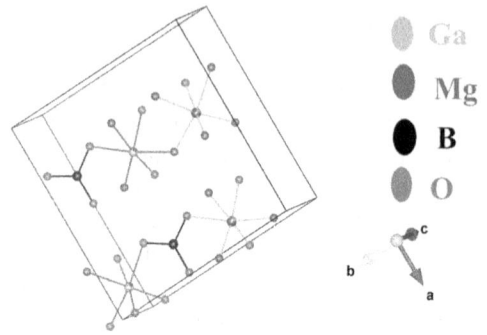

FIGURE 1.5 The structure of MgGaBO$_4$ along [001].

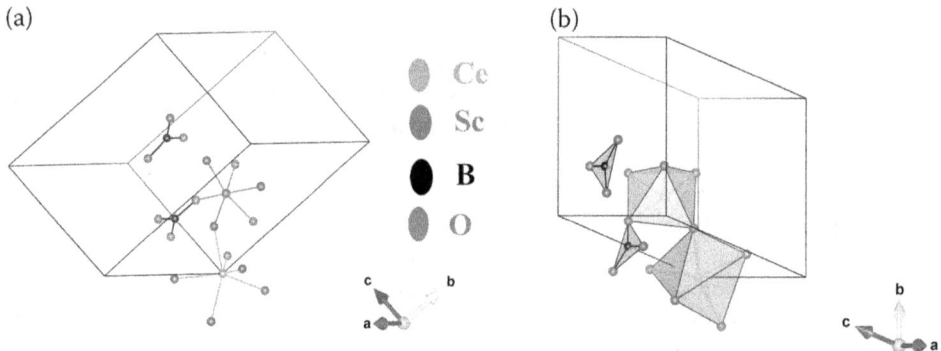

FIGURE 1.6 (a) Unit-cell contents of CeSc$_3$(BO$_3$)$_4$. (b) Sc$_6$O polyhedral connections in CeSc$_3$(BO$_3$)$_4$.

The compound CeSc$_3$(BO$_3$)$_4$ adopts the normal trigonal huntite structure shown by Figure 1.6 with Ce-centred distorted trigonal prisms and Sc-centred distorted octahedral dispersed between planes of BO$_3$ triangles that extend parallel to (001). Dissimilar polyhedra share only vertices, resulting in isolation of the CeO$_6$ prisms and BO$_3$ triangles; the ScO$_6$ octahedra share edges to form helices extending along (001).

ix. Mineral Boracite

The boracite represents a family of compounds having the general formula M$_3$B$_7$O$_{13}$X where M = Mg, Cr, Mn, Fe, Co, Ni, Cu, Zn. Cd and X = Cl, Br, I. The term 'boracite' is due to the mineral Mg$_3$B$_7$O$_{13}$Cl [36]. At present, more than 25 isomorphous boracite compounds exist. A complete structural investigation of high and low boracite Mg$_3$B$_7$O$_{13}$Cl was reported by Ito et al. [37,38]. High boracite was described as cubic and low boracite as orthorhombic. Complete structural analysis was given for Sr$_2$B$_5$O$_9$Cl is given in Figure 1.7.

1.4.2 CRYSTAL CHEMISTRY OF BORATES

In recent years, more and more research has been concentrated on borate crystals. Most of the obtained crystals contain independent orthoborate BO$_3$ groups in their structures. The structural chemistry of anhydrous borates is characterized by a variety of discrete and condensed BO$_3$ and BO$_4$ anions [39]. Predominant among these structures, especially binary and more complex borates are isolated BO$_3$ triangles (65%). Of the rest, almost one-half is represented by framework structures containing three-dimensional B$_n$O$_m$ polyanions. In this case, BO$_3$ and BO$_4$ groups join by sharing common O atoms. Then, it is followed by insular (pyroborates and ring metaborates), network and chain-forming structures.

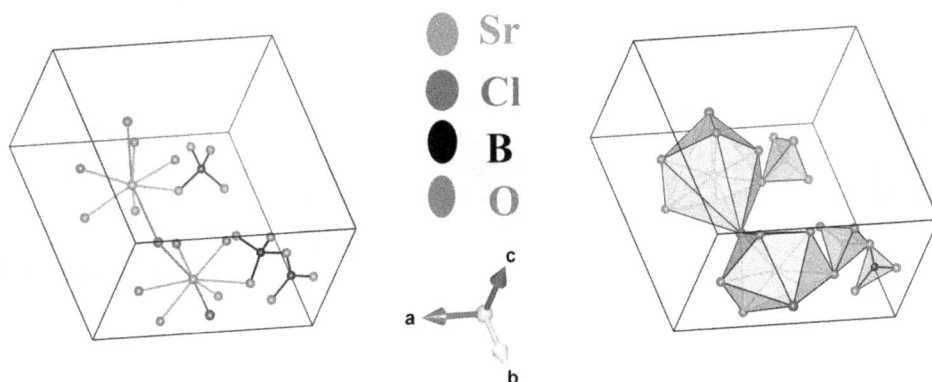

FIGURE 1.7 Unit cell of $Sr_2B_5O_9Cl$ showing coordination of Sr at various sites.

There are more than 50 types of boron-oxygen anions and polyanions in anhydrous borates [40]. However, all of them consist exclusively of three basic types of structural units of different composition. The first set of these basic types identify with BO_3 triangles (Δ) and BO_4 tetrahedra ($\square$) as fundamental structural units (FSU), which are shown in Figure 1.8(a). They are isolated in orthoborates and can be merged in pyroborates, metaborates and polyborates.

The second basic type of structural units is usually built up of several FSU (from two to five) joined by sharing common O atoms. There is a rather limited number of these anionic groups. They are shown in Figure 1.8(b) and designated below as combined basic structural units (CSU). Only two of them, 2Δ and $3\square$, can be isolated in the structure of anhydrous pyroborates and some metaborates, respectively. The rest of the structural units of this type are components of various one-, two-, and three-dimensional B_nO_m polyanions by sharing their free O atoms.

Finally, in polyanions of anhydrous borates, it is expedient to distinguish one or more types of structural units. Structural elements of this category are, as a rule, more complex and characterize the borate structures on the whole. It is appropriate to designate these anionic groups as complete radicals of polyanions (CRP). Some of the typical CRP are demonstrated in Figure 1.8(c) and (d). The CRP may be visualized as one or more of the CSU and additional FSU (BO_3 triangles and/or BO_4 tetrahedra) joined by sharing common O atoms. Thus, it represents a full repeating fragment of a polyanion corresponding in composition aliquot to the B-O anion part in the structural formula of the borate. It is useful to note that the CRP identifies themselves with the CSU in the anhydrous orthoborate structures, and with the CSU in the pyroborates, metaborates and in some more complex borate compounds.

1.4.3 APPLICATIONS OF BORATE HOST MATERIALS AS PHOSPHORS

Solid-state inorganic borates have become a focus of technological interest due to a variety of physical and chemical features exhibited by these compounds [41]. Owing to possible three-, or fourfold coordination of borate atoms, borates form a great number of compounds having diverse structures. Borates intrinsically possess characteristics that are advantageous for optical materials [42], which include wide transparency range, large electronic bandgap, good thermal and chemical stability, low preparative temperature, optical stability with good nonlinear characteristics and exceptionally high optical damage threshold [43–45]. The unique crystal structure of borates determines their enhanced UV transparency, good nonlinearity and relatively high resistance against laser-induced damage. Recent research on inorganic borates has been focused on the synthesis and characterization of compounds with potential application as optical material [46]. Borate compounds currently have been of considerable interest to the scientific community owing to their wide range of applications: Laser, NLO material, phosphor material and scintillator material, etc. Most of the borates are polyfunctional materials with nonlinear optical, piezo-electrical and

(a)   (b)

(c)

(d)

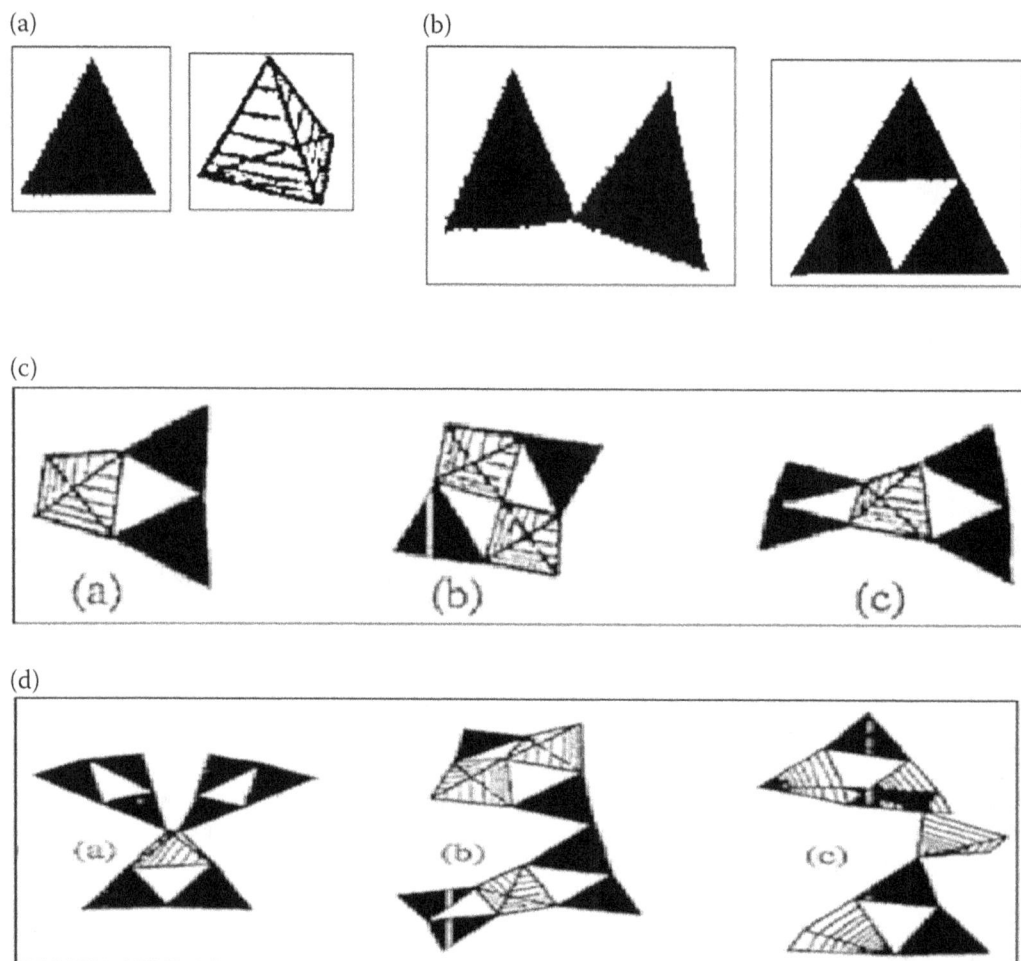

FIGURE 1.8 (a) Configuration of some basic B-O structural units in anhydrous borates: Fundamental structural units (Δ and $\square$ isolated forms occur in orthoborate structure only). (b) Combined basic structural units as insular anions in pyroborate (a) and metaborate (b). (c) Combined basic structural units forming network and framework polyanions: (a) $2\Delta + 1\square$ (b) $2\Delta + 2\square$ (c) $4\Delta + 1\square$. (d) Combined radicals of polyanions in network and framework polyborate structure: (a) $(2\Delta + 1\square) + 2(3\Delta)$ (b) $(4\Delta + 1\square) + (2\Delta + 2\square)$ (c) $(2\Delta + 1\square) + (2\Delta + 2\square) + 1\square$.

acousto-electrical properties. Some borates are also suitable as a laser material for miniature laser [47]. Borate crystals with the structure of the naturally occurring mineral huntite $CaMg_3(CO_3)_4$ are widely known as polyfunctional materials having device potential due to their good thermal and chemical stabilities. The general formula of huntite borate is $LnM_3(BO_3)_4$, Ln = rare earth element and M = Al, Ga, Cr, Fe, Sc. Among them, rare earth aluminium borates $LnAl_3(BO_3)_4$ have attracted considerable attention for their luminescence properties and possible application as single crystal mini-laser [48]. In the following sections, the applications of inorganic borate compounds as LASER and NLO material and phosphor material are discussed.

1.4.3.1 Inorganic Borate as NLO and LASER Material

The NLO materials are materials used to generate new LASER sources of frequencies that cannot be obtained directly from available lasers. Efficient frequency conversion depends on crystal properties such as the effective nonlinear coefficient, refractive indices, phase-matching conditions and damage threshold. The borates with NLO properties form a large family of compounds.

The first NLO borate described was $KB_5O_8.4H_2O$ [49] in 1975, but intense research on NLO borate materials began with the advent of low-temperature BaB_2O_4 and LiB_3O_5 which are the most frequently used borate materials. A subsequent search for new borate compounds led to the continued discovery of new borate compounds with improved NLO properties.

Borate materials are also useful as a self-frequency doubling active laser sources. A commercial self-frequency doubling laser source is $YAl_3(BO_3)_4$: Nd^{3+}. Many new borates have been recently reported. Borophosphates which contain the borate group and the phosphate group as basic structural units has also drawn attention recently, as new functional materials for application in laser technologies [50]. Complex boron oxide materials attract considerable attention because they can be used as multifunctional optical materials in which the laser effect and nonlinear optical phenomena occur simultaneously. Borotungstate with the general formula Ln_3BWO_9, Ln = La, Pr, Nd, Sm – Ho are the promising complex borates [51]. Lanthanide calcium oxoborate $LnCa_4O(BO_3)_3$, Ln = La – Lu and Y, constitute a family of compounds intensively studied since 1997 due to their good NLO properties. These crystals could be doped with rare earth and can be used for self-frequency doubling and self-frequency mixing [52].

1.4.3.2 Inorganic Borate as Phosphor Material

During the past few years, a number of borate materials have been studied extensively due to their unique combinations of large electronic bandgaps, strong nonlinear optical properties, chemical and environmental stabilities and mechanical robustness. Due to their large electronic bandgaps, borate materials are excellent host lattices for luminescent ions. Rare earth borate compounds normally have high UV transparency and exceptional optical damage threshold, which makes them attractive for numerous practical applications such as in lamps and display applications. There are many excellent phosphors in the borate family, for example, YBO_3: Eu^{3+}, $GdBO_3$: Eu^{3+} and $LnMgB_5O_{10}$, etc. which have been applied in the region of UV-excited phosphors and integrated optics [53]. Rare earth borate compounds are an interesting class of luminescent materials. The rare earth borate phosphors were first introduced by R.I. Smirnova et al. [54] and have been of little or no interest for two decades. More recently, due to rising demand for new efficient phosphors for various applications such as lamp phosphor, FPD, PDP, etc. rare earth borates have attracted attention [55,56]. Rare earth orthoborates, RBO_3 (R = Y, La, Gd), doped with rare earth ions (Eu^{3+} and Tb^{3+}) are interesting luminescent materials (56). Rare earth borates doped with Eu^{3+} are potential red-emitting phosphors. Gadolinium borate phosphors ranging from orthoborates to pentaborates have proved to be potential candidates for practical applications in fluorescent lamps due to their high efficiency [57–59].

Haloborates activated by Ce^{3+} ions have been recently shown to be a promising material for detecting thermal neutrons. Strontium haloborates doped with Eu^{2+} are well-known X- ray storage phosphors.

Plasma display panels (PDP) are regarded as the most promising candidate for large-sized flat panel displays (FPDs). Phosphors for the application in PDP are required to have high conversion efficiency by VUV radiation of 147 and/or 172 nm from the Xe gas plasma. The inorganic borate compounds have strong absorption in the VUV region [60] and therefore widely used as host lattices of phosphors for PDP applications. $LnBO_3$ doped with Eu^{3+} has been widely used as a luminescent material in plasma display panels (PDPs) due to the high quantum efficiency and good color coordinates under 147 nm VUV excitation [61]. At present, the most widely used red-emitting phosphor for PDP is $(YGd)BO_3$: Eu^{3+} [62]. The doped $YAl_3(BO_3)_4$ can be utilized as red PDP phosphor (Tables 1.3 and 1.4).

- Lamp Phosphors
- Led Phosphors
- UV-Based Photo Therapy Phosphors
- PDP Phosphors
- Radiation Dosimetery Phosphors
- Mechno or Layo-luminescent Phosphors
- Nir Quantum Cutting Phosphors
- Neutron Radiography

TABLE 1.3
Borate host inorganic NLO and LASER materials

Sr. no.	Inorganic borate	Applications	Ref.
1	CsB_3O_5	NLO material to generate deep UV	[63,64]
2	$CsLiB_6O_{10}$	NLO material for SHG and Higher order harmonic generation	[65–67]
3	$YAl_3(BO_3)_4:Nd^{3+}$	Self-frequency doubling LASER material	[68–77]
4	LiB_3O_5	NLO material for SHG	[78–82]
5	$NaBe_2BO_3F_2$ $KBe_2BO_3F_2$	VUV NLO material used in LASER chemistry and LASER medical science	[65,83–85]
6	$BaZn_2(BO_3)_2$	NLO material for SHG	[86]
7	β-BaB_2O_4	NLO material	[87–89]
8	$Ca_4GdO(BO_3)_3$	NLO material	[90]
9	$Ca_4YO(BO_3)_3$	NLO material for THG	[91]
10	$K_2Al_2B_2O_7$	New NLO material for THG	[92]
11	$YAl_3(BO_3)_4:Yb^{3+}$	Excellent self-frequency doubling (SFD) material	[93]
12	BiB_3O_6	NLO material for SHG and THG	[94]
13	$YAl_3(BO_3)_4:Cr^{3+}$	Tunable LASER gain medium	[95]
14	$Ca_4NdO(BO_3)_3$	Mini-laser material	[96]
15	$Sr_3Sc(BO_3)_3:Cr^{3+}$	LASER material	[97,98]
16	$LaBa_3(BO_3)_4:Nd^{3+}$	Miniature LASER material	[99]
17	$NdAl_3(BO_3)_4$	LASER material	[100–106]
18	$Ca_4GdO(BO_3)_3:Nd^{3+}$	Green-emitting micro-chip LASER	[107–110]
19	$Ca_4YO(BO_3)_3:Yb^{3+}$	Self-frequency doubling LASER material	[111]
20	$LaSC_3(BO_3)_4$	Tunable LASER material	[112]
21	$LaSC_3(BO_3)_4:Nd^{3+}$	Efficient LASER material	[113–115]
22	$LaB_3O_6:Pr^{3+}$	Micro-chip LASER material	[116]
23	$Ba_3Y(BO_3)_3:Nd^3$	LASER gain medium	[117]
24	$YbAl_3(BO_3)_4$	Stoicheometric NLO LASER material	[118]
25	$YAl_3(BO_3)_4:Pr^{3+}$	LASER material in UV region	[119]
26	$(Gd_{0.5}Y_{0.5})Ca_4O(BO_3)_4:Yb^{3+}$	Diode pumping LASER host material	[120]
27	$Sr_3Y(BO_3)_3:Yb^{3+}$	Femto-second and tunable LASER	[121]
28	$Sr_3La_2(BO_3)_4:Yb^{3+}$	LASER material	[122]
29	$Sr_3Gd(BO_3)_3:Nd^{3+}$	LASER material	[123]

Neutron radiography (NR) is a special application of neutron beams that developed neutron beam applications. The use of neutron beams for radiographic purposes is a relatively new method of non-destructive testing. Neutron imaging has expanded rapidly as a means of non-destructive testing (NDT) of materials. It is possible to develop scintillators, thermo-luminescent, optically stimulable, materials from the borates. These form the basis for the various methods developed for neutron radiography. Various neutron scintillators, phosphors for imaging plates, based on borates are reviewed in detail in chapter 10. Some important applications such as nuclear fuel inspection, bioimaging, dentistry and oral surgery, inspection of mechanical parts, etc. are discussed.

1.4.4 MIXED BORATE PHOSPHORS

Apart from these simple borates, several complex compositions exist involving mixed anions as well as double metal borates. Some of these complex borates show interesting luminescence

TABLE 1.4

Borate host inorganic luminescent materials

Sr. no.	Inorganic borate	Applications	Ref.
1	(Ce Gd)MgB$_5$O$_{10}$	Red-emitting phosphor in special deluxe lamps	[124]
2	(Ce Gd)MgB$_5$O$_{10}$: Tb^{3+}	Green-emitting phosphor in tricolor lamps	[124,125]
3	Ca$_4$GdO(BO$_3$)$_3$:Eu^{3+}	Potential red lamp phosphor	[126–128]
4	GdB$_3$O$_6$: Ce^{3+},Tb^{3+}	Potential green lamp phosphor	[129]
5	SrBPO$_5$: Eu^{3+}	Storage phosphor	[130,131]
6	CaLaB$_7$O$_{13}$: Ce^{3+},Tb^{3+}	Green-emitting phosphor in low-pressure Hg vapor lamps	[132]
7	CaLaB$_7$O$_{13}$: Eu^{3+}	Red-emitting phosphor in low-pressure Hg vapor lamps	[132]
8	SrB$_4$O$_7$: Sm^{2+}	Optical pressure gauge	[133]
9	(Sr$_{0.89}$Na$_{0.05}$)BPO$_5$: Ce$_{0.05}$, Tb$_{0.01}$	Green-emitting phosphor in tricolor lamps	[134]
10	Sr$_2$B$_5$O$_9$Br: Ce^{3+}	Potential storage phosphor for thermal neutron	[135–137]
11	SrB$_4$O$_7$: Eu^{2+}	Commercial UV-emitting phosphor in medical lamps	[138,139]
12	Sr$_2$B$_5$O$_9$Cl: Eu^{2+}	Blue component of daylight phosphor	[140–144]
13	Sr$_2$B$_5$O$_9$Cl: Eu^{2+}**(Thin Film)**	Blue component in FED	[145–147]
14	Ba$_2$B$_5$O$_9$Br: Eu^{2+}	X-ray storage phosphor	[148]
15	SrB$_2$Si$_2$O$_8$: Eu^{2+}	Blue-emitting phosphor	[149]
16	InBO$_3$: Tb^{3+}	Green-emitting phosphor in CTV screens	[150]
17	La(BO$_3$,PO$_4$): Ce^{3+} Gd^{3+} Tb^{3+}	Green-emitting phosphor in high-quality tricolor lamps	[151]
18	Ba$_2$B$_5$O$_9$Cl:Tb^{3+}**(Thin Film)**	Green-emitting phosphor in FED	[152]
19	Ba$_2$B$_5$O$_9$Cl:Eu^{2+}**(Thin Film)**	Blue-emitting phosphor in flat panel display (FPD)	[153–155]
20	YBO$_3$: Eu^{3+}	Red-emitting VUV phosphor	[156–159]
21	YAl$_3$(BO$_3$)$_4$:Ho,Yb	Upconversion phosphor	[160]
22	CaBPO$_5$: Tb^{3+}	Green-emitting VUV phosphor	[161]
23	Y$_3$BO$_6$: Eu^{3+}	Red phosphor in vacuum discharge lamps or screens	[162]
24	(YGd)BO$_3$:Tb^{3+}	Green component in PDP	[163]
25	(YGd)BO$_3$:Eu^{3+}	Red component in PDP	[164–173]
26	GdAl$_3$(BO$_3$)$_4$:Eu^{3+}	Red component in PDP	[174,175]
27	(Y Gd)Al$_3$(BO$_3$)$_4$:Eu^{3+}	Red component in PDP	[176]
28	GdAl$_3$(BO$_3$)$_4$:Tb^{3+}	Green-emitting VUV phosphor	[177]
29	BaZr(BO$_3$)$_2$: Eu$_{3+}$	Red component in PDP	[178,179]

properties and are covered in this chapter. These include double borates containing rare earth RM$_3$(BO$_3$)$_4$, pentaborates LaMgB$_5$O$_{10}$, M$_3$R$_2$(BO$_3$)$_4$, where M is alkaline earth, mixed anion borates like aluminoborate SrAl$_2$B$_2$O$_7$, silicate-borates like pekovite, SrB$_2$Si$_2$O$_8$, haloborates, M$_2$B$_5$O$_9$X, where M is alkaline earth and X is halogen, phosphate borates, MBPO$_5$, where M is alkaline earth. Phosphors based on these compositions find use in various applications like fluorescent lamps, colour TV, plasma display panels, high-intensity discharge lamps based on xenon, optically pumped solid-state lasers, eye-safe lasers and X-ray imaging. This is briefly illustrated in chapter 11.

REFERENCES

[1] Seitz, F., Gorton, R., and Schulman, J. H. (1948). Luminescent Crystals. In: *Preparation and characteristics of Solid Luminescent Materials*. Pub: John Wiley & Sons, inc, NY, 1, 28. 10.1063/1.3066240

[2] Blasse, G., and Grabmaier, B. C. (1994). How Does a Luminescent Material Absorb Its Excitation Energy?. In: *Luminescent Materials*. Springer, Berlin, Heidelberg. 10.1007/978-3-642-79017-1_2

[3] Seitz, F., (1948). In: 'Luminescent Crystals', Preparation and characteristics of Solid Luminescent Materials, Symp. Cornell Univ., John Wiley & Sons, INC, NY, 9

[4] Johnson, R. P. (1939). Luminescence of sulphide and silicate phosphors. *J. Opt. Soc. Am.*, 29, 387–391. 10.1364/JOSA.29.000387

[5] Johnson, R. P. (1939). Decay of Willemite and zinc sulphide phosphors. *Phys. Rev.*, 55, 881. 10.1103/PhysRev.55.881

[6] Blasse G., and Grabmaier B. C. (1994) A General Introduction to Luminescent Materials. In: *Luminescent Materials*. Springer, Berlin, Heidelberg. 10.1007/978-3-642-79017-1_1

[7] Justel, T., Nikol, H., and Ronda, C. (1998). New developments in the field of luminescent materials for lighting and displays. *Angew. Chem. Int. Ed.*, 37, 3084–3103. 10.1002/(SICI)15213773(19981204)37:22 <3084::AID-ANIE3084>3.0.CO;2-W

[8] Mitschke, U., and Bauerle, P. (2000). The electroluminescence of organic materials. *J. Mater. Chem.*, 10, 1471–1507. 10.1039/A908713C

[9] Li, H., Inoue, S., Machida, K., and Adachi, G. (1999). Preparation and luminescence properties of organically modified silicate composite phosphors doped with an europium(III) β-diketonate complex. *Chem. Mater.*, 11, 3171–3176. 10.1021/cm990251a

[10] Antic-Fidance, E., Lemaitre-Blaise, M., Chaminade, J., and Porcher, P. (1992). Luminescence and crystal field calculation of aragonite-type $LaBO_3:Eu^{3+}$. *J. Alloys Compd.*, 180, 223–228. 10.1016/0925-8388(92)90384-L

[11] Levin, E. M., Roth, R. S., and Martin, J. B. (1961). Polymorphism of ABO_3 type rare earth borates. *Am. Mineral.*, 46 (9-10), 1030–1055.

[12] Weidelt, J., and Bambauer, H. U. (1968). Ein mit Aragonit isotypes Cerborat. Die Naturwissenschaften, 342–342. 10.1007/BF00600458

[13] Chadeyron, C., El-Ghozzi, M., Mahiou, R., Arbus, A., and Cousseins, J. C. (1997). Revised structure of the orthoborate YBO_3. *J. Solid State Chem.*, 128, 261–266. 10.1006/jssc.1996.7207

[14] Chaminade, J. P., Garcia, A., Pouchard, M., Fouassier, C., Jacquier, B., Perret-Gallix, D., and Gonzalez-Mestres, L. (1990). Crystal growth and characterization of $InBO_3: Tb^{3+}$. *J. Cryst. Growth*, 99, 799–804. 10.1016/S0022-0248(08)80029-2

[15] Cox, J. and Keszler, D. A. (1994). $InBO_3$. *Acta. Cryst.*, C50, 1857–1859. 10.1107/S0108270194003999

[16] Frondel, C., Biedl, A., and Ito, J. (1966). New type of ferric iron tourmaline1. *Am. Mineral.*, 51 (9–10), 1501–1505.

[17] Burns, P. C., Hawthorne, F. C., Macdonald, D. J., Dellaventura, G., and Parodi, G. C. (1993). The crystal structure of stillwellite. *Can. Mineral.*, 31, 147–152.

[18] Rulmant, A., and Tarte, P. (1988). Lanthanide borogermanates LnBGeO5: Synthesis and structural study by X-ray diffractometry and vibrational spectroscopy. *J. Solid State Chem.*, 75, 244–250. 10.1016/0022-4596(88)90163-6

[19] Chi, L., Chen, H., Zhuang, H., and Huang, J. (1997). Crystal structure of LaBSiO5. *J. Alloys and Compd.*, 252 (1-2), L12–L15. 10.1016/S0925-8388(96)02625-4

[20] Barbier, J., Penin, N., and Cranswick, L. M. (2005). Melilite-type borates $Bi_2ZnB_2O_7$ and $CaBiGaB_2O_7$. *Chem. Mater.*, 17, 3130–3136. 10.1021/cm0503073

[21] Krzhizhanovskaya, M. G., Bubnova, R. S., Filatov, S. K., Meyer, D. C., and Paufler, P. (2006). Crystal structure and thermal behaviour of $(Rb,Cs)BSi_2O_6$ solid solutions. *Cryst. Res. Technol.*, 41, 285–292. 10.1002/crat.200510575

[22] Palaspagar, R. S., Gawande, A. B., Sonekar, R. P., Omanwar, S. K. (2015). $Eu^{3+}{\rightarrow}Eu^{2+}$ reduction in $BaAl_2B_2O_7$ phosphor in oxidizing environment, Optik, 126, 5030–5032. 10.1016/j.ijleo.2015.09.056

[23] Miclos, D., Smrcok, L., Durovic, S., Gypesova, D., and Handlovic, M. (1992). Refinement of the structure of boroleucite, $K(BSi_2O_6)$. *Acta. Cryst.*, C41, 1831–1832. 10.1107/S0108270192002270

[24] Mazza, D., and Lucco-Borlera, M. (1997). On the substitution of Fe and B for Al in the pollucite $(CsAlSi_2O_6)$ structure. *J. Eur. Ceramic Soc.*, 17, 1767–1772. 10.1016/S0955-2219(96)00245-2

[25] Bubnova, R. S., Levin, A. A., Stepanov, N. K., Belger, A., Meyer, D. C., Polyakova, I. G., Filatov, S. K., and Paufler, P. (2002). Crystal structure of $K_{1-x}Cs_xBSi_2O_6$ (x = 0.12, 0.50) boroleucite solid solutions and thermal behaviour of $KBSi_2O_6$ and $K_{0.5}Cs_{0.5}BSi_2O_6$. *Zeitschrift fur Kristallographie Crystalline Materials*, 217 (2), 55–62.

[26] Voldan, J. (1979). Crystallization of a three-component compound in the system $K_2O\text{-}SiO_2\text{-}B_2O_3$. *Silikaty*, 23, 133–135.

[27] Voldan, J. (1981). Crystallization of $Rb_2O \cdot B_2O_3 \cdot {}_4SiO_2$. *Silikaty*, 25, 165–167.

[28] Satava, V., Klouzkova, A., Lezal, D., and Novotna, M. (2002). *Ceramics-Silikity*, 46, 37–40.

[29] Liebau, F. (1985). Crystal Chemical Classification of Silicates: Special Part. In: *Structural Chemistry of Silicates*. Springer, Berlin, Heidelberg. 10.1007/978-3-642-50076-3_7

[30] Palmer, D. C., Dove, M. T., Ibberson, R. M., and Powell, B. M. (1997). Structural behavior, crystal chemistry, and phase transitions in substituted leucite: High-resolution neutron powder diffraction studies. *Am. Mineral.*, 82 (1-2), 16–29. 10.2138/am-1997-1-203

[31] Bigi, S., Brigatti, M. F., and Capedri, S. (1991). Crystal chemistry of Fe- and Cr-rich warwickite. *Am. Mineral.*, 76 (7-8), 1380–1388.

[32] Balaev, A. D., Bayukov, O. A., Vasil'ev, A. D., and Ovchinnikov, S. G., Abd-Elmeguid, M., and Rudenko, V. V. (2003). *J. Exp. Theoretical Phys.*, 79, 989.

[33] Liang, J. K., Xu, T., Yang, Z., Chen, X. L., and Lan, Y. C. (2001). Phase relations in the MgO-Ga_2O_3-B_2O_3 system and crystal structure of $MgGaBO_4$. *J. Alloys Compd.*, 319, 247–252. 10.1016/S0925-8388(01)00871-4

[34] Reynolds, T. A., Peterson, G. A., and Keszler, D. A. (2000). Stoichiometric, trigonal huntite borate $CeSc_3(BO_3)_4$. *Int. J. Inorg. Mater.*, 2, 101–106. 10.1016/S1466-6049(00)00011-8

[35] Couwenberg, I., Binnemans, K., De Leebeeck, H., and Gorller- Walrand, C. (1998). Spectroscopic properties of the trivalent terbium ion in the huntite matrix $TbAl_3(BO_3)_4$. *J. Alloys Compd.*, 274, 157–163. 10.1016/S0925-8388(98)00549-0

[36] Peters, T. E., and Baglio, J. (1970). Luminescence and structural properties of alkaline earth chloroborates activated with divalent europium. *J. Inorg. Nuclear Chem.* 32, 1089–1095. 10.1016/0022-1902(70)80103-8

[37] Ito, T., Morimoto, N., and Sadanaga, R. (1951). The crystal structure of boracite. *Acta Crystallogr.*, 4, 310–316. 10.1107/S0365110X51001033

[38] Campa- Malina, J. Ulloa-Godinez, S., Barrera, A. Bucio, L., and Mata, J. (2006). Nano and micro reoriented domains and their relation with the crystal structure in the new ferroelectric boracite $Zn_3B_7O_{13}Br$. *J. Phys. Condens. Mater.*, 18, 4827. 10.1088/0953-8984/18/20/007

[39] Leonyuk, N. I. (1997). Structural aspects in crystal growth of anhydrous borates. *J. Crystal Growth*, 174, 301–307. 10.1016/S0022-0248(96)01164-5

[40] Leonyuk, N. I., and Leonyuk, L. I. (1983). *Crystal Chemistry of Anhydrous Borates.*

[41] Cheng, W. D., Zhang, H., Zheng, F. K., Chen, J. T., Zhang, Q. E., and Pandey, R. (2000). Electronic structures and linear optics of $A_2B_2O_5$ (A = Mg, Ca, Sr) pyroborates. *Chem. Mater.*, 12, 3591–3594. 10.1021/cm000188l

[42] Corbel, G., Rotoux, R., and Leblanc, M. (1998). Ab initio structure determination of new rare earth fluoride borates $Ln_3(BO_3)_2F_3$(Ln = Sm, Eu, and Gd). *J. Solid State Chem.*, 139, 52–56. 10.1006/jssc.1998.7800

[43] Chaminade, J. P., Gravereau, P., Jubera, V., and Fouassier, C. (1999). A new family of lithium rare-earth oxyborates, $LiLn_6O_5(BO_3)_3$ (Ln = Pr–Tm): Crystal structure of the gadolinium phase $LiGd_6O_5(BO_3)_3$. *J. Solid State Chem.*, 146, 189–196. 10.1006/jssc.1999.8331

[44] Keszler, D. A. (1996). Borates for optical frequency conversion. *Curr. Opin. Solid State Mater. Sci.*, 1, 204–211. 10.1016/S1359-0286(96)80085-4

[45] Becker, P. (1998). Borate materials in nonlinear optics. *Adv. Mater.*, 10, 979–992. 10.1002/(SICI)1521-4095(199809)10:13<979::AID-ADMA979>3.0.CO;2-N

[46] Park, H., and Barbier, J. (2000). Crystal structures of new gallo-borates $MGa_2B_2O_7$, M = Sr, Ba. *J. Solid State Chem.*, 154, 598–602. 10.1006/jssc.2000.8901

[47] Xia, H. R., Jiang, H. D., Gua, M., Wang, J. Y., Wei, J. Q., Hu, X. B., and Liu, Y. G. (2001). Absorption and emission properties of erbium calcium oxyborate crystals. *Opt. Commun.*, 188, 233–238. 10.1016/S0030-4018(00)01168-8

[48] Hinatsu, Y., dai, Y., Ito, K., Wakeshima, M., and Alemi, A. (2003). Magnetic and calorimetric studies on rare-earth iron borates $LnFe_3(BO_3)_4$ (Ln = Y, La–Nd, Sm–Ho). *J. Solid State Chem.*, 172, 438–445. 10.1016/S0022-4596(03)00028-8

[49] Dewey, C. F., Jr., Cook, W. R., Jr., Hodgson, R. T., and Wynne, J. J. (1975). Frequency doubling in $KB_5O_8 \cdot 4H_2O$ and $NH_4B_5O_8 \cdot 4H_2O$ to 217.3 nm. *Appl. Phys. Lett.*, 26, 714. 10.1063/1.88047

[50] Duan, C. J., Li, W. F., Wu, X. Y., Chen, H. H., Yang, X. X., and Zhao, J. T. (2006). Syntheses and X-ray excited luminescence properties of $Ba_3BP_3O_{12}$, $BaBPO_5$ and Ba_3BPO_7. *J. Lumin.*, 117, 83–89. 10.1016/j.jlumin.2005.03.016

[51] Maczka, M., Tomaszewski, P., Stepien Damm, J., Majchrowski, A., Macalik, L., and Hanuza, J. (2004). Crystal structure and vibrational properties of nonlinear Eu_3BWO_9 and Nd_3BWO_9 crystals. *J. Solid State Chem.*, 177, 3595–3602. 10.1016/j.jssc.2004.06.016

[52] Antic- Fidancev, E., Lupai, A., Caramanian, A., and Aka, G. (2004). Spectroscopic study of europium doped RCOB host lattices: Evidence of local perturbations. *J. Alloys Compd.*, 380, 141–145. 10.1016/j.jallcom.2004.03.041

[53] Zhang, Y., and Li, Y. D. (2004). Red photoluminescence properties and crystal structure of sodium rare earth oxyborate. *J. Alloys Compd.*, 370, 99–103. 10.1016/j.jallcom.2003.09.035

[54] Smirnova, R. I., Konopko, L. A., and Ivanenko, L. A. (1969). *Otkrytiya Izobreteniya Bulletin*, 46, 74.

[55] Mordkovich, V. (2000). Photo and cathodoluminescence in cerium-activated yttrium-aluminium borates. *MRS Proceedings*, 621, Q5.1.1. 10.1557/PROC-621-Q5.1.1

[56] Tukia, M., Holsa, J., Lastusaari, M., and Niittykoski, J. (2005). Eu^{3+} doped rare earth orthoborates, RBO_3 (R = Y, La and Gd), obtained by combustion synthesis. *Opt. Mater.*, 27, 1516–1522. 10.1016/j.optmat.2005.01.017

[57] Zhiran, H., and Blasse, G. (1985). Energy transfer phenomena in luminescent materials based on GdB_3O_6. *Mater. Chem. Phys.*, 12, 257–274. 10.1016/0254-0584(85)90096-3

[58] Hair, J. Th. de., and Canijnendijk, W. L. (1980). The intermediate role of Gd^{3+} in the energy transfer from a sensitizer to an activator (Especially Tb^{3+}). *J. Electrochem. Soc.*, 127, 161. 10.1149/1.2129608

[59] Jagannathan, R., Rao, R. P., and Kutty, T. N. R. (1989). Sensitized luminescence of Sm^{3+} in $GdBO_3$. *Mater. Chem. Phys.*, 23, 329–333. 10.1016/0254-0584(89)90075-8

[60] Lee, K. G., Yu, B. Y., Pyun, C. H., and Mho, S. I. (2002). Vacuum ultraviolet excitation and photoluminescence characteristics of $(Y,Gd)Al_3(BO_3)_4/Eu^{3+}$. *Solid State Commun.*, 122, 485–488. 10.1016/S0038-1098(02)00195-3

[61] Wen, F., Li, W., Liu, Z., Kim, T., Yoo, K., Shin, S., Moon, J. H., and Kim, J. H. (2005). Hydrothermal synthesis of Sb^{3+} doped and (Sb^{3+}, Eu^{3+}) co-doped YBO3 with nearly white light luminescence. *Solid State Commun.*, 133, 417–420. 10.1016/j.ssc.2004.12.013

[62] Wang, Y., Endo, T., He, L., and Wu, C. (2004). Synthesis and photoluminescence of Eu^{3+} doped $(Y,Gd)BO_3$ phosphors by a mild hydrothermal process. *J. Cryst. Growth*, 268, 568–574. 10.1016/j.jcrysgro.2004.04.093

[63] Wu, Y., Sasaki, T., Nakai, S., Yokoyani, A., Tang, H., and Chen, C. (1993). CsB_3O_5: A new nonlinear optical crystal. *Appl. Phys. Lett.*, 62, 2614. 10.1063/1.109262

[64] Kagebayashi, Y., Deki, K., Morimoto, Y., Miyazawa, S., and Sasaki, T. (2000). Development of new NLO borate crystals. *Japanese J. Appl. Phys.*, 39 L, 1093. 10.1142/S0218863501000589

[65] Mori, Y., Kuroda, I., Nakajima, S., and Nakai, S. (1995). New nonlinear optical crystal: Cesium lithium borate. *Appl. Phys. Lett.*, 67, 1818. 10.1063/1.115413

[66] Zhang, T., and Yonemura, M. (1997). Properties of $CsLiB_6O_{10}$ and $KBe_2BO_3F_2$ crystals for second-harmonic generation with ultrashort laser pulses. *Japanese J. Appl. Phys.*, 36, 6353. 10.1143/JJAP.36.6353

[67] Mori, Y., Kuroda, I., Nakajima, S., Sasaki, T., and Nakai, S. (1995). Growth of a nonlinear optical crystal: Cesium lithium borate. *Japanese J. Appl. Phys.*, 34, L 296–309. 10.1016/0022-0248(95)00309-6

[68] Nelmes, R. I., (1974). Structural studies of boracites. A review of the properties of boracites. *J. Phys. C: Solid State Phys.*, 7, 3840. 10.1088/0022-3719/7/21/008

[69] Pan, H. F., Liu, M. G., Xue, J., and Lu, B. S. (1990). The spectra and sensitisation of laser self-frequency-doubling $Nd_xY_{1-x}Al_3(BO_3)_4$ crystal. *J. Phys: Condens. Mater.*, 2, 4525. 10.1088/0953-8984/2/19/018

[70] Jaque, D., Capmany, J., and Sole, G. (1997). Optical bands and energy levels of Nd^{3+} ion in the $YAl_3(BO_3)_4$ nonlinear laser crystal. *J.Phys: Condens. Mater.*, 9, 9715. 10.1088/0953-8984/9/44/024

[71] Jaque, D., Capmany, J., Sole, G., DLuo, Z., and Jiang, A. D. (1998). Continuous-wave laser properties of the self-frequency-doubling $YAl_3(BO_3)_4$:Nd crystal. *J. Opt. Soc. Am. B*, 15, 1656–1662. 10.1364/JOSAB.15.001656

[72] Jaque, D., Capmany, J., and Garcia Sole, J. (1999). Red, green, and blue laser light from a single $Nd:YAl_3(BO_3)_4$ crystal based on laser oscillation at 1.3 μm. *Appl. Phys. Lett.*, 75, 325. 10.1063/1.124364

[73] Ji, Y., Liang, J., Xie, S., and Wu, X. (1994). Phase relations in the system $BaO-Nd_2O_3-B_2O_3$ ($B_2O_3 \geqslant 50$ mol%) and crystal growth of $NdBaB_9O_{16}$. *J. Cryst. Growth*, 137, 521–527. 10.1016/0022-0248(94)90993-8

[74] Luo, Z. D., Zhang, H., Huang, Y. D., Qiu, M. W., Huang, Y. C., Tu, C. Y., and Jiang, A. D. (1989). Study of $Li_6Y(BO_3)_3:Nd^{3+}$ crystal – A new laser crystal. *Chinese Phys. Lett.*, 6, 440. 10.1002/crat.21 70260125

[75] Nikolov, V., and Peshev, P. (1994). A new solvent for the growth of $Y_{1-x}Nd_xAl_3(BO_3)_4$ single crystals from high-temperature solutions. *J. Cryst. Growth*, 144, 187–192. 10.1016/0022-0248(94)90454-5

[76] Jung, S. J., Choi, D., Kang, J. K., and Chung, S. J. (1995). Top-seeded growth of Nd: YAI3(BO3)4 from high temperature solution. *J. Cryst. Growth*, 148, 207–210. 10.1016/0022-0248(94)00873-6

[77] Jaque, D. (2001). Self-frequency-sum mixing in Nd doped nonlinear crystals for laser generation in the three fundamental colours: The NYAB case. *J. Alloys Compd.*, 323–324, 204–209. 10.1016/S0925-8388(01)01111-2

[78] Scripsick, M. P., Fang, X. H., Edwards, G. J., and Halliburton, L. E. (1993). Point defects in lithium triborate (LiB_3O_5) crystals. *J. Appl. Phys.*, 73, 1114. 10.1063/1.353275

[79] Gua, R., Markgraf, S. A., Furukawa, Y., Sato, M., and Bhalla, A. S. (1995). Pyroelectric, dielectric, and piezoelectric properties of LiB_3O_5, *J. Appl. Phys.*, 78, 7234. 10.1063/1.360435

[80] Yangyeng, J., Shuqing, Z., Yujing, H., Hongwu, Z., Ming, L., and Chaoen, H. (1991). Growth of lithium triborate (LBO) single crystal fiber by the laser-heated pedestal growth method. *J. Cryst. Growth*, 112, 283–286. 10.1016/0022-0248(91)90929-Y

[81] Katsumata, T., Ohshima, K., Oe, K., Hisamoto, M., Ohtaki, T., Konoura, H., Nakagawa, H., and Takahashi, K. (1992). Non-wetting container material for growing lithium and barium borate crystals. *J. Cryst. Growth*, 125, 270–280. 10.1016/0022-0248(92)90340-O

[82] Lim, A. R. (1993). Local structure of LiB_3O_5 single crystal from ^7Li nuclear magnetic resonance. *J. Appl. Phys.*, 94 (2003) 5095. 10.1063/1.1607512

[83] Mei, L., Wang, Y., Chen, C., and Wu, B. (1993). Nonlinear optical materials based on $MBe_2BO_3F_2$ (M = Na, K). *J. Appl. Phys.*, 74, 7014. 10.1063/1.355060

[84] Mei, L., Wang, Y., and Chen, C. (1994). Crystal structure of sodium beryllium borate fluoride. *Mater. Res. Bull.*, 29, 81–87. 10.1016/0025-5408(94)90108-2

[85] Keszler, D. A. (1996). Borates for optical frequency conversion. *Curr. Opin. Solid State Mater. Sci.*, 1, 204. 10.1016/S1359-0286(96)80085-4

[86] Smith, R. W., and Keszler, D. A. (1992). The noncentrosymmetric orthoborate $BaZn_2(BO_3)_2$. *J. Solid State Chem.*, 100, 325–330. 10.1016/0022-4596(92)90107-7

[87] Guaitieri, D. M., and Chai, B. H. T. (1989). Growth of β-barium borate from $NaCl-Na2O$ solutions. *J. Cryst. Growth*, 97, 613–616. 10.1016/0022-0248(89)90562-9

[88] Bordui, P. F., Caivert, G. D., and Blachman, R. (1993). Immersion-seeded growth of large barium borate crystals from sodium chloride solution. *J. Cryst. Growth*, 129, 371–374. 10.1016/0022-0248 (93)90469-D

[89] Polgar, K., and Peter, A. (1993). Etching study on beta barium metaborate (β-BaB_2O_4) single crystals. *J. Cryst. Growth*, 134, 219. 10.1016/0022-0248(93)90129-K

[90] Aka, G., Mougel, F., Auge, F., Kahn- Harari, A., Vivien, D., Benitez, J. M.., Salin, F., Pelenc, D., Balembois, F., Georges, P., Brun, A., Linain, N., and Jacquet, M. (2000). Overview of the laser and non-linear optical properties of calcium-gadolinium-oxo-borate $Ca_4GdO(BO_3)_3$, *J. Alloys Compd.*, 303–304, 401–408. 10.1016/S0925-8388(00)00648-4

[91] Iwai, M., Kuroda, I., Nakajima, S., Sasaki, T., and Nakai, S. (1997). Crystal growth and optical characterization of rare-Earth (Re) calcium oxyborate $ReCa_4O(BO_3)_3$ (Re = Y or Gd) as new nonlinear optical material. *Japanese J. Appl. Phys.*, 36, L276. 10.1143/JJAP.36.L276

[92] Hu, Z. G., Higashiyama, T., Yoshimura, M., Yap, Y. K., Mori, Y., and Sasaki, T. (1998). A new nonlinear optical borate crystal $K_2Al_2B_2O_7$ (KAB). *Japanese J. Appl. Phys.*, 37, L1093. 10.1143/JJAP.37.L1093

[93] Jiang, H., Li, J., Wang, J., Hu, X. B., Liu, H., Teng, B., Zhang, C. Q., Dekker, P., and Wang, P. (2001). Growth of Yb: $YAl_3(BO_3)_4$ crystals and their optical and self-frequency-doubling properties, *J. Cryst. Growth*, 233, 248–252. 10.1016/S0022-0248(01)01562-7

[94] Teng, B., Wang, J., Wang, Z., Jiang, H., Hu, X., Song, R., Liu, H., Liu, Y., Wei, J., and Shao, Z. (2001). Growth and investigation of a new nonlinear optical crystal: Bismuth borate BiB_3O_6, *J. Cryst. Growth*, 224, 280–283. 10.1016/S0022-0248(01)00975-7

[95] Wang, G., Hna, T. P. J., Gallagher, H. G., and Henderson, B. (1995). Novel laser gain media based on Cr^{3+} doped mixed borates $RX_3(BO_3)_4$. *Appl. Phys. Lett.*, 67, 3906. 10.1063/1.115313

[96] Norrestam, R., Nygren, M., and Bovin, J. O. (1992). Structural investigations of new calcium rare earth (R) oxyborates with the composition $Ca_4RO(BO_3)_3$, *Chem. Mater.*, 4, 737–743. 10.1021/cm00021a044

[97] Thompson, P. D., and Keszler, D. A. (1989). The new strontium scandium borate $Sr_3Sc(BO_3)_3$. *Chem. Mater.*, 1, 292–294. 10.1021/cm00003a004

[98] Thompson, P. D., and Keszler, D. A. (1994). Structure of $Sr_3Sc(BO_3)_3$. *Chem. Mater.*, 6, 2005–2007. 10.1021/cm00047a020

[99] Yan, J. F., and Yang, H. Y. P. (1987). Crystal structure of a new mini-laser material, $Nd_2Ba_3(BO_3)_4$. *Mater. Res. Bull.*, 22, 1347–1353. 10.1016/0025-5408(87)90299-6

[100] Ha, K., and Shi, S. (1988). Optical phonons and fluorescence quenching in new laser crystal nab. *J. Lumin.*, 40–41, 698–699. 10.1016/0022-2313(88)90396-1

[101] Sardar, D. K., and Powell, R. C. (1981). Time-resolved site-selection spectroscopy studies of $NdAl_3(BO_3)_4$ crystals. *J. Lumin.*, 22, 349–358. 10.1016/0022-2313(81)90012-0

[102] Hong, H. Y. P., and Dwight, K. (1974). Crystal structure and fluorescence lifetime of $NdAl_3(BO_3)_4$, a promising laser material. *Mater. Res. Bull.*, 9, 1661–1665. 10.1016/0025-5408(74)90158-5

[103] Wang, G., He, M., and Luo, Z. (1991). Structure of β-$NdAl_3(BO_3)_4$(NAB) crystal. *Mater. Res. Bull.*, 26, 1085–1089. 10.1016/0025-5408(91)90092-Z

[104] Jaque, D., Enguita, O., J., Garcia Sole., Jiang, A. D., and Luo, Z. D. (2000). Infrared continuous-wave laser gain in neodymium aluminum borate: A promising candidate for microchip diode-pumped solid state lasers. *Appl. Phys. Lett.*, 76, 2176. 10.1063/1.126289

[105] Luo, Z. D., Jiang, A. D., Huang, Y., and Qiu, M. (1991). Studies on the self-activated laser crystal:neodymium aluminium borate $NdAl_3(BO_3)_4$. *Sci. China A*, 34, 762–768. 10.1360/ya1991-34-6-762

[106] Chinn, S. R., and Hong, H. Y. P. (1975). CW laser action in acentric $NdAl_3(BO_3)_4$ and $KNdP_4O_{12}$. *Opt. Commun.*, 15, 345–350. 10.1016/0030-4018(75)90242-4

[107] Mougel, F., Auge, F., Aka, G., Kahn-Harari, A., Vivien, D., Balenbois, F., Georges, P., and Brun, A. (1998). New green self-frequency-doubling diode-pumped $Nd:Ca_4GdO(BO_3)_3$ laser. *Appl. Phys. B*, 67, 533–535. 10.1007/s003400050540

[108] Brenier, A., and Kityk, I. V. (2001). Spectroscopic properties of Pr^{3+} doped $Ca_4GdO(BO_3)_3$ (GdCOB). *J. Appl. Phys.*, 90, 232. 10.1063/1.1379779

[109] Aka, G., Reino, E., Loiseau, P., Vivien, D., Ferrand, B., Fulbert, L., Pelenc, D., Lucas-Leclin, G., and Georges, P. (2004). $Ca_4REO(BO_3)_3$ crystals for green and blue microchip laser generation: From crystal growth to laser and nonlinear optical properties. *Opt. Mater.*, 26, 431. 10.1016/j.optmat.2004.02.009

[110] Mougel, F., Aka, G., Kahn-Harari, A., and Vivien, D. (1999). CW blue laser generation by self-sum frequency mixing in $Nd:Ca_4GdO(BO_3)_3$ (Nd:GdCOB) single crystal. *Opt. Mater.*, 13, 293–297. 10.1016/S0925-3467(99)00081-6

[111] Jiang, H., wang, J., Zhang, H., Hu, X., Burns, P., and Piper, J. A. (2002). Spectral and luminescent properties of Yb^{3+} ions in $YCa_4O(BO_3)_3$ crystal. *Chem. Phys. Lett.*, 361, 499–503. 10.1016/S0009-2614(02)00989-2

[112] Long, X., Lin, Z., Hu, Z., Wang, G., and Van, T. P. J. (2002). Optical study of Cr^{3+} doped $LaSc_3(BO_3)_4$ crystal. *J. Alloys Compd.*, 347, 52–55. 10.1016/S0925-8388(02)00785-5

[113] Chen, W., Wang, G., Lin, Z., and Hu, Z. (1999). Spectral parameters of Nd^{3+} ion in β-$Nd^{3+}:LaSc_3(BO_3)_4$ crystal. *Opt. Commun.*, 162, 49–52. 10.1016/S0030-4018(99)00080-2

[114] Chen, W., Wang, G., Lin, Z., and Hu, Z. (1999). Optical transition probability of Nd^{3+} ions in a α-$Nd^{3+}:LaSc_3(BO_3)_4$ crystal. *Phys. Rev.*, B 60, 15469. 10.1103/PhysRevB.60.15469

[115] Meyn, J. P., Jensen, T., and Hubber, G. (1994). Spectroscopic properties and efficient diode-pumped laser operation of neodymium-doped lanthanum scandium borate. *IEEE J. Quantum Electron.*, 30, 913–917. 10.1109/3.291362

[116] Lin, X., Xiong, F., Chen, Y., Gong, X., Liao, J., and Huang, Y. (2005). Growth and spectral properties of $Pr^{3+}:LaB_3O_6$ single crystal. *J. Cryst. Growth*, 280, 113–117. 10.1016/j.jcrysgro.2005.03.007

[117] Pan, S., Hu, Z., Lin, Z., and Wang, G. (2003). Growth and X-ray diffraction of Nd^{3+} doped Ba_3Y $(BO_3)_3$ crystal. *J. Cryst. Growth*, 247, 452–456. 10.1016/S0022-0248(02)02052-3

[118] Xu, Y., Gong, X., Chen, Y., Huang, M., Luo, Z., and Huang, Y. (2003). Crystal growth and optical properties of YbAl3(BO3)4: A promising stoichiometric laser crystal, *J. Cryst. Growth*, 252, 241. 10.1016/S0022-0248(03)00863-7

[119] Leonyuk, N. I., Kaporulina, E. V., Maltsev, V. V., Pilipenko, O. V., Melekhova, M. D., and Mokhov, A. V. (2004). Crystal growth and characterization of YAl$_3$(BO$_3$)$_4$ doped with Sc, Ga, Pr, Ho, Tm, Yb. *Opt. Mater.*, 26, 443–447. 10.1016/j.optmat.2003.09.010

[120] Zhang, Y., Lin, Z., Hu, Z., and Wang, G. (2005). Growth and spectroscopic properties of Yb^{3+}:Gd$_{0.5}$Y$_{0.5}$Ca$_4$O(BO$_3$)$_3$ crystal. *J. Alloys Compd.*, 390, 194–196. 10.1016/j.jallcom.2004.07.058

[121] Zhao, D., Hu, Z., Hu, Z., Lin, Z., and Wang, G. (2005). Growth and spectral properties of Er^{3+}/Yb^{3+} codoped Sr$_3$Y(BO$_3$)$_3$ crystal. *J. Cryst. Growth*, 277, 401–405. 10.1016/j.jcrysgro.2005.01.081

[122] Pan, J., Lin, Z., Hu, Z., Zhang, L., and Wang, G. (2006). Crystal growth and spectral properties of Yb^{3+}:Sr$_3$La$_2$(BO$_3$)$_4$ crystal. *Opt. Mater.*, 28, 250–254. 10.1016/j.optmat.2004.12.019

[123] Pan, J., Wu, S., and Wang, G. (2006). Crystal growth and spectral properties of Nd^{3+}:Sr$_3$Gd(BO$_3$)$_3$ crystal. *Opt. Mater.*, 28, 391–394. 10.1016/j.optmat.2004.12.022

[124] Smet, B. M. J. (1987). Phosphors based on rare-earths, a new era in fluorescent lighting. *Mater. Chem. Phys.*, 16, 283–299. 10.1016/0254-0584(87)90103-9

[125] Ronda, C. R., Justel, T., and Nikol, H. (1998). Rare earth phosphors: Fundamentals and applications. *J. Alloys Compd.*, 275–277, 669–676. 10.1016/S0925-8388(98)00416-2

[126] Aron, A., Tigreat, P. Y., Caramanian, A., and Antic- Fidancev, E. (2000). Infrared and visible emission of Pr^{3+}, Eu^{3+}, Yb^{3+}/Er^{3+} in Ca$_4$Gd(BO$_3$)$_3$O (GdCOB). *J. Lumin.*, 87–89, 611–613. 10.1016/S0022-2313(99)00328-2

[127] Duan, C. J., Chen, H. H., Yang, X. X., and Zhao, J. T. (2006). Luminescence properties of Eu^{3+}, Tb^{3+} or Tm^{3+} activated Ca$_4$GdO(BO$_3$)$_3$ under X-ray and UV excitation. *Opt. Mater.*, 28, 956–961. 10.1016/j.optmat.2005.05.005

[128] Zhang, Y., and Li, Y. (2005). Photo-luminescent properties of Eu^{3+} ion-doped Ca$_4$REO(BO$_3$)$_3$ (RE double bond La,Y,Gd). *J. Lumin.*, 113, 45–49. 10.1016/j.jlumin.2004.08.051

[129] Guo, F., and Peng, Y. (1988). Photoluminescence of terbium ion in the binary system of GdB$_3$O$_6$ – CeB$_3$O$_6$. *J. Lumin.*, 40–41, 175–176. 10.1016/0022-2313(88)90144-5

[130] Nakamura, T., Takeyama, T., Takahashi, N., Jagannathan, R., Karthikeyani, A., Smith, G. M., and Riedi, P. C. (2003). High-frequency EPR investigation of X-ray storage SrBPO$_5$:Eu phosphor. *J. Lumin.*, 102–103, 369–372. 10.1016/S0022-2313(02)00537-9

[131] Karthikeyani, A., and Jagannathan, R. (2000). Eu^{2+} luminescence in stillwellite-type SrBPO$_5$ – A new potential X-ray storage phosphor. *J. Lumin.*, 86, 79–85. 10.1016/S0022-2313(99)00171-4

[132] Risheng, C., Jinggen, H., Ling, L., and Yan, X. (1988). Study of the new boron-rich calcium rare earth borate CaLnB$_7$O$_{13}$. *Mater. Res. Bull.*, 23, 1699–1704. 10.1016/0025-5408(88)90178-X

[133] Datchi, F., Le Toyllec, R., and Loubeyre, P. (1997). Improved calibration of the SrB$_4$O$_7$:Sm^{2+} optical pressure gauge: Advantages at very high pressures and high temperatures. *J. Appl. Phys.*, 81, 3333. 10.1063/1.365025

[134] Lu, C. H., Godbole, S. V., and Natarajan, V. (2005). Luminescence characteristics of strontium borate phosphate phosphors. *Mater. Chem. Phys.*, 94, 73–77. 10.1016/j.matchemphys.2005.04.011

[135] Dotsenko, V. P. (2000). Rapid synthesis of nanoscale colloidal metal clusters by microwave irradiation. *J. Mater. Chem.*, 10, 9. 10.1039/B002232M

[136] Dotsenko, V. P., berezovskaya, I. V., Efryushina, N. P., Voloshinovskii, A. S., Dorenbos, P., and van Eijk, C. W. E. (2001). Luminescence of Ce^{3+} ions in strontium haloborates, *J. Lumin.*, 93, 137–145. 10.1016/S0022-2313(01)00179-X

[137] Sidorenko, A. V., Bos, A. J. J., Dorenbos, P., van Eijk, C. W. E., Rodnyi, P. A., berezovskaya, I. V., Dotsenko, V. P., Guillot- Noel, O., and Gourier, D. (2004). Radiation induced defects in Sr$_2$B$_5$O$_9$Br:Ce^{3+} storage phosphor, *J. Phys: Condens. Mater.*, 16, 4131. 10.1088/0953-8984/16/23/027

[138] Z., Pei, Zeng, Q., and Su, Q. (2000). The application and a substitution defect model for Eu^{3+} → Eu^{2+} reduction in non-reducing atmospheres in borates containing BO$_4$ anion groups. *J. Phys. Chem. Solids*, 61, 9–12. 10.1016/S0022-3697(99)00237-1

[139] Keszler, D. A. (1999). Synthesis, crystal chemistry, and optical properties of metal borates. *Curr. Opin. Solid State Mater. Sci.*, 4, 155–162. 10.1016/S1359-0286(99)00011-X

[140] Mc Carthy, G. J., and White, W. B. (1970). On the stabilities of the lower oxides of the rare earths. *J. Less Common Metals*, 22, 409–417. 10.1016/0022-5088(70)90128-1

[141] Machida, K., Adachi, G., and Shiokava, J. (1979). Luminescence properties of Eu(II)-borates and Eu^{2+} activated Sr-Borates. *J. Lumin.*, 21, 101–110. 10.1016/0022-2313(79)90038-3

[142] Peters, T. E., and Baglio, J. (1970). *Inog J. Nucl. Chem.*, 32, 1083.

[143] Diaz, A., and Keszler, D. A. (1996). Red, green, and blue Eu^{2+} luminescence in solid-state borates: A structure-property relationship. *Mater. Res. Bull.*, 31, 147–151. 10.1016/0025-5408(95)00182-4

[144] Pei, Z. Zeng, Q., and Su, Q. (1999). A study on the mechanism of the abnormal reduction of Eu^{3+} → Eu^{2+} in Sr$_2$B$_5$O$_9$Cl prepared in air at high temperature. *J. Solid State Chem.*, 145, 212–215. 10.1006/jssc.1999.8246

[145] Zhao, J., and Cocivera, M. (2001). Luminescent characteristics of blue-emitting Sr$_2$B$_5$O$_9$Cl:Eu thin-film phosphors. *Appl. Phys. Lett.*, 79, 740. 10.1063/1.1391410

[146] Jiang, Y. D., Zhang, F., Summers, C. J., and Wang, Z. I. (1999). Synthesis and properties of Sr$_2$CeO$_4$ blue emission powder phosphor for field emission displays. *Appl. Phys. Lett*, 74, 1677. 10.1063/1.123652

[147] Kumar, D., Sankar, J., Cho, K. G., Cracium, V., and Singh, R. K. (2000). Enhancement of cathodoluminescent and photoluminescent properties of Eu:Y$_2$O$_3$ luminescent films by vacuum cooling. *Appl. Phys. Lett.*, 77, 2518. 10.1063/1.1318938

[148] Meijrink, A., and Blasse, G. (1991). Photostimulated luminescence and thermally stimulated luminescence of some new X-ray storage phosphors. *J. Phys. D: Appl. Phys.*, 24, 626. 10.1088/0022-3727/24/4/016

[149] Verstgen, J. M. P. J., Ter Vrugt, J. W., and Wanmaker, W. L. (1972). Luminescence of Eu^{2+}-activated SrB$_2$Si$_2$O$_8$. *J. Inorg. Nucl. Chem.*, 34, 3588.

[150] Welkar, T. (1991). Recent developments on phosphors for fluorescent lamps and cathode-ray tubes. *J. Lumin.*, 48–49, 49–56. 10.1016/0022-2313(91)90075-7

[151] Diang, S. J., Zhang, D. W., Wang, P. F., and Wang, J. J. (2001). Preparation and photoluminescence of the Ce, Tb and Gd-doped lanthanum borophosphate phosphor. *Mater. Chem. Phys.*, 68, 98–104. 10.1016/S0254-0584(00)00300-X

[152] Hao, J., Gao, J., and Cocivera, M. (2015). Stimuli responsive upconversion luminescence nanomaterials and films for various applications. *Appl. Phys. Lett.*, 44, 1585–1607. 10.1039/C4CS00171K

[153] Hao, J., Gao, J., and Cocivera, M. (2003). *Appl. Phys. Lett.*, 82, 2778.

[154] Hao, J., and Cocivera, M. (2001). *Appl. Phys. Lett.*, 79, 740.

[155] Hao, J., and Cocivera, M. (2001). *Appl. Phys. Lett.*, 81, 4154.

[156] Wei, Z. G., Sun, L. D., Jiang, X. C., Liao, C. S., and Yan, C. H. (2003). Correlation between size-dependent luminescent properties and local structure around Eu^{3+} ions in YBO$_3$:Eu nanocrystals: An XAFS study. *Chem. Mater.*, 15, 3011–3017. 10.1021/cm0341888

[157] Yang, Z., Ren, M., Lin, J. H., Su, M. Z., Tao, T., and Wang, W. (2004). Ordered Nanosheet-based YBO$_3$:Eu^{3+} assemblies: Synthesis and tunable luminescent properties. *J. Phys. Chem. B*, 108, 11, 3387–3390. 10.1021/jp037301q

[158] Wei, Z. G., Sun, L. D., Liao, C. S., Jiang, X. C., and Yan, C. H. (2003). Size dependence of luminescent properties for hexagonal YBO$_3$:Eu nanocrystals in the vacuum ultraviolet region. *J. Appl. Phys.*, 93, 9783. 10.1063/1.1575496

[159] Jiang, C., Yan, C. H., Sun, L. D., Wei, Z. G., and Liao, C. S. Hydrothermal homogeneous urea precipitation of hexagonal YBO$_3$:Eu^{3+} nanocrystals with improved luminescent properties. *J. Solid State Chem.*, 175 (2003) 245-151. 10.1016/S0022-4596(03)00276-7

[160] Li, J., Wang, J., Tan, H., Cheng, X., Song, F., Zhang, H., and Zhao, S. (2003). Growth and optical properties of Ho,Yb:YAl$_3$(BO$_3$)$_4$ crystal. *J. Cryst. Growth*, 256, 324–327. 10.1016/S0022-0248(03)01370-8

[161] Liang, H., Zeng, Q., Tao, Y., Wang, S., and Su, Q. (2003). VUV-UV excited luminescent properties of calcium borophosphate doped with rare earth ions. *Mater. Sci. Eng.*, B98, 213–219. 10.1016/S0921-5107(03)00034-5

[162] Boyer, D., Bertrand-Chadeyran, G., Mahiou, R., Brioude, A., and Mugnier, J. (2003). Synthesis and characterization of sol–gel derived Y$_3$BO$_6$:Eu^{3+} powders and films. *Opt. Mater.*, 24, 35–41. 10.1016/S0925-3467(03)00102-2

[163] Kwon, I. E., Yu, B. Y., Bae, H., Hwang, Y. J., Kwon, T. W., Kim, C. H., Pyun, C. H., and Kim, S. J. (2000). Luminescence properties of borate phosphors in the UV/VUV region. *J. Lumin.*, 87–89, 1039–1041. 10.1016/S0022-2313(99)00532-3

[164] Kang, Y. C., and Park, S. B. (1999). Morphology of $(Y_xGd_{1-x})BO_3$:Eu phosphor particles prepared by spray pyrolysis from aqueous and colloidal solutions. *Japanese J.Appl. Phys.*, 38, L 1541. 10.1143/JJAP.38.L1541

[165] Ronda, C. R. (1997). Recent achievements in research on phosphors for lamps and displays. *J. Lumin.*, 72-74, 49–54. 10.1016/S0022-2313(96)00374-2

[166] Kim, D. S., and Lee, R. Y. (2000). Synthesis and photoluminescence properties of (Y,Gd)BO$_3$: Eu phosphor prepared by ultrasonic spray. *J. Mater. Sci.*, 35, 4777–4782. 10.1023/A:1004864426980

[167] Justel, T., Krupa, J. C., and Wiechert, D. U. (2001). VUV spectroscopy of luminescent materials for plasma display panels and Xe discharge lamps. *J. Lumin.*, 93, 179–189. 10.1016/S0022-2313(01)00199-5

[168] Yu, Z., Huang, X., Zhung, W., Cui, X., and Li, H. (2005). Crystal structure transformation and luminescent behavior of the red phosphor for plasma display panels. *J. Alloys Compd.*, 390, 220–222. 10.1016/j.jallcom.2004.07.065

[169] Sohn, J. R., Kang, Y. C., Park, H. D., and Yoon, S. G. (2002). (YGd)BO$_3$:Eu phosphor particles prepared from the solution of polymeric precursors by spray pyrolysis. *Japanese J. Appl. Phys.*, 41, 6007. 10.1143/JJAP.41.6007

[170] Wei, Z. G., Sun, L. D., Liao, C. S., Cheng Jiang, X. C., and Yan, C. H. (2002). Synthesis and size dependent luminescent properties of hexagonal (Y,Gd)BO$_3$:Eu nanocrystals. *J. Mater. Chem.*, 12, 3665–3670. 10.1039/B207103G

[171] Jeoung, B. W., Hong, G. Y., Yoo, W. T., and Yoo, J. S. (2004). Preparation of spherical phosphor (Y,Gd)BO$_3$:Eu by polymeric-aerosol pyrolysis. *J. Electrochem. Soc.*, 151, H213. 10.1149/1.1790513

[172] Gupta, S. K., Agrawal, D. C., and Mahapatra, Y. N. (2004). Photoluminescence and morphological studies of $(Y_{0.5}Gd_{0.5})BO_3$: Eu phosphor powders prepared by the urea hydrolysis route. *J. Electrochem. Soc.*, 151, H239. 10.1149/1.1804814

[173] Sano, H., Matsumoto, T., Matsumoto, Y., and Koinuma, H. (2005). *Appl. Phys. Lett.*, 86, 21124.

[174] Park, W., Lee, R. Y., Summer, C. J., Do, Y. R., and Yang, H. G. Photoluminescence properties of $Al_3GdB_4O_{12}$:Eu phosphors, *Mater. Sci. Eng.*, B78, 28–31. 10.1016/S0921-5107(00)00509-2

[175] Wang, Y., Uheda, K., Takizawa, H., Mizumoto, U., and Endo, T. (2001). Synthesis of $Gd_{1-x}Eu_xAl_3(BO_3)_4$ $(0 < x \leq 1)$ and its photoluminescence properties under UV and vacuum ultraviolet regions. *J. Electrochem. Soc.*, 148, G430. 10.1149/1.1383778

[176] Lee, K. G., Yu, B. Y., Pyun, C. H., and Mho, Sun-II. (2002). Vacuum ultraviolet excitation and photoluminescence characteristics of $(Y,Gd)Al_3(BO_3)_4/Eu^{3+}$. *Solid State Commun.*, 22, 485–488. 10.1016/S0038-1098(02)00195-3

[177] Hongpeng, Y., Hong, G., Zeng, X., Pyun, C. H., Yu, B. Y., and Bae, H. S. (2000). VUV excitation properties of $LnAl_3B_4O_{12}$:Re (Ln = Y, Gd; Re = Eu, Tb). *J. Phys. Chem. Solids*, 61, 1985–1988. 10.1016/S0022-3697(00)00192-X

[178] Tian, L., You, B. Y., Pyun, C. H., Park, H. L., and Mho, Sun-il. (2004). New red phosphors BaZr$(BO_3)_2$ and $SrAl_2B_2O_7$ doped with Eu^{3+} for PDP applications. *Solid State Commun.*, 129, 43–46. 10.1016/j.ssc.2003.09.012

[179] Tian, L., Kim, S. J., Park, H. L., and Mho, S. I. (2006). Variation of the photoluminescence and vacuum ultraviolet excitation characteristics of $BaZr(BO_3)_2$:Eu^{3+} by the incorporation of Al^{3+}, La^{3+}, or Y^{3+} into the lattice. *Mater. Res. Bull.*, 41, 29–37. 10.1016/j.materresbull.2005.07.039

2 Borate Phosphor
Synthesis and Characterization

K. A. Koparkar

CONTENTS

2.1   INTRODUCTION

The inorganic compounds are composed of host materials doped with suitable impurity ions. The synthesis of borate-based phosphor materials may appear straightforward as the host materials are well-known, but in practice the synthesis of borate-based phosphors with desired characteristics is quite multifarious and complicated.

DOI: 10.1201/9781003207757-2

In this chapter, we are describing the methods of synthesis and mechanisms that are involved in the preparation of borate-based phosphors useful for various applications. We have employed some traditional, modified and novel methods for synthesis of various borate-based materials.

2.2 SYNTHESIS TECHNIQUES

2.2.1 CONVENTIONAL METHOD

2.2.1.1 Solid State Diffusion (SSD) Method

In the SSD method, the precursors are made to react through a diffusion process. The intermixing of precursors at an atomic level assisted by high temperature without melting the constituents is expected to achieve. Reaction time and the temperature bear a sort of reciprocal relation. Although many researchers frequently used this technique, it operates at high temperatures (~1600°C) and takes a long time for a reaction process (even two to three days). Due to insufficient mixing and the low reactivity of raw materials, several intermediate phases easily exist in the products and the possibility of obtaining a desired single-phase product becomes difficult, which also restricts the use of doping concentration and hence strongly affects luminescence performance of phosphors. Repeated grinding and calcinations are required to eliminate these intermediate compounds, thus increasing the consumption in time as well as energy. It is difficult to control over particles size and morphology so that the phosphor suffers for long time calcinations at a very high temperature with the result that its particles have a large size (~5–10 μm) with wide distribution range and heterogeneous morphology, which results in unfavorable luminescent properties in the applied devices for its industrial aspects [1,2].

In our experimental study, we modified the step-wise SSD method, which required less time compared to the high-temperature SSD method [3]. The general presentation of the SSD method is as shown in Figure 2.1.

Advantage

- The only advantage of this method is that the precursors are readily available.

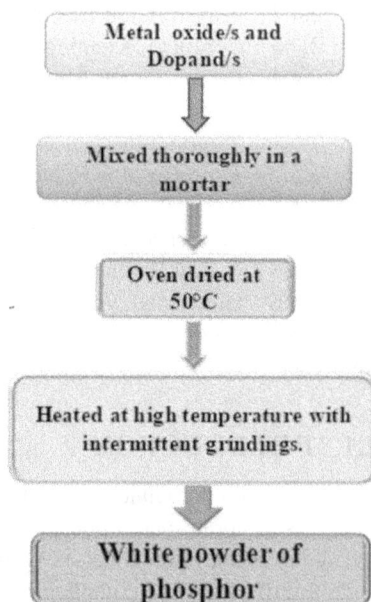

FIGURE 2.1 Flow chart of SSD method.

Disadvantages

- Coarse grain size, powder agglomerates withnon-homogeneous properties due to limited diffusion lengths.
- Repeated cycles of heating and cooling followed after crushing the material in between these cycles.
- This added processing time as well as impurity during crushing.
- Powders produced by this method are not suited for the fabrication of coatings and fibers because of 22 non-uniform-sized particles [4,5].

In Table 2.1, there are several reports on the solid-state synthesis of borate-based phosphors doped with rare earth ions.

Therefore, it becomes necessary to explore simple, easy and low-cost methods to overcome the disadvantages associated with the solid-state diffusion process.

2.2.2 NON-CONVENTIONAL METHOD

2.2.2.1 Co-Precipitation Method (CP)

The co-precipitation technique is one reliable method and recently employed for the formation of salts from oxides. In this method, two or more salts simultaneously undergo precipitation, as precipitate obtained from the clear, homogeneous mixture of host and dopant activators. It ensures uniform distribution of activators at the atomic level in host material particles, leading to homogeneity of product, due to its many advantages, such as low cost, low reaction temperature, simple equipment and less consumption of energy; the red phosphors prepared by the co-precipitation method have better luminous performance than other methods [21]. The general flow chart of CP is

TABLE 2.1

Various borate-based phosphors synthesized through solid state diffusion method

Sr. no.	Phosphors	Particle size/ range	Morphology	Required temperature for synthesis	Synthesis time	Ref.
1	$YBO_3:Eu^{3+}$	30–50 nm	Spherical	1100°C	5 h	[6]
2	$YBO_3:Eu^{3+}$	NR	NR	1300°C	3 h	[7]
3	$(Y,Gd)BO_3:Sb^{3+}$	NR	NR	800°C–1150°C	2–4 h	[8]
4	$YBO_3:Eu^{3+}$	NR	NR	1250°C	4–5 h	[9,10]
5	$YBO_3:Dy^{3+}, Bi^{3+}$	0.6 µm	Regular-agglomerated	900°C	8 h	[11]
6	$Ba_2Gd_5B_5O_{17}:Ce^{3+}/Tb^{3+}$	NR	NR	1200°C	8 h	[12]
7	$GdAl_3(BO_3)_4:Cr^{3+}$	1 µm	Hexagonal shape	1100°C	6 h	[13]
8	$NaSr_4(BO_3)_3:Pb^{2+}$	NR	NR	850°C	2 h	[14]
9	$LiSr4(BO_3)_3: Pb^{2+}$	NR	NR	800°C	2 h	[15]
10	$KCaBO_3:Eu^{3+}$	NR	NR	950°C	21 h	[16]
11	$BaB_8O_{13}:Gd^{3+}$	2 to 10 µm	Agglomeration particles	800°C	24 h	[17]
12	$BaB_4O_7:Ce$	~2 mm	Irregular shape	700°C	8 h	[18]
13	$MgB_4O_7:Mn,Tb$	20–25 µm	Irregular shape	800°C	5 h	[19]
14	$LiCaBO_3:Tm^{3+}$	100 µm	NR	750°C	3 h	[20]

*NR = Not reported.

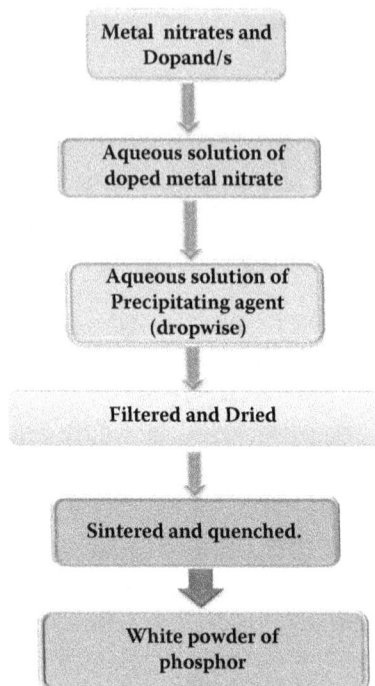

FIGURE 2.2 Flow chart of the CP method.

as shown in Figure 2.2. But if two salts undergo precipitation, it may happen that their rate of precipitation is different. In the precipitation method only one compound undergoes precipitation; even if the desired compound is water soluble, it might be prepared by dissolving the appropriate constituents in a common solvent, followed by re-crystallization, which can be achieved by boiling off the solvent. However, the solvent molecules may remain as contaminants and this can adversely affect the luminescence performance of the phosphor, conditions of pH, temperature and reactant concentration further eliminate the traces of unwanted insoluble particles from the precipitate and repeated washing with plenty of water has to be done.

The CP method is highly preferred to have particles with well-controlled size and morphology. Obviously, the precipitant concentration and surfactant will be the key factors to be monitored and hence the role of these factors is to be understood in the right spirit. The general flow chart of CP is as shown in Figure 2.2.

In the precipitation process, the solution is either seeded or undergoes auto-precipitation. The problem can arise when two or more components are co-precipitated; different species do not always deposit from solution at the reaction in a specific pH.

The CP method is used to prepare ceramic oxides through formation of intermediate precipitates, usually hydrous oxides or oxalates, so that an intimate mixture of components is formed during precipitation, and chemical homogeneity is maintained on calcinations [22,23].

The precipitate was washed, dried, calcined and milled to get the required oxide product. This method has limitations to obtain phosphors with high luminescence efficiency. In CP, careful controls of solution conditions are required for precipitation of all cations and thus maintain chemical homogeneity on the molecular scale. The method is frequently used for the manufacture of high-performance phosphors.

There are several reports on the co-precipitation synthesis of borate-based phosphors doped with RE ions, as tabulated in Table 2.2.

TABLE 2.2
Various Borate based phosphors synthesized through the CP method

Sr. no.	Phosphors	Particle size/range	Morphology	Required temperature for synthesis	Synthesis time	Ref.
1	α-Zn(BO$_2$)$_2$:Eu^{3+}	530 to 850 nm	Monodispersed nanoparticles	800°C	1 h	[24]
2	YBO$_3$:Eu^{3+}	20–40 nm	Spherical-like	800°C	2 h	[25]
3	(Y, GdBO$_3$:Eu^{3+}	NR	NR	950°C	2 h	[26]
4	YBO$_3$:Eu^{3+}	250 to 400 nm	NR	900°C	2 h	[27]
5	Y$_2$O$_3$:Eu^{3+}	30–125 nm	Spherical shape	800°C	1 h	[28]
6	NaBaBO$_3$:Eu^{3+}	150 nm to 5 μm	tabular particles	700°C	8 h	[29]
7	YBO$_3$:Sm^{3+}	96 to 102 nm	spherical particles	1200°C	1 h	[30]
8	CaB$_4$O$_7$:Mn	9 nm	Spherical shapes	80°C	24 h	[31]
9	NaBaPO$_4$:Gd^{3+}	NR	NR	800°C	3 h	[32]
10	α-Zn(BO$_2$)$_2$:Eu^{3+}	30 nm	Nanosheets morphology	60°C	7 h	[33]

*NR = Not reported.

Advantages

- It gives crystalline size in the nano range, which depends on the precipitating agent used in the reaction.
- The crystallite size and morphology of the phosphor prepared using this method can be well controlled.

Disadvantages

- The precipitate needs to be repeatedly washed, dried and calcined to get the pure phase of phosphor. It will be a very slow and time-consuming process.
- This method requires a precipitating agent for the formation of precipitate.

2.2.2.2   Sol-Gel Method

The sol-gel method is one of the traditional and simplest wet chemical syntheses of oxide by the molecular chemical process. The molecular motions and hence the chemical reactions may proceed very swiftly in the liquid state. Colloidal solute particles are suspended by hydrolysis, gelatin agent to obtain the gel. Generally citric acid and ethylene glycol are used as the gelation agents to obtain gel; formation of allkoxide, ceramic powders and silica hybrid materials are obtained successfully by this method [34–36]. A polymeric fiber linking monomers and slurry-like mixtures were obtained, which further dried by vacuum-rotation dryer and calcination. This method can produce a high homogeneity and morphology of the chemical composition of the materials. But dryness of the obtained gel is difficult and requires sophisticated dryer techniques and this deliberately slow process takes long time; further, the precursor powders obtained by grinding the dry gels are later required to be repeatedly calcined in the range of 800–1500°C and then milled several times [37,38], otherwise flaws and cracks within phosphor pellets were obtained. The general flow chart of the sol-gel method is shown in Figure 2.3.

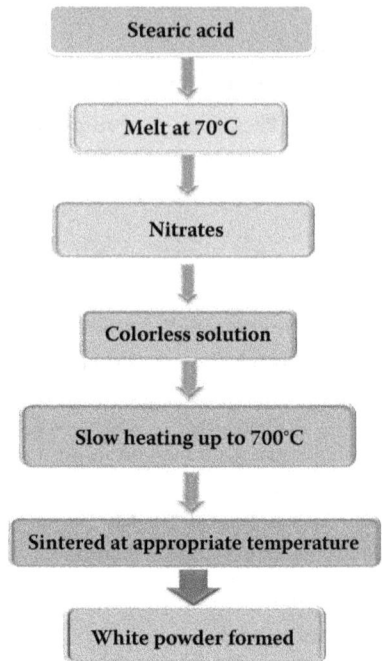

FIGURE 2.3 Flow chart of sol-gel method.

Advantages

- Good chemical homogeneity due to mixing components at the colloidal level.
- Lower reaction temperatures required than conventional powder mixing method.
- As it involves handling liquid feeds, small amounts of dopants can be readily introduced, while lower crystallization temperatures enable preparation of phases that are unstable at high temperatures.
- Because many alkoxides are liquids or volatile solids, they can be purified to form an extremely pure oxide source, which is important for the electro-ceramic synthesis.
- The improved homogeneity is associated with lower crystallization temperatures than the use of colloids, often between 400°C and 800°C for gel-to-oxide conversion.

Disadvantages

- Alkoxides are relatively expensive compared to precursors for sol-gel processing of colloids.
- Using solvent-based rather than water-based process.
- High purity and high cost chemicals are required for the gel formation (Table 2.3).

2.2.2.3 Combustion Synthesis (CS) Method

Combustion synthesis method could be the best option driven to overcome the solid-state diffusion method. The attempts are made to eliminate the diffusion control problems of the solid-state diffusion method [49]. Strongly exothermic reactions are observed to sustain themselves and propagate in the form of a combustion wave until the reactants are completely consumed. Philpot et al. [50] and Booth et al. [51] were the first to report such observations and to lay out the mathematical foundation, which is nowadays referred to as the self-propagating high-temperature synthesis (SHS), or CM [52].

This has led to a low cost, energetically efficient method for the development of advanced materials. The CS method is a promising method to prepare high-purity, small-sized particle phosphors because the starting raw materials are intermixed at the atomic level and homogeneously mixed in liquid phases, and the high temperature generated instantly by the exothermic reaction can volatilize low boiling point

TABLE 2.3
Various Borate based phosphors synthesized through the sol-gel method

Sr. no.	Phosphors	Particle size/range	Morphology	Required temperature for synthesis	Synthesis time	Ref.
1	$REAl_3(BO_3)_4:Eu^{3+}/Tb^{3+}$ (RE = Y, Gd)	200–300 nm	Agglomerated particles	950°C	3 h	[39]
2	Dy-doped strontium borates	NR	NR	950°C	3 h	[40]
3	$Sr_3RE_2(BO_3)_4:Tb^{3+}$	40 nm	Agglomerated particles	900°C	3 h	[41]
4	$YBO_3:Tb^{3+}$ and $Y,GdBO_3:Tb^{3+}$	0.1–2 μm	Plate-like structure	1200°C	4 h	[42]
5	$Li_2O–B_2O_3$ system	60 nm to 0.5 μm	Hexagonal platelets	500°C	5 h	[43]
6	$Sr_3Y_2(BO_3)_4:Eu^{3+}$	100–300 nm	Agglomerated particles	900°C	3 h	[44]
7	$Gd_{0.9}Eu_{0.1}BO_3$	150–250 nm	Highly facetted and smooth surfaces	1200°C	4 h	[45]
8	$RE_3BO_6: Eu^{3+}/Tb^{3+}$ (RE = Y, Gd)	200 nm	Foamy shape of material	1000°C	3 h	[46]
9	$β-LaB_5O_9$	NR	NR	870°C	NR	[47]
10	$Ca_3La_3(BO_3)_5: Eu^{3+}$	NR	A rounded shape	900°C	3 h	[48]

*NR = Not reported.

impurities, leading to purer products. In addition, this method results in products with narrow particle distribution because of the decrease in reaction time (a few seconds during the combustion reaction) [53]. When heated rapidly at 550°C, the solution containing a stoichiometric amount of redox mixture boils and dehydrates, followed by decomposition, generating combustible gases; the volatile combustible gases ignite and burn with a flame. The large amount of escaping gases dissipates heat and thereby prevents material from sintering and thus provides conditions for formation of a nanocrystalline phase. Also, as gases escape they leave voluminous, foaming and crystalline fine powder, occupying the entire volume of the container and have no chance of forming agglomerations, unlike in the other conventional processes. Therefore, in combustion synthesis, instantaneous and very high temperature, combined with the release of a large volume of volatiles from a liquid mixture results in the production of nanoparticles.

The process simply involves highly exothermic redox chemical reactions between metals and non-metals; a metathetical (exchange) reaction between reactive compounds or reactions involving redox compounds/mixtures. This process is nothing but the 'combustion' consisting of flaming (gas phase) and smoldering (heterogeneous) as well as explosive reactions.

The CS method was successfully used in the preparation of a large number of technologically useful oxide (refractory oxides, magnetic, dielectric, semiconducting, insulators, catalysts, sensors, phosphors, etc.) and non-oxide (carbides, borides, silicides, nitrides, etc.) materials. Important parameters that control the CS method are the particle size and shape of the reactants, ignition techniques and stoichiometric ratio, processing of reactant particles and adiabatic temperature, which is a measure of the exothermicity of the reaction [54–56]. It is an effective, low-cost method for production of various industrially useful materials.

The good fuel not only acts as a complexing agent for metal ions but also reacts non-violently without producing toxic gasses [57].

Various fuels and oxidizers, which are used in the combustion synthesis, are listed in Tables 2.4 and 2.5 with their valancies, respectively.

The reaction temperatures depend on the type of fuel and oxidizer used in the stoichiometric molar ratio of fuel to oxidizers used in the synthesis. The role of fuel is to burn and the oxidizer is used to supply the oxygen in proper quantities or in excess.

The temperatures achieved during the combustion are indicated by the color of the flame witnessed. The effect of the fuel-to-oxidizer ratio on flame temperature and flame duration time are tabulated in Table 2.6. In combustion, the unwanted ingredients are released in the form of gases that actually make the product in small sized particles. The more the gases released the more dispersion

TABLE 2.4

Various fuels used in the CS method

Sr. no.	Fuel	Formula	Valency
1	Urea	NH_2CONH_2	+6
2	Tetraformal trisazine	$C_4H_{16}N_6O_2$	+28
3	Carbohydrazide	CH_6N_4O	+8
4	Glycine	$C_2H_5NO_2$	+9
5	Malonodihydrazide	$C_3H_8N_4O_2$	+16
6	3 Methyl Pyrazole-5 One	$C_4H_6N_2O$	+20
7	Dofirmyl hydrazine	$C_2H_4N_2O_2$	+8
8	Oxalyl dihydrazide		+10
9	Citric acid		+18
10	Oxaldohydrazide		+9

TABLE 2.5

Various oxidizers used in the CS method

S.N.	Oxidizers	Valency
1	Ammonium perchlorate	−5
2	Ammonium nitrate	−2
3	$M(NO_3)$	−10
4	$M(NO_3)_2$	−15
5	$M(NO_3)_3$	−20
6	$M(NO_3)_4$	−25

TABLE 2.6

Effect of fuel-to-oxidizer ratio on flame temperature and flame temperature duration time

Sr. no.	Ammonium nitrate	Urea	Molar ratio	Flame colour	Duration of flame in sec.
1	12	4	1:0.33	Fent-Yellow	12
2	12	8	1:0.66	Yellow	20
3	12	10	1:0.83	Dark-Yellow	30
4	12	12	1:1	Fent-Orange	35
5	12	14.4	1:1.2	Orange	35
6	12	18	1:1.5	Orange-Red	36
7	12	24	1:2	Orange-Red	40

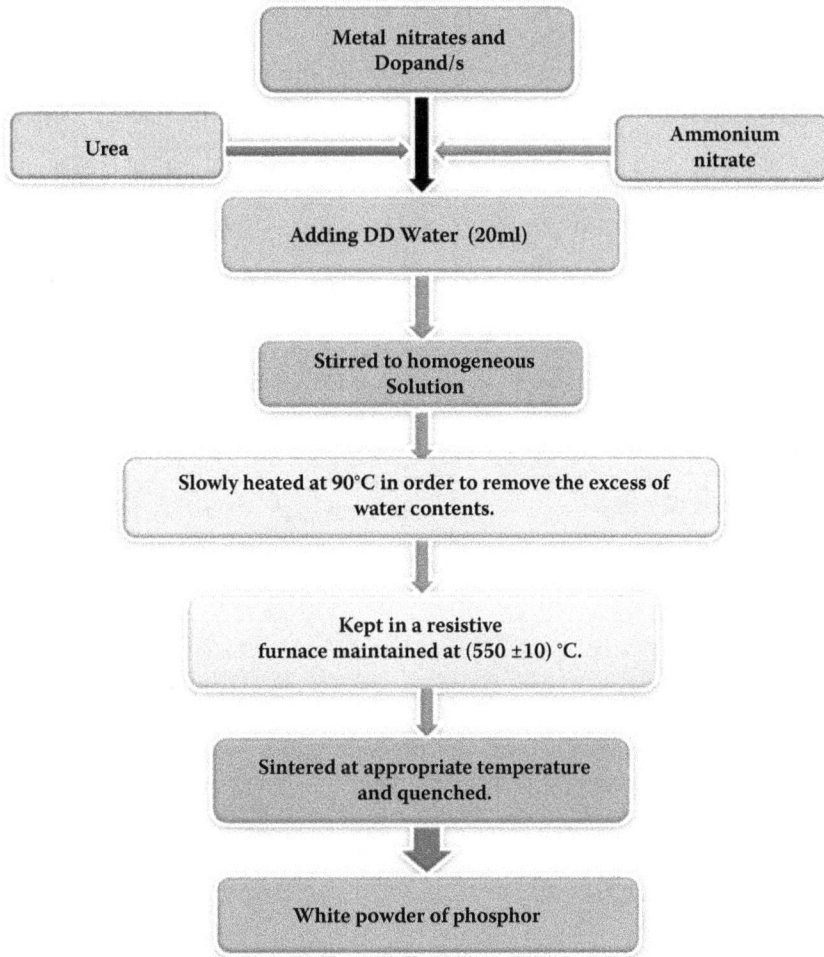

FIGURE 2.4 The flow chart of CS method.

and hence the size of the product particle will be reduced. The duration of the flame decides the size of the particles. Therefore, the fuel-to-oxidizer ratio has to be accurately monitored for the desired particle size of the product. Table 2.6 mentions the effect of the fuel-to-oxidizer ratio on flame colour while preparing phosphors. From this table it can be concluded that when the fuel-to-oxidizer ratio is greater, the orange to red flame produced more heat, which can persist for a long time during the CS method and hence this method requires less annealing temperature for preparation of phosphor materials. The CS method involves a ratio of ammonium nitrate, keeping constant as an oxidizer and different moles of urea as a fuel. The general CS method is shown in Figure 2.4.

In Table 2.7, there are several reports on the CS method of borate-based phosphors doped with rare earth ions. There are several reports indicating that the CS route has been successfully used for preparation of nanomaterials and is accomplished in a number of countries [58].

Advantages

- Simple, low-cost, energy-saving method.
- Homogeneously mixed in liquid phases and high temperature generated instantly by exothermic reaction.

TABLE 2.7

Various Borate based phosphors synthesized through the CS method

Sr. no.	Phosphors	Particle size/ range	Morphology	Required temperature for synthesis	Synthesis time	Ref.
1	$YBO_3:Tb^3$	Micrometer	Irregular	1000°C	1 h	[59]
2	$(Y,Gd)BO_3:Bi^{3+}, Pr^{3+}$	4–10 μm	Irregular	800°C	4 h	[60]
3	$Sr_3Y(BO_3)_3:Ce^{3+}$	70–90 nm	Slightly agglomerated	800°C	5 h	[61]
4	$M_3Y_2(BO_3)_4:Eu^{3+}$ (M=Ba, Sr)	1 μm	Rosette morphology	900°C	2 h	[62]
5	$YCa_4O(BO_3)_3:RE^{3+}$ (RE = Eu^{3+}, Tb^{3+})	1–2 μm	Rosette morphology	900°C	90 min	[63]
6	$KSr_4(BO_3)_3:Dy$	1–5 μm	Agglomerated	750°C	2 h	[64]
7	$KCa_4(BO_3)_3:Pb^{2+}$	4–12 μm	Irregular grains	800°C	3 h	[65]
8	$Sr_2Mg(BO_3)_2:Pb^{2+}, Gd^{3+}$	4–8 μm	Irregular grains	800 ± 10 °C	3 h	[66]
9	$BaAl_2O_4:Cr^{3+}$	Micrometer	Platelets	500 °C	5 min	[67]
10	$YAl3(BO3)4:Sm^{3+}$ and $LaMgB5O10:Ce^{3+}$	NR	NR	750°C	90 min	[68]

*NR = Not reported.

- The combustion-derived powders have a narrow size distribution.
- The combustion synthesis as a preparation process to produce homogeneous, very fine crystalline, un-agglomerated, multi-component oxide ceramic powders without the intermediate decomposition and/or calcining steps.

Disadvantages

- Evolution of a large amount of toxic gases (NH_3, H_2O, CO_2).
- It is very difficult to maintain the fuel/oxidizer (F/O) ratio in CM.

2.2.2.4   Hydrothermal (HT) Synthesis

The hydrothermal technique is widely used in industrial processes for the dissolution of bauxite prior to the precipitation of gibbsite in the Bayer process and for the preparation of aluminosilicate zeolites. Many researchers have shown an interest in this method for the synthesis of materials but it is not as popular as the sol-gel process. This is surprising because hydrothermal synthesis offers a low-temperature, direct route to submicrometer-sized powders with a narrow distribution, avoiding the calcination step required in sol-gel processing.

The hydrothermal method, which uses autogenous pressure developed at temperatures above the boiling point of water, has been used especially in the synthesis of various ceramic oxide powders. The advantage of the hydrothermal method is that ceramic materials can be synthesized at relatively low temperatures (100–300°C) without milling or calcinations [69]. The particle size and shape can also be controlled by various processing variables such as temperature, pH and the addition of surfactants or mineralizers. The reaction is controlled by dissolution/precipitation of reactants in an aqueous medium. Therefore, the above processing variables are thought to have a significant influence on the dissolution or precipitation behavior, even though the exact roles or effects are not fully understood and differ in various systems [70]. In Table 2.8, there are several reports on the HT method of borate-based phosphors doped with rare earth ions.

TABLE 2.8

Various Borate based phosphors synthesized through the HT method

Sr. no.	Phosphors	Particle size/range	Morphology	Required temperature for synthesis	Synthesis time	Ref.
1	$(Y,Gd)BO_3:Eu^{3+}$	1.6 μm	Uniform plate morphology	300°C	3 h	[71]
2	$GdBO_3:Eu^{3+}$	1 μm and width 20 nm	Rodlike	180°C	8 h	[72]
3	$Ca2B_2O_5:RE$ (RE = Eu^{3+}, Tb^{3+}, Dy^{3+})	1–5 μm and 100 nm	Rodlike	190°C	1 h	[73]
4	$NH_4HB_4O_7 \cdot 3H_2O$: Eu^{3+}/ Tb^{3+}/Dy^{3+}	100 nm to 1 μm	Flower-like, microsphere, silkworm-chrysalis-like morphology	150°C	12 h	[74]
5	$GdBO_3:Eu^{3+}$	1.6 μm	Plate morphology	300°C	3 h	[75]
6	$YBO_3:Eu$	3–5 μm	Spherical morphology	300°C	6 h	[76]
7	$Y_{1-x}BO_3:xTb^{3+}$	3–8 μm	Flake-like morphology	260°C	6 h	[77]
8	YBO_3: Sb^{3+}, Eu^{3+}	200 nm	Flower and hexagonal flakes	200°C	3 days	[78]
9	$Lu_{1-x-y}Ce_xTb_y BO_3$	100–150 nm	Nanofibers or nanoflakes	200°C	24–72 h	[79]
10	$LnPO_4:Tb,Bi$ (Ln=La,Gd)	0.8 μm	Spherical morphology	240°C	6 h	[80]

2.3 PHYSICAL, CHEMICAL AND SURFACE CHARACTERIZATION

2.3.1 Diffraction and Scattering Methods

2.3.1.1 Single-Crystal X-Ray Diffraction

Single-crystal structure determination is an important characterization in materials science, due to its physical, chemical and biological functions that are used to know crystal structures. Single-crystal X-ray diffraction is a non-destructive analytical technique that gives information about the internal lattice of crystalline substances and it provides details about the crystal data and various phases of material samples. In this method, microcrystalline powder sample is used for treatment [81,82]

Single-crystal diffractometers use either 3- or 4-circle goniometers. These circles ask the four angles (2θ, χ, φ and Ω) that outline the connection between the space lattice, the incident ray and detector. Crystalline material samples are put onto goniometer heads. For cantering, the crystal sample within the X-ray beam by adjusting the X, Y and Z orthogonal directions.

X-rays leave the collimator and are directed at the crystal. X-ray beams are transmitted, diffracted or reflected from or to the crystal sample. A beam ray stop is directly opposite the collimator to dam transmitted rays and stop burn-out of the detector. The detector detects a reflected ray by some angle involved. Diffracted rays at the right orientation for the configuration are then collected by the detector. Figure 2.5 shows the geometry of single-crystal X-Ray diffraction.

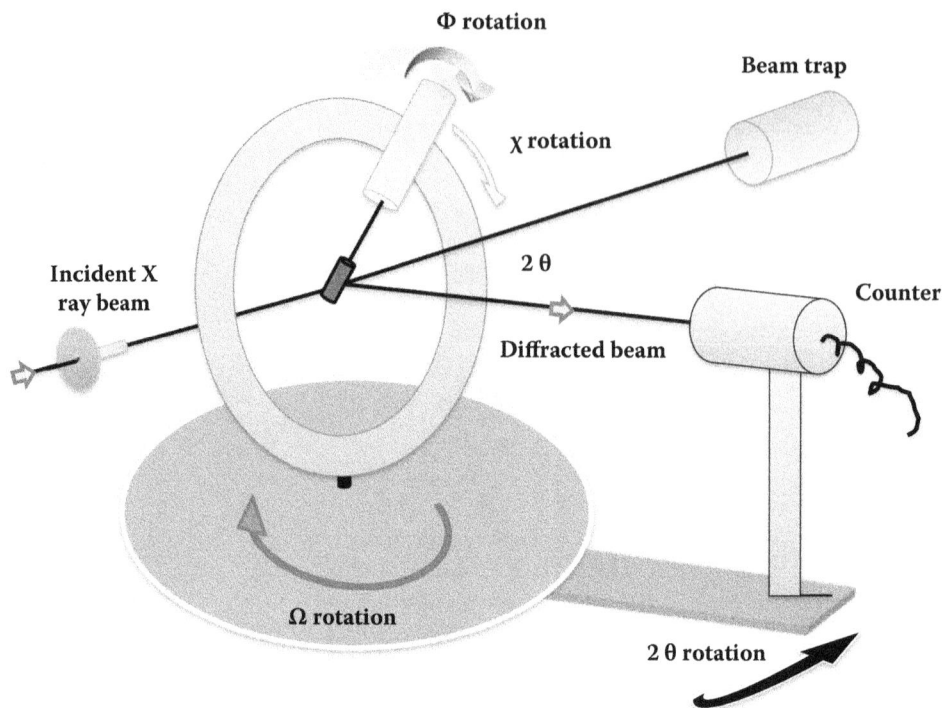

FIGURE 2.5   Schematic diagram of 4-circle diffractometer; the angles between the incident ray, the detector and the sample.

2.3.1.2   Powder X-Ray Diffraction (XRD)

X-rays are electromagnetic radiation with typical photon energies in the range of 100 eV to 100 keV. For diffraction applications, only short-wavelength X-rays in the range of a few angstroms to 0.1 angstrom (1 keV to 120 keV) are used. Because the wavelength of X-rays is comparable to the size of atoms, they are ideally suited for probing the structural arrangement of atoms and molecules in a wide range of materials. The energetic X-rays can penetrate deep into the materials and provide information about the bulk structure [83].

X-rays primarily interact with electrons in atoms, collide and some photons from the incident beam are deflected away from the original. If the wavelength of these scattered X-rays did not change, the process is called elastic scattering, where only the momentum transfer takes place. These diffracted X-rays are measured to extract information about the material, since they carry information about the electron distribution in materials. Diffracted waves from different atoms can interfere with each other and the resultant intensity distribution is strongly modulated by this interaction. If the atoms are arranged in a periodic fashion as in crystals, the diffracted waves will consist of sharp interference maxima with the same symmetry as in the distribution of atoms [84]. Measuring the diffraction pattern therefore allows us to deduce the distribution of atoms in a material. When certain geometric requirements are met, X-rays scattered from a crystalline solid can constructively interfere, producing a diffracted beam. In 1912, W. L. Bragg gave the relation, which elucidates the condition for constructive interference [85].

$$n\lambda = 2d\sin\theta$$

The above equation is called Bragg's equation.

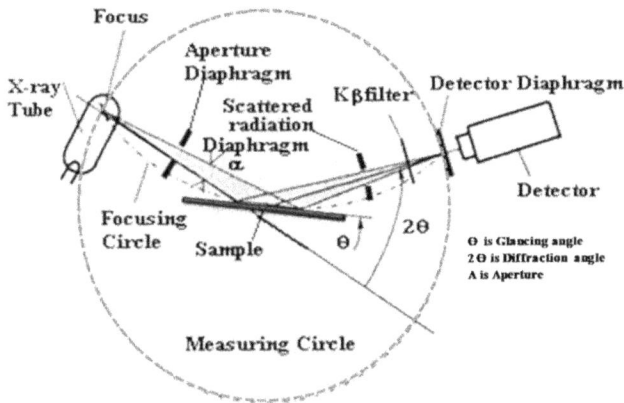

FIGURE 2.6 Schematic of the system for XRD measurements.

Where n denotes the order of diffraction, λ represents the wavelength, d is the interplanar spacing and θ signifies the scattering angle. The distance between similar atomic planes in a crystal is known as d spacing and is measured in angstroms. The angle of diffraction is called θ angle and is measured in degrees. For practical reasons, the diffractometer measures an angle twice that of the θ angle [86]. The schematic diagram of the XRD system is shown in Figure 2.6.

2.3.1.3 Small-Angle Scattering

Figure 2.7 depicts the geometry of small angle X-ray scattering. [It consist of an X-ray tube, detector and sample holder. It consist of an aerosol generator, switching module, pure gas system, inlet, inner flow tube X-ray source, conductive kapton windows and SAXS detector]. It makes glancing angle, diffraction angle and aperture. As the name suggests, scattering is not a diffraction method; small-angle scattering is used for low-order condensed matter materials, metals alloys, liquid crystals, porous materials, powders, ceramics, synthetic polymers in solution and in bulk and biological macromolecules in solution. This method provides structural information on particle size and size distributions, shape and orientation distributions. Moreover, it also provides the internal low-resolution molecular structure of biological systems [87].

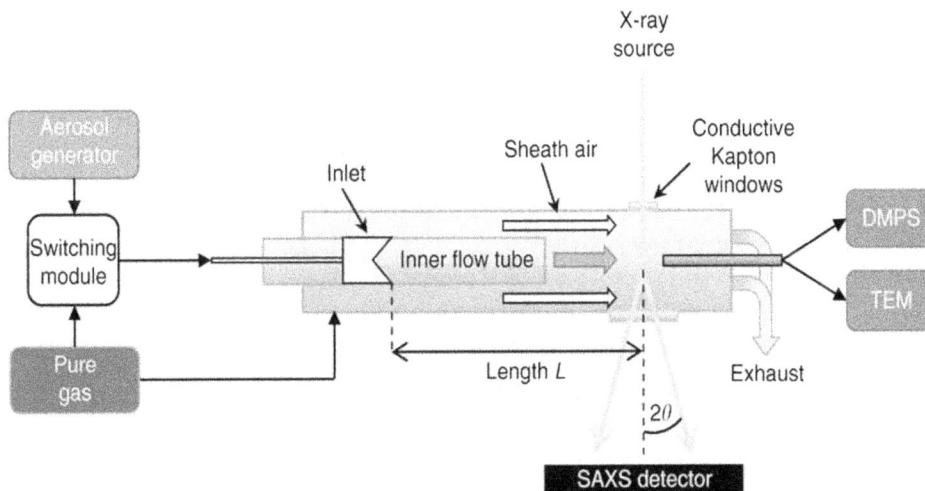

FIGURE 2.7 Illustration of small-angle X-ray scattering [88].

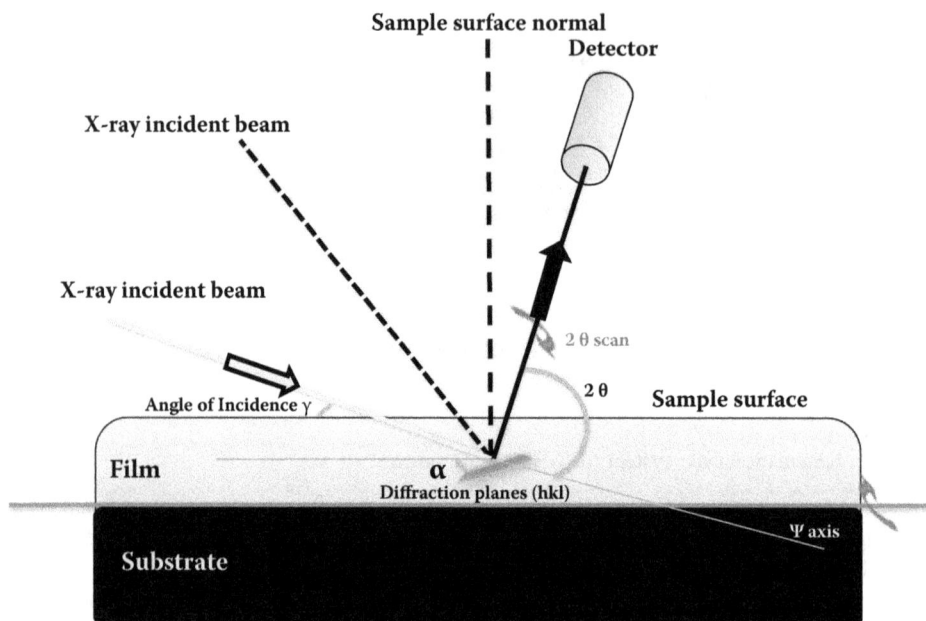

FIGURE 2.8 Illustration of grazing incidence X-ray diffraction process.

2.3.1.4 Grazing Incidence X-Ray Diffraction

For materials such as thin films, it is very difficult to do an analysis due to their small diffracting volume, which results in low diffracted intensities. The synchrotron technique of grazing incidence X-ray diffraction is a powerful tool for studying surfaces. The intense radiation from synchrotron and the increased path length of the incident X-ray beam passing through the film results in resonance of intensity of the diffracted/scattered beam so that conventional analysis can be acquired. This diffraction method provides the limited depth penetration of the evanescent wave setup when there is total reflection of X-rays incident at the grazing angle to the surface. The completely externally reflected wave penetrates only a few nanometres into the material, giving extremely high surface sensitivity below the critical angle (typically between 0.25° and 0.5°) [89]. Figure 2.8 shows the illustration of the grazing incidence X-ray diffraction process.

2.3.1.5 Electron Diffraction

Electron diffraction is the same as X-ray and neutron diffraction. Mostly it is used for phase identification, and disorder and defect identification [90]. The electron diffraction characterizations are performed in a transmission electron microscope (TEM). In this process, before the interaction of electrons with the sample, they are accelerated by an electrostatic potential. Through this interaction of the atomic potential of the object and the incident electron beam transmitted through it, diffracted electrons are generated. These are focused into a regular arrangement of diffraction spots by an electromagnetic objective lens. The experimental setup maintained in such a way that it allows projection and collection of the electron diffraction pattern. Figure 2.9 depicts the diagram of electron diffraction geometry.

2.3.1.6 Fiber Diffraction

Fibre diffraction is one of the most powerful and relevant method to available for the study of macromolecular structure. This is high degree of regularity in helical structures and the availability of information on covalent bond geometry from X-ray single crystal studies of model compounds [91]. The fiber diffraction method is a subarea of scattering intermediate between the single-crystal diffraction and powder diffraction method in the fiber diffraction method, molecular structure is

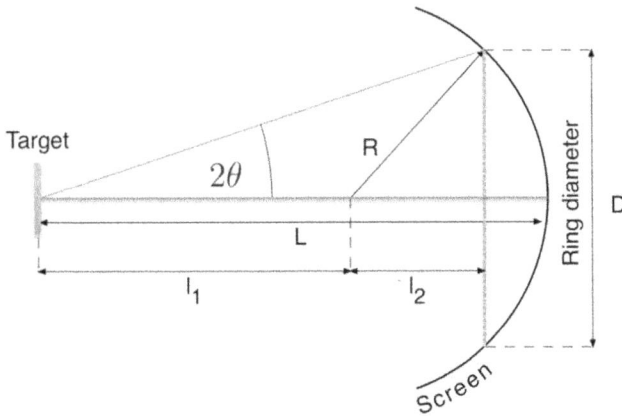

FIGURE 2.9 Illustration of electron diffraction geometry.

determined from scattering data (usually of X-rays, electrons or neutrons). In materials science considers fiber symmetry a simplification, because it gives almost all the information is in a single two-dimentional diffraction pattern and the result is on photographic film (or detector).

The prepared crystal sample is typically an extruded fiber, with a well-defined crystal axis aligned along the fiber axis (In the ideal pattern the fiber axis is called the meridian), and cylindrical averaging about that axis (the perpendicular direction is called the equator) [92].

2.4 MICROSCOPY

2.4.1 SCANNING ELECTRON MICROSCOPY (SEM)

Scanning electron microscopy (SEM) has become the most powerful and flexible tool for micro- and nanosize particles for morphological characterization. SEM produces material sample images by scanning the sample surface with a focused beam of electrons. The electrons interact with atoms in the phosphor material, producing different signals that contain information about the surface topography phosphor material. The electron beam is scanned in a raster scan pattern, and the detector detected the signal to produce an image. In the Everhart-Thornley detector (secondary detector) of SEM mode. After detecting a sample of phosphor material, the intensity of the signal depends on the other things and specimen topography [93].

In SEM, the electron beam is illuminated on the sample and then scans in a raster pattern. Initially, the source of electrons is generated electron at the top of the column. They are then accelerated and attracted by the positively charged anode.

We can see that in Figure 2.10 [It shows Electron source, Anode, Condenser lenses, objective lens, BSE detector, X-ray detector, sample, SE detector, amplifier, scan generator and computer for showing a graph]. The entire electron column should be under vacuum. For all component parts of an SEM, the electron source is inside a special chamber to maintain vacuum and protect it against contamination, vibrations and noise due to these users allowed to acquire a high-resolution image. Without vacuum, other atoms and molecules can be present in the column and interact with electrons causes the electron beam to deflect and reduces the image quality. On the other hand, a high vacuum also increases the collection efficiency of electrons by the detectors that are in the column.

For controlling the path of electrons, optical microscopes and lenses are used to control the pathway of the electrons. Because electrons cannot pass through glass, the lenses are used. They simply consist of coils of wires inside metal pole pieces. The magnetic field is generated by applying for current passes through the coils. An electron has its own magnetic field so that electrons are very sensitive to the magnetic field. The path of electrons inside the microscope column can be controlled by these electromagnetic lenses

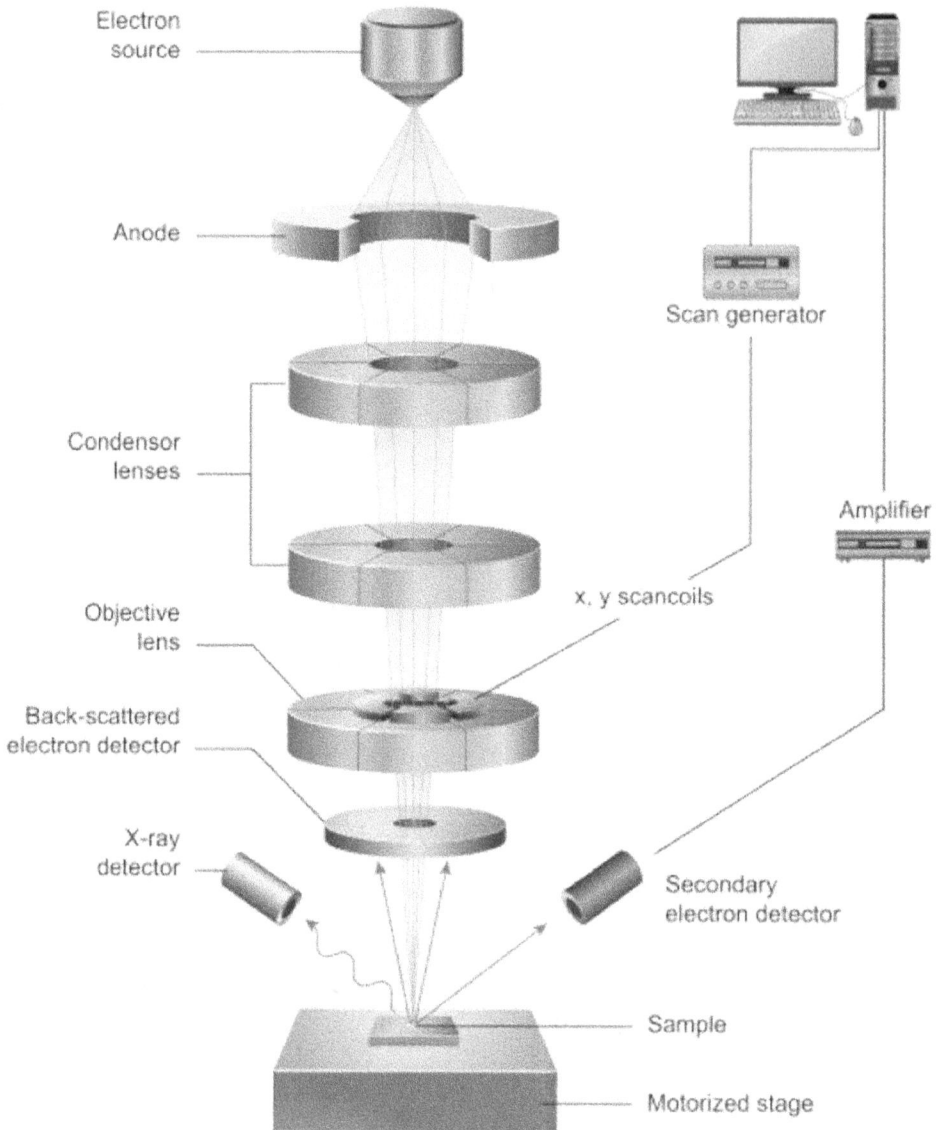

FIGURE 2.10   Mechanism of image formation by SEM [96].

simply by adjusting the current that is applied to them. When electrons are exposed on a sample, it can generate many different types of electrons, photons or irradiations. There are two types of electrons used for imaging of SEM that are backscattered (BSE) and secondary electrons (SE) (Figure 2.10).

BSEs belong to the primary electron beam and primary electron are reflected back after elastic interactions between the beam and the sample. On the other hand, secondary electrons originate from the atoms of the sample. Different types of signals used by an SEM and the area from which they originate [94]. For the detection of BSEs, solid-state detectors are positioned above the sample, concentrically to the electron beam, to maximize BSE collection. For the detection of SEs, the Everhart-Thornley detector is mainly used. It consists of a scintillator inside a Faraday cage, which is positively charged and attracts the SEs. SEs accelerates the electrons and convert them into light then reached to a photomultiplier for amplification. The SE detector is detecting the electron then used to form a 3D image shown on a PC monitor [95]. The interaction of high energetic electrons with matter is shown in Figure 2.11.

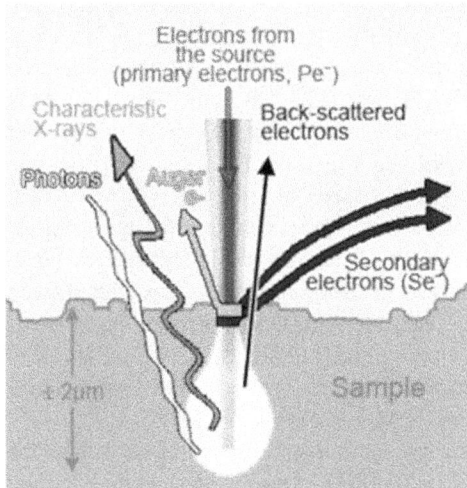

FIGURE 2.11 Interaction of highly energetic electrons with matter [96].

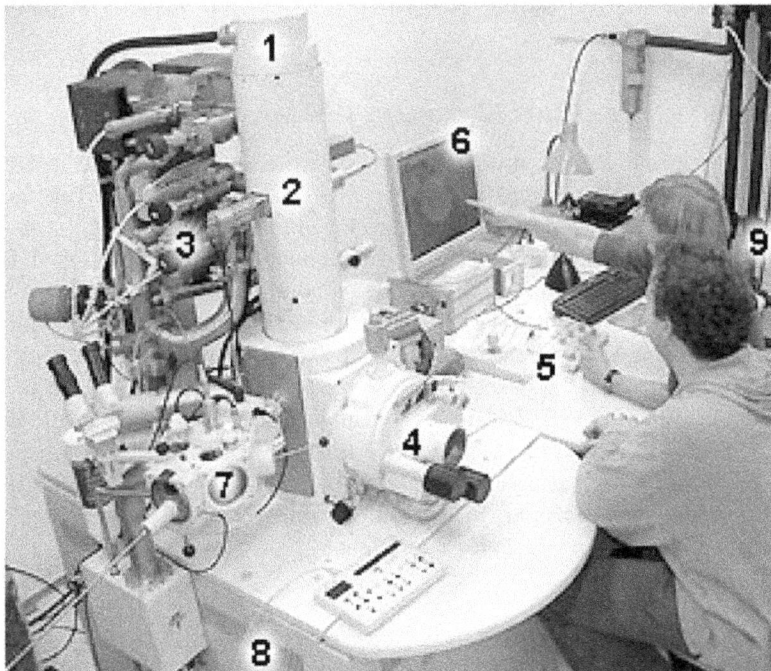

FIGURE 2.12 1: Electron cannon in the upper part of the column (here, a so-called field-emission source). 2: Electromagnetic lenses to direct and focus the electron beam inside the column. 3: Vacuum pumps system. 4: Opening to insert the object into the high-vacuum observation chamber in conventional SEM mode. 5: Operation panel with focus, alignment and magnification tools and a joystick for positioning of the sample. 6: Screen for menu and image display. 7: Cryo-unit to prepare (break, coat and sublimate) frozen material before insertion in the observation chamber in Cryo-SEM mode. 8: Electronics stored in cupboards under the desk. 9: Technicians Mieke Wolters-Arts and Geert-Jan Janssen discussing a view [96].

The SEM is limited to a very low working distance, extremely shallow depth of field and the SEM does not provide details of nonconductive soft materials and takes long scan times for large sample areas. A complete assembly of SEM is displayed in Figure 2.12.

2.4.2 Transmission Electron Microscopy (TEM)

TEM is considered to be the most popular technique in characterizing non-material in electron microscopy. It gives information up to the atomic level. The first TEM was demonstrated by Max Knoll and Ernst Ruska in 1931, with this group developing the first TEM with a resolution greater than that of light in 1933 and the first commercial TEM in 1939. In 1986, Ruska was awarded the Nobel Prize in Physics for the development of transmission electron microscopy [97].

TEM is a microscopy technique in which a beam of electrons is transmitted through a foil of specimen is transformed into elastically or inelastically scattered electrons when the electron beam interacts with the phosphor material specimen to form an image. The specimen is less than 100 nm thin. An image is formed from the interaction of the electrons with the sample as the beam is transmitted through the specimen. The image is then magnified and focused onto an imaging device, such as a fluorescent screen, a layer of photographic film, or a sensor such as a scintillator attached to a charge-coupled device. The mechanism of TEM is shown in Figure 2.13.

The advantage of TEM is to show the size, degree of aggregation and dispersion, as well as the heterogeneity of nanophosphor. When compared to TEM with SEM in providing spatial resolution in good quality and analytical measurements. The precise particle size of brightfield images as well as darkfield images are provided by the TEM [98,99]. A complete setup of TEM is displayed in Figure 2.14.

2.4.3 Confocal Scanning Laser Microscopy (CSLM)

The CSLM technique has become an important tool in materials sciences than traditional optical microscopy [101]. The principle of confocal microscopy was patented by Marvin Minsky in 1957 but it took a few good years before it was fully developed to incorporate a laser scanning process. CSLM is one of the most significant advances in optical microscopy ever developed. By following M. David Egger and Mojmir Petran [102] fabricated a multiple-beam confocal microscope in the late 1960s that utilized a spinning (Nipkow) disk for examining unstained brain sections and ganglion cells. Continuing in this arena, Egger went on to develop the first mechanically scanned confocal laser microscope, and published the first recognizable images of cells in 1973 [103]. During the late 1970s and the 1980s, advances in computer and laser technology, coupled to new algorithms with digital manipulation of images, led to a growing interest in confocal microscopy [104]. The technique essentially scans an object point-by-point using a focused laser beam to allow for a three-dimension transformation. In a traditional microscopy light can penetrate while a confocal microscope images one depth level at a time.

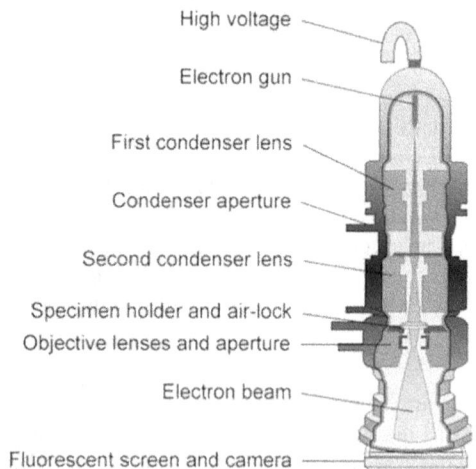

High voltage

Electron gun

First condenser lens

Condenser aperture

Second condenser lens

Specimen holder and air-lock
Objective lenses and aperture

Electron beam

FIGURE 2.13 Mechanism of image formation by TEM [100].

Fluorescent screen and camera

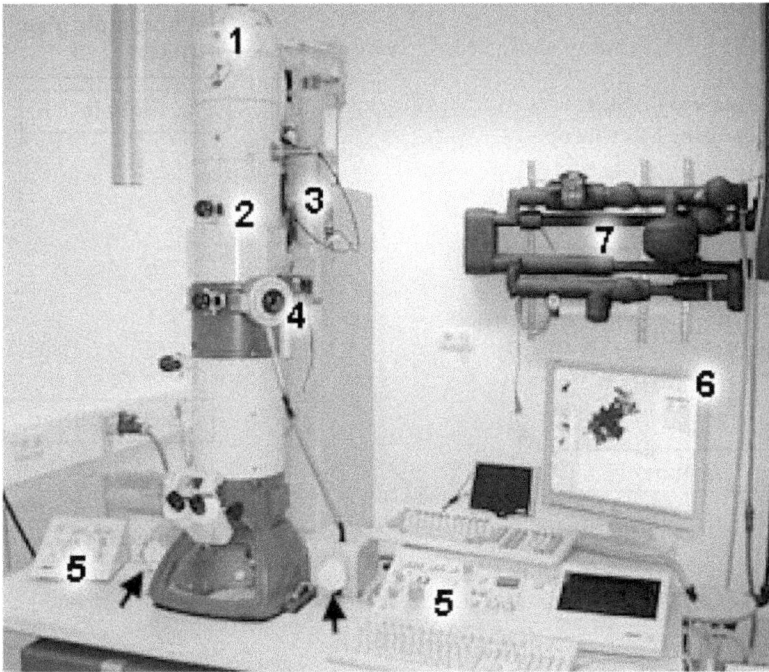

FIGURE 2.14 1: Electron cannon in the upper part of the column. 2 Electromagnetic lenses to direct and focus the electron beam inside the column. 3: Vacuum pumps system. 4: Opening to insert a grid with samples into the high-vacuum chamber for observation. 5: Operation panels (left for alignment; right for magnification and focussing; arrows for positioning the object inside the chamber). 6: Screen for menu and image display. 7: Water supply to cool the instrument [96].

In CSLM technology, a laser beam is focussed onto the sample instate of lamp (Figure 2.15). The intensity of the laser light is illuminated with neutral density filters to scanning mirrors. One mirror is set to move toward the X direction and other Y direction. After that on the sample, the beam is brought to the back focal plane of the objective lens. If your sample is fluorescent, it emits light. The part of the light will pass back into the objective lens. The direction of the sample light path is the same path as the laser light. The result is to produce a spot of light that is not scanning but standing still. Then this light passes via a semi-transparent mirror toward the detection system.

The first object in the detection system is the pinhole aperture. If this is a fluorescent light, then it will be a different colour from the laser light. The various filters used for the different wavelengths of light collected and passed through a polarizer. To increase light intensity, the photomultiplier tube (PMT) is used. The output of PMT is an electrical signal with amplitude that is converted to a series of digital numbers by an analog-to-digital converter in the computer. Finally, it shows the image of the sample on a computer.

2.4.4 ATOMIC FORCE MICROSCOPY (AFM)

Since the introduction of the Nobel Prize–winning scanning tunneling microscope (STM) and then the invention of the atomic force microscopy (AFM) from the landmark publication by Binnig, Quate and Gerber, the field of scanning probe microscopy has exploded well beyond using interatomic forces to image topography on the nanometre scale. The ability to measure intermolecular forces and see atoms is scientifically tantalizing [106].

The AFM is an instrument for the study of surface properties for both conductive and non-conductive samples [107]. AFM is able to build three-dimensional maps of surface properties in both

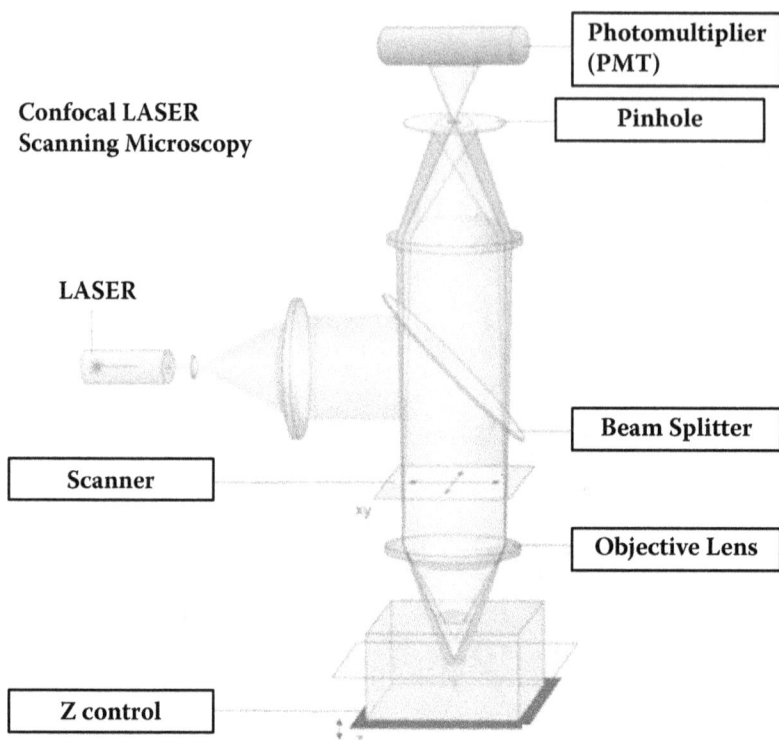

FIGURE 2.15   Schematic diagram of confocal scanning laser microscope [105].

air and liquid environments, with high lateral (<25 nm), vertical (0.1 A°), and force (1 pN) resolution. Due to such powerful capabilities, AFM is an indispensible technique in tribology to understand surface topography, adhesion, lubrication and wear at the atomic scale [108,109]. AFM is a very-high-resolution type of scanning probe microscopy (SPM), with demonstrated resolution on the order of fractions of a nanometer, more than 1,000 times better than the optical diffraction limit.

The AFM is a kind of scanning probe microscope in which a topographical image of the sample surface can be achieved based on the interactions between a tip and a sample surface. The AFM was invented by Gerd Binning et al. in 1986 at IBM Zurich based on the STM (scanning tunneling microscope) already presented in 1981. While the latter depends on the conductive samples, the AFM allows also the use of non-conductive samples. In 1987, the inventors were awarded the Nobel Prize in Physics for the achievements.

A typical AFM consists of a cantilever with a small tip (probe) at the free end, a laser, a four-quadrant photodiode and a scanner. The surface characteristics can be explored with very accurate resolution in a range of 100 μm to less than 1 μm. A schematic representation of AFM is shown in Figure 2.16.

2.5   SPECTROSCOPY

2.5.1   PHOTOLUMINESCENCE SPECTROSCOPY

Photoluminescence is the most popular type of luminescence because a large selection of reliable and inexpensive excitation sources are available and also because the effect can often be observed with the naked eye. Usually an excitation source emits in UV/VUV and the PL occurs in visible or NIR. Therefore, only a part of the absorbed energy is transformed into luminescent light. The rest of it ends up as molecular vibrations, or simply as heat.

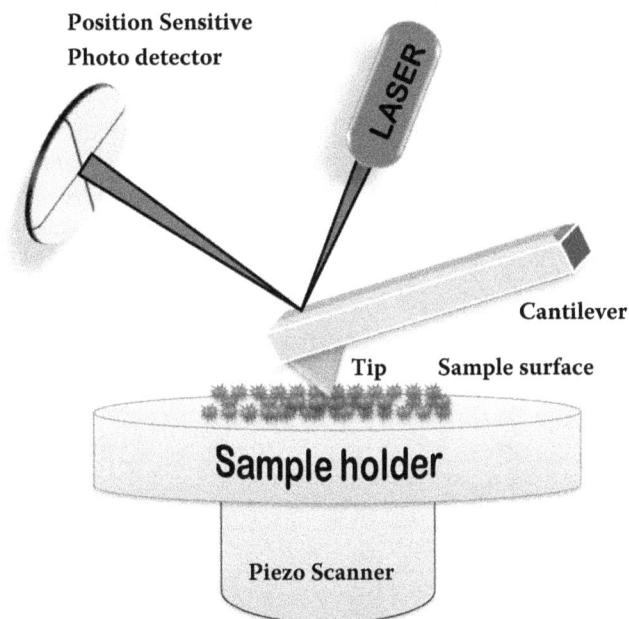

FIGURE 2.16 Schematic diagram of AFM.

There is always a delay between the moment the material has absorbed the higher-energy photon and the moment the secondary lower-energy photon is re-emitted. This delay is defined by the lifetime of excitation states, or simply by how long atoms or molecules are able to stay in excited high-energy conditions. Delay time can vary many orders of magnitude for different materials. Based on practical observations, two types of photoluminescence were historically established: 'fluorescence' and 'phosphorescence'.

Technically, delay time is the only difference between them. It is shorter for fluorescence (10^{-12} to 10^{-7} s) and much longer for phosphorescence (up to a few hours and even days).

Fluorescence is a 'fast' photoluminescence. The effect is widely used in such everyday practical applications as industrial and residential lightning (neon and fluorescent lamps) as an analytical technique in science and as a quality and process control method in industry.

Phosphorescence is a 'slow' photoluminescence. In contrast to fluorescence, it demonstrates itself as a glowing that lasts long after the excitation light is gone. Phosphorescent materials are usually called 'glow-in-the-dark'. This effect is generally used by the Department of Transportation to attract drivers' attention to road signs, in advertising campaigns to produce glowing stickers and promotional materials, as well as in numerous industries to notify people of potential hazards and dangers.

Figure 2.17 show that the photoluminescence spectrometer. It consists of a xenon lamp. It emits ultraviolet to visible light, a monochromator to select the specific wavelength for excitation and emission. A second monochromator is coupled with a photomultiplier tube (PMT) to examine the photoluminescence signal. The filters in the excitation and emission monochromator is used to disperse light from 200 to 900 nm. The slit's width of emission and excitation control the number of photon spread (band pass) of the light. The xenon lamp is illuminated to an elliptical mirror and reflected toward the entrance slit of the excitation monochromator. The excitation monochromator selects the specific wavelengths of excitation light and then it fell on the sample. The sample absorbs a specific wavelength of light. Some part of the incident light is then emitted from the sample and some of the incident light is emitted by means of photoluminescence light. The emitted photoluminescence light is transmitted in a narrow range centered about the specified emission of wavelength and detector captures the emission wavelength signal. The signal is amplified and

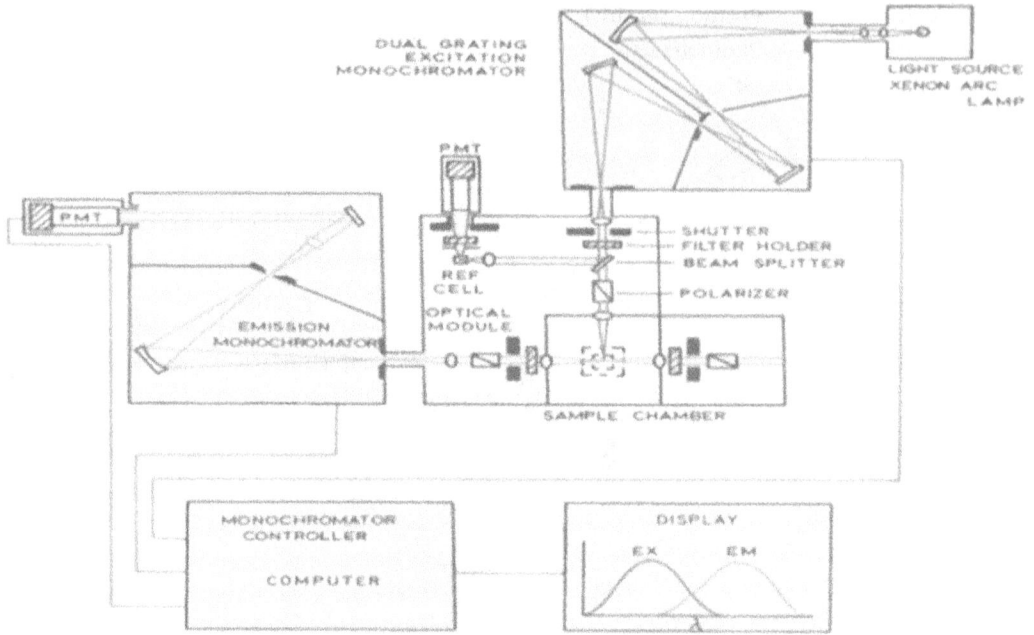

FIGURE 2.17 Schematic diagram of a spectrofluorometer.

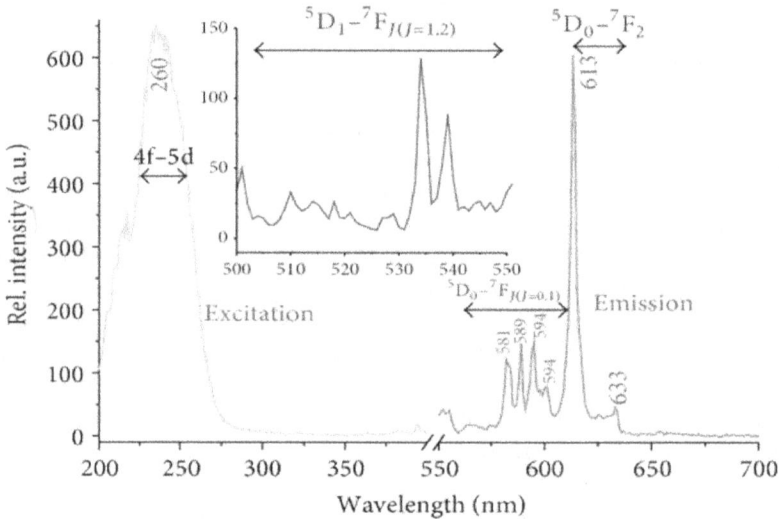

FIGURE 2.18 Photoluminescence Excitation and emission spectra of $K_2Y_2B_2O_7$:Eu^{3+} phosphor [112].

generates a voltage that is proportional to the measured emitted intensity. Finally, emission spectra scanning are excited by specified wavelength recorded on computer software between wavelength and intensity [110,111]. The photoluminescence excitation and emission spectra of phosphor is displayed in Figure 2.18.

2.5.2 TIME-RESOLVED PHOTOLUMINESCENCE SPECTROSCOPY

The luminescent system generally consists of a host lattice and a luminescent centre, often called an 'activator'. In general, the host needs to be transparent to the radiation source used for the excitation process. The activator absorbs the exciting radiation and is raised to an excited state. The excited state returns to the ground state by emission of radiation or by non-radiative decay. It is necessary to suppress this non-radiative process. In some materials, the excitation radiation is not absorbed by the activator but the other ion may absorb the exciting radiation and subsequently transfer it to the activator. In this case, the absorbing ion is called a 'sensitizer'. In many cases, the host lattice transfers its excitation energy to the activator so that the host lattice acts as the sensitizer. High-energy excitation always excites the host lattice. Direct excitation of an activator is only possible with ultraviolet and visible radiation [113]. Time resolved photoluminescence (TRP) spectroscopy is the tool of choice to study fast electronic deactivation processes that result in the emission of photons, a process called fluorescence. The lifetime of a molecule in its lowest excited singlet state usually ranges from a few picoseconds (ps) up to nanoseconds (ns) [114].

For the excited state (singlet or triplet state), the lifetime can be extremely different, varying from ms to s. In the first case, we generally refer to the emission process as fluorescence, while in the second case as phosphorescence (ps to ns). Both phenomena are generally summarized under the term 'luminescence'.

An examination of the decay properties of the luminescent materials indicates that they fall in to two broad classifications [115]:

In the first type, the decay in equation 2.1 is given by:

$$I(t) = I_0 \, \exp(-\alpha t) \tag{2.1}$$

where, I_0 is the initial intensity, $I(t)$ is the intensity at time t and α is a constant.

This resembles closely the process governing the progress of monomolecular reaction. This behaviour suggests that in these cases the luminescence takes place by simple excitation with subsequent optical emission in the active centre, the excitation energy remaining closely localised in the centre between excitation and emission. The decay constant is independent of temperature and is small.

Most of the luminescent materials, which are valued for their long decay characteristics, obey a decay in equation (2.2) of a second type:

$$I(t) = I_0/(\beta t + 1)n \tag{2.2}$$

where, I_0 is the initial intensity, $I(t)$ is the intensity at time t and β and n are constants.

Equation (2.2) is similar to the rate equation for the bimolecular reaction. The constant β is dependent on temperature. The atoms or clusters of atoms become ionised during the excitation and the luminescent radiation is emitted during recombination of the free electrons and the ionised centres. Johnson [116] has suggested that essentially all centres become ionized during excitation and that a majority of free electrons are recaptured into a state, which has a very long lifetime (of the order of milliseconds), because the optical transition to the ground state is forbidden. These electrons contribute an exponential component to the decay curve. The remaining electrons are captured at the trapping centres and are released over a period of time that is long compared to the lifetimes of the excited state of the fluorescing centre [117,118]. The time resolved spectroscopy spectra of borate phosphors is shown in Figure 2.19.

2.5.3 INFRARED SPECTROSCOPY (IR)

IR dpectroscopy detects the wavelength of infrared light that is absorbed by a molecule. Luminescence materials absorb the particular wavelength, which is the wavelength of the vibration of bonds in the molecule of materials.

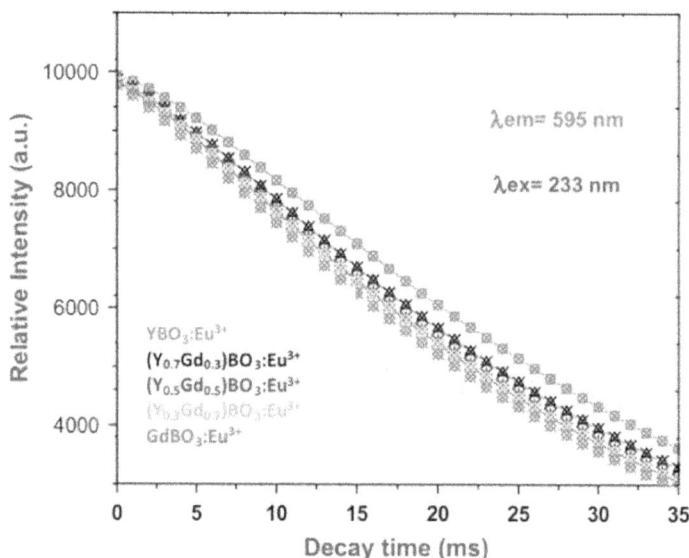

FIGURE 2.19 Time decay diagram of borate phosphors [119].

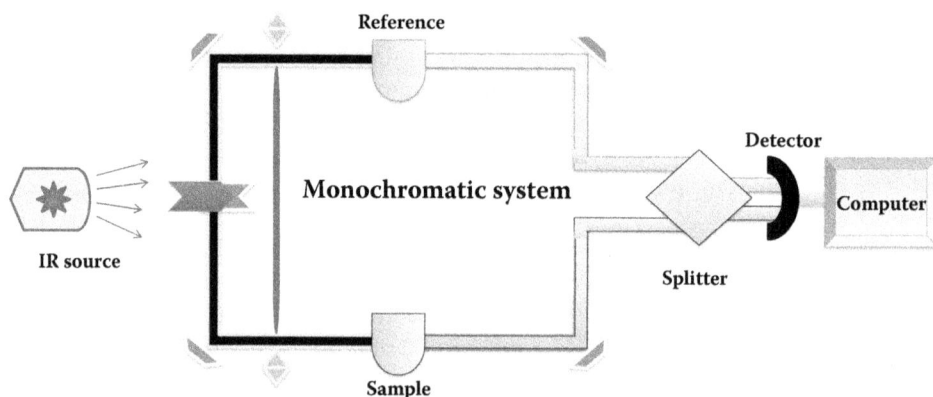

FIGURE 2.20 Schematic diagram of IR spectroscopy.

In the IR region, energy is required to excite the bonds and vibrate with more amplitude. The area of partial positive and negative charge in a molecule allows exciting vibrational energy of the molecule by the electric field component of the electromagnetic (EM) wave. The change in the vibrational energy leads to another corresponding change in the dipole moment of the given molecule in the material. The intensity of the absorption depends on the polarity of the bond. Symmetrical nonpolar bonds in $N\equiv N$ and $O=O$ do not absorb radiation, as they cannot interact with an electric field.

Most of the bands that indicate what functional group is present are found in the region from $4,000 \text{ cm}^{-1}$ to $1,300 \text{ cm}^{-1}$. Their bands can be identified and used to determine the functional group of an unknown compound. The instrumentation of IR spectroscopy is shown in Figure 2.20. A beam of IR light from this is split into two ways and passed through the reference and the sample, respectively. Then, both of these beams are reflected to pass through a splitter through a detector. At last, the result is monitored on a computer through the detector.

Figure 2.21 show the classification of different regions of IR spectra. A graph is plotted with the IR light absorbed on the y-axis against wavelength on the x-axis.

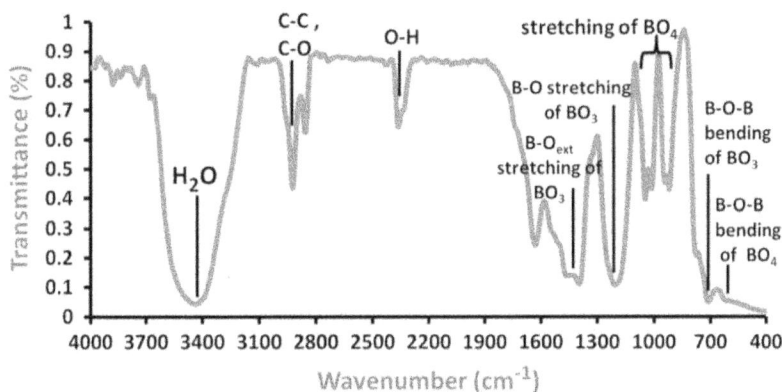

FIGURE 2.21   Graph of typical infrared absorption wavelength in IR [120].

2.5.4   RAMAN SPECTROSCOPY

Raman spectroscopy is an analytical technique, in which scattered light is used to measure the vibrational energy modes of a sample. This phenomenon was first observed by K. S. Krishnan in 1928. After that, Indian scientist C. V. Raman worked with K. S. Krishnan [121]. Raman spectroscopy provides chemical and structural information of material, in addition to the identification of substances through their characteristic Raman 'fingerprint' [122].

When light falls on material, then light is scattered by molecules, as shown in Figure 2.22; the oscillating EM field of a photon induces a polarisation of the molecular electron. The energy transfer from a lower to a higher state (energy of the photon transferred to the molecule).

The energy transfer from molecules to scattering photons is the inelastic scattering process [123]. If the molecule gains energy from the photon during the scattering, then the scattered photon loses energy and its wavelength increases, which is called Stokes Raman scattering. Inversely, if the molecule loses energy by relaxing to a lower vibrational level, the scattered photon gains the corresponding energy and its wavelength decreases, which is called anti-Stokes Raman scattering. Quantum mechanically, Stokes and anti-Stokes are equally likely processes. However, with an ensemble of molecules, the majority of molecules will be in the ground vibrational level (Boltzmann distribution) and Stokes scatter is the statistically more probable process. As a result, the Stokes Raman scatter is always more intense than the anti-Stokes and for this reason, it is nearly always the Stokes Raman scatter that is measured in Raman spectroscopy [124]. Figure 2.23 shows the Jablonski diagram.

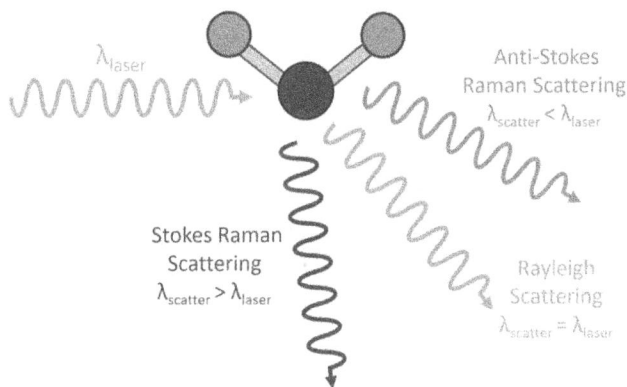

FIGURE 2.22   Three types of scattering processes that can occur when light interacts with a molecule.

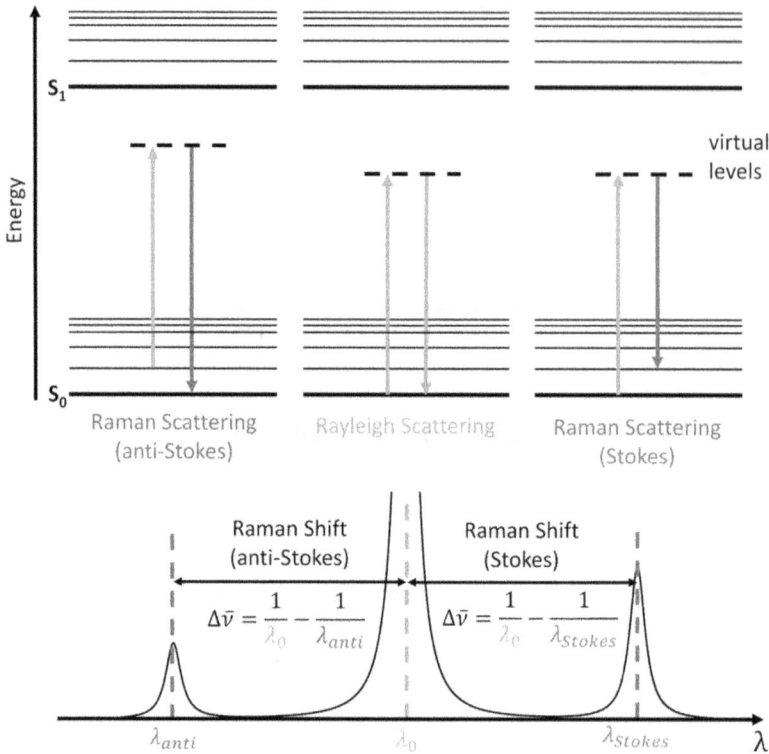

FIGURE 2.23 Jablonski diagram showing the origin of Rayleigh, Stokes and anti-Stokes Raman Scatter [125].

2.5.5 X-Ray Photoelectron Spectroscopy (XPS)

XPS is used to detect photoelectrons (all elements with the exception of hydrogen and helium through the detection of the binding energies of the photoelectrons) emitted from a sample of luminescence material by single-energy X-ray photons. It is a simple and routine technique for the compositional and chemical state analysis of surfaces. The basic principle of XPS is the photoelectric effect discovered by Hertz in 1887 [126,127] and extended to surface analysis by K. Siegbahn in the 1960s. Siegbahn won the Nobel Prize in Physics in 1981 for his work in XPS [128].

> Chemical states identify small variations in binding energies of the photoelectron lines as well as Auger lines, satellite peaks. XPS is initiated by irradiating a sample with monoenergetic soft X-rays, most commonly Mg Kα (1,253.6 eV with a line width ≈ 0.7 eV) or Al Kα (1,486.6 eV with a line width ≈ 0.85 eV). In many modern instruments, the Al Kα X-ray line is further narrowed (to ≈ 0.35 eV) using a monchrometer. Schematic representations of the photoemission and Auger processes are shown in Figures 2.24 and 2.25 [129,130].

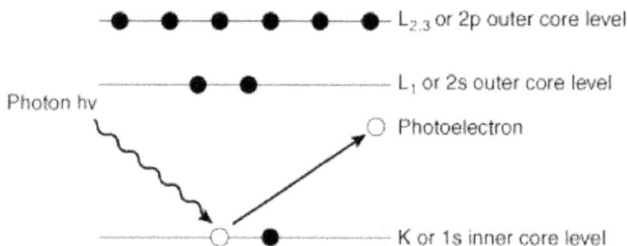

FIGURE 2.24 Schematic diagram of photoemission process used for XPS [131].

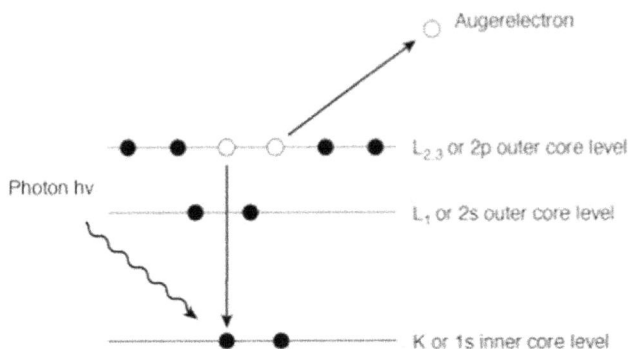

FIGURE 2.25 Schematic diagram of the Auger process. Note that the energy of the Auger electron is independent of the energy of the photon creating the core-level vacancy [131].

2.5.6 ULTRAVIOLET-VISIBLE SPECTROSCOPY

Ultraviolet–visible spectroscopy or ultraviolet-visible spectrophotometry (UV-Vis or UV/Vis) refers to absorption spectroscopy or reflectance spectroscopy in the ultraviolet-visible spectral region. This means it uses light in the visible and adjacent (near-UV and near-infrared [NIR]) ranges. The absorption or reflectance in the visible range directly affects the perceived color of the chemicals involved. In this region of the electromagnetic spectrum, molecules undergo electronic transitions. This technique is complementary to fluorescence spectroscopy, in that fluorescence deals with transitions from the excited state to the ground state, while absorption measures transitions from the ground state to the excited state [132,133].

Ultraviolet-visible (UV-vis) spectroscopy is used to obtain the absorbance spectra of a compound in solution or as a solid. What is actually being observed spectroscopically is the absorbance of light energy or electromagnetic radiation, which excites electrons from the ground state to the first singlet excited state of the compound or material. The UV-vis region of energy for the electromagnetic spectrum covers 1.5–6.2 eV, which relates to a wavelength range of 800–200 nm. The Beer-Lambert law in equation (2.3) is the principle behind absorbance spectroscopy. For a single wavelength, A is absorbance (unitless, usually seen as arb. units or arbitrary units), ε is the molar absorptivity of the compound or molecule in solution ($M^{-1}cm^{-1}$), b is the path length of the cuvette or sample holder (usually 1 cm) and c is the concentration of the solution (M).

$$A = \varepsilon \ b \ c \qquad (2.3)$$

Molecules containing π-electrons or non-bonding electrons (n-electrons) can absorb the energy in the form of ultraviolet or visible light to excite these electrons to higher anti-bonding molecular orbitals [134]. The principle of UV spectroscopy is basically related to the interaction of light with matter. As light is absorbed by matter, the result is an increase in the energy content of the atoms or molecules. When ultraviolet radiations are absorbed, this results in the excitation of the electrons from the ground state towards a higher energy state. Molecules containing π-electrons or non-bonding electrons (n-electrons) can absorb energy in the form of ultraviolet light to excite these electrons to higher anti-bonding molecular orbitals. The more easily excited the electrons, the longer the wavelength of light it can absorb. There are four possible types of transitions (π–π^*, n–π^*, σ–σ^* and n–σ^*), and they can be ordered as follows: σ–$\sigma^* >$ n–$\sigma^* > \pi$–$\pi^* >$ n–π^*. The absorption of ultraviolet light by a chemical compound will produce a distinct spectrum that aids in the identification of the compound (Figure 2.26).

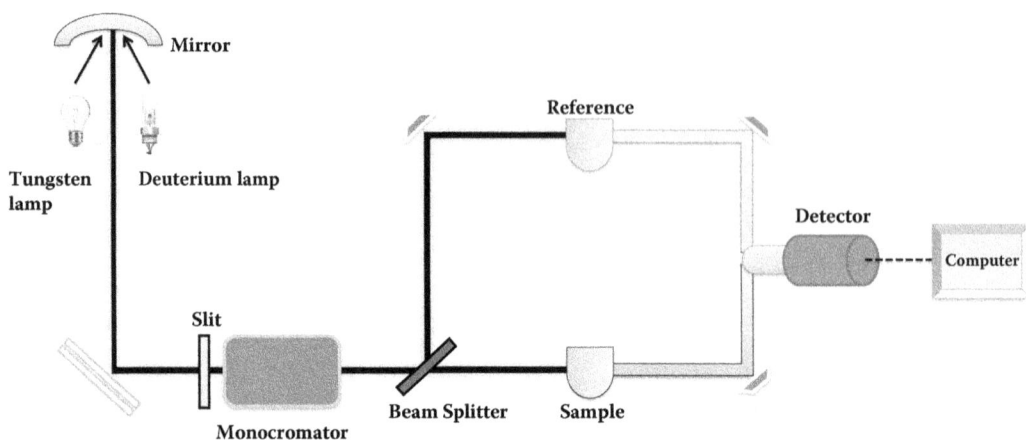

FIGURE 2.26 Schematic representation of UV spectroscopy.

2.5.7 THERMAL ANALYSIS (TA)

In material science research, thermal analysis has been used to understand synthesis temperature and nature of material with respect to temperature. There are three main groups of TA (Figure 2.27) to measure physical parameters, such as mass, thermogravimetry (TG), differential thermogravimetry (DTG), differential scanning calorimetry (DSC), differential thermal analysis (DTA), thermo and dynamic mechanical analysis, thermooptical analysis (TOA) and thermo-sonomentry analysis (TS), as shown in Figure 2.27.

The information of borate materials by various techniques of thermal analysis gives instruction for understanding borate materials and new phases of boron materials. TA follows the temperature of a sample upon heating and cooling, i.e. to record dependencies T = f(T).

In TA, the introduction of a thermally inert reference substance is subjected to the same heating or cooling program as the sample. Then, TA is transformed into DTA by developing a change in measuring setup [135]. The modified form of the equation is $\Delta T = f(T)$.

When measuring more than one parameter simultaneously in one device, then the combination DTA-TG is useful to measure. Simply, a DTA measuring cell is assembled onto a sensitive balance that was realized at first with the derivatograph (derivatograph was the trademark of a measuring device, not the name of a method. It is a hint to the first-time derivative of the TG curve) [136].

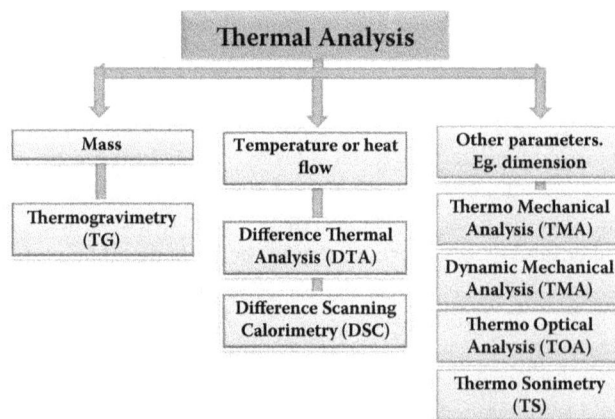

FIGURE 2.27 Classification of thermal analysis.

REFERENCES

[1] Kingsley, J. J., and Patil, K. C. (1988). A novel combustion process for the synthesis of fine particle α-alumina and related oxide materials. *Mater. Lett.*, 6, 427–432. 10.1016/0167-577X(88)90045-6

[2] Westesen, K., and Wehler, T. (1993). Investigation of the particle size distribution of a model intravenous emulsion. *J. Pharm Sci.*, 82, 12, 1237–1244. 10.1002/jps.2600821211

[3] Koparkar, K. A., Bajaj, N. S., and Omanwar, S. K. (2015). Aldo-keto synthesis effect on Eu3+ fluorescence in YBO_3 compared with solid state diffusion. *J. Rare Earths*, 33, 5, 486–490. 10.1016/S1002-0721(14)60445-2

[4] Li, C. M., Lambert, N., Chen, Y. N., Chen, G. Y., and Nori, F. (2012). Witnessing Quantum Coherence: From solid-state to biological systems. *Sci. Rep.*, 2012, 2, 885. 10.1038/srep00885

[5] Bu, J., Wang, P., Ai, L., Sang, X., and Li, Y. (2011). A new method for preparation and luminescence of $YBO_3:Eu^{3+}$ phosphor. *Adv. Mater. Res.*, 287, 1460. 10.4028/www.scientific.net/AMR.287-290.1460

[6] Wei, Z., Sun, L., Liao, C., Yan, C., and Huang, S. (2002). Fluorescence intensity and color purity improvement in nanosized $YBO_3:Eu$. *Appl. Phys. Lett.*, 80, 1447. 10.1063/1.1452787

[7] Wanga, L., and Wang, Y. (2007). Enhanced photoluminescence of $YBO_3:Eu^{3+}$ with the incorporation of Sc^{3+} Bi^{3+} and La^{3+} for plasma display panel application. *J. Lumin.*, 122, 921. 10.1016/j.jlumin.2006.01.327

[8] Chen, L., Luo, A., Deng, X., Xue, S., Zhang, Y., Liu, F., Zhu, J., Yao, Z., Jiang, Y., and Chen, S. (2013). Luminescence and energy transfer in the Sb^{3+} and Gd^{3+} activated YBO_3 phosphor. *J. Lumin.*, 143, 670. 10.1016/j.jlumin.2013.06.014

[9] Dubey, V., Kaur, J., Agrawal, S., and Suryanarayana, N. S. (2013). Synthesis and characterization of Eu^{3+} doped YBO_3 phosphor. *Int. J. Lumin. Appl.*, 3, 98.

[10] Dubey, V., Kaur, J., Agrawal, S., Suryanarayana, N. S., and Murthy, K. V. R. (2014). Effect of Eu^{3+} concentration on photoluminescence and thermoluminescence behavior of $YBO_3:Eu^{3+}$ phosphor. *Superlattices Microstruct.*, 67, 156. 10.1016/j.spmi.2013.12.026

[11] Zhang, W., Liu, S., Hu, Z., Liang, Y., Feng, Z., and Sheng, X. (2014). Preparation of $YBO_3:Dy^{3+}$, Bi^{3+} phosphors and enhanced photoluminescence. *Mater. Sci. Eng. B*, 187, 108. 10.1016/j.mseb.2014.05.006

[12] Vijayakumar, R., Devakumar, B., and Huang, X. (2021). Energy transfer induced color-tunable emissions from $Ba_2Gd_5B_5O_{17}:Ce^{3+}/Tb^{3+}$ borate phosphors for white LEDs. *J. Lumin.*, 229, 117685. 10.1016/j.jlumin.2020.117685

[13] Huang, D., Zhu, H., Deng, Z., Yang, H., Hu, J., Liang, S., Chen, D., Ma, E., and Guo, W. (2021). A highly efficient and thermally stable broadband Cr^{3+}-activated double borate phosphor for near-infrared light-emitting diodes. *J. Mater. Chem. C*, 9, 1, 164–172. 10.1039/D0TC04803H

[14] Chauhan, A. O., Koparkar, K. A., Bajaj, N. S., and Omanwar, S. K. (2016). Synthesis and photo-luminescence properties of Pb^{2+} doped inorganic borate phosphor $NaSr_4(BO_3)_3$. *AIP Conference Proceedings*, 1728, 020470. 10.1063/1.4946521

[15] Chauhan, A. O., Sawala, N. S., Palan, C. B., and Omanwar, S. K. (2015). Synthesis and PL study of UV emitting phosphor $LiSr_4(BO_3)_3:Pb^{2+}$. *Int. J. Lumin. Appl.*, 5, 4, 465–467.

[16] Das, S., Reddy, A. A., Babu, S. S., and Prakash, G. G. (2011). Controllable white light emission from Dy^{3+}–Eu^{3+} co-doped $KCaBO_3$ phosphor. *J. Mater. Sci.*, 46, 7770-7775. 10.1007/s10853-011-5756-5

[17] Tamboli, S., Nair, G. B., Dhoble, S. J., and Burghate, D. K. (2018). Energy transfer from Pr^{3+} to Gd^{3+} ions in BaB_8O_{13} phosphor for phototherapy lamps. *Phys. B: Condens. Matter*, 535, 232–236. 10.1016/j.physb.2017.07.042

[18] Gavhane, K. H., Bhadane, M. S., Bhoir, A. S., Kulkarni, P. P., Patil, B. J., Bhoraskar, V. N., Dhole, S. D., and Dahiwale, S. S. (2020). Tm-Tstop analysis and dosimetric properties of Ce doped BaB_4O_7 phosphor. *J. Alloys Compd.*, 817, 152805. 10.1016/j.jallcom.2019.152805

[19] Sahare, P. D., Singh, M., and Kumar, P. (2015). Synthesis and TL characteristics of $MgB_4O_7:Mn,Tb$ phosphor. *J. Lumin.*, 160, 158–164. 10.1016/j.jlumin.2014.11.042

[20] Anishia, S. R., Jose, M. T., Annalakshmi, O., Ponnusamy, V., and Ramasamy, V. (2010). Dosimetric properties of rare earth doped $LiCaBO_3$ thermoluminescence phosphors. *J. Lumin.*, 130, 10, 1834–1840. 10.1016/j.jlumin.2010.04.019

[21] Ali, A., Zafar, H., Zia, M., Ul Haq, I., Phull, A. R., Ali, J. S., and Hussain, A. (2016). Synthesis, characterization, applications, and challenges of iron oxide nanoparticles. *Nanotechnol. Sci. Appl.*, 9, 49–67. 10.2147/NSA.S99986

[22] Shao, Z., Zhou, W., and Zhu, Z. (2012). Advanced synthesis of materials for intermediate-temperature solid oxide fuel cells. *Prog. Mater. Sci.*, 57, 804. 10.1016/j.pmatsci.2011.08.002

[23] Chen J., Shi Y., and Shi J. (2004). Synthesis of $(Y,Gd)_2O_3$:Eu nanopowder by a novel co-precipitation processing. *J. Mater. Res.* 19, 3586–3591. 10.1557/JMR.2004.0477

[24] Jadhav A. P., Pawar A., Kim C. W., Cha H. G., Pal U., and Kang Y. S. (2009). Effect of different additives on the size control and emission properties of $Y2O_3$:Eu^{3+} nanoparticles prepared through the coprecipitation method. *J. Phys. Chem. C*, 113, 16652. 10.1021/jp9059399

[25] Yadav, S. R., Dutta, R. K., Kumar, M., and Pandey, A. C. (2009). Improved color purity in nano-size Eu^{3+}-doped YBO_3 red phosphor. *J. Lumin.*, 129, 1078. 10.1016/j.jlumin.2009.04.032

[26] Hu, Y., Tao, Y., Huang, Y., Yu, X., Zhang, C., Liang, T., and Yu, J. (2011). Luminescent properties of $(Y, Gd)BO_3$:Eu^{3+} under VUV excitation for PDP prepared by co-precipitation method. *Adv. Mater. Rapid Commun.*, 5, 348.

[27] Koparkar, K. A., Bajaj, N. S., and Omanwar, S. K. Effect of partially replacement of Gd^{3+} ions on fluorescence properties of YBO_3:Eu^{3+} phosphor synthesized via precipitation method. *Opt. Mater.*, 39, 74–80. 10.1016/j.optmat.2014.11.001

[28] Jadhav, A. P., Kim, C. W., Cha, H. G., Pawar, A. U., Jadhav, N., A.Pal, U., and Kang, Y. S. (2009). Effect of different surfactants on the size control and optical properties of Y_2O_3:Eu^{3+} nanoparticles prepared by coprecipitation method. *J. Phys. Chem. C*, 113, 31, 1932–7447. 10.1021/jp903067j

[29] Qiao, X., Cheng, Y., Qin, L., Qin, C., Cai, P., Kim, S. I., and Seo, H. J. (2014). Coprecipitation synthesis, structure and photoluminescence properties of Eu^{3+}-doped sodium barium borate. *J. Alloys Compd.*, 617, 946–951. 10.1016/j.jallcom.2014.08.050

[30] Lu, H. Y., and Chen, Y. S. (2015). The influence of codopant aluminum ions (Al^{3+}) on the optical characteristics of YBO_3: Sm^{3+} phosphors. *J. Nanomater.*, 16, 1, 1. 10.1155/2015/615863

[31] Haghiri, M. E., Saion, E., Abdullah, W. S. W., Soltani, N., Hashim, M., Navasery, M., and Shafaei, M. A. (2013). Thermoluminescence studies of manganese doped calcium tetraborate (CaB4O7:Mn) nanocrystal synthesized by co-precipitation method. *Radiat. Phys. Chem.*, 90, 1–5. 10.1016/j.radphyschem.2013.05.009

[32] Singh, V., Mahamuda, S., Rao, A. S., Rao, J. L., and Irfan, M. (2020). Luminescence and EPR properties of UVB emitting Gd doped $NaBaPO_4$ phosphor prepared by co-precipitation method. *Optik*, 206, 164086. 10.1016/j.ijleo.2019.164086

[33] Zheng, Y., Qu, Y., Tian, Y., Rong, C., Wang, Z., Li, S., Chen, X., and Ma, Y. (2009). Effect of Eu^{3+}-doped on the luminescence properties of zinc borate nanoparticles. *Colloids Surf. A Physicochem. Eng. Asp.*, 349, 19–22.

[34] Lu, C. H., and Saha, S. K. (2000). Synthesis of ferroelectric nanocrystalline $SrBi_2Ta_2O_9$ powder by colloid-gel process. *Mater. Res. Bull.*, 35, 13, 2135–2143. 10.1016/S0025-5408(00)00422-0

[35] Kahoul, A., Nkeng, P., Hammouche, A., Naamoune, F., and Poillerat, G. (2001). A sol+gel route for the synthesis of $Bi_2Ru_2O_7$ pyrochlore oxide for oxygen reaction in alkaline medium. *J. Solid State Chem.*, 161(2001), 379. 10.1006/jssc.2001.9346

[36] Campero, A., Cardoso, J., and Pacheco, S. (1997). Ethylene glycol-citric acid-silica hybrid organic-inorganic materials obtained by the sol-gel method. *J. Sol-Gel Sci. Technol.*, 8, 535–539. 10.1007/BF02436895

[37] Xue, J., Wu, T., Dai, Y., and Xia, Y. (2019). Electrospinning and electrospun nanofibers: Methods, materials, and applications. *Chem. Rev.*, 119, 8, 5298–5415. 10.1021/acs.chemrev.8b00593

[38] Hsu, W. T., Wu, W. H., and Lu, C. H. (2003). Synthesis and luminescent properties of nano-sized Y3Al5O12:Eu3+ phosphors. *Mater. Sci. and Eng. B*, 104, 40. 10.1016/S0921-5107(03)00268-X

[39] Yan, B., and Wang, C. (2008). Synthesis and luminescence properties of $REAl_3(BO_3)_4$:Eu^{3+}/Tb^{3+} (RE=Y, Gd) phosphors from sol–gel composition of hybrid precursors. *Solid State Sci.*, 10, 1, 82–89.

[40] Santiago, M., Marcazzó, J., Grasselli, C., Lavat, A., Molina, P., Spano, F., and Caselli, E. (2011). Thermo- and radioluminescence of undoped and Dy-doped strontium borates prepared by sol-gel method. *Radiat. Meas.*, 46, 12, 1488–1491. 10.1016/j.radmeas.2011.01.006

[41] Shyichuk, A., and Lis, S. (2013). Green-emitting nanoscaled borate phosphors $Sr_3RE_2(BO_3)_4$:Tb^{3+}. *Mater. Chem. Phys.*, 140, 2–3, 447–452. 10.1016/j.matchemphys.2013.03.019

[42] Rao, R. P. (2003). Tb3+ activated green phosphors for plasma display panel applications. *J. Electrochem. Soc.*, 150, 8, H165. 10.1149/1.1583718

[43] Bengisu, M., Yilmaz, E., Farzad, H., and Reis, S. T. (2008). Borate, lithium borate, and borophosphate powders by sol–gel precipitation. *J. Sol-Gel Sci. Technol.*, 45, 3, 237–243. 10.1007/s10971-008-1681-8

[44] Shyichuk, A. A., and Lis, S. (2011). Photoluminescence properties of nanosized strontium-yttrium borate phosphor $Sr_3Y_2(BO_3)_4$:Eu^{3+} obtained by the sol-gel Pechini method. *J. Rare Earths*, 29, 12, 1161–1165. 10.1016/S1002-0721(10)60617-5

[45] Seraiche, M., Guerbous, L., Kechouane, M., Potdevin, A., Chadeyron, G., and Mahiou, R. (2017). VUV excited luminescence of $Gd_{0.9}Eu_{0.1}BO_3$ nanophosphor prepared by aqueous sol-gel method. *J. Lumin.*, 192, 404–409. 10.1016/j.jlumin.2017.07.012

[46] Wang, C., and Yan, B., Sol–gel synthesis and photoluminescence of RE_3BO_6:Eu^{3+}/Tb^{3+} (RE=Y, Gd) microcrystalline phosphors from hybrid precursors, *J. Non-Cryst. Solids.*, 354 (10–11,) 962–969. 10.1016/j.jnoncrysol.2007.08.029

[47] Yang, R., Sun, X., Jiang, P., Gao, W., Cong, R., and Yang, T. (2018). Sol-gel syntheses of pentaborate β-LaB_5O_9 and the photoluminescence by doping with Eu^{3+}, Tb^{3+}, Ce^{3+}, Sm^{3+}, and Dy^{3+}. *J. Solid State Chem.*, 258, 212–219. 10.1016/j.jssc.2017.10.022

[48] Zhang, Y., Li, Y., and Yin, Y. (2005). Red photoluminescence and morphology of Eu3+ doped $Ca3La_3(BO_3)_5$ phosphors. *J. Alloys Compd.*, 400, 1–2, 222–226. 10.1016/j.jallcom.2005.04.001

[49] Patil, K. C., Aruna, S. T., and Ekambaram, S. (1997). Combustion synthesis. *Curr. Opin. Solid State Mater. Sci.*, 2, 158–165. 10.1016/S1359-0286(97)80060-5

[50] Philpot, K. A., Munir, Z. A., and Holt, J. B. (1987). An investigation of the synthesis of nickel aluminides through gasless combustion. *J. Mater. Sci.*, 22, 159–169. 10.1007/BF01160566

[51] Booth, F. (1953). The theory of self-propagating exothermic reactions in solid systems. *Trans. Faraday Soc.*, 49, 1953, 272–281. 10.1039/TF9534900272

[52] Moore, J. J., and Feng, H. J. (1995). Combustion synthesis of advanced materials, Part I. Reaction parameters. *Prog. Mater. Sci.*, 39, 243–273. 10.1016/0079-6425(94)00011-5

[53] Ekambaram, S., Patil, K. C., and Maaza, M. J. (2005). Synthesis of lamp phosphors: Facile combustion approach. *J. Alloys Compd.*, 393, 81–92. 10.1016/j.jallcom.2004.10.015

[54] Merzhanov, A. G., and Borovinskaya, I. P. (1993). Historical retrospective of SHS: An autoreview. *Int. J. Self-Propagating High-Temp. Synth.*, 6, 1993, 119.

[55] Subrahmanyam, J., and Vijaykumar, M. (1992). Self-propagating high-temperature synthesis. *J. Mater. Sci.*, 27, 6249–6273. 10.1007/BF00576271

[56] Koparkar, K. A., Bajaj, N. S., and Omanwar, S. K. (2014). Synthesis and characterization of $MgY_2B_2O_7$:Eu(III) phosphors. *Int. J. Chem Tech Res.*, 6, 3287–3290.

[57] Li, X., Liu, H., Wang, J., Zhang, X., and Cui, H. (2004). Preparation and properties of YAG nanosized powder from different precipitating agent. *Opt. Mater.*, 25, 4, 407–412.

[58] Koparkar, K. A., Bajaj, N. S., and Omanwar, S. K. (2015). Combustion synthesis and photoluminescence properties of Eu3+ activated Y2Zr2O7 nano phosphor. *Indian J. Phys.*, 89, 295–298. 10.1007/s12648-014-0554-y

[59] Onani, M. O., Okil, J. O., and Dejene, F. B. (2014). Solution–combustion synthesis and photoluminescence properties of YBO3:Tb3+ phosphor powders. *Physica B*, 439, 133–136. 10.1016/j.physb.2013.10.056

[60] Gawande, A. B., Sonekar, R. P., and Omanwar, S. K. (2014). Combustion synthesis and energy transfer mechanism of $Bi^{3+} \rightarrow Gd^{3+}$ and $Pr^{3+} \rightarrow Gd^{3+}$ in YBO_3. *Combust. Sci. Technol.*, 186, 785–791. 10.1080/00102202.2013.878708

[61] Hargunani, S. P., Sonekar, R. P., Singh, A., Khosla, A., and Arya, S., (2020) Structural and spectral studies of Ce^{3+} doped $Sr_3Y(BO_3)_3$ nano phosphors prepared by combustion synthesis, *Materials Technology*, 1–12. 10.1080/10667857.2020.1859052

[62] Ingle, J., Sonekar, R., Omanwar, S., Wang, Y., and Zhao, L. (2014). Combustion synthesis and luminescent properties of metal yttrium borates M_3Y_2 $(BO_3)_4$:Eu^{3+} (M=Ba, Sr) for PDPs applications. *Solid State Sci.*, 33, 19–24. 10.1016/j.solidstatesciences.2014.04.007

[63] Ingle, J., Gawande, A., Sonekar, R., Omanwar, S., Wangd, Y., and Zhao, L. (2013). Combustion synthesis and optical properties of oxy-borate phosphors $YCa_4O(BO_3)_3$:RE^{3+} (RE = Eu^{3+}, Tb^{3+}) under UV, VUV excitation. *J. Alloys Compd.*, 585, 633–636. 10.1016/j.jallcom.2013.09.178

[64] Bajaj, N., and Omanwar, S. (2013). Combustion synthesis and characterization of phosphor $KSr_4(BO_3)_3$:Dy^{3+}. *Opt. Mater.*, 35, 1222–1225. 10.1016/j.optmat.2013.01.025

[65] Gawande, A. B., Sonekar, R. P., and Omanwar, S. K. (2014). Synthesis and PL study of UV emitting phosphor $KCa_4(BO_3)_3$:Pb^{2+}. *J. Lumin.*, 149, 200–203. 10.1016/j.jlumin.2014.01.044

[66] Bhagat, S. P., Gawande, A. B., and Omanwar, S. K. (2015). Photoluminescence study of a novel UV emitting phosphor $Sr_2Mg(BO_3)_2$:Pb^{2+},Gd^{3+}. *Opt. Mater.*, 40, 36–40. 10.1016/j.optmat.2014.11.043

[67] Singh, V., Chakradhar, R. P. S., Rao, J. L., and Zhu, J.-J. (2008). Studies on red-emitting Cr^{3+} doped barium aluminate phosphor obtained by combustion process. *Mater. Chem. Phys.*, 111, 1, 143–148.

[68] Pekgözlü, İ. (2013). Photoluminescence properties of $M_2Mg(BO_3)_2$:Sm^{3+} (M:Sr and Ba). *J. Lumin.*, 134, 8-13. 10.1016/j.jlumin.2012.09.031

[69] Dawson, W. J. (1988). Hydrothermal synthesis of advanced ceramic powder. *Am. Ceram. Soc. Bull.*, 67, 1673–1678.

[70] Kim, T., and Kang, S. (2005). Hydrothermal synthesis and photoluminescence properties of nanocrystalline $GdBO_3$:Eu^{3+} phosphor. *Mater. Res. Bull.*, 40(11), 1945–1954. 10.1016/j.materresbull.2005.06.001

[71] Wang, Y., Endo, T., He, L., and Wu, C. (2004). Synthesis and photoluminescence of Eu^{3+}-doped (Y,Gd)BO_3 phosphors by a mild hydrothermal process. *J. Crystal Growth*, 268, 568–574. 10.1016/j.jcrysgro.2004.04.093

[72] Szczeszak, A., Grzyb, T., Barszcz, B., Nagirnyi, V., Aleksei, K., and Lis, S. (2013). Hydrothermal synthesis and structural and spectroscopic properties of the new triclinic form of $GdBO_3$:Eu^{3+} nanocrystals. *Inorg. Chem.*, 52(9), 4934–4940. 10.1021/ic302525k

[73] Zheng, J., Wu, X., Ren, Q., Bai, W., Ren, Y., Wang, M., and Hai, O. (2020). Investigation of luminescence properties and energy transfer in Sm^{3+} and Eu^{3+} co-doped $Sr_3Y(BO_3)_3$ red phosphors. *Optics Laser Technol.*, 122, 105857. 10.1016/j.optlastec.2019.105857

[74] Liang, P., Liu, J. W., and Liu, Z. H. (2016). Controllable hydrothermal synthesis of Eu^{3+}/Tb^{3+}/Dy^{3+} activated $Zn_8[(BO_3)_3O_2(OH)_3]$ micro/nanostructured phosphors, energy transfer and tunable emissions. *RSC Adv.*, 6 (92) 89113–89123. 10.1039/C6RA19101K

[75] Wang, Y., Endo, T., He, L., and Wu, C. Synthesis and photoluminescence of Eu^{3+}-doped (Y,Gd)BO_3 phosphors by a mild hydrothermal process. *J. Crystal Growth*, 268, 3–4, 568–574. 10.1016/j.jcrysgro.2004.04.093

[76] He, L., and Wang, Y. H. (2005). Control of morphology of YBO_3:Eu by a mild hydrothermal process. *Mater. Sci. Forum*, 475, 1829–1832. 10.4028/www.scientific.net/MSF.475-479.1829

[77] Wang, Y. H., Wu, C. F., and Zhang, J. C. (2006). Hydrothermal synthesis and photoluminescence of novel green-emitting phosphor $Y_{1-x}BO_3$:xTb^{3+}. *Mater. Res. Bull.*, 41, 8, 1571–1577. 10.1016/j.materresbull.2005.05.031

[78] Wen, F., Li, W., Liu, Z., Kim, T., Yoo, K., Shin, S., Moon, J.-H., and Kim, J. H. (2005). Hydrothermal synthesis of Sb^{3+} doped and (Sb^{3+}, Eu^{3+}) co-doped YBO_3 with nearly white light luminescence. *Solid State Commun.*, 133, 7, 417–420. 10.1016/j.ssc.2004.12.013

[79] Shmurak, S. Z., Kedrov, V. V., Kiselev, A. P., Fursova, T., and Shmytko, I. (2017). Spectral and structural characteristics of $Lu_{1-x-y}Ce_xTb_yBO_3$ orthoborates prepared by the hydrothermal synthesis method. *Phys. Solid State*, 59, 2017, 1171–1182. 10.1134/S1063783415080326

[80] Wang, Y., Wu, C., and Wei, J. (2007). Hydrothermal synthesis and luminescent properties of $LnPO_4$:Tb,Bi (Ln=La,Gd) phosphors under UV/VUV excitation. *J. Lumin.*, 126, 2, 503–507. 10.1016/j.jlumin.2006.09.006

[81] https://serc.carleton.edu/research_education/geochemsheets/techniques/SXD.html

[82] Ma, T., Kapustin, E. A., Yin, S. X., Liang, L., Zhou, Z., Niu, J., Li, Hua, L., Wang, Y., Su, J., Li, J., Wang, X., Wang, W. D., Wang, W., Sun, J., and Yaghi, O. M. (2018). Single-crystal x-ray diraction structures of covalent organic frameworks. *Science*, 361, 48–52. 10.1126/science.aat7679

[83] Nazarov, M., and Noh, D. Y., (2011) . New Generation of Europium- and Terbium-Activated Phosphors, From Syntheses to Applications, CRC Press, 02-Sep-2011 - Science - 300 pages 10.4032/9789814364058

[84] Patil, A., (2010). Study of electrical, structural and gas sensing characteristics of ZnO thick solid films with different dopants. August, 2010 http://hdl.handle.net/10603/2233

[85] Daniel, R., and Musil, J., (2014). Novel Nanocomposite Coatings, Advances and Industrial Applications, ISBN 9789814411172, CRC Press, 16-Dec-2014 - Science - 344 pages).

[86] X-Ray and Neutron Diffraction, Materials Characterization (2019). SN 978-1-62708-213-6.

[87] Rice, S. A. (1956). Small angle scattering of X-rays. A. Guinier and G. Fournet. Translated by C. B. Wilson and with a bibliographical appendix by K. L. Yudowitch. Wiley, New York, 1955. 268 pp. J. Polym. Sci., 19, 594–594. 10.1002/pol.1956.120199326

[88] Bauer, P. S., Amenitsch, H., Baumgartner, B., Köberl, G., Rentenberger, C., and Winkler, P. M. (2019). In-situ aerosol nanoparticle characterization by small angle X-ray scattering at ultra-low volume fraction. *Nat. Commun.*, 10, 2041–1723. 10.1038/s41467-019-09066-4

[89] Tanner, B., Hase, T., Lafford, T., and Goorsky, M. (2004). Grazing incidence in-plane X-ray diffraction in the laboratory. *Powder Diffr.*, 19, 1, 45–48. 10.1154/1.1649319

[90] Cowley, J. M., (1992) . Electron Diffraction Techniques, International Union of Crystallography Science – 2, 423 pages. ISBN 0-19-855733-7 10.1002/crat.2170290319

[91] Fuller, W., and Mahendrasingam, A. (2017). X-Ray Diffraction, Studies of Fibres and Films. Editor (s): John C. Lindon, George E. Tranter, David W. Koppenaal, *Encyclopedia of Spectroscopy and Spectrometry*. 3rd ed., Academic Press, 667–675. ISBN 9780128032244.

[92] Arnott, S., and Wonacott, A. J. (1966). The refinement of the molecular & crystal structures of polymers using X-ray data and stereochemical constraints. *Polymer*, 7, 157–166. 10.1016/0032-3861(66)90009-7

[93] von Ardenne, M. (1938). Das Elektronen-Rastermikroskop. Theoretische Grundlagen. *Zeitschrift für Physik (in German)*, 109(9-10), 553–572, Bibcode:1938ZPhy.109.553V. 10.1007/BF01341584

[94] Pendse, D. R., and Chin. A. K. (2001). Cathodoluminescence and Transmission Cathodoluminescence, Editor(s): K. H. Jürgen Buschow, Robert W. Cahn, Merton C. Flemings, Bernhard Ilschner, Edward J. Kramer, Subhash Mahajan, Patrick Veyssière, *Encyclopedia of Materials: Science and Technology*, Elsevier, 2001, 1–7, ISBN 9780080431529, 10.1016/B0-08-043152-6/00190-X

[95] Xu, R. (2015). Light scattering, a review of particle characterization applications. *Particuology*, 18, 11–21. 10.1016/j.partic.2014.05.002

[96] https://www.vcbio.science.ru.nl/en/fesem/info/principe/

[97] The Nobel Prize in Physics 1986, Perspectives – Life through a Lens. nobelprize.org.

[98] Su, D. (2017). Advanced electron microscopy characterization of nanomaterials for catalysis. *Green Energy Environ.*, 2, 2, 70–83. 10.1016/j.gee.2017.02.001

[99] Kumar, P. S., Pavithra, K. G., and Naushad, M. (2019). Chapter 4 - Characterization Techniques for Nanomaterials, Editor(s): Sabu Thomas, El Hadji Mamour Sakho, Nandakumar Kalarikkal, Samuel Oluwatobi Oluwafemi, Jihuai Wu, *Nanomaterials for Solar Cell Applications*, Elsevier, 2019, 97–124.

[100] http://getdrawings.com/drawing-tag/microscope

[101] Pentassuglia, S., Agostino, V., Tommasi, T., (2018). Electroactive Biofilm, A Biotechnological Resource, Editor(s): Klaus Wandelt, *Encyclopedia of Interfacial Chemistry*, Elsevier, 2018, 110–123, ISBN 9780128098943. https://doi.org/10.1016/B978-0-12-409547-2.13461-4

[102] Egger, M. D., and Petran, M. (1967). New reflected-light microscope for viewing unstained brain and ganglion cells. *Science*, 157, 305–307. 10.1126/science.157.3786.305

[103] Davidovits, P., and Egger, M. D. (1973). Photomicrography of Corneal Endothelial Cells in vivo. *Nature*, 244, 366–367. 10.1038/244366a0

[104] Amos, W. B., and White, J. G. (2003). How the confocal laser scanning microscope entered biological research. *Biol. Cell*, 95, 335–342. 10.1016/s0248-4900(03)00078-9

[105] https://bitesizebio.com/19958/what-is-confocal-laser-scanning-microscopy/

[106] Ohnesorge, F. (1993). True atomic resolution by atomic force microscopy through repulsive and attractive forces. *Science*, 260, 5113, 1451–1456. 10.1126/science.260.5113.1451PMID 17739801. S2CID 27528518 Bibcode:1993Sci.260.1451O

[107] Binnig, G., Quate, C. F., and Gerber, C. H. (1986). Atomic force microscope. *Phys. Rev. Lett.*, 56, 930–933. 10.1103/PhysRevLett.56.930

[108] Adams, J., Hector, L., Siegel, D., Yuand, H., and Zhong, J. (2001). Adhesion, lubrication and wear on the atomic scale. *Surf. Interface Anal.*, 31, 619–626. 10.1002/sia.1089

[109] Bhushan, B., Israelachvili, J. N., and Landman, U. (1995). Nanotribology: Friction, wear and lubrication at the atomic scale. *Nature*, 374, 607–616. 10.1038/374607a0

[110] Sanz-Medel, A., and Costa-Fernandez, J. M. (2005). PHOSPHORESCENCE | Principles and Instrumentation, Editor(s): Paul Worsfold, Alan Townshend, Colin, Pool. *Encyclopedia of Analytical Science*. 2nd ed., Elsevier, 149–157, ISBN 9780123693976 .

[111] Gooijer, C., Ariese, F., and Velthorst, N. H.(2005). FLUORESCENCE | High-Resolution Techniques, Editor(s): Paul Worsfold, Alan Townshend, Colin Poole. *Encyclopedia of Analytical Science*. 2nd ed., Elsevier, 119–128, ISBN 9780123693976.

[112] Koparkar, K. A., Bajaj, N., and Omanwar, S., (2014) A potential candidate for lamp phosphor, Eu^{3+} activated $K_2Y_2B_2O_7$. *Adv. Opt. Technol.*, 1–5. 10.1155/2014/706459

[113] Blasse, G., and Grabmaier, B. C. (1994). *Luminescent materials*, Springer-Verlag, Berlin, Heidelberg, 11, ISBN: 978-3-540-58019-5 .

[114] Lakowicz, J. R. (2006). *Principles of Fluorescence Spectroscopy*. 3rd ed., Springer, London, UK.

[115] Seitz, F.(1948). 'Luminescent Crystals', Preparation and characteristics of Solid Luminescent Materials, Symp. Cornell Univ., John Wiley & Sons, INC, NY p 9.

[116] Johnson, R. P. (1939). Luminescence during intermittent optical excitation. *J. Opt. Soc. Amer.*, 29, 7, 283–290. 10.1364/JOSA.29.000283

[117] Lakowicz, J. R., Laczko, G., Cherek, H., Gratton, E., and Limkeman, M. (1984). Analysis of fluorescence decay kinetics from variable-frequency phase shift and modulation data. *Biophys. J.*, 46, 463–477. 10.1016/S0006-3495(84)84043-6

[118] Wagnieres, G. A., Star, W. M., and Wilson, B. C. (1998). In vivo fluorescence spectroscopy and imaging for oncological applications. *Photochem. Photobiol.*, 68, 5, 603–632. 10.1111/j.1751-1 097.1998.tb02521.x

[119] Koparkar, K. A., Bajaj, N. S., and Omanwar, S. K. (2015). Effect of partially replacement of Gd^{3+} ions on fluorescence properties of $YBO_3:Eu^{3+}$ phosphor synthesized via precipitation method. *Opt. Mater.*, 39, 74-80. 10.1016/j.optmat.2014.11.001

[120] Solgi, S., Tafreshi, M. J., and Ghamsari, M. S. (2019). A facile route for synthesis of highly pure α-CaB_4O_7 compound. *Mater. Res. Express*, 6, 026205–026209. 10.1088/2053-1591/aaeec5

[121] Raman, C. V., and Krishnan, K. S. (1928). A new type of secondary radiation. *Nature*, 121, 501–502. 10.1038/121501c0

[122] Raman, C. V. (1928). A new radiation. *Indian J. Phys.*, 2, 387–398. http://www.jstor.org/stable/24101519

[123] Smith, E. , and Dent, G. (2005). *Modern Raman Spectroscopy, A Practical Approach*. 1st ed. Wiley, ISBN 0-471-49668-5

[124] Bumbrah, G., and Sharma, R. (2015). Raman spectroscopy – Basic principle, instrumentation and selected applications for the characterization of drugs of abuse. *Egypt. J. Forensic Sci.*, 6, 209–215. 10.1016/j.ejfs.2015.06.001

[125] Raman, C. V., and Krishnan, K. S. (1928). A new type of secondary radiation. *Nature*, 121, 501–502. 10.1038/121501c0

[126] Briggs, D., and Seah, M. P. (1990). Practical Surface Analysis. *Angewandte Chemie*, John Wiley, Chichester. 10.1002/ange.19951071133

[127] Siegbahn, K. (1966). Alpha- beta- and gamma-ray spectroscopy. *Am. J. Phys.*, 34, 1966, 275. 10.1119/1.1972911

[128] Siegbahn, K. (1982). Electron-spectroscopy for atoms, molecules, and condensed matter. *Science*, 217, 111–121. 10.1126/science.217.4555.111

[129] Moulder, J. F., Chastain, J., and King, R. C. (1995). Handbook of X-ray photoelectron spectroscopy: A reference book of standard spectra for identification and interpretation of XPS data, Physical Electronics, Eden Prairie, MN (1995).

[130] Zhu, H., Zhang, L., Zuo, T., Gu, X., Wang, Z., Zhu, L., and Yao, K. (2008). Sol–gel preparation and photoluminescence property of $YBO_3:Eu^{3+}/Tb^{3+}$ nanocrystalline thin films. *Applied Surf. Sci.*, 254(20), 6362–6365. 10.1016/j.apsusc.2008.03.183

[131] Baer, D. R., and Thevuthasan, S. (2010). Chapter 16 - Characterization of Thin Films and Coatings, Editor(s): Peter M. Martin, *Handbook of Deposition Technologies for Films and Coatings*. 3rd ed., William Andrew Publishing, NY, 13815 United States,749–864.

[132] Kohli, R. (2019). *Developments in Surface Contamination and Cleaning, Chapter 3 - Methods for Assessing Surface Cleanliness*, Editor(s): Rajiv Kohli, K. L. Mittal, 12, Elsevier, 23–105.

[133] Skoog, D., Holler, F., and Crouch, S. (2007). *Principles of Instrumental Analysis*. 6th ed. Thomson Brooks/Cole, Belmont, CA, 169–173. ISBN 978-0-495-01201-6 .

[134] Metha, A. (13 December 2011). *"Principle". PharmaXChange.info.*

[135] Lewis, M. J., (1996). Sensible and latent heat changes: Physical Properties of Foods and Food Processing Systems, Pages 220–245, ISBN 9781855732728. 10.1533/9781845698423.220

[136] Paulik, J., Paulik, F., and Arnold, M. (1983). Thermogravimetric examination of the dehydration of calcium nitrate tetrahydrate under quasiisothermal and quasi-isobaric conditions. *J. Thermal Anal.*, 27, 409-418. 10.1007/BF01914678

3 Borate Phosphors
Lamp Phosphors

N. S. Sawala and P. A. Nagpure

CONTENTS

DOI: 10.1201/9781003207757-3

3.1   HISTORY AND INVENTION OF FLUORESCENT LAMPS

French physicist Alexandre E. Becquerel in 1857 invented the phenomena of fluorescence and phosphorescence and hypothesized the development of fluorescent tubes similar to those made today. Alexandre Becquerel tried to coat the electric discharge tubes with luminescent materials, a process was further developed in fluorescent lamps in later years [1,2].

About half a century later, with the speculation of fluorescent lamps, Peter Cooper Hewitt patented (U.S. patent 889,692) the first low-pressure mercury vapour lamp in 1901. In this lamp, electric energy excites mercury vapour to create luminescence. In the same period, German physicist Julius Plucker and Heinrich Geissler succeeded in producing light by passage of an electric current through a glass tube containing a small amount of gas. Peter Hewitt also succeeded in producing intense light from mercury-filled tubes in the late 1890s, but this bluish-green light was quite unpleasant. Hence, it is not suitable for lighting at home, but can be used in photographic studios and industry [3,4].

Edmund Germer is honoured by some historians as being the inventor of the first true fluorescent lamp. However, it can be argued that fluorescent lamps have a long history of development before Germer [5,6].

The team of scientists of General Electric Company led by George Inman designed the first practical and viable fluorescent lamp (U.S. Patent No. 2,259,040) that was first sold in 1938. Thus, GE became the first company to introduce the fluorescent lamp into the market.

Several other inventors patented versions of the fluorescent lamp, including Thomas Edison. He filed a patent (U.S. Patent 865,367) on May 9, 1896, for a fluorescent lamp that was never sold. However, he did not use mercury vapour to excite the phosphor. His lamp used X-rays [7,8].

As the demand for the fluorescent light grew, several advancements were made. In 1980, when Philips designed the first line of screw-in fluorescents for magnetic ballasts this made a revolution in lighting technology. The reasonableness, along with the performance of these lamps, made them well-liked in both commercial and residential lighting. In the 1990s, a T5 lamp was introduced that offered an even more efficient solution. Today, fluorescent lighting is one of the most widely sold options in the market because it is still an affordable and pleasurable lighting solution [9].

3.2   PHOSPHORS FOR FLUORESCENT LAMPS

In the beginning, naturally occurring fluorescent material such as willimite (Zn_2SiO_4: Mn^{2+}) was used as phosphor in fluorescent lamps [10,11]. The invention of calcium halo-phosphate phosphor in the 1940s led to a significant breakthrough in fluorescent lighting ($Ca_5(PO_4)_3(Cl, F):Sb^{3+}, Mn^{2+}$).

The blue band peaking near 480 nm is due to the emission of Sb^{3+} activator ions at 254 nm excitation. The excitation energy is also transferred from Sb^{3+} to Mn^{2+} resulting in the orange-red Mn^{2+} emission peaking at about 580 nm. The optimum white light can be obtained by adjusting the ratio of the orange to blue emissions; for example, increasing the concentration of Mn^{2+} enhances the orange emission and suppresses the blue emission. Thus proper white light can therefore be obtained from a single phosphor material. The color rendition index (CRI) of this 'cool white' fluorescent lamp is between 50 and 60. The low CRI in this case is due to the deficiency of the red spectral region in the emission spectrum. It shows that the two complementary emission bands of halo-phosphate-based phosphor do not cover the entire visible region of the spectrum. The CRI is improved by using the blend of halo-phosphate phosphor with a red-emitting strontium ortho-phosphate Sn^{2+}-activated phosphor. These are known as the 'deluxe' lamps which display higher CRI, typically 85, due to the continuous emission over the entire visible region from the resultant blend. The improvement in CRI is accompanied by a considerable fall in the luminous efficacy of the lamp to 50 lm/W [12].

3.3   RARE-EARTH TRI-PHOSPHORS: LITERATURE REVIEW

Due to the low CRI and less luminous efficacy of the halo-phosphate lamps, although acceptable in many applications, is not good enough to gratify the consumers to install such lamps in the living areas of their home; hence these lamps are mostly used in the basements, parking lots, warehouses, etc. Then it has been thought that high CRI and high luminous efficacy can also be obtained through the use of rare-earth phosphors where white light is produced by blending three (red-, green- and blue-emitting) phosphors. Theoretical modelling had also predicted the possibility of high CRI and high luminous efficacy with a blend of three phosphors in the ratio 60:30:10; having emission bands at 610 nm (red), 550 nm (green) and 450 nm (blue), respectively [12].

The modern generation of low-pressure mercury fluorescent lamps with high CRI (nearly 80) as well as high light output (above 80 lm/W) were developed in the 1970s. Such fluorescent lamps with the blend of three rare earth–activated phosphors were made commercially available. The tri-color or tri-phosphor blend used in the lamps consists of red-emitting $Y_2O_3:Eu^{3+}$, green-emitting $CeMgAl_{11}O_{19}:Tb^{3+}$ and blue-emitting $BaMgAl_{10}O_{17}:Eu^{2+}$ [13]. The use of these tri-colour phosphors in the lamps has made a reduction in lamp diameter possible. This ultimately resulted in the development of the compact fluorescent lamps (CFLs) with various sizes and shapes. A reduction of the internal tube diameter from 36 mm (wall load of 270 w/m^2) to 25 mm (330 w/m^2), and even to 10 mm (1,100 w/m^2) became possible without any significant loss in the output during the lamp's life [14].

3.3.1 $Y_2O_3:Eu^{3+}$

All the characteristics of this phosphor, such as the emission spectrum, quantum efficiency, lumen maintenance, etc. of this phosphor are ideal for fluorescent lamps.

In general, electronic transitions of Eu^{3+} are influenced by the symmetry of the ion in the host lattice. If inversion symmetry is present, the electric-dipole transitions are forbidden, so that the emission is restricted to only magnetic-dipole $^5D_0 \rightarrow {}^7F_1$ transitions. If inversion symmetry is absent, electric-dipole transitions are most probable [15,16].

The commercial $Y_2O_3:Eu^{3+}$ phosphor contains relatively high Eu^{3+} concentrations (3–5 mole percent). At lower Eu^{3+} concentrations, emission on the higher energy side are due to transitions from $^5D_1$ and $^5D_2$ enhances at the expense of the red $^5D_0 \rightarrow {}^7F_2$ emission. The $Y_2O_3:Eu^{3+}$ phosphor absorbs the 254 nm mercury discharge emission through a charge transfer transition involving the Eu^{3+} ion and the neighbouring O^{2-} ions.

The $Y_2O_3:Eu^{3+}$ phosphor is still the most expensive component of the tri-phosphor blend. Research aimed at replacing this phosphor with less-expensive alternatives have not met with success.

3.3.2 $CeMgAl_{11}O_{19}:Tb^{3+}$

Verstegen et al . in 1974 has found the green emission in Tb^{3+} in $CeMgAl_{11}O_{19}$ the tri-colour lamp phosphor which has the magneto-plum bite type (hexagonal) structure. But, Tb^{3+} cannot be excited directly as the charge transfer band of Tb^{3+} lies well above the predominant mercury line at 254 nm in the lamp. In order to absorb the 254 nm radiation efficiently, a sensitizer, Ce^{3+} which has an intense absorption in this region is used.

Often there is a considerable contribution to the emission from the higher level $^5D_3 \rightarrow {}^7F_j$, mainly in the blue. The Ce^{3+} emission of $CeMgAl_{11}O_{19}$ has an emission maximum at 330 nm and the first excitation maximum at 270 nm with a large stokes shift (~8,000 cm^{-1}). The energy transfer between Ce^{3+} ions does not occur and concentration quenching is absent. The $Ce^{3+} \rightarrow Tb^{3+}$ transition is a one-step transfer process which requires high Tb^{3+} concentration for effective quenching of Ce^{3+} emission [17]. The optimum phosphor composition $(Ce_{0.67}Tb_{0.33})MgAl_{11}O_{19}$ exhibits high quantum efficiency and excellent lumen output and maintenance during lamp operation.

3.3.3 BaMgAl$_{10}$O$_{17}$:Eu^{2+}

The blue-emitting phosphor $BaMgAl_{10}O_{17}:Eu^{2+}$ [18], popularly known as BAM, is commonly used in tri-colour fluorescent lamps.

As in all Eu^{2+}-activated samples, the intense excitation spectrum is found throughout the UV spectral region, and is associated with 4f → 5d electronic transitions. The synthesis of BAM is generally accomplished under a reducing atmosphere [19–21]. The blue phosphors represent only a minor weight fraction of the tri-phosphor blend (about 10% for colour temperature of 4,100 K). However, blends designed for higher colour temperatures, say 6,500 K, require higher amounts of the blue emitting component.

The list of the red, green and blue phosphor materials commonly used for tri-colour lamp phosphors is tabulated in Table 3.1.

3.4 REQUIREMENT FOR LAMP PHOSPHORS

Since the fluorescent lamp converts electrical energy to optical radiation, the conversion efficiency should be as high as possible. There are many factors affecting the conversion efficiency of fluorescent lamps but it mostly depends on the phosphors that are used. In order to have good conversion efficiency, the phosphor used should have the following qualities.

3.4.1 ADOPTABILITY OF AOPANT (ACTIVATOR) BY THE PHOSPHOR HOST

The phosphor host should be compatible with the luminescent centres i.e. accommodate the luminescent ions and allow them to be involved in the luminescence process. Further, they should not react with the luminescent ions to form compounds [23].

TABLE 3.1
Tri-colour lamp phosphors [22]

Phosphor	Colour	Q.E.	Emission (nm)
Y_2O_3: Eu^{3+}	Red	97	613
$(CeGd)MgB_5O_{10}$: Mn^{2+}		90	620
$(CeGd)MgB_5O_{10}$: Tb^{3+} (CBT)	Green	93	545
$CeMgAl_{11}O_{19}$: Tb^{3+} (CAT)		90	545
$(La, Ce)PO_4$: Tb^{3+}		93	545
$Sr_5(PO_4)_3Cl$: Eu^{3+}	Blue	90	445
$BaMgAl_{10}O_{17}$: Eu^{2+} (BAM)		90	450
$Sr_2Al_6O_{11}$: Eu^{2+} (SAL)		90	460
$Sr_4Al_{14}O_{25}$: Eu^{2+} (SAE)		90	490
$BaAl_8O_{13}$: Eu^{2+} (BAE)		90	480

3.4.2 Efficient Absorption of UV (185 and 254 nm)

The efficient absorption of UV radiations and its conversion into visible light by the phosphor is indicated by its quantum efficiency. The phosphors should have quantum efficiency above 80%. A factor often used to index this property of the lamp is the visual sensitivity defined by

$$Q = \frac{\int E(\lambda) V(\lambda) \, d\lambda}{E(\lambda) \, d\lambda}$$

where E is the energy of the emitted photon and V is the sensitivity of the eye. The lamp output is also often expressed in terms of light emitted per unit energy consumed. Theoretically, the maximum output can be 310 lumen/watt (Elenbaas 1959). Lamp and phosphor performances have been characterized using various other indices also [23].

3.4.3 Transparency in the Visible Range

The phosphor should be transparent over entire range visible spectrum, so that phosphor will not reabsorb the emitted visible radiation.

3.4.4 Good Persistence, Long Lifetime (Phosphorescence)

The phosphors continue to emit light when the source of excitation has been removed: this afterglow, which continues for hours in the case of some sulphide phosphors, is only a fraction of a second with other phosphors [24,25]. It is important for fluorescent lamp phosphors to have a bright, short afterglow to minimise the stroboscopic effect due to flickering of light emitted by the lamp, which occurs when discharge lamps are operated on alternating voltage supplies.

3.4.5 Phosphor Maintenance

The maintenance of the phosphor efficiency during the long life of lamp has perhaps greater importance than high initial intensity. It is well known that the light output of a fluorescent lamp decreases during life, particularly in the first 100 hours. The foremost decrease in efficiency is due to the mercury that deposits a thin film on the tube coating which prevents the UV radiation from reaching the phosphor. This deposition on the phosphor surface is also aggravated by traces of alkali in the phosphor or on the glass surface of the lamp. The poor efficiency maintenance by the phosphor may also be due to the close association of phosphor and mercury which makes possible a chemical reaction and results in a change of valency of the activator [26,27].

3.4.6 Stability under Industrial Handling

Phosphors are subjected to various thermal, mechanical and chemical treatments when they are applied on the inner walls of the lamp. For dispersing the phosphor uniformly in a coating solution, various steps such as milling in dry or liquid medium, baking etc. are necessary. They must retain initial PL properties after undergoing these processes. They should have minimal losses after heating treatments. Only a few phosphors get through this test [27].

3.4.7 Concentration Quenching

In order to obtain high emission efficiency, it would seem obvious to make the activator concentration as high as possible. In many cases, however, it is found that the emission efficiency decreases if the activator concentration exceeds a specific value known as the critical concentration. This effect is called concentration quenching. The concentration quenching can be explained in a number of cases as follows.

If the concentration of the activator becomes so high that the probability of energy transfer exceeds that for emission, the excitation energy repeatedly goes from the one-activator ion to the other [28–30].

If the host lattice is not perfect, that is it contains all kinds of sites such as at the surface, at dislocations, impurities, etc. where the excitation energy may lost, in some ambiguous way. While traversing the lattice, the excitation energy will encounter such a site and dissipate as heat. It makes no contributions to the luminescence. The efficiency then decreases, in spite of the increase of the activator concentration [28–30].

3.4.8   PARTICLE SIZE AND MORPHOLOGY PHOSPHOR

The proper particle size of the phosphor material is very important in increasing its luminous efficacy and reducing the cost of the lamp. The increasing surface-to-volume ratio by decreasing the particle size of the phosphor material reduces the quantity of the phosphor material to cover the inner surface of the lamp, thereby reducing the cost of the lamp. Also, backscattering and back reflection of light by phosphor particles can be minimized by optimizing phosphor particle sizes and morphology [31,32].

All of these qualities are not found in a single phosphor. This is quite understandable as many of the qualities are mutually exclusive. Making compromises on some of the qualities, the lamp phosphors are chosen in order get optimum outcome.

3.5   BORATE-BASED FLUORESCENT LAMP PHOSPHORS

Inorganic borates have become a focus of technological interest due to a variety of physical and chemical features exhibited by these compounds [33]. Owing to possible three-, or fourfold co-ordination of borate atoms, borates form a great number of compounds having diverse structures. Borates fundamentally possess characteristics that are advantageous for its use as lamp phosphor, which include wide transparency range, large electronic band gap, good thermal and chemical stability, low preparative temperature, optical stability with good nonlinear characteristics and exceptionally high optical damage threshold. Recent research on inorganic borates has been focused on the synthesis and characterization of compounds with potential application as phosphor material for fluorescent lamps [34–36].

In this section of the chapter, the reports on the some inorganic borate hosts which have the potential as lamp phosphor are discussed.

3.5.1   RED-EMITTING BORATE PHOSPHORS: LITERATURE REVIEW

Most of the red-emitting borate host phosphors utilize the Eu^{3+} ion as an activator due to strong UV absorption and narrow emission in the red region of Eu^{3+}. The characteristic emission lines of Eu^{3+} originate from the $^5D_0 \rightarrow ^7F_j$, (j = 0, 1, 2, 3, 4) transitions of Eu^{3+} ions that are in the orange-red region. Specifically the $^5D_0 \rightarrow ^7F_2$ transitions in Eu^{3+} corresponds to red emission around 610 nm, which is known to be ideal for tri-colour lamps. The electric dipole allowed $^5D_0 \rightarrow ^7F_2$ transitions are significant when Eu^{3+} occupies non-centrosymmetric positions in the host lattices and the contributions from the other transitions $^5D_0 \rightarrow ^7F_j$ (j = 0, 1, 3, 4.....) are negligibly weak.

3.5.1.1   REBO$_3$: Eu^{3+}

The ortho-borates (REBO$_3$, where RE = Rare Earths Y, La, Gd and Lu) possess the hexagonal vaterite-type structure and the activator Eu^{3+} ion can occupy the RE^{3+} site with the point symmetry S6. Therefore, the orange emission at 592 nm from the magnetic dipole $^5D_0 \rightarrow ^7F_1$ transition is dominant in the rare ortho-borates [37–39].

3.5.1.2   YCaBO₃: Eu³⁺

The excitation spectrum for the 611 nm emission mainly consists of broadband peaking near 254 nm. It is in agreement with that of the commercial red phosphors Y_2O_3:Eu^{3+} [40]. The prominent emission line 610 nm is due to the $^5D_0 \rightarrow ^7F_2$ transition of Eu^{3+}, which indicates that the Eu^{3+} ions occupy a non-centro-symmetric positions in the $YCaBO_4$ lattice.

3.5.1.3   La₂CaB₁₀O₁₉: Eu³⁺

The excitation spectrum consists of a strong broad CT band peaking around 250 nm and the emission spectrum under 254 nm excitation consists of a series of sharp lines at 576, 588, 597, 611 and 616 nm [41,42].

3.5.1.4   Li₂Y₂B₂O₇: Eu³⁺

The excitation spectra (λ_{em} = 611 nm) consist of a single broadband peaking at 234 nm. The emission spectra under UV (λ_{ex} = 254 nm) excitation consist of a group of lines between 600 and 660 nm corresponding to $^5D_0 \rightarrow ^7F_j$ (j = 1, 2) transitions [43].

3.5.1.5   Na₃Y(BO₃)₂: Eu³⁺

The excitation spectra consist of a broadband peaking around 250 nm and the emission spectrum consists of an intense red emission line at 611 nm corresponding to ($^5D_0 \rightarrow ^7F_2$) transition of Eu^{3+} [44].

3.5.2   GREEN-EMITTING BORATE PHOSPHORS

Trivalent terbium Tb^{3+} ion is used as a green-emitting centre in a variety of commercial phosphors. The electronic configuration of Tb^{3+} ion is $4f^8$. In the case of the Tb^{3+} ion, the absorption is usually due to the allowed f-d transition. From the excited state of the $4f^75d^1$ configuration, the electron loses energy to lattice and comes to ^5D_j. $^5D_3 \rightarrow ^7F_j$ emission is in UV and blue region while $^5D_4 \rightarrow ^7F_j$ emission is predominantly green.

Trivalent cerium Ce^{3+} ion can strongly absorbed UV radiation and efficiently transfer its energy to Tb^{3+}. Hence, Ce^{3+} ion acts as sensitize for Tb^{3+} ion. The sensitization of green emission of Tb^{3+} ion by Ce^{3+} to Tb^{3+} energy transfer process has been extensively investigated in a variety of inorganic borate host materials [45]. The conditions for the energy transfer from the broad band emitter (Ce^{3+}) to the line emitter (Tb^{3+}) are,

   i. the absorbing ions should be nearest neighbours in the crystal lattice so that there is an orbital wave function overlap between the sensitizer and activator and
   ii. an overlap between the energy level of the sensitizer and activator.

3.5.2.1   GdMgB₅O₁₀: Ce,Tb

A new class of lanthanum pentaborate ($LnMgB_5O_{10}$), first time synthesised and characterised by B. Saubat et al. The Ln ions are 10 coordinated with Mg ions in a distorted octehedra [45]. The rare earth polyhedra share edges to form isolated zigzag chains. The shortest inter and intra Ln-Ln distances 0.4 nm and 0.6 nm, respectively. In $GdMgB_5O_{10}$, the Gd^{3+} ion assists in the transport of energy from sensitizer Ce^{3+} to activator Tb^{3+} ions. Since the Ce^{3+} levels are resonant with the lowest Gd^{3+} excited state (^6P_J), Ce^{3+} efficiently transfer the energy to Gd^{3+} ions. This energy is then trapped by the Tb^{3+} ions to give green luminescence. The quantum efficiency for this green-emitting phosphor is high and also phosphor shows excellent stability in fluorescent lamps. The synthesis methods, such as solid-state diffusion, combustion synthesis and sol-gel were employed [46,47].

The most prominent emission peaks were found around 543 nm due to $^5D_4 \rightarrow ^7F_5$ transition of Tb^{3+}. The excitation observed around 270 nm corresponds to absorption of Ce^{3+} ion.

3.5.2.2 $Ce_{0.2}Gd_{0.6}Tb_{0.2}Mg_{0.9}Mn_{0.1}B_5O_{10}$

An efficient energy transfer is also observed from Gd^{3+} to Mn^{2+} ions. In this case, in the addition to the Tb^{3+} emission, there is a broad emission peaking at about 616 nm, which can be attributed to the Mn^{2+} ion. This type of phosphor can be synthesised via solid-state diffusion, combustion synthesis and sol-gel [48]. The high-intensity emission peak/band: around 543 nm due to the $^5D_4 \rightarrow {}^7F_5$ transition of Tb^{3+} and 616 nm due to the transition in Mn^{2+}. An excitation around 270 nm corresponds to absorption of Ce^{3+} ions.

3.5.2.3 $(La_{0.5} Ce^{3+}_{0.3} Tb^{3+}_{0.2}) BaB_9O_{16}$

Intense and strong green emission of Tb^{3+} ions at 541 nm can be obtained in $LaBaB_9O_{16}$ by co-doping the host lattice with Ce^{3+} and Tb^{3+} ions. Ce^{3+} ions act as sensitizers and Tb^{3+} ions act as activators. The excitation energy is absorbed by Ce^{3+} ions and transfers it to the Tb^{3+} ions, which give intense green emission at 541 nm.

The excitation spectrum for $^5D_4 \rightarrow {}^7F_5$ transition (541 nm) of Tb^{3+} ions consists of intense broadband, which corresponds to the 4f-5d transitions of Ce^{3+} ions. The phosphor can be prepared by solid-state diffusion combustion synthesis [49]. The intense dominant emission peak was found at 541 nm due to the $^5D_4 \rightarrow {}^7F_5$ transition of Tb^{3+} when excitation energy around 260 nm corresponds to absorption of Ce^{3+} ions.

3.5.2.4 $Na_2La_2B_2O_7$: Tb^{3+}

The emission peaks around 544 nm due to $^5D_4 \rightarrow {}^7F_5$ transition of Tb^{3+}, while the excitation around 252 nm due to the transition between the $4f^8$ and $4f^7 5d^1$ states of Tb^{3+} ion. The combustion synthesis method is ideal for this phosphor [50].

3.5.2.5 $Sr_3RE_2(BO_3)_4$: Tb^{3+} (RE = Y, La and Gd)

By using the Pechini sole-gel method $Sr_3RE_2(BO_3)_4$:Tb^{3+} (RE=Y, La and Gd) was prepared. The emission band around 544 nm due to $^5D_4 \rightarrow {}^7F_5$ transition of Tb^{3+} ion was observed at an excitation energy around 240 nm due to the transition between the $4f^8$ and $4f^7 5d^1$ states of Tb^{3+}. The excitation spectra consist of f-d band and f-f bands of Tb^{3+} as well as f-f transitions of Gd^{3+} in the case of $Sr_3Gd_2(BO_3)_4$:Tb^{3+} [50,51].

3.5.3 BLUE-EMITTING BORATE PHOSPHORS

There are very few reports on the blue-emitting borate host phosphors for fluorescent lamps in the literature. Some results on emissions in the blue region by Eu^{2+}-, Ce^{3+}- and Sb^{3+}-activated borate host phosphors are presented in this section.

3.5.3.1 BaB_8O_{13}: Eu^{2+}

The Eu^{2+} ion is commonly employed as blue-emitting centre in most of the blue-emitting phosphors. However, there are several reports of Eu^{2+} doped compounds, which shows the emission colour of Eu^{2+} varies from ultraviolet to red. The emission and absorption spectra of Eu^{2+} ions usually consists of a broadband due to transitions between the $^8S_{7/2}$ ($4f^7$) ground state and the crystal field components of the $4f^6 5d^1$ excited states [52–56].

It consists of a broad excitation band in the wavelength range of 220 nm to 340 nm and emission band in the wavelength range of 380 nm to 460 nm, peaking at 408 nm.

3.5.3.2 $A_6 M_{1-x} Ce_x M'(BO_3)_6$ (A = La,Sr; M = La, Y, Gd, Lu; and M' = Mg, Al, Ga)

The blue emission is also associated with Ce^{3+} ions in several inorganic phosphors. The Ce^{3+} ion shows allowed optical transitions in absorption and emission, which are of the f $\rightarrow$ d type. The Ce^{3+} ion shows an efficient broadband luminescence in inorganic hosts due to its 4f $\rightarrow$ 5d parity allowed electric dipole transitions [57,58].

Novel blue-emitting activated borate phosphors have the formula $A_6MM'(BO_3)_6:Ce^{3+}$, wherein A = La, Sr; M = La, Y, Gd, Lu; and M' = Mg, Al, Ga are found to emit a broadband (340–480 nm) with a peak in the range 410–420 nm [59].

Synthesis method attempted: solid-state diffusion.

3.5.3.3 $LaBaB_9O_{16}: Sb^{3+}$

The excitation spectrum monitored at 445 nm emission consists of two overlapping broadbands peaking at 235 nm and 261 nm, which corresponds to the $^1S_0 \rightarrow ^3P_1$ transition of hte Sb^{3+} ion. Under UV 254 nm excitation, $LaBaB_9O_{16}: Sb^{3+}$ shows deep blue emission. The emission spectrum consists of a single broad blue band peaking at 445 nm, corresponding to the $^1P_{1,0} \rightarrow ^1S_0$ transitions of iosn. Similar emission is observed for 235 nm excitation. The Sb^{3+} ion is therefore a good activator for obtaining blue phosphor materials [40]. The combustion synthesis methods were attempted for phosphor preparation.

3.6   APPLICATIONS OF FLUORESCENT LAMPS

Fluorescent lamps can be used in industrial, commercial and residential lighting. Fluorescent lamps provide a shadow-free lighting in all applications.

1. Fluorescent lamps can provide high light output for lighting of large areas and are well suited for detailed task lighting in industrial applications.
2. Fluorescent lamps provide a uniform light level for office space and are also well suited for detailed task lighting in commercial applications.
3. Fluorescent lamps can also be used in residential applications as up-lighting from cabinets, beams or coves and under-cabinet lighting. They are effective lighting for kitchens, valences, cimices and fascias.
4. Smaller fluorescent lamps are used in sign and display applications such as jewellery stores.
5. Ceiling fixtures tend use a circling fluorescent light, which can be found in residential applications.
6. Fluorescent lamps are used in decorating commercial stores.
7. Fluorescent lamps are used in commercial advertisements.

3.7   LIMITATIONS AND CHALLENGES

3.7.1   Cost

The development of high efficacy and high CRI fluorescent lamps would not have been possible without the application of rare-earth phosphors. The main disadvantage of the rare-earth phosphors is their high cost, which is more than double the cost of some fluorescent lamps. As a compromise, between phosphor cost and performance, a double-coating scheme is widely used in the fluorescent lamp industry.

Cost may also be reduced by decreasing the phosphor particle size. A covering of four layers of phosphor particles requires proportionally less material with small particle (nano) size. One more way of reducing the cost is to decrease in the required rare-earth ion concentration or, in some cases, by completely eliminating rare-earth ions.

3.7.2   Environmental and Energy Issues

There is an increasing concern that mercury, which is an essential part of the operation of fluorescent lamps, could percolate into groundwater supplies when the discarded lamps are thrown in dumping lands. Although only about 0.1 mg of mercury is required to generate the desired 5–10 mtorr of mercury vapour pressure, chemical binding of mercury by glass, phosphors and electrodes during lamp operation requires dosing of the fluorescent lamps with several milligrams of mercury to ensure the availability of sufficient free mercury to sustain the desired vapour pressure. Hence, significant efforts are required to reduce the amount of mercury in fluorescent lamps [60,61].

The maximum discharge efficiency of mercury is only 70%; that means only 70% of supplied electric energy can be converted into electromagnetic radiation. Thus, 30% electric energy is lost in the first step of fluorescent lighting. Also, it has been said earlier that the phosphor absorbs the ultraviolet radiation from the mercury discharge and emits visible light. Unfortunately, the mechanism of the change itself causes reductions in energy, since the energy corresponding to an absorbed photon of UV radiation of wavelength 253.7 nm is 4.8 eV and the energy of emitted visible photons of wavelength, say 550 nm, is 2.2 eV. In addition to this, the phosphors are not 100% the absorbed UV photon to visible photon [61].

3.8   FUTURE PROSPECTS AND SCOPE

Changing lamp phosphor requirements are presently creating new problems in phosphor research. The major problems that are being addressed at the present time are as follows:

1. Improving the control of stoichiometry in order to reduce defect concentrations.
2. Reducing the cost of rare-earth phosphors, particularly the terbium and europium components.
3. Development of alternative phosphors with high quenching temperatures and possessing high stability.
4. Obtaining a better understanding of the various mechanisms controlling phosphor stability so that the choice of useful lamp phosphors may be broadened.

3.9   MERCURY-FREE FLUORESCENT LAMPS

Cold light, such as that emitted by a fluorescent lamp, surrounds us and gives support to our comfortable life. Nevertheless, a shadow of destruction has been cast over present fluorescent lamps. Waste electrical and electric equipment (WEEE) and restriction of the use of certain hazardous substances in electrical and electric equipment directives restrict the use of hazardous substances (lead, mercury, cadmium and hexavalent chromium, etc.) in the manufacturing of electrical and electronic equipment. After July 2006, electrical and electronic equipment containing such harmful chemicals cannot be introduced to the European market. Although fluorescent lamps based on mercury discharge are not presently restricted, the use of mercury may be prohibited in the near future. Consequently, demand for new materials for modern lighting technology has therefore increased. White LEDs and organic LEDs are possible alternatives, but both technologies are faced with tough problems for lighting applications. The present white LED is a pseudo-white light that consists of yellow photoluminescence of a cerium-doped yttrium aluminium garnet (YAG) phosphor and blue emission of a blue LED. This light colour is not true white and its light directivity is too high for interior lighting and backlighting. In contrast, the organic EL is a flat panel lighting that is suitable for interior lighting but the EL has a limited life span for practical application. Major improvements will therefore require a conceptually new approach, which leads to mercury-free fluorescent lamps [62].

The mercury-free fluorescent lamp is fundamentally a Xe discharge lamp with a phosphor layer inside the flat glass. A difference between the present mercury-based fluorescent lamp and the mercury-free fluorescent lamp is the radiation wavelength of the gas discharge. The mercury-free fluorescent lamp contains a mixture of Xe and Ne gases. The generated internal radiation is mostly 147 nm (8.4 eV), which is the resonance wavelength of the Xe discharge. The other (molecular Xe state) is distributed in 173 nm (7.2 eV). The molecular state becomes important as the gas pressure increases. Vacuum ultraviolet (VUV) irradiation from the Xe gas is converted into visible light by the VUV phosphors. Highly efficient, new VUV phosphors are required for mercury-free fluorescent lamps because the main emission peak of this discharge is located at a shorter wavelength than that of the conventional mercury discharge (254 nm) [62,63].

Replacement of mercury discharge–based fluorescent lamps by a Xe discharge mercury-free fluorescent lamp does not resolve environmental and energy issues at the same time. The discharge efficiency of Xe gas is much less than that of mercury (emission of 254 nm or 4.9 eV radiations from mercury discharge with 50% energy loss and emission of 172 nm or 7.2 eV radiatons with 70% energy loss) [64].

Hence, to actually make the Xe discharge-based fluorescent lamps more energy efficient than the currently used Hg-based tubes, quantum efficiencies of the phosphors need to be above 150%. The quantum-cutting phosphor therefore serves this need. Quantum cutting provides a means to obtain two or more photons for each photon absorbed. The development of such VUV excitable visible quantum cutting phosphors would lead toward the realization of a highly efficient and environmentally benign lighting technology [65].

There are several rare earth–activated borate host phosphors reported in the literature that can be effectively excited by VUV. The focus is now to shift to the investigation of the combination of proper activators and sensitizesr that will provide quantum efficiency greater than unity, through the phenomenon of visible quantum cutting in these borate host phosphors [66–71].

REFERENCES

[1] Gribben, J. (2004). *The scientists; a history of science told through the lives of its greatest inventors.* Random House, New York, 424–432.

[2] Harrison, G. R., and Forbes, G. S. (1925). Spectral energy characteristics of the mercury vapor lamp. *J. Opt. Soc. Am.*, 10, 1–17.

[3] Blasse, G., and Grabmaier, B. C. (1994). *Luminescent materials*, Springer-Verlag, Berlin, Heidelberg, 10–11.

[4] Gaster, L., and Dow, J. S. (1915). *Modern illuminants and illuminating engineering*, Whittaker & Co, SY16 2AB UK,, 107–111.

[5] Edison, T. A. (1880). Electric-Lamp. No. 223,898. Patented Jan. 27, U. S. patent 865367 Fluorescent Electric Lamp.

[6] Moore's Etheric Light. 1896. The Young Newark Electrician's New And Successful Device. *New York Times*, Retrieved 2008-05-26.

[7] Justel, T., Nikol, H., and Ronda, C. (1998). *Angew. Chem. Int. Ed.*, 37, 3084–3103.

[8] Claude, G. (1913). The development of neon tubes. *Eng. Mag.*, 271–274.

[9] Van Dulken, S. (2002). *Inventing the 20th century: 100 inventions that shaped the world: from the airplane to the zipper*, New York University Press, New York, 42.

[10] Cao, L. M., Bu, W. B., Ren, C. A., Xu, X., and Li, Q. (2008). Preparation and luminescent properties of $Zn_2SiO_4:Mn^{2+}$. *Key Eng. Mater.*, 368–372, 375–377.

[11] Lojpur, V., Nikolić, M. G., Jovanović, D., Medić, M., Antić, Ž., and Dramićanin, M. D. (2013). Luminescence thermometry with $Zn_2SiO_4:Mn^{2+}$ powder. *Appl. Phys. Lett.*, 103, 141912, 1–5.

[12] Srivastava, A. M., and Soules, T. F. (2000). *Phosphors*, John Wiley & Sons, Inc., Hoboken, New Jersey.

[13] Stevels, A. L. N. (1976). Recent developments in the application of phosphors. *J. Lumin.*, 12/13, 97.

[14] Josephine, S., MaxWei, S., and Sohn, M. D. (2016). A retrospective analysis of compact fluorescent lamp experience curves and their correlations to deployment programs. *Energy Policy*, 98, 505–512.

[15] Blasse, G., Bril, A., and Nieuwpoort, W. C. (1966). On the Eu^{3+} fluorescence in mixed metal oxides: Part I- The crystal structure sensitivity of thr intensity ratio of electric and magnetic dipole emission. *J. Phys. Chem. Solids*, 27, 1587–1592.

[16] Adam, J., Metzger, W., Koch, M., Rogin, P., Coenen, T., Atchison, J. S., and König, P. (2017). Light emission intensities of luminescent Y_2O_3:Eu and Gd_2O_3:Eu particles of various sizes. *Nanomaterials*, 7, 26–30.

[17] Sommerdijk, J. L., and Verstegen, J. M. P. J. (1974). Concentration dependence of the Ce^{3+} and Tb^{3+} luminescence of $Ce_{1-x}Tb_xMgAl_{11}O_{19}$. *J. Lumin.*, 9, 415–419.

[18] Kim, K.-B., Kim, Y.-I., Chun, H.-G., Cho, T.-Y., Jung, J.-S., and Kang, J.-G. (2002). Structural and optical properties of $BaMgAl_{10}O_{17}$ Eu^{2+} phosphor. *Chem. Mater.*, 14, 5045–5052.

[19] Chandar Rao, P., Ravinder, G., Anusha, S., and Sreelatha, C. J. (2021). Synthesis and characterization of $BaMgAl_{10}O_{17}$: Eu^{2+} phosphor for efficient displays. *AIP Conf. Proc.*, 2317, 030017, 1–6.

[20] Ekambaram, S., and Patil, K. C. (1995). Synthesis and properties of rare earth doped lamp phosphors. *Bull. Mater. Sci.*, 18, 921–930.

[21] Welkar, T. (1991). An Interdisciplinary Journal of Research on Excited State Processes in Condensed Matter. *J. Lumin.*, 48, 49, 49- and 50.

[22] Ya, Z., and Brgoch, J. (2021). Opportunities for next-generation luminescent materials through artificial intelligence. *J. Phys. Chem. Letters*, 12, 764–772.

[23] Allen, J. W. (1991). Impact processes in electroluminescence. *J. Lumin.*, 48–49, 49–50.

[24] Matossi, F., and Nudelman, S. (1959). Luminescence. *Methods Exp. Phys.*, 6, 293–334.

[25] Kim, D., Kim, H.-E., and Kim, C.-H. (2020). Enhancement of long-persistent phosphorescence by solid-state reaction and mixing of spectrally different phosphors. *ACS Omega*, 5, 10909–10918.

[26] Illuminating Engineering Society (2020). *ANSI/IES LM-86-20, approved method: Measuring luminous flux and color maintenance of remote phosphor components*, IES, New York.

[27] M. S. Ibrahim, W. K. C. Yung and J. Fan (2018). Lumen Maintenance Lifetime Prediction for Phosphor-converted White LEDs with a Wiener Process based Model *IEEE Xplore* 20th International Conference on Electronic Materials and Packaging 1–4.

[28] Lakowicz, J. R. (2006). Mechanisms and dynamics of fluorescence quenching. *Principles of Fluorescence Spectroscopy*, 331–351.

[29] Tian, B., Chen, B., Tian, Y., Sun, J., Li, X., Zhang, J., Zhong, H., Cheng, L., and Hua, R. (2012). Concentration and temperature quenching mechanisms of Dy^{3+} luminescence in $BaGd_2ZnO_5$ phosphors. *J. Phys. Chem. Solids*, 73, 1314–1319.

[30] Meza, O., Villabona-Leal, E. G., Diaz-Torres, L. A., Desirena, H., Rodríguez-López, J. L., and Pérez, E. (2014). Luminescence concentration quenching mechanism in $Gd_2O_3:Eu^{3+}$. *J. Phys. Chem. A*, 118, 1390–1396.

[31] Wei-Ning, W., Widiyastuti, W., Ogi, T., Lenggoro, I. W., and Okuyama, K. (2007). Correlations between crystallite/particle size and photoluminescence properties of sub micrometer phosphors. *Chem. Mater.*, 19, 1723–1730.

[32] Park, H. K., Oh, J. H., and Do, Y. R. (2012). Toward scatter-free phosphors in white phosphor-converted light-emitting diodes. *Opt. Express*, 20, 10218–10228.

[33] Thompson, P. D., Huang, J. F., Smith, R. W., and Keszler, D. A. (1991). The mixed orthoborate pyroborates Sr2Sc2B4O11 and Ba2Sc2B4O11: pyroborate geometry. *J. Solid State Chem.*, 95, 126–135.

[34] He, M., Kienle, L., Simon, A., Chen, X. L., and Duppel, V. (2004). Re-examination of the crystal structure of $Na_2Al_2B_2O_7$: stacking faults and twinning. *J. Solid State Chem.*, 177, 3212–3218.

[35] Wu, L., Chen, X. L., Li, H., He, M., Xu, Y. P., and Li, X. Z. (2005). Structure determination and relative properties of novel cubic borates $MM'_4(BO_3)_3$ (M = Li, M′ = Sr; M = Na, M′ = Sr, Ba). *Inorg. Chem.*, 44, 6409–6414.

[36] Cai, G. M., Sun, Yu, Li, H. K., Fan, X., Chen, X. L., Zheng, F., and Jin, Z. P. (2011). Crystal structure and photoluminescence of Tb^{3+}-activated $Ba_3InB_3O_9$. *Mater. Chem. Phys.*, 129, 761–768.

[37] Das, K., Marathe, A., Zhang, X., Zhao, Z., and Chaudhuri, J. (2016). Superior white light emission and color tunability of tri-doped $YBO_3:Tb^{3+}$, Eu^{3+} and Dy^{3+} for white light emitting diodes. *RSC Adv.*, 6, 95055–95061.

[38] Murthy, K. V. R., Sai Prasad, A. S., and Rao, M. R. (2012). Luminescence characteristics of Eu and Tb doped $YGdBO_3$ phosphor. *Phys. Procedia*, 29, 70–75.

[39] Zhang, J., Yang, M., Jin, H., Wang, X., Zhao, X., Liu, X., and Peng, L. (2012). Self-assembly of $LaBO_3:Eu$ twin microspheres synthesized by a facile hydrothermal process and their tunable luminescence properties. *Mater. Res. Bull.*, 47, 247–252.

[40] Ingle, J. T., Gawande, A. B., Sonekar, R. P., Nagpure, P. A., and Omanwar, S. K. (2013). Synthesis and photoluminescence of inorganic borate host red emitting VUV phosphor $YCaBO_4:Eu^{3+}$. *AIP Conf. Proc.*, 1536, 895–899.

[41] Lin, H., Hou, D., Li, L., Tao, Y., and Liang, H. (2013). Luminescence and site occupancies of Eu^{3+} in $La_2CaB_{10}O_{19}$. *Dalton Trans.*, 42, 12891–12897.

[42] Pązik, R., Zawisza, K., Watras, A., Maleszka-Bagińska, K., Boutinaud, P., Mahiou, R., and Dereń, P. J. (2012). Temperature induced emission quenching processes in Eu^{3+}-doped $La_2CaB_{10}O_{19}$. *J. Mater. Chem.*, 22, 22651–22657.

[43] Koparkar, K. A., Bajaj, N. S., and Omanwar, S. K. (2015). Effect of Eu^{3+} doping on fluorescence properties of novelphosphor $Li_2Y_2B_2O_7$. *Optoelectr. Adv. Mater. – Rapid Comm.*, 9, 7–8, 915–918.

[44] Zeng, H., You, F., Peng, H., and Huang, S. (2015). Energy transfer from Ce^{3+} to Tb^{3+}, Dy^{3+} and Eu^{3+} in $Na_3Y(BO_3)_2$. *J. Rare Earth*, 33, 1051–1055.

[45] Saubat, B., Vlasse, M., and Foussier, C. (1980). Synthesis and structural study of the new rare earth magnesium borates LnMgB5O10 (Ln = La, ..., Er). *J. Solid State Chem.*, 34, 271.

[46] Leskel, M., Saakes, M., and Blasse, G. (1984). Energy transfer phenomena in $GdMgB_5O_{10}$. *Mater. Res. Bull.*, 19, 151–159.

[47] Lin, C. K., Yu, M., Pang, M. L., and Lin, J. (2006). Photoluminescent properties of sol-gel derived (La,Gd)MgB$_5$O$_{10}$:Ce^{3+}/Tb^{3+} nanocrystalline thin films. *Opt. Mater.*, 28, 913–918.

[48] Johan de Hair, and Johan Van Kemenade (1983). New Tb3+ and Mn2+ activated phosphors and their applications in "Deluxe" lamps Conference: 3rd Int. Conf. Science and Technology of Light Sources, At: Toulouse.

[49] Sonekar, R. P., Omanwar, S. K., Moharil, S. V., Muthal, P. L., Dhopte, S. M., and Kondawar, V. K. (2009). Luminescence in LaBaB$_9$O$_{16}$ prepared by combustion synthesis . *J. Lumin.*, 129, 624–628.

[50] Andrii, S., and Lis, S. (2013). Green-emitting nanoscaled borate phosphors Sr$_3$RE$_2$(BO$_3$)$_4$:Tb^{3+}. *Mater. Chem. Phys.*, 140, 447–452.

[51] Sankar, R. (2008). Efficient green luminescence in Tb^{3+} activated borates, A$_6$MM′(BO$_3$)$_6$. *Opt. Mater.*, 31, 268–275.

[52] Zhang, X., Zhang, J., Xu, J., and Su, Q. (2005). Luminescent properties of Eu2+-activated SrLaGa3S6O phosphor. *J. Alloys Compd.*, 389, 247-252.

[53] Jiang, Y., Ning, L., Xia, S., Yin, M., and Tanner, P. A. (2004). Third-order contributions to the 8S7/2 → 6P7/2, 6P5/2 two-photon transitions of Eu2+ in KMgF3. *J. Phys: Condens Matter*, 16, 2773–2778.

[54] Lucas, F., Jaulmes, S., Quarton, M., Le Mercier, T., Guillen, F., and Fouassier, C. (2000). Crystal structure of SrAl$_{12}$B$_2$O$_7$ and Eu^{2+} luminescence. *J Solid State Chem.*, 150, 404–409.

[55] Jee, S. D., Park, J. K., and Lee, S. H. (2006). Photoluminescence properties of Eu^{2+} activated Sr$_3$SiO$_5$ phosphors. *J. Mater. Sci.*, 41, 3139–3141.

[56] Sonekar, R. P., Omanwar, S. K., and Moharil, S. V. (2009). Combustion synthesis and photo-luminescence of Eu^{2+} doped BaB$_8$O$_{13}$. *Ind. J. Pure Appl. Phys.*, 47, 441–443.

[57] Meza-Rochaa, A. N., Camarillob, I., Moralesc, R. L., and Caldiño, U. (2017). Reddish-orange and neutral/warm white light emitting phosphors: Eu^{3+}, Dy^{3+} and Dy^{3+}/Eu^{3+} in potassium mzinc phosphate glasses. *J. Lumin.*, 183, 341–347.

[58] Caldino, U., Speghini, A., and Bettinelli, M. (2006). Optical spectroscopy of zinc metaphosphate glasses activated by Ce^{3+} and Tb^{3+} ions . *J. Phys: Condens. Matter*, 18, 3499–3505.

[59] Yury Yu, D., Deyneko, D. V., Spassky, D. A., Lazoryak, B. I., and Stefanovich, S. Yu. (2021). A novel high color purity blue-emitting Tm^{3+} doped β-Ca$_3$(PO$_4$)$_2$ type phosphor for WLED application. *Optik*, 227, 166027–166032.

[60] Sha Chen, and Jiaxing Zhang. 2016. Environmental Impacts of Compact Fluorescent Lamps and Linear Fluorescent Lamps in China, International Conference on Civil, Transportation and Environment (ICCTE 2016) ISBN 978-94-6252-185-8.

[61] Lim, S. R., Kang, D., Ogunseitan, O. A., and Schoenung, J. M. (2013). Potential environmental impacts from the metals in incandescent, compact fluorescent lamp (CFL) and light - emitting diode (LED) bulbs. *Environ. Sci. Technol.*, 47, 1040–1047.

[62] Weber, L. F. (2000). *Colour plasma displays, the electrical engineering handbook*, edited by Richard C. Dorf, CRC Press, Boca Raton, Florida, 1997.

[63] Kenji, T. (2006). Recent research and development of VUV phosphors for a mercury free lamp. *J. Alloys Compd.*, 408–412, 665–668.

[64] Kenji, T., Yu Ichiro, I., Takashi, N., Kazuyoshi, U., and Mineo, S. (2003). New VUV phosphor, NaLnGeO$_4$:Eu^{3+} (Ln = Rare Earth). *Chem. Lett.*, 32, 346–347.

[65] Ronda, C. R. (1995). Phosphors for lamps and displays: an application view. *J. Alloys Compd.*, 225, 534–538.

[66] Mildren, R. P., and Carman, R. J. (2001). Enhanced performance of a dielectric barrier discharge lamp using short pulsed excitation. *J. Phys. D: Appl. Phys.*, 34, L1–L3.

[67] Sommerdijk, J. L., Bril, A., and Jager, A. W. (1974). Two photon luminescence with ultraviolet excitation of trivalent praseodymium. *J. Lumin.*, 8, 341–343.

[68] Srivastava, A. M., and Doughty, D. A. (1996). Photon cascade luminescence of Pr^{3+} in LaMgB$_5$O$_{10}$. *J. Electrochem. Soc.*, 143, 4113–4116.

[69] Chen, Y., Shi, C., Yan, W., Qi, Z., and Fu, Y. (2006). Energy transfer between Pr^{3+} and Mn^{2+} in SrB$_4$O$_7$: Pr, Mn. *Appl. Phy. Lett.*, 88, 061906, 1–3.

[70] Depp, S. W., and Howard, W. E. (1993). Flat-panel displays: recent advances in microelectronics and liquid crystals make possible video screens that can be hung on a wall or worn on a wrist. *Sci. Am.*, 268, 90–97.

[71] Pappalardo, R. (1976). Calculated quantum yields for photon-cascade emission (PCE) for Pr^{3+} and Tm^{3+} in fluoride hosts. *J. Lumin.*, 14, 159–193.

4 Borate Phosphors for Solid-State Lighting

Devayani P. Awade

CONTENTS

DOI: 10.1201/9781003207757-4

4.1   HISTORICAL BACKGROUND

4.1.1   THE ADVENT OF LIGHT

Artificial lighting has always fascinated humankind, as it is helpful to lengthen the day. That increased overall working hours, making times without natural lighting more productive, secure and delightful. It was among the first inventions of humankind. The first use of artificial light sources was thousands of years back. The early man created light by burning wood. It was a very inefficient way to produce light, but the heat produced helped our predecessors meet their fundamental needs [1,2].

Each subsequent improvement in lighting led to significant lifestyle improvements. Each invention improved the energy efficiency of the light source as well. Hence, it is evident that modern society's progression is mainly dependent on human competence to produce light.

A new era began in lighting technology with the introduction of LEDs. This technology has the potential to alter lighting as we perceive it. At present, we are advancing towards the age of LED lighting technology that is economical, energy-saving and environmentally friendly [1].

4.1.2   THE GENERATIONS OF LIGHTING

The evolution of lighting technology can be categorically divided into four generations [3]. These four generations are discussed in detail in the next section.

The first evolution is everything that is related to open fire, i.e. burning of wood, kerosene, candles, etc. In this, most of the energy is dissipated in heat and only a small part of the energy is converted into light. Hence, these light sources were highly inefficient. However, they were adopted for the longest period. Many developing countries are still facing a short supply of electricity and hence for an artificial light they are still dependent on the traditional system of burning oil or wood [4].

The second evolution of light sources was started when T. A. Edison discovered the incandescent lightbulb in 1879 [5]. His first commercial product, using carbonized bamboo fibers, operated at about 60 watts for about 100 hours and had an efficacy of approximately 1.4 lm/W. Further improvements over time have raised the efficacy of the currently used 120-volt, 60-watt incandescent lamp to about 15 lm/W for products with an average lifetime of 1,000 hours. Although the incandescent lightbulb produces a colour-rendering index (CRI) close to sunlight, it wastes power by emitting too much light outside of the visible spectrum, which is not for illumination purposes. Approximately 95% of the power consumed is emitted as heat, rather than as visible light [1].

The third evolution of light sources began with the introduction of fluorescent tubes by General Electric (GE) in 1939 [6]. Meyer et al. filed the patent in 1927 [7]. In the fluorescent tubes, UV radiation emitted by the low-pressure mercury plasma is used to excite the phosphor layer inside the tube. The emission from the blend of phosphors coated controls the colour of the light produced and prevents harmful UV light from escaping. Fluorescent lamps are more efficient than incandescent lightbulbs since they convert a smaller proportion of the energy into heat. However, the CRI produced is lower than an incandescent lightbulb. Moreover, **mercury toxicity** is the major issue with fluorescent lamps. Each compact fluorescent bulb has approximately 5 mg of mercury. Therefore, disposal of these on large scale will lead to mercury contaminating soil and underground water unless proper precaution is taken. However, fluorescent tubes are widely used for general lighting applications [1] at present.

All these light sources – incandescent and fluorescent lamps – have evolved to their present performance levels over the last 60 to 120 years of research and development.

Industry researchers have studied all aspects of improving the efficiency of these sources, and while marginal incremental improvements are possible, there is little room for significant, paradigm-shifting, efficacy improvements. Hence, current research is focused on the evolving fourth-generation LED technology, which has the potential of high efficiency, small size and

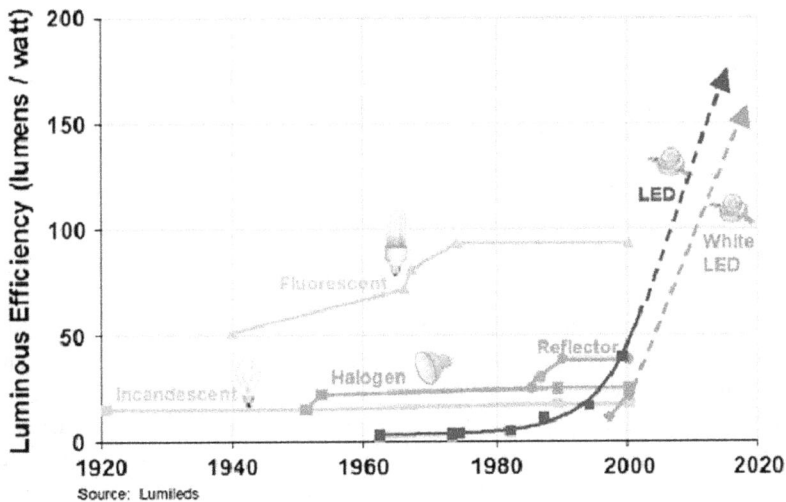

FIGURE 4.1 Historical and predicted efficacy of light sources [8].

environmentally friendly. The exponential projection is illustrated for light-emitting diodes (LEDs) in Figure 4.1 [8].

The fourth evolution of light sources began in 1991 with the invention of blue-emitting InGaN-based LED by S. Nakamura [9]. Later, various scientists contributed to increasing the efficiency of blue LEDs [10–12], which became the basis for the fabrication of white LEDs [13–16]. In other words, the invention of blue LEDs has started the era of solid-state lighting. The white LEDs are going to replace conventional light sources such as incandescent bulbs and fluorescent tubes in the near future. Moreover, white LEDs are already a dominant technology in various fields such as backlighting units of LCDs in mobile phones, handheld devices, computer monitors and TVs [17]. With the increasing efficiency of the blue LEDs, the new luminescent materials that can convert blue light to green, yellow, orange and red became necessary to obtain white light.

4.2 FUNDAMENTALS OF LEDS

4.2.1 OVERVIEW OF LEDS

"Solid-State Lighting (SSL) refers to a type of lighting that utilizes light-emitting diodes (LEDs), organic light-emitting diodes (OLED), or polymer light-emitting diodes as sources of illumination rather than electrical filaments or gas" [18]. Here the term "solid-state" indicates that the light-emitting source is a solid block of semiconductor. In traditional incandescent and fluorescent lamps, glass or vacuum tubes were used. Solid-state devices can withstand the vibrations and shock more efficiently, which increases their life span. Moreover, in SSL, visible light is produced with no heat dissipation [1,19].

As per the 2019 report of U.S. Department of Energy, "LEDs have surpassed, or matched, all conventional lighting technologies in terms of energy efficiency, lifetime, versatility, and colour quality, and, due to their increasing cost competitiveness, LEDs are successfully competing in a wide variety of lighting applications. Going forward, LED technology is expected to continue to improve, with increasing efficacy and decreasing prices while enabling new opportunities for lighting design and energy savings" [19].

For OLEDs, the technology is lagging behind that of LEDs; however, the industry experts project that the price and performance improvements will accelerate in the near future, positioning OLEDs at comparable price and performance levels with LED in 2027 [20,21].

4.2.2 MERITS AND OPPORTUNITIES

Solid-state lighting is an economical, environmentally friendly and efficient type of lighting. It has various advantages compared to conventional lighting sources such as incandescent bulbs and fluorescent tubes. The merits and opportunities of LED technology are listed below:

1. Good Efficacy

The energy efficiency of different lighting sources is expressed in terms of luminous efficacy or useful lumens. Luminous efficacy describes the amount of light emitted per unit of power (watts) consumed by the bulb. In traditional light sources, most of the energy consumed is wasted and only a small portion is converted into desired visible light. In the case of the incandescent lightbulb, only 5% of the energy supplied is converted into visible light, whereas fluorescent tubes convert 25% of the energy consumed [21]. Compared to these conventional sources, LEDs are extremely energy efficient. Two major reasons for this are 1) they waste very little energy in the form of infrared radiation 2) and they emit light directionally relative to every other commercially available lighting source [22]. LEDs generally consume very low amounts of power. Because of their high lumen output per watt, LEDs are capable of turning about 70% of their energy into light and there is the scope of improvement further if associated with perfect material and devices [22].

2. Allows a Wide Variety of Lighting

With the help of LEDs, we can have artificial lighting similar to daylight. We can have more control over the colour as well as on intensity of the lighting. The optical properties of light sources like CRI, CCT and émission spectrum, etc. can be tuned depending upon the application area. That widens the application areas, enabling their use in a variety of applications, including general illumination, display technology, medical treatment, transportation, communications, agriculture and biotechnology [21].

3. Higher Durability

LEDs provide 50,000 hrs of life compared to 1,000 hrs of incandescent lightbulbs and a nearly 10,000 hrs lifetime of fluorescent tubes [21]. Recent advancements increased the life span of LEDs to 100,000 operating hours or more. Making it multiple times longer than metal halides and even the sodium vapor lights. A longer lifetime will reduce the frequency of replacement. That will make it more economical as ultimately it will reduce the maintenance cost of the lighting [22].

4. Instantaneous Response

LEDs switch on rapidly, even when cold, and this is a particular advantage for certain applications such as vehicle brake lights. Moreover, frequent switching on/off does not cause degradation of this device [22].

5. Cool Lightning/No Heat Dissipation

In incandescent bulbs, heat is produced in the form of infrared radiation, which makes bulbs hot to the touch. Imposing restrictions on its installment. LEDs are free from such heating effect hence can be positioned in locations where heating from conventional sources would cause a particular problem e.g. illuminating food or textiles [23]. However, precaution need to be taken to limit the driving current of LED within the rated value as increasing it beyond the rated value will increase the junction temperature within the semiconductor leading to permanent damage [23]. Nowadays some LED lamps are designed with series resistors to limit the operating current, resulting in no cold filament current variation [24].

6. Reliable Device

LEDs are solid-state devices. Hence, they can withstand vibrations and impact better. Moreover, they do not have a filament and other fragile parts that might break. They can be safely used in all-weather conditions and withstand temperature fluctuations to a certain degree [25].

7. Small Size

LEDs are small in size – typical high-brightness LED chips to measure 0.3 mm by 0.3 mm, while high-power devices can be 1 mm × 1 mm or larger [24]. That enables their application in mobile phone handsets, backlight keypads and liquid-crystal display (LCD) screens. Moreover, due to their small size, LEDs are easily coupled to light pipes and thereby light can be efficiently and flexibly distributed. They can be placed on floors, walls, ceilings or furniture, and hence interesting decorative designs are possible [25].

8. Environmentally Friendly

Replacing a single 60-watt lightbulb with an LED results in a reduction of approximately 160 kg of CO_2 emissions per year. If you replaced 10 lamps in your home with LEDs, that would represent a reduction of 1,599 kg CO_2 emissions annually [26]. The fossil fuel power stations release pollutants like sulfur oxides, nitrogen oxides, carbon dioxide, fly ash, etc. that may cause environmental hazards like acid rain [21]. LEDs do not contain mercury as in the case of fluorescent and mercury vapor lamps. Now, as per the European Union directives, lead-free material is preferred for soldering while fixing LEDs to circuit boards [26]. In addition, LEDs are power-saving devices and hence more environmentally friendly.

4.2.3 SEMICONDUCTOR PHYSICS OF LEDS

LEDs are semiconductor devices that convert electricity into light through a *p-n* junction. N and p-type semiconductors form this p-n junction are shown in Figure 4.2.

In the case of LEDs, an electron is excited to a higher energy state and the relaxation of this excited electron back to the lower energy state after combining with hole generates light. The

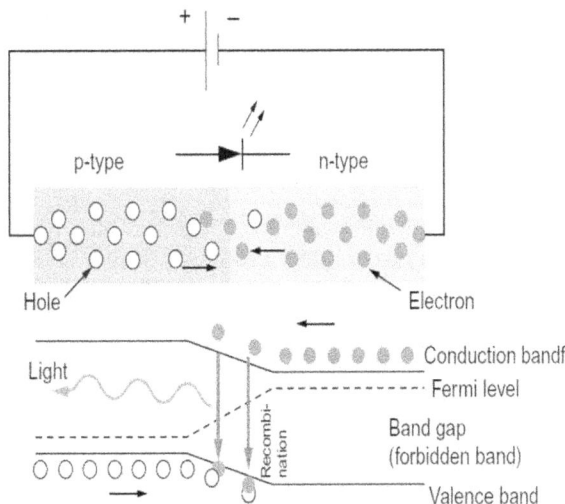

FIGURE 4.2 Schematic diagram of PN junction in forward bias mode [27].

energy of the photon (light) released in the recombination process, is equivalent to the bandgap energy (hv = Eg). The wavelength or colour emitted by the LED depends on the materials used to make the diode, which means that the emission colour of LEDs can be tuned from infrared to ultraviolet by varying either the composition or the thickness of the quantum well [28].

The semiconductor material can have either the direct band structure or the indirect band structure. In the direct band structure, the relaxation of the excited electrons from the conduction band into the valence band can occur under momentum conservation, while for the indirect band structure, the relaxation process needs the assistance of photons to achieve momentum conservation. Direct bandgap semiconductors such as gallium arsenide, are used to make optical devices such as LEDs and semiconductor lasers [29].

The recombination's rate of electrons with holes is proportional to the density of electrons, the density of holes and a proportionality factor, which is a measure of the probability of the recombination.

In addition to the band-to-band transition, various other types of transition may take place, like band to impurity level transition, donor states to acceptor states transitions and Auger recombination process.

The first two transitions are similar to the band-to-band transition but the Auger recombination's process is undesirable. For Auger recombination, the energy released during electron-hole pair recombination may be transferred to other carriers and then gets dissipated instead of generating photons [30]. With the help of bandgap energy Eg, the center emission wavelength of an LED can be obtained by using the formula $\lambda = \frac{hc}{Eg}$. The line widths of the emission spectrum can be increased either by using high doping levels or by using graded substrates. The extraction of light from the semiconductor material is very less as compared to the light generation inside the LED.

The reason for this is the mismatch of the refractive index of semiconductors, which is in the range of 3–3.5 to the refractive index of the surrounding medium that is in the range of 1–1.5. As a result, strong total internal reflection normally occurs at the interface of the semiconductor and the surroundings that prevent the light from coming out from the semiconductor material. This restricts the improvement of the efficiency of light-emitting diodes (LEDs) [31]. To improve the light-extraction efficiency various methods are adopted like fabricating narrow ridges on the surface of semiconductor materials. Recently it is reported that if these ridges are coated with a thin film made of a material (e.g. SiO2) whose refractive index is lower than that of the semiconductor, then light-extraction efficiency increases significantly (1.5 times higher than the previously discovered method) [32].

4.3 PHOSPHOR CONVERTED WHITE LED

It is evident that one LED at a time can produce a single colour with a restricted amount of light, whereas for white light generation, the entire visible spectrum is required to be spanned in appropriate proportion. In general, white light can be created with LEDs in two primary methods [21,33–35].

4.3.1 MULTI-LED CHIP APPROACH

In this light, coming from three monochromatic red, green and blue (RGB) LEDs is mixed, resulting in white light to human eyes (Figure 4.3). Potentially very high colour rendering possible with this method but there are certain constraints. The instability of colour temperature exists for RGB LEDs as the degradation of different colour LEDs or variations of driving current. Moreover, RGB LEDs require different driving currents for different colour LEDs; hence, their fabrication becomes more complex.

4.3.2 PHOSPHOR CONVERSION OR WAVELENGTH CONVERSION

In this, a GaN-based blue or UV LED chip is coated with a yellow or multi-chromatic phosphor. The mixing of light from the phosphor and the LED chip appear white to human eyes.

(a)

RGB LED

Green LED

Blue LED   +   +   Red LED

400   500   600   700 nm

(b)   Blue LED + Yellow phosphor

Blue LED   +   Yellow phosphor

400   500   600   700 nm

(c)   Near UV LED + RGB phosphor

UV-LED   +   +   RGB Phosphor

400   500   600   700 nm

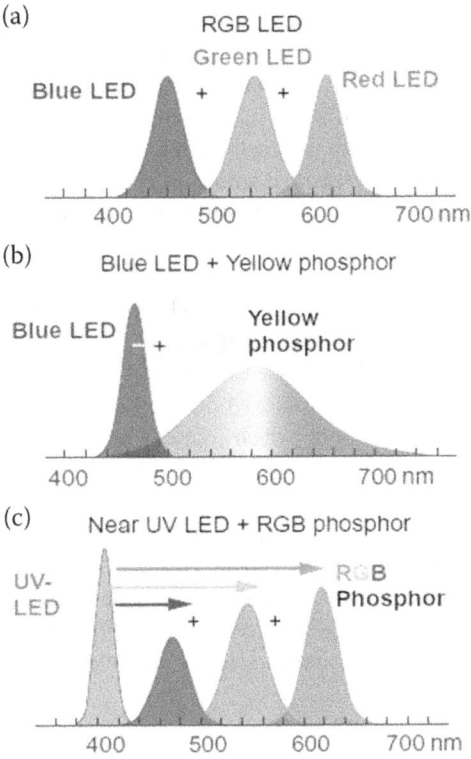

FIGURE 4.3 Three methods of generating white light from LEDs: (a) red + green + blue-LEDs, (b) blue-LED + yellow phosphor, (c) UV-LED + RGB phosphors [27].

Wavelength conversion involves converting some or all of the LED's output into visible wavelengths. Methods used to accomplish this feat include:

1. Blue LED and yellow phosphor – This method is considered the least expensive method for producing white light. Blue light from an LED is used to excite a phosphor which then re-emits yellow light. This balanced mixing of yellow and blue lights results in the appearance of white light.
2. Blue LED and several phosphors – In this method, a blue LED chip is coated with a blend of different phosphors, as shown in Figure 4.4. The process involved is similar with

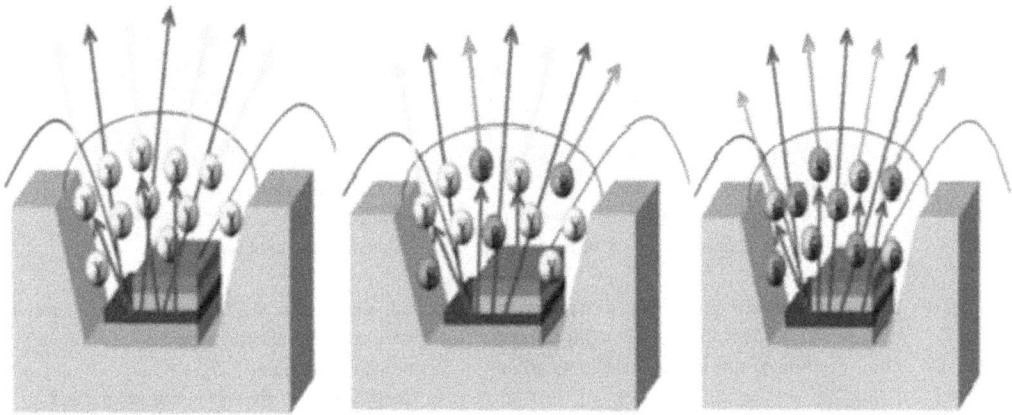

FIGURE 4.4 (a) A blue LED + yellow phosphors, (b) a blue LED + yellow phosphors + red phosphors, (c) a blue LED + green phosphors + red phosphors [35].

yellow phosphors, except that now each excited phosphor re-emits a different colour. These different light colours are combined with the originating blue light from the LED chip to create white light. The resulting light, however, has a richer and broader wavelength spectrum and produces a higher colour quality light, albeit at an increased cost.

3. Ultraviolet (UV) LED and red, green and blue phosphors – In this method, the UV light is used to excite the different phosphors, which are doped at measured amounts. The emitted light colours from the phosphors are mixed, resulting in white light with the richest and broadest wavelength spectrum.

The first phosphor that was combined with blue-emitting LED to produce white light was YAG:Ce, which was originally developed for flying-spot cathode-ray tubes [36]. However, due to the deficiency of intensity in the red spectral region, such a white LED possessed low CRI values (poor colour reproducibility) and high CCTs; thus, the generated white light looked bluish. Nevertheless, people are used to incandescent bulbs generating yellowish-white light, which creates a pleasant and cozy atmosphere. Therefore, the intensity in the red spectral region needs to be increased to meet the requirements for general illumination. Furthermore, near-UV LED-based white light sources recently emerged as an alternative to the blue ones. The advantage of near-UV LED is that there are much more phosphors that can be excited in near-UV than in the blue spectral region. However, complex blending of the three RGB phosphors in appropriate amounts is a complex task. At the same time, the near-UV LEDs are not that efficient yet, but the progress in efficiency enhancement is significant and even deep UV LEDs recently were reported [37–41]. Table 4.1 summarizes the advantages and disadvantages of white light based on different technology.

4.3.2.1 White Light Sources via Blue LED Chip

White light can be produced from the blue LED chip by various methods, by either coating a single phosphor or employing a blend of several phosphors. The simplest technique is single phosphor conversion. In this method, yellow-emitting phosphor, like YAG:Ce, is solely applied on top of a blue-emitting LED chip [42–45]. Though the fabrication of white LED by this method is comparatively easy, this approach results in high CCTs and low CRI values (Ra). The reason is the lack of emission in the red spectral region. The CRI defines the ability of a light source to illuminate an object and how its illumination makes the colour appear to the human eye compared to the sun (for

TABLE 4.1

Comparison of white LED technology [21,33–35]

Types	Advantages	Disadvantages
Blue LED+ Phosphor	• Most mature technology • High volume manufacturing processing • Relatively low luminous flux • Relatively high efficacy • Comparatively low cost	• High CCT(cool/ blue appearance) • Low CRI (typically in the 70s) • Color variability in the beam
Near-UV LED+ Phosphor	• Higher colour rendering • Warmer colour temperature possible • Colour appearance less affected by chip variation	• Less mature technology • Relatively low efficiency • Relatively low light output
RGB LEDS	• Potentially very high colour rendering • Color flexibility both in multi-colour displays and different shades of white	• Individual coloured LEDS respond differently to drive current, operating temperature, dimming and operating time. • Controls needed for colour consistency add expenses

6,500 "D65" and 5,000 K "D50") or black body radiators (for 4,200; 3,450; 2,950; 2,700 K) as a light source. The CRI of the sunlight is defined as 100 [2].

Another way is to adopt a combination of two phosphors like yellow and red, or green and red phosphors with blue LED. In two-phosphor, converted LEDs the CCT and CRI values can be adjusted according to the requirements in different fields i.e. in street lighting CRI is not important, however, in museums, for instance, the CRI values should be close or equal to 100 for the best colour reproducibility. Hence, this technique is superior to single phosphor LEDs.

The review of the recent articles indicates that Eu^{2+} and Ce^{3+} ions are most suitable for white light sources via blue LED chip [33–35]. Both of them possess optically allowed 4f to 5d transitions. The optical transitions involve 5d orbitals, which participate in chemical bonds of the activator ions to their chemical surrounding. Therefore, the position of the lowest excited 5d-levels is sensitive to the local environment of the ion and can be tuned by changing the host lattice (e.g. silicates, aluminates, nitrides, etc.) or modifying the host lattice itself (e.g. replacing Y by Lu in YAG:Ce). Thus, the absorption band can be shifted to the near-UV or blue spectral region where the respective LED can efficiently excite the ions.

4.3.2.2 White Light Sources via Near-UV LED Chip

There are two reasons for which near-UV LED chips are preferred for white LED fabrications. The first is, recently the efficiency of near-UV LEDs is considerably increased and the second, with near-UV LEDs there is a wider choice of phosphors. The emission band of near-UV LEDs lies in the range of 380 to 420 nm, so the suitable activators that can be used are Eu^{3+}, Pr^{3+}, Mn^{4+}, Tb^{3+}, Sm^{2+}, Sm^{3+} [46–57], etc. in addition to Eu^{2+} and Ce^{3+} [58,59] ions.

Generation of white light from near-UV LED is based on the RGB concept. The white light is generated by mixing red, green and blue colours. There are two popular approaches for this; one method employs broadband red-emitting phosphor and another method line-emitting red phosphor. The second approach in terms of luminous efficacy of the device is more preferable because the emission at wavelengths longer than 650 nm is practically of no use and according to Žukauskas et al., it is considered "waste" [1].

Nowadays, a single-phase white-emitting phosphor concept is emerging as an alternative to the RGB concept. In this, a single host is doped with more than one activator that shows emission spectra in more than one band resulting from the energy transfer between the activators like Eu^{2+} or Ce^{3+} that emit blue light and coactivator ions like Mn^{2+} that emit red light. Colour tuning is possible by choosing the appropriate concentration of the activator and coactivators [60–64].

4.3.3 CHARACTERISTICS OF LED PHOSPHORS

In the case of phosphor-converted white LEDs, the conversion phosphor is used to down-convert the light of UV or blue LEDs into visible light. Hence, the performance of fabricated LEDs is highly dependent on the phosphors used. Although there are a huge number of phosphors, only a few of them are suitable for use in LEDs. This depends on the requirements for LED phosphors. Only those phosphors that meet the requirements can be potentially applied in white LEDs. These selection rules or requirements help phosphor researchers or engineers design, develop, and utilize phosphors correctly.

With the increasing demand for warm white LEDs, the need for materials that can convert the LED output light to cyan, green, yellow, orange and red arose. However, such materials need to meet quite high requirements to be applied in white LEDs. Such requirements are [65] as discussed.

4.3.3.1 Temperature Quenching

For the practical application of phosphors in the fabrication of white LEDs, its thermal stability is gauged by thermal quenching. To achieve longer life and higher efficiency of LEDs, the conversion phosphor used should have more thermal stability [66]. In other words, the less thermal quenching means more life span of LED lamps. Thermal quenching relies significantly on the crystal structure and chemical composition of the host lattice.

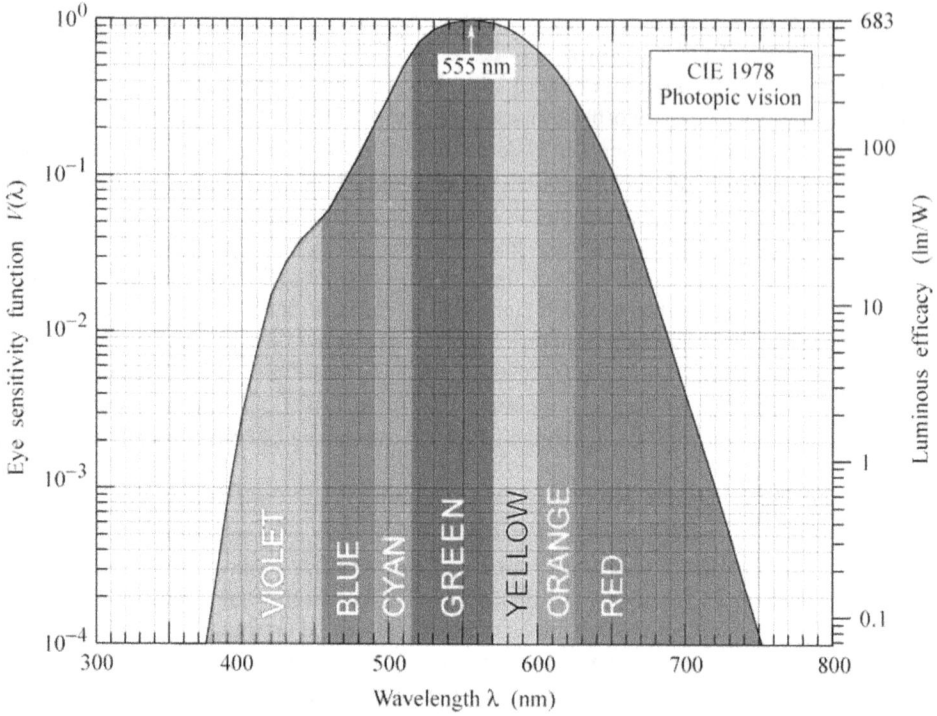

FIGURE 4.5 Eye-sensitive function, V(λ) (left ordinate) and luminous efficacy measured in lumens per watt of optical power (right ordinate). V(λ) is highest at 555 nm [67].

4.3.3.2 Luminous Efficacy

Luminous efficacy is the measure of the energy efficiency of a light source. "The luminous efficacy of a light source is defined as the ratio of the total luminous flux (lumens) emitted by the light source to the power absorbed (watts or equivalent) either by its radiant flux or its electrical power consumption" [65]. It is expressed in lumens/watt (lm/W).

The response of the human eye is not uniform for the visible spectrum that ranges from 380 nm to 780 nm. The eye sensitivity peaks at 555 nm and declines on both sides (Figure 4.5). Therefore, if 100% of electrical power were converted into monochromatic green radiation of 555 nm, then the theoretical maximum value of efficacy for that light source would be 683 lm/W. However, white LED sources are supposed to span the entire visible spectrum hence the efficacy of actual white sources is well below this ideal value [34]. The luminous efficiency and luminous efficacy are the terms related to one another.

Luminous efficiency (η_v) is the ability of the light source to convert the consumed power (P) into visible radiation(Φ_v) flux [1].

$$\eta_v = \eta_e \times K \equiv \frac{\Phi_v}{P} \tag{4.1}$$

It is expressed in terms of the radian efficiency of the source (η_e) and the luminous efficacy (K).

Radiant efficiency of a light source (η_e) is the ability of the light source to convert consumed power into radiant flux i.e. conversion efficiency of the source from the electrical energy to electromagnetic radiation. It is a unitless quantity [1].

$$\eta_e = \frac{\Phi_e}{P}(W/W) \tag{4.2}$$

Here, (Φ_e) is the radiation flux and P is the consumed power.

From equations 4.1 and 4.2, the luminous efficacy K is the ratio of luminous flux (Φ_v) to radiant flux, given below [1].

$$K = \frac{\Phi_v}{\Phi_e} \tag{4.3}$$

$$K = \frac{\Phi_v}{\Phi_e} = \frac{683lm}{W} \times \frac{\int_{380}^{780} V(\lambda)S(\lambda)d\lambda}{\int_0^\infty S(\lambda)d\lambda} \tag{4.4}$$

In equation 4.3,

$V(\lambda)$ – is the 1924 CIE relative luminous efficiency function for the photonic vision that is defined in the visible range of 380–780 nm, as shown in Figure 4.10.

$S(\lambda)$ – is the spectra power distribution (SPD), which describes the power per unit area per unit wavelength of illumination, or more generally, the per-wavelength contribution to radiant flux:

$$S(\lambda) = \frac{d\Phi_e}{d\lambda} \tag{4.5}$$

Chhajed et al. [68] and Mirhosseini et al. [69] proposed the calculations of the spectral power distribution of multichip white LEDs as well as phosphor-converted white LEDs. The SPD of an LED can be described by using the Gaussian function:

$$S(\lambda) = P \times \frac{1}{\sigma\sqrt{2\pi}} \times exp\left[-\frac{1}{2}\left(\frac{\lambda - \lambda_{peak}}{\sigma}\right)^2\right] \tag{4.6}$$

$$\sigma = \frac{\lambda_{peak}^2 E}{2hc\sqrt{2ln2}} \tag{4.7}$$

where

ΔE – is the full width at half maximum (FWHM) of the emission spectrum and

P – is the optical power of the LED.

For trichromatic white LEDs, the SPD can be given as,

$$S_{white}(\lambda) = S_{blue}(\lambda) + S_{green}(\lambda) + S_{red}(\lambda) \tag{4.8}$$

For phosphor-converted bi-chromatic white LEDs, by considering the overlap between the absorption spectrum of the phosphor and the emission spectrum of a blue LED, the number of photons absorbed by the phosphor per second is determined as follows:

$$photon_{abs} = \int \frac{S_{blue}(\lambda)}{hc/\lambda}[1 - exp(-S_{abs}(\lambda)t)d\lambda] \tag{4.9}$$

where
t – denotes the thickness of phosphor, and
$S_{abs}(\lambda)$ – is the absorption spectrum of phosphor.
Conversion of pumped LED power into yellow power is

$$S_{yellow}(\lambda) = photon_{abs} \times \eta \times S_{ems}(\lambda) \tag{4.10}$$

where
η – the quantum efficiency of the phosphor,
and $Sems(\lambda)$ – is the emission spectrum of the phosphor.
Therefore, the SPD of the phosphor-converted dichromatic white LEDs is given as [21]

$$S_{white}(\lambda) = S_{emblue}(\lambda) + S_{yellow}(\lambda) \tag{4.11}$$

where
$S_{emblue}(\lambda)$ – is the blue power transmitted through the phosphor.

4.3.3.3   Colour Rendering Index (CRI)

This is a figure of merit of a light source. It indicates how well a light source renders colours of illuminated people and objects on a scale of 0–100. It is related to the idea that if a red object is illuminated with red light it appears red, but it appears grey or black if it is illuminated with blue light. The colour of the object illuminated will change with the colour of the source used to illuminate it. The reference source used is the sun (or a black body), and by definition, it has a colour rendering index (CRI) of 100 as it covers the entire visible spectrum.

The *colour rendering index* (CRI) of a light source is evaluated based on 'Ra' values. The 'Ra' value is obtained by averaging the values of the special colour rendering indices of eight standard colour samples [68].

$$R_a = \frac{1}{8} \sum_{i=1}^{8} R_i \tag{4.12}$$

Here, Ri – is the special colour rendering indices for each sample, and is given by [68]

$$R_i = 100 - 4.6\Delta E_i (i = 1, \ldots, 14) \tag{4.13}$$

where
ΔEi is the colour differences of 14 selected Munsell samples when illuminated by a reference source and when illuminated by a given source.

The light source is expected to cover the entire visible spectrum. Hence, a good source should have a high CRI value. If the CRI value for a light source is more than 70, then it is acceptable. The CRI value of the convention incandescent lamp is 95 but luminous efficacy is very low. The cool, white fluorescent lamp has a CRI of 62, whereas in the case of a fluorescent lamp coated with tri-colour emitting phosphors CRI increases to more than 80. There is always a trade-off relationship between luminous efficacy and the colour-rendering index. The increased CRI is at the cost of a decrease in luminous efficacy. The human eye is most sensitive at 555 nm. Hence, luminous efficacy of radiation is highest for monochromatic radiation of 555 nm wavelength. However, for good CRI values, broadband spectral distribution throughout the visible spectrum is desirable. Hence, generally, balance is adopted between the two depending on the application areas [21,65,68,69].

4.3.3.4 CIE Chromaticity Coordinates

The Commission Internationale de l'Eclairage (CIE) 1931 system is commonly adopted to express the composition of any colour in terms of three primary colours. CIE chromaticity coordinates (x,y) are computed from the spectral power distribution of the light source and the colour-matching functions (Figure 4.6).

Tri-stimulus values X, Y and Z are obtained by integrating the spectrum with the standard colour-matching functions $x(\lambda)$, $y(\lambda)$ and $z(\lambda)$ by adopting the following equations [21]:

$$X = \int \bar{x}(\lambda) S(\lambda) d\lambda$$

$$Y = \int \bar{y}(\lambda) S(\lambda) d\lambda$$

$$Z = \int \bar{z}(\lambda) S(\lambda) d\lambda$$

The tri-stimulus values X, Y and Z specify the lightness, hue and saturation of the colour.

The 1931 CIE chromaticity coordinates (x, y and z) of any light source having spectrum $S(\lambda)$ are calculated by using the following equations [71]:

$$x = \frac{X}{X + Y + z}$$

$$y = \frac{Y}{X + Y + z}$$

$$z = \frac{Z}{X + Y + z} \equiv 1 - x - y$$

From the above equations, it is clear that quoting two values of chromaticity coordinated (x, y) is sufficient to define the colour since the addition of three (x, y and z) equals one.

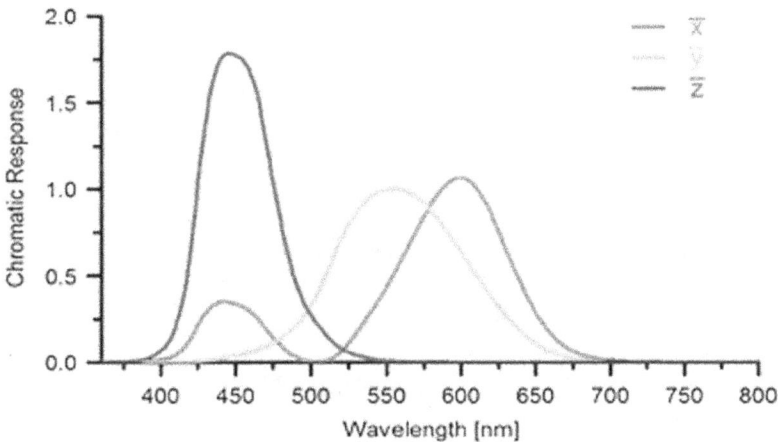

FIGURE 4.6 CIE 1931 XYZ colour-matching functions [70].

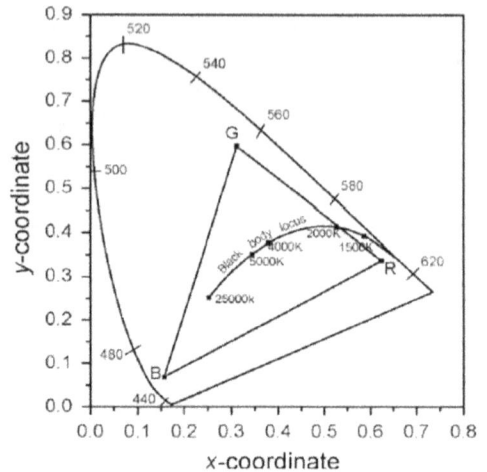

FIGURE 4.7 1931 CIE chromaticity diagram and CCT [34].

Using x, y as the coordinates, a two-dimensional chromaticity diagram (the CIE 1931 colour space diagram) is plotted as shown in Figure 4.7. The blackbody locus represents the chromaticities of black bodies having various (colour) temperatures.

4.3.3.5 High Chemical Stability

LEDs have a phosphor coating, which is continuously exposed to UV or blue radiation depending upon the chip used. Hence, the chemical composition, as well as the crystal structure of the phosphor material, should be strong enough to withstand these continuous UV/blue radiation exposures. Low chemical stability of the phosphor material will lead to a complex production process that will increase the cost. It will also decrease the luminous efficacy and degrade the LEDs faster, resulting in a reduction of their lifetime. The phosphors should be stable to ambient humidity change and should not react chemically with the atmospheric gases [65,72].

4.3.3.6 Correlated Colour Temperature (CCT)

Planck's blackbody radiator will change the colour with the increase in temperature. Its colour changes to red, orange, yellow, white and finally to blue with the temperature rise. So each colour is associated with a particular value of temperature expressed on the Kelvin scale. Colour temperature is the term used to express the numerical value of the temperature in the Kelvin scale for that colour. The conventional incandescent bulb emits light when the filament is heated that resembles closely the blackbody emission. Hence, its colour temperature is the temperature of filament for that particular colour. For the light sources that do not emit light on heating and adopt other ways than thermal radiation to produce light like CFL or LEDs, the terminology correlated colour temperature (CCT) is used in place of colour temperature [34]. Correlated colour temperature (CCT) is a relative term used with reference to theoretical blackbody emission. "*Correlated colour temperature* (CCT) is an extrapolation of the colour of the light source to the colour of a blackbody radiator of a given colour temperature such that they appear the same colour to the human eye [21]". Single phosphor-converted white LEDs fabricated by coating blue LED with YAG:Ce phosphor generally shows high correlated colour temperatures (above 5,000 K) and hence produce "cold" bluish-white light. So a multi-phosphor approach is adopted that helps in tuning the correlated colour temperatures (from 2,700 to 6,500 K) by using different combinations of the suitable phosphors. Furthermore, daylight has varying colour temperatures from dawn to dusk and that relates to human feelings. So, depending upon the need, now it is possible to select the lighting system from warm/cool colour light. However, the LEDs designed for general illumination purposes should have CCT on or near the blackbody locus [34].

4.3.3.7   High Quantum/Conversion Efficiency of Phosphors

The phosphors used for LED applications are required to have high conversion efficiency. The conversion phosphors should convert the maximum amount of UV or blue light used for excitation into required visible emission and there should be a minimum loss in the luminescence process. Hence, these phosphors are required to have strong absorption in UV (350–410 nm) or blue (440–480 nm) light region, a broad excitation spectrum that covers the entire UV/blue region and useful emission in the visible region for good quantum efficiency [21]. Another factor that affects quantum efficiency is concentration quenching. The luminescence efficiency of a phosphor increases with the increase inactivator concentration in the host. However, this is observed up to a certain specific value. This specific value is termed 'critical concentration'. When the activator concentration exceeds a critical concentration, the luminescence efficiency of the phosphor decreases. This effect is called 'concentration quenching'. With very high concentration, the probability of energy transfer exceeds that for emission, then the excitation energy repeatedly goes from one activator to the other and eventually lost at the surface, dislocations or impurities [73]. Thus, it does not contribute to luminescence. The efficiency then decreases despite the increase of the activator concentration. In addition to this, the quantum efficiency is affected by parameters such as particle size and particle size distribution [74].

4.4   BORATE-BASED PHOSPHOR-CONVERTED LEDS

The borates possess excellent properties as host structures of phosphors due to the inherent attributes of the large bandgap and covalent bond energy. Moreover, they possess the advantages of low synthesizing temperature and high chemical and physical stability. A variety of borate host materials doped with rare earth and other ions have been reported as phosphor materials for different applications [75–79].

4.4.1   Red/Orange-Emitting Phosphors

Red phosphor plays a key role in improving the CRI of the white LED system. Therefore, efficient red phosphors are required. Hence, researchers are trying to find an efficient red phosphor that will meet the requirement [80–86]. Rare earth–doped sulfides and oxysulfides red phosphors [87–89] are efficient but cannot be used widely because these compounds are sensitive to moisture moreover their synthesis methods are not easy. Recently, nitrides and oxynitrides phosphors have been reported as good potential red phosphors because of their good thermal stability [90–92]. However, the production cost of these phosphors is very high because the synthesis of the phosphors requires very high firing temperatures and high nitrogen pressures. Borates host phosphors have good chemical stability, are environmentally friendly, have convenient synthesis conditions with low cost and comparatively low synthesis temperatures (around 850°C in the air); hence, they have attracted much attention [93–98]. Here we are discussing recently reported red-emitting borate host phosphors.

1. S. Yuan et al. reported red-emitting borate host phosphor $Ba_2Mg (BO_3)_2:Eu^{2+} Mn^{2+}$ [77]. A solid-state reaction method was adopted for the synthesis of this phosphor. Its excitation band extends from 250–450 nm, which is suitable for near-ultraviolet LED chips (350–420 nm). When excited by 365 nm light, the phosphor exhibits strong red emission centred at 615 nm.

2. Another red-emitting phosphor $ZnB_2O_4:Bi^{3+}$, Eu^{3+} [99] was reported by W. R. Liu et al. In this red phosphor, ZnB_2O_4 host was co-activated with Eu^{3+} and Bi^{3+} ions. This was prepared by a solid-state reaction method. The composition-optimized $(Zn_{0.9}Eu_{0.1}) B_2O_4$ phosphor exhibits a dominant emission peak at 610 nm ($^5D_0–^7F_2$) with CIE coordinates of (0.63, 0.36) under the excitation at 393 nm. It is reported by the authors that co-doping with Bi^{3+} ions in $ZnB_2O_4:Eu^{3+}$, the emission intensity and quantum efficiency was enhanced by an increment of 14% and 6%, respectively.

3. Y. Zhang et al. reported Eu^{3+}-activated $KSr_4 (BO_3)_3$ [100] red borate host phosphor. The synthesis method used was the solid-state reaction method. $KSr_4 (BO_3)_3:Eu^{3+}$ exhibits a dominant emission peak at 612 nm (5D0-7F2) with CIE coordinates of (0.64, 0.35) under the excitation at 394 nm. They studied co-doping of M^+ ions (M = Li, Na, and K) in KSr_4 $(BO_3)_3:Eu^{3+}$ for charge compensation. It was observed that the intensities of emission spectra at 612 nm were increased greatly, but the CIE coordinates were not changed.

4. To obtain novel phosphors, researchers have explored many Sr-B-based inorganic compounds as host lattices and discovered various strontium borate phosphors as well. Among the strontium borates, the crystal structure of $NaSrBO_3$ was reported by Wu et al. [96]. Later, F. Yang et al. [101] prepared the $NaSrBO_3:Eu^{3+}$ phosphor by solid-state reaction method in which prolonged heating of 8 hours was required at 850°C. In our lab, we prepared $NaSrBO_3:xEu^{3+}$ phosphor by a simple, time-saving, economical method of modified solution combustion synthesis at a comparatively lower temperature using urea as fuel [102]. Figure 4.8 shows the photoluminescence excitation and emission spectrum of $NaSr_{1-x}BO_3: xEu^{3+}$ (for x = 0.09). The excitation spectrum shows broadband from 250 to 350 nm due to O–Eu charge-transfer band (CTB) transition and the peaks in the range of 350–400 nm originated from f–f transitions of Eu^{3+} ions. The emission spectrum consists of five peaks in the range of 560–720 nm, that are due to $^5D_0-^7F_J$ (J = 0, 1, 2,3 and 4) transitions of Eu^{3+} ions, respectively. The main peak at 614 nm corresponds to electron dipole $^5D_0-^7F_2$ transition and the peak at 592 nm ascribes to magnetic dipole $^5D_0-^7F_1$ transition of Eu^{3+} ions.

Figure 4.9 shows the variation of PL intensity of $NaSr_{1-x}BO_3: xEu^{3+}$ phosphor for different doping concentrations of Eu^{3+} ions [102].

5. J. Li et al. [103] reported Sm^{3+}-doped $CaBi_2B_2O_7$ orange-red-emitting phosphor. This was synthesized by the solid-state reaction method. $CaBi_2B_2O_7:Sm^{3+}$ phosphor is effectively excited by 404 nm near-ultraviolet (NUV) light and features orange-red emission centred at 560, 605 and 649 nm corresponding to the $^4G_{5/2} \rightarrow {}^6H_J$ (J = 5/2, 7/2, 9/2) transitions of Sm^{3+} ions, respectively. The optimum concentration of Sm^{3+} ions in $CaBi_2B_2O_7$ is found to be 5.0 mol%.

6. S. N. Bajaj et al. [104] reported highly intense reddish-orange emitting borate phosphors $YAl_3(BO_3)_4:Sm^{3+}$. The synthesis method deployed was the combustion synthesis method using urea as a fuel. As reported by the authors, the emission spectrum of Sm^{3+} in $YAl_3(BO_3)_4$ phosphor monitored for 402 nm excitation composed of 564, 600 and 648 nm emission peaks

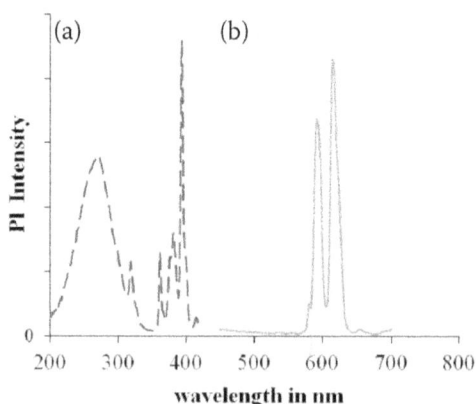

FIGURE 4.8 PLE (a) and PL (b) curve for $NaSr_{1-x}BO_3:xEu^{3+}$ for x = 0.06 [102].

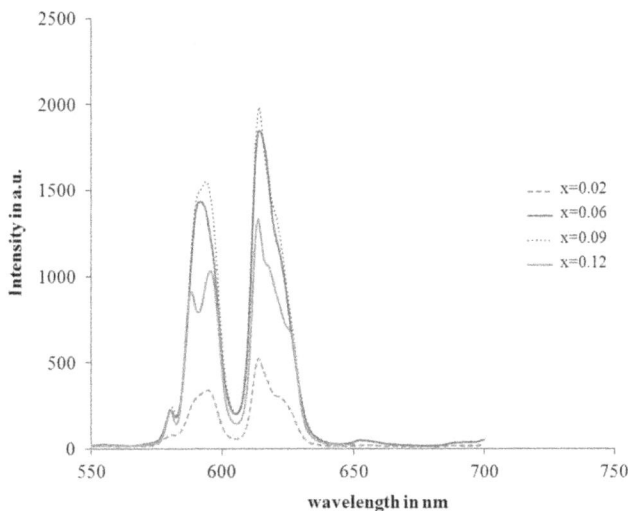

FIGURE 4.9 PL of NaSr1-xBO3:xEu3+ for different 'x' values for an excitation wavelength of 394 nm [102].

and the excitation spectra monitored at 600 nm emission shows peaks at 345, 362, 375, 402, 421 and 445 nm.

7. S. Yuan et al. [105] reported a new red phosphor, $Ba_2Mg(BO_3)_2$: Eu, Mn borate phosphor. The synthesis method adopted was the solid-state reaction method. The excitation band for this phosphor is extending from 250–450 nm, which is adaptable to the emission band of near-ultraviolet LED chips (350–420 nm). Upon the excitation of 365 nm light, the phosphor exhibits strong red emission centred at 615 nm.

4.4.2 YELLOW-EMITTING PHOSPHORS

Blue LED and the yellow phosphor is considered the least expensive method for producing white light. Blue light from the LED is used to excite a phosphor, which then re-emits yellow light. This balanced mixing of yellow and blue lights results in the appearance of white light. Presently, the yellow-emitting YAG:Ce^{3+} phosphor is the commercially widely used phosphor. However, the attempts to find new phosphors that will replace YAG: Ce^{3+} to give warmer white light are driving more and more researchers to this field. Here we discuss a few recently reported borate host yellow-emitting phosphors.

1. One of the efficient borate host phosphors reported is the Ca_2BO_3Cl:xEu^{2+} phosphor. The crystal structure of Ca_2BO_3Cl was first reported by J. Majling [106] et al. and after that several researchers tried to develop this phosphor in various aspects. We prepared this yellow-emitting phosphor by conventional solid-state reaction method in a slightly reductive atmosphere produced by charcoal at 1,173 K [107].

In this chloroborate host, Eu^{2+} ions show broad yellow emission peaking at 560 nm. The photoluminescence excitation spectrum monitored at 560 nm shows a broad excitation band from 300 nm to 500 nm, peaking at 380 nm. There are several small peaks superimposed. The excitation band shows an excellent spread in both the NUV and blue regions of the spectrum, as shown in Figure 4.10. This makes it a potential candidate for LED applications.

The emission spectrum monitored at 380 nm as well as 450 nm shows an emission wavelength of 560 nm corresponding to $4f^65d^1{\rightarrow}4f^7$ transition of Eu^{2+} ions. However, the comparative intensity is less in the latter case.

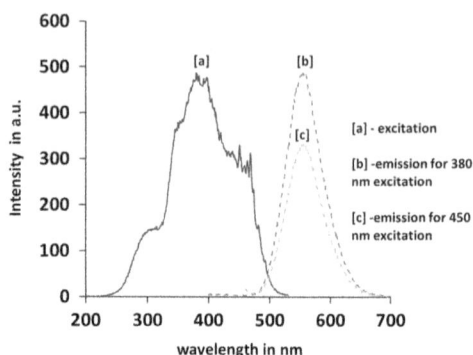

FIGURE 4.10 PLE (a) and PL (b) curve for $Ca_2BO_3Cl:xEu^{2+}$ for x = 0.04 [107].

2. To obtain novel yellow-emitting borate phosphors, researchers have tried many Sr-B-based inorganic compounds as host lattices and attained various phosphors with broadband emissions. W.-S. Song et al. reported the $Sr_3B_2O_6:Eu^{2+}$ phosphor [108] that shows 578 nm emission at 400 nm excitation. They tried to obtain the efficient yellow emission by varying host composition, Eu concentration and firing temperature. The 1,300°C-fired Eu^{2+}-doped $Sr_3B_2O_6$ was found to be the most efficient candidate and fabricated white LED. The warm white lights with CIE chromaticity coordinates of (0.340–0.372, 0.287–0.314) and colour rendering indices of 75–77 under the forward currents of 5–40 mA.

3. Y. Xue et al. reported $(Sr, Ca)_3B_2O_6:Eu$ [109] yellow–green-emitting phosphor synthesized using precursors prepared via a facile sol-gel route. This phosphor shows emission peak centred at 540 nm under near-UV excitation.

4.4.3 GREEN-EMITTING PHOSPHORS

Recently, there are a few reports on the borate host green-emitting phosphors. The activators used are Tb^{3+}, Eu^{2+} and Ce^{3+} in various borate hosts. There are few reports of green/white LED fabrication by using these phosphors singly/blending with other phosphors by coating UV/blue LED chips. Here we are discussing recently reported green-emitting phosphors borate host phosphors:

1. R. Wang et al. reported a Tb^{3+}-doped $Sr_3B_2O_5$ phosphor [110] that was synthesized by the solid-state reaction method. This phosphor can be effectively excited by ultraviolet (UV) 376 nm and exhibits bright green emission centred at 545 nm.

2. Novel green phosphor, $LiCaBO_3:Tb^{3+}$ was reported by Z. Wang et al. [111]. The synthesis method used was the solid-state reaction method. The authors reported that the excitation band is extended from 220 to 400 nm for the phosphor. It exhibited a strong green emission located at 544 nm with chromatic coordination (0.25, 0.58). The variation of emission intensities of $LiCaBO_3$: Tb^{3+} phosphor by varying Tb^{3+} concentration, and co-doping by charge compensator Li^+, Na^+ and K+ ions is also reported.

3. G. M. Cai et al. reported crystal structure and photoluminescence of Tb^{3+}-activated $Ba_3InB_3O_9$ phosphor [112]. This phosphor exhibited green emission at about 543 nm with 370 nm light excited.

4. W. Geng et al. reported a new borate host green-emitting phosphor $NaBaY(BO_3)_2$: Ce^{3+}, Tb^{3+} (NBY: Ce^{3+}, Tb^{3+}) [113] for LED applications. As reported by the authors NBY: Ce^{3+} and Tb^{3+} present a wide excitation band ranging from 310 to 400 nm and efficient green emission at 542 nm, with a full-width at half-maximum of 3.3 nm. Blending this green-emitting phosphor along with $BaMaAl_{10}O_{17}$: Eu^{2+} (BAM: Eu^{2+}), and $CaAlSiN_3$: Eu^{2+} a white LED device was fabricated by the authors using a 370 nm UV chip. The colour coordinate, correlated colour temperature and colour rendering index of the manufactured LED device were reported as (0.335, 0.347), 5,511 K and 80.16, respectively.

5. C. Guoet et al. reported the brightest green-emitting phosphor $Na_2Gd_2B_2O_7$:Ce^{3+}, Tb^{3+} suitable for the application of near-UV LEDs [114]. In this phosphor, the efficient non-radiative energy transfer from Ce^{3+} to Tb^{3+} is reported. The authors claim that Tb^{3+} can give a bright green-emitting light by a strong excitation band of Ce^{3+}, centred at about 358 nm, which indicates that it can serve as a potential green-emitting phosphor for n-UV LEDs.

6. Panase et al. reported Terbium-doped $Sr_2B_2O_5$ green phosphor that was synthesized by combustion synthesis method at about 550°C [115]. The excitation and emission spectra specify that the prepared phosphor can be effectively excited by 353 nm and reveal bright green emission centred at 545 nm due to the D → F transition of Tb^{3+} ions.

7. Green-emitting Terbium-doped $LiBaB_9O_{15}$ phosphor was reported by V. Singh et al. [116]. This phosphor shows a dominant and strong green luminescent peak at 542 nm that is due to the d–f transition. As reported by the authors, this phosphor exhibited a broad excitation band centred at 230 nm and several weak peaks in the wavelength range of 300–400 nm.

8. Narrow-band green-emitting borate phosphor $NaBaB_9O_{15}$:Eu^{2+} was reported by Zhuo et al. [117]. This phosphor shows the emission spectrum with a maximum centred at 515 nm and a full width at half-maximum (fwhm) of approximately 61 nm when excited using a blue excitation source (λex = 430 nm). The NBBO:Eu,Na emission spectra measured using typical UV and near-UV excitation sources (λex = 330 nm, 365 nm and 395 nm) all show the same narrow green emission.

9. Zhen et al. reported a novel green phosphor, $Sr_2MgB_2O_6$:Tb^{3+}, Li^+ [118] for white light-emitting diodes. It was prepared by the solid-state reaction method. The excitation and emission spectra reported for this phosphor indicate that this phosphor can be effectively excited by ultraviolet 368 nm radiation and exhibited bright green emission centred at 545 nm.

4.4.4 Blue-Emitting Phosphors

1. Blue-emitting $NaSrBO_3$:Ce^{3+} phosphor with high efficiency and high colour purity for near-UV light-emitting diodes was reported by W.-R. Liu et al. [119]. This phosphor shows bright blue emission centred at 400 nm under 365 nm excitation.

2. X. Li et al. reported an ideal blue $Sr_3B_2O_6$: Ce^{3+} [120] phosphor prepared by the sol-combustion method. The samples display two excitation peaks at 292 and 343 nm, and also two emission peaks at 432 and 468 nm.

3. In 2012, A. Reddy et al. [121] reported the luminescence of Eu^{3+} in the $KCaBO_3$ host matrix. We reported Ce^{3+} luminescence in $KCaBO_3$ borate host [122]. The photo-luminescence excitation and emission spectrum of $KCaBO_3$:xCe^{3+} for (x = 0.02) is shown in Figure 4.11. The excitation spectrum monitored at 427 nm shows a small excitation peak at 287 nm and a broad excitation band peaking at 360 nm. The emission spectrum exhibits intense blue emission at 427 nm when excited with a 365 nm wavelength.

4. Nakamura et al. reported the luminescence of sintered $Ca_2B_5O_9Cl$: Eu^{2+} excited by UV light or X-rays [123]. The phosphors of Eu^{2+} ions doped alkaline earth halo-borates $M_2B_5O_9X$ (M -Ca, Sr, Ba; X- Cl, Br) have gained special attention because of their possible applications as storage phosphors for X-ray imaging and thermal neutron detection. There are few reports on its potential application for white light-emitting diodes [124].

5. We prepared $Ca_2B_5O_9Cl$: Eu^{2+} [125] phosphor in our lab by the conventional solid-state reaction method. The emission spectrum monitored at 360 nm and 396 nm excitation shows an emission wavelength of 452 nm corresponding to $4f^65d^1 \rightarrow 4f^7$ transition of Eu^{2+} ions with FWHM of 37 nm (Figure 4.12). This phosphor shows intense blue emission with colour coordinates (0.145, 0.0368), as shown in Figure 4.13.

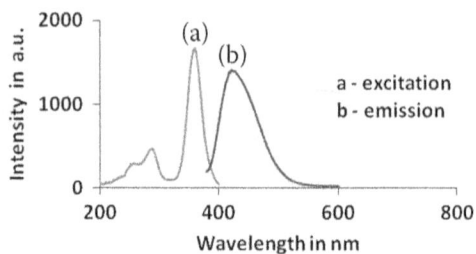

FIGURE 4.11 PLE (a) and PL (b) curve for KCaBO$_3$: xCe^{3+} [122].

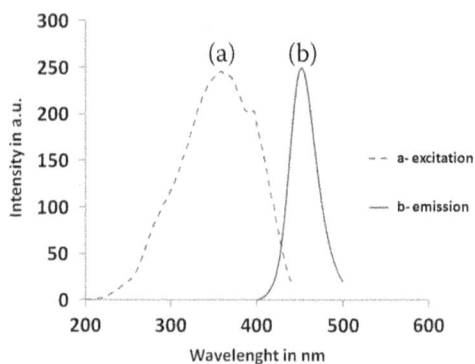

FIGURE 4.12 PLE (a) and PL (b) curve for Ca$_2$B$_5$O$_9$Cl: xEu^{2+} for x = 0.08 [125].

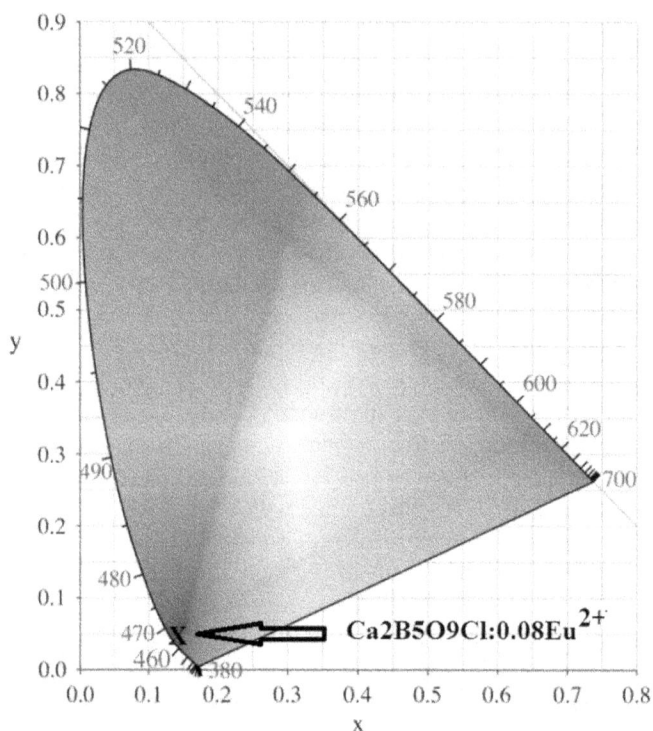

FIGURE 4.13 CIE colour coordinates for the as-synthesised Ca$_2$B$_5$O$_9$Cl:0.08Eu^{2+} [125].

CIE colour coordinates 1931 for the as-synthesized phosphor is (0.145, 0.0368), indicating better colour purity with FWHM of 37 nm (Figure 4.13).

6. S. N. Bajaj et al. [83] reported a $LaMgB_5O_{10}:Ce^{3+}$ borate host blue-emitting phosphor that exhibits symmetrical band centred on 430 nm. The excitation spectra monitored at 430 nm emission shows peaks at 272, 326 and 372 nm.

4.4.5 WHITE-EMITTING PHOSPHORS

In some borate hosts, two or more activators were used to produce white light. Here we are discussing a few recently reported such phosphors.

1. Q. Waan et al. reported potential colour tunable white-emitting phosphor, $LiSr_4(BO_3)_3$: Ce^{3+}, Eu^{2+} suitable for ultraviolet light-emitting diodes [126]. The co-doping results in the resonant energy transfer between Ce^{3+} and Eu^{2+} ions in this phosphor. White light emitted from this phosphor shows chromaticity coordinate (0.34, 0.30).
2. L. Panlai et al. reported a potential single-phase white-emitting $LiBaBO_3:Ce^{3+}$, Eu^{2+} [127] phosphor for white LEDs. This Ce^{3+}/Eu^{2+} co-doped $LiBaBO_3$ phosphor was synthesized by a high-temperature solid-state reaction method. Under UV excitation, white light was generated by coupling blue and yellow-green emission bands attributed to Ce^{3+} and Eu^{2+} emissions, respectively. The luminous efficacy of $LiBaBO_3:1\%Ce^{3+}$, $2\%Eu^{2+}$ phosphor was reported to be 290 lm/W by the authors.
3. S. Das et al. reported controllable white light emission from Dy^{3+} and Eu^{3+} co-doped $KCaBO_3$ phosphor [128].
4. White-light-emitting phosphors reported, like $Sr_3Y_2(BO_3)_4:Dy^{3+}$ when excitation by 370 nm shows the emission of 488 nm, 575 nm and 665 nm with CIE coordinates of (0.2997, 0.3142) [129].
5. $GdAl_3(BO_3)_4:Dy^{3+}$, Ce^{3+} phosphor with excitation by 400 nm shows the emission of 480 nm, 575 nm and 620 nm with CIE coordinates of (0.31, 0.33), CCT 6480 K and the CRI reported of 78 [130].
6. Another borate host phosphor $Sr_3B_2O_6:Ce^{3+}$, Eu^{2+} when excited with 351 nm, shows an emission of 434 nm and 574 nm, and reported CIE coordinates are (0.31, 0.24) [131].
7. $Ba_2 Ca(BO_3)_2:Ce^{3+}$,Mn^{2+} [132] phosphor with excitation by 345 nm shows an emission of 420 nm, 480 nm and 625 nm with CIE coordinates (0.33, 0.31) and CCT 5861. The warm white lights with Ra of 95.43 ($\lambda ex = 345$ nm) and of 86.37 ($\lambda ex = 395$ nm) are obtained by combining selected phosphors with UV or violet LEDs.
8. The phosphor $Ba_2 Ca(B_3O_6)_2:Eu^{2+}$,$Mn^{2+}$ with excitation of 330 nm shows emission of 459 nm and 608 nm with CIE coordinates (0.37, 0.25) [133].
9. The chloroborate $Ca_2BO_3Cl:Ce^{3+}$, Eu^{2+} with excitation of 349 nm shows emission of 421 nm and 575 nm with CIE coordinates of (0.326, 0.334) [134,135].

There are a few recently reported single-phase and colour tuneable phosphors for white-light-emitting applications [136,137]. Some researchers also report rare-earth-free borate host phosphors [138].

4.5 APPLICATIONS

LEDs are the most popular light source nowadays because of their efficiency, availability of a range of colours and structural flexibility. Hence, they have numerous applications.

4.5.1 LEDs FOR LIGHTING APPLICATIONS

LEDs are widely used for decorative lighting, night lighting and outdoor lighting. The reason for this popularity is that though the initial installation costs for this lighting system is on the higher side, it is reliable and almost maintenance-free. LEDs are now a common solution for theatrical stage lighting as well [139].

4.5.1.1 Household Lighting

Earlier, the only purpose of household lighting was ambient lighting, leading to a uniform amount of lighting throughout a room for general vision and orientation. Thanks to LED lighting, now it is possible to choose different lighting for every room, depending upon the purpose and mood [140]. Typical requirements for household lighting are ambient lighting, highlighting a particular area and decorative lighting [141]. LEDs provide solutions for each type. Recessed downlights, cove lighting or pendant-hung LED fixtures are available for ambient lighting [142]. The decorative lighting fixtures should have an essentially pretty look; for example, chandeliers and wall sconces. Nowadays it is possible to choose a mix of LED light sources in the living room that can be adjusted depending upon the occasion. Adjustable spotlight LED lamps are available that can be pointed at the walls to illuminate a particular artwork, family photos or simply to reduce unwanted glare [143]. These spotlight lamps can be used on walls or on ceilings to eliminate unwanted shadows. New LED three-way lamps can be used in all three-way table lamps you may have in your living room. Dimmable overhead fixtures like LED chandelier bulbs that offer soft to neutral tones for the dining room are now available in beautiful colours.

So the SSL technology helps us to choose the light that complements the interior design, furniture, wall colours and other decorations in the house in addition to trustworthy general illumination. Some of the latest designs include a two-in-one design i.e. LED bulb with bluetooth speaker music lightbulb B22 LED white + RGB light ball bulb colourful lamp with remote control for home. The Halonix Hexa 1,200 mm ceiling fan with built-in six-colour LED light and remote (white) is now available on the market [144]. In addition, portable flexible USB LED light, solar and electric rechargeable portable LED emergency light, etc. are already on the market. Soon we may expect to have smart houses enabled with LiFi technology.

4.5.1.2 Street Lighting

The world has become small since we have good roads connecting across. Since there are different types of roads, different street lightings are required. In addition, lighting should be competent to match different slots of the day like dusk, dawn or changing weather conditions [145]. Street lighting should help people to safely continue their travels during the daytime and should reduce the risk of nighttime accidents. The street lighting requires high maintenance costs that includes lamp replacement and high-energy consumption. Hence, energy-efficient, reliable, long-life LEDs are rapidly replacing conventional light sources. They are small in size, more flexible and optically control the light better. Companies like Wipro have launched skyline plus series, especially for street lighting [146]. In addition, LEDs offer all the different colour temperature options combined with great colour rendering. Some leading companies like LEDiL are coming up with different types of LED packages (high-power, mid-power, CSP, COB, etc.) to have a variety of street light luminaires [147].

4.5.1.3 Automobile Lighting

LEDs are cool devices i.e. they do not emit any heat. This property makes them a great choice for the automobile lighting industry [148]. The most common application is for signage, along with many other uses [149]. Since they operate on very small power, LEDs are popularly used for small applications like flashlights, strobe lights and mobile phone cameras [150]. They are widely used in signals or indicators; for example, in traffic lights and exit signs, vehicle brake lights, etc. Nowadays, LEDs are used in vehicle headlights as well due to their long lifetime. LEDs are utilized for impressive styling in both rear and forward lighting [151]. They are widely used in interior lighting applications in cars for better illumination as well as to create a pleasant ambiance [152].

4.6 CURRENT TRENDS AND INNOVATIONS

As per Haitz's law: "Every decade, the amount of light generated by LEDs increases by a factor of 20, and the cost per lumen falls by a factor of 10." This we could observe till 2016, literally [153]. However, after that, this technology is continuously improving, leading to an increase in efficacy and reduction in production costs. Nowadays, the cost of LEDs has become comparable to that of compact fluorescent lamps, and hence they are widely accepted.

- Flexible, in-molded LED foil using a roll-to-roll type process is reported by a Finland research center.
- Earlier LED lighting systems used to have external circuit board drivers. Now there are DC LED chips that handle all the power requirements for the LED directly on the chip. Hence, the packaging became more compact.
- Recently, WiFi-enabled smart LED bulbs have launched on the market by various leading companies (Figure 4.14).
- Like Wi-Fi, we are expecting a **Li-Fi** system. Harald Haas from the University of Edinburgh, Scotland invented Li-Fi in 2011 [154]. In this system, light from LEDs is modulated at high rates, creating a data network (Figure 4.15). It is a high-speed, bidirectional, and fully networked wireless communication of data using light. In the Li-Fi system, LEDs will play the role of the router. Thus, by flashing LED lights on and off at very high speeds, data will be sent to a receiver in a binary code. This will enable common household LED lightbulbs to enable data transfer, thereby boosting the speed up

SMART DECORATIVE LIGHTING

RGB Lights
16 million colors

Brightness Control
10–100%

Set a Schedule or Timer

Group Devices

FIGURE 4.14 Wipro WiFi-enabled smart LED bulb B22 9-Watt (16 million colors + warm white/neutral white/white) (Compatible with Amazon Alexa and Google Assistant) [144].

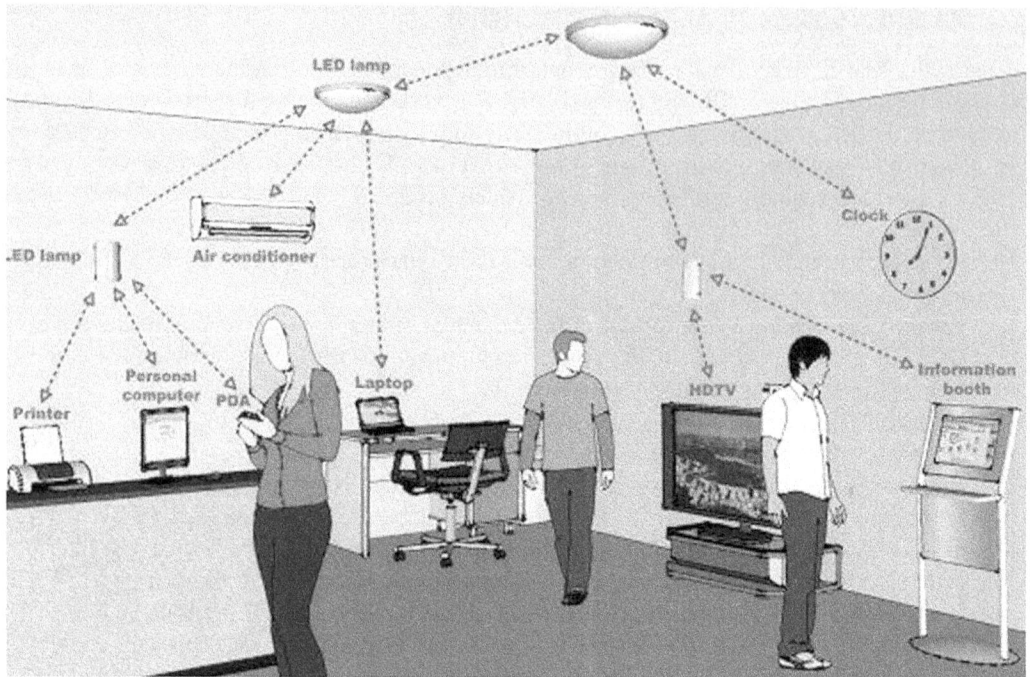

FIGURE 4.15 Typical Li-Fi environment [158].

to 224 GB per second [155]. So instead of transmitting information via one data stream, this visible light would make it possible to transmit the same information using thousands of data streams simultaneously. This Li-Fi system has various merits like more secure wireless communication and energy savings as it will combine and eliminate health concerns. Indonesia is among the first countries to pilot Li-Fi in an education center. In India, Akrund and Navanagar villages in the Aravalli district of Gujarat have become the first smart villages with Li-Fi-based internet connectivity [156].

- In the automobile sector, a new advanced forward lighting system has launched. In this system, an array of LEDs will be controlled by forward-sensing technology. A microcontroller is used to selectively flash the required LEDs while dimming the others. For example, in the array of LEDs, it will dim those LEDs that are producing disturbing glare to an oncoming driver while keeping high beams on for other areas of the roadway. That will add to driver safety. These systems will be adopted by technologically advanced automobile companies [157].

4.7 LIMITATIONS AND CHALLENGES

We have discussed the merits and advantages of LEDs in section 4.2.2. LEDs indeed offer various benefits but at the same time, some limiting factors restrict their wide acceptance for every lighting-related application. The challenges faced by this technology are listed below [159]:

1. **Higher cost**

Currently, LED prices are higher than other light sources available. This is due to the high manufacturing costs associated with these semiconductor devices. But this can be compensated with better efficiency, lower energy consumption, lower maintenance costs and a longer lifetime.

2. RGB LEDs and potential colour shift

RGB LEDs have inconsistent colour temperature and light output. The output light from LEDs change with time, temperature and driving current also varies from device to device. This may result in variation of colour and intensity of output light.

3. Require constant driving current

Driving LEDs above their rated current causes the junction temperature to rise to levels where permanent damage may occur. In addition, a transformer compatibility issue needs to be addressed.

4. Complex fabrication

Fabrication of LEDs involves the growth of crystalline layers across the surface of a semiconductor wafer at high temperatures. The precision in laying these layers decides the properties of the LED. So the quality may differ from wafer to wafer, or batch to batch. Proper designing and use of an effective driver is the key to obtain all the benefits of LEDs. Extracting light from LEDs is a separate science with its limitations.

5. Standardization is not streamlined

Since this is an emerging technology coming up with new improvements every day, it has become difficult to set the standards unanimously. So there are various standards relating to LEDs in domestic lighting, automotive lighting and street lighting. This lack of standard standardization measures is a current issue in the LED field. The organizations like CIE, NEMA and IES are trying to come up with common standards.

6. Poor quantum efficiency

Although LEDs have high efficiency and consume a small amount of power, the devices produce a small total number of lumens.

4.8 FUTURE PROSPECTS AND SCOPE

LEDs have the potential to replace conventional light sources, with good economic and environmental savings. Phosphors required for the said purpose should have properties such as high luminescence efficiency, stability against the environment, long shelf life, ease of preparation, low production cost, large-scale batch processing, batch uniformity, etc [65]. All these properties are seldom found in one phosphor and usually compromise has to make on one or more parameters.

So there is scope to undertake synthesis, characterisation and study the novel luminescent materials suitable for solid-state lighting.

- Conventional synthesis methods can be modified to make them more economical and environment friendly that will produce cost-effective and intense emitting phosphors. There is a lot of scope to synthesize phosphors that will be agglomeration-free, nanosize particles with pre-designed morphology.
- New borate hosts with suitable morphology need to be reported for LED application.
- There is a scope in investigating the epoxy suitable for borate host phosphors and improving the phosphor blending techniques for LED applications.
- Red-emitting phosphor plays a key role in improving the CRI of white LEDs. High-intensity

red-emitting borate host phosphor with a wide excitation band is required for LED applications and has not yet been reported.
- There is a scope to develop intense emitting borate host phosphor by exploring various doping ions like rare-earth ions, transition metal ions and other ions by deploying unconventional synthesis methods.
- It will be interesting to study energy transfer by co-doping the phosphors with various ion combinations by investigating their energy band structure.

In conclusion, I would like to state that LED lighting and its further applications are still at an early, exciting and highly innovative stage. It would be my utmost pleasure if any of my readers contribute to this new era of efficient and eco-friendly lighting.

REFERENCES

[1] Shur, M. S., and Zukauskas, R. (2005). Solid-state lighting: toward superior illumination. *Proceedings of the IEEE*, 93(10), 1691–1703. 10.1109/JPROC.2005.853537
[2] Žukauskas, A., Vaicekauskas, R., Ivanauskas, F., Vaitkevičius, H., and Shur, M. S. (2008). Spectral optimization of phosphor-conversion light-emitting diodes for ultimate colour rendering. *Applied Physics Letters*, 93(5), 051115. 10.1063/1.2966150
[3] A. Katelnikovas, A. Kareiva, H. Winkler, T. Jüstel, ICL'11 – 16th International Conference on Luminescence, Ann Arbor, MI, USA, June 26–July 1 (2011) 60.
[4] Johnstone, B. (2007). *Brilliant!: Shuji Nakamura and the Revolution in Lighting Technology.* Prometheus Books, Amherst, New York, USA.
[5] T.A. Edison, U.S. patent No. 223, (1879) 898.
[6] Thayer, R. N., and Barnes, B. T. (1939). The basis for high efficiency in fluorescent lamps. *Journal of the Optical Society of America*, 29(3), 131–134. https://www.osapublishing.org/josa/abstract.cfm?URI=josa-29-3-131
[7] F. Meyer, H. Spanner, and E. Germet, US 2 182(1939) 732.
[8] Tarashioon, S., Baiano, A., Van Zeijl, H., Guo, C., Koh, S. W., Van Driel, W. D., and Zhang, G. Q. (2012 May 1). An approach to "Design for Reliability" in solid state lighting systems at high temperatures. *Microelectronics Reliability*, 52(5), 783–793. 10.1016/j.microrel.2011.06.029
[9] Nakamura, S. (1991). GaN growth using GaN buffer layer. *Japanese Journal of Applied Physics*, 30(10A), L1705. 10.1143/jjap.30.11705
[10] Nakamura, S., Mukai, T., and Senoh, M. (1994). Candela-class high-brightness InGaN/AlGaN double-heterostructure blue-light-emitting diodes. *Applied Physics Letters*, 64(13), 1687–1689. 10.1063/1.111832
[11] Wierer, J. J., Steigerwald, D. A., Krames, M. R., O'shea, J. J., Ludowise, M. J., Christenson, G., ... and Stockman, S. A. (2001). High-power AlGaInN flip-chip light-emitting diodes. *Applied Physics Letters*, 78(22), 3379–3381. 10.1063/1.1374499
[12] Zakheim, D. A., Smirnova, I. P., Roznanskii, I. V., Gurevich, S. A., Kulagina, M. M., Arakcheeva, E. M., ... and Itkinson, G. V. (2005). High-power flip-chip blue light-emitting diodes based on AlGaInN. *Semiconductors*, 39(7), 851–855. 10.1134/1.1992647
[13] Narukawa, Y., Ichikawa, M., Sanga, D., Sano, M., and Mukai, T. (2010). White light emitting diodes with super-high luminous efficacy. *Journal of Physics D: Applied Physics*, 43(35), 354002. 10.1088/0022-3727/43/35/354002
[14] Mueller-Mach, R., Mueller, G. O., Krames, M. R., and Trottier, T. (2002). High-power phosphor-converted light-emitting diodes based on III-nitrides. *IEEE Journal of Selected Topics in Quantum Electronics*, 8(2), 339–345. 10.1109/2944.999189
[15] Mueller-Mach, R., Mueller, G. O., and Krames, M. R. (2004, January). Phosphor materials and combinations for illumination-grade white PCLEDs. In Third International Conference on Solid State Lighting (Vol. 5187, pp. 115–122). International Society for Optics and Photonics. 10.1117/12.514074
[16] Nakamura, S., and Fasol, G. (2013). *The Blue Laser Diode: GaN Based Light Emitters and Lasers.* Springer Science & Business Media,Berlin, Germany.
[17] Jüstel, T., Möller, S., Winkler, H., and Adam, W. (2012). *Luminescent Materials in Ullmann's Encyclopedia of Industrial Chemistry*, Vol. A1-28. 1–75
[18] https://www.lrc.rpi.edu/programs/solidstate/sslwhat.asp

[19] https://www.energy.gov/sites/default/files/2020/02/f72/2019_ssl-energy-savings-forecast.pdf
[20] https://www.idtechex.com/en/research-report/oled-lighting-opportunities-2017-2027-forecasts-technologies-players/526
[21] Xie, R. J., Li, Y. Q., Hirosaki, N., and Yamamoto, H. (2011). *Nitride Phosphors and Solid-state Lighting*. Boca Raton, FL: Taylor & Francis. 10.1201/b10939
[22] https://www.stouchlighting.com/blog/top-15-advantages-of-led-lighting
[23] https://innovationtoronto.com/2017/03/flexible-led-displays-containing-printed-electronics-ready-for-roll-to-roll-manufacturing/
[24] https://www.ledsmagazine.com/architectural-lighting/retail-hospitality/article/16696265/benefits-and-drawbacks-of-leds
[25] https://worldwidescience.org/topicpages/l/led+accent+lighting.html
[26] https://www.thelightbulb.co.uk/resources/ultimate-guide-led-lights-advantages-leds/
[27] Khan, Q., Khan, S. A., and Bao, Q. (2020 Jan 1). Light-emitting devices. In *2D Materials for Photonic and Optoelectronic Applications* (pp. 175–197). Woodhead Publishing, Sawston, Cambridge, 10.1016/B978-0-08-102637-3.00007-3
[28] Humphreys, C. J. (2008). Solid-state lighting. *MRS Bulletin*, 33(4), 459–470. 10.1557/mrs2008.91
[29] https://www.doitpoms.ac.uk/tlplib/semiconductors/direct.php
[30] Fu, H., and Zhao, Y. 2018). Efficiency Droop in GaInN/GaN LEDs. In *Nitride Semiconductor Light-Emitting Diodes (LEDs)*, Jan 1 (pp. 299–325). Woodhead Publishing, Sawston, Cambridge.
[31] Huang, J. J., Kuo, H. C., and Shen, S. C. (2017 Oct 24). *Nitride Semiconductor Light-Emitting Diodes (LEDs): Materials, Technologies, and Applications*. Woodhead Publishing, Sawston, Cambridge, 10.1016/C2016-0-01551-6
[32] https://www.led-professional.com/technology/light-generation/successful-extraction-of-light-from-semiconductors-with-the-highest-efficiency
[33] Steigerwald, D. A., Bhat, J. C., Collins, D., Fletcher, R. M., Holcomb, M. O., Ludowise, M. J., ... and Rudaz, S. L. (2002). Illumination with solid state lighting technology. *IEEE Journal of Selected Topics in Quantum Electronics*, 8(2), 310–320. 10.1109/2944.999186
[34] Ye, S., Xiao, F., Pan, Y. X., Ma, Y. Y., and Zhang, Q. Y. (2010 Dec 1). Phosphors in phosphor-converted white light-emitting diodes: recent advances in materials, techniques and properties. *Materials Science and Engineering: R: Reports*, 71(1), 1–34. 10.1016/j.mser.2010.07.001
[35] Chen, L., Lin, C. C., Yeh, C. W., and Liu, R. S. (2010). Light converting inorganic phosphors for white light-emitting diodes. *Materials*, 3(3), 2172–2195. 10.3390/ma3032172
[36] Blasse, G., and Bril, A. (1967). A new phosphor for flying-spot cathode-ray tubes for colour television: yellow-emitting Y3Al5O12–Ce3+. *Applied Physics Letters*, 11(2), 53–55. 10.1063/1.1755025
[37] Hirayama, H., Tsukada, Y., Maeda, T., and Kamata, N. (2010). Marked enhancement in the efficiency of deep-ultraviolet AlGaN light-emitting diodes by using a multiquantum-barrier electron blocking layer. *Applied Physics Express*, 3(3), 031002 10.1016/j.jcrysgro.2003.12.030
[38] Wang, H. X., Li, H. D., Lee, Y. B., Sato, H., Yamashita, K., Sugahara, T., and Sakai, S. (2004). Fabrication of high-performance 370 nm ultraviolet light-emitting diodes. *Journal of Crystal Growth*, 264(1–3), 48–52. 10.1016/j.jcrysgro.2003.12.030
[39] Lee, S. N., Paek, H. S., Kim, H., Kim, K. K., Cho, Y. H., Jang, T., and Park, Y. (2008). Growth and characterization of the AlInGaN quaternary protective layer to suppress the thermal damage of InGaN multiple quantum wells. *Journal of Crystal Growth*, 310(16), 3881–3883. 10.1016/j.jcrysgro.2008.05.056
[40] Liu, Y., Egawa, T., Ishikawa, H., and Jimbo, T. (2003). Growth and characterization of high-quality quaternary AlInGaN epilayers on sapphire. *Journal of Crystal Growth*, 259(3), 245–251. 10.1016/j.jcrysgro.2003.07.014
[41] McIntosh, F. G., Boutros, K. S., Roberts, J. C., Bedair, S. M., Piner, E. L., and El-Masry, N. A. (1996). Growth and characterization of AlInGaN quaternary alloys. *Applied Physics Letters*, 68(1), 40–42. 10.1063/1.116749
[42] Park, B. K., Park, H. K., Oh, J. H., Oh, J. R., and Do, Y. R. (2012). Selecting morphology of $Y_3Al_5O_{12}$: Ce_{3+} phosphors for minimizing scattering loss in the pc-LED package. *Journal of the Electrochemical Society*, 159(4), J96. 10.1149/2.031204jes
[43] Kottaisamy, M., Thiyagarajan, P., Mishra, J., and Rao, M. R. (2008). Color tuning of Y3Al5O12: Ce phosphor and their blend for white LEDs. *Materials Research Bulletin*, 43(7), 1657–1663. 10.1016/j.materresbull.2007.09.005
[44] Shen, C. Y., Li, K., Hou, Q. L., Feng, H. J., and Dong, X. Y. (2010). White LED based on YAG: Ce, Gd phosphor and CdSe–ZnS core/shell quantum dots. *IEEE Photonics Technology Letters*, 22(12), 884–886. 10.1109/LPT.2010.2046724

[45] Uchida, Y., Setomoto, T., Taguchi, T., Nakagawa, Y., and Miyazaki, K. (2000, June). Characteristics of high-efficiency InGaN-based white LED lighting. In *Display Technologies Iii* (Vol. 4079, pp. 120–126). International Society for Optics and Photonics. 10.1117/12.389397

[46] Adachi, S., Abe, H., Kasa, R., and Arai, T. (2012). Synthesis and Properties of Hetero-Dialkaline Hexafluorosilicate Red Phosphor KNaSiF6:Mn4+. *Journal of the Electrochemical Society*, 159, J34– J37.

[47] Adachi, S., Abe, H., Kasa, R., and Arai, T. (2011). Synthesis and properties of hetero-dialkaline hexafluorosilicate red phosphor KNaSiF6: Mn4+. *Journal of the Electrochemical Society*, 159(2), J34. 10.1149/2.064202jes

[48] Arai, Y., and Adachi, S. (2011). Optical properties of Mn4+-activated Na2SnF6 and Cs2SnF6 red phosphors. *Journal of Luminescence*, 131(12), 2652–2660. 10.1016/j.jlumin.2011.06.042

[49] Kasa, R., and Adachi, S. (2012). Red and Deep Red Emissions from Cubic K2SiF6:Mn4+ and Hexagonal K2MnF6 Synthesized in HF/KMnO4/KHF2/Si Solutions,*Journal of the Electrochemical Society*, 159, J89–J95.

[50] Kasa, R., and Adachi, S. (2012). Red and deep red emissions from cubic K2SiF6: Mn4+ and hexagonal K2MnF6 synthesized in HF/KMnO4/KHF2/Si solutions. *Journal of the Electrochemical Society*, 159(4), J89. 10.1149/2.005204jes

[51] Arai, Y., and Adachi, S. (2011). Optical Transitions and Internal Vibronic Frequencies of MnF62 - Ions in Cs2SiF6 and Cs2GeF6 Red Phosphors,*Journal of the Electrochemical Society*, 158, J179–J183.

[52] Arai, Y., and Adachi, S. (2011). Optical Transitions and Internal Vibronic Frequencies of MnF 6 2- Ions in Cs2SiF6 and Cs2GeF6 Red Phosphors. *Journal of the Electrochemical Society*, 158(6), J179. 10.1149/1.3576124

[53] Wang, J., Jing, X., Yan, C., Lin, J., and Liao, F. (2006). Influence of fluoride on f–f transitions of Eu3+ in LiEuM2O8 (M = Mo, W). *Journal of Luminescence*, 121(1), 57–61. 10.1016/j.jlumin.2005.10.003

[54] Guo, C., Gao, F., Xu, Y., Liang, L., Shi, F. G., and Yan, B. (2009). Efficient red phosphors Na5Ln (MoO4) 4: Eu3+ (Ln = La, Gd and Y) for white LEDs. *Journal of Physics D: Applied Physics*, 42(9), 095407. 10.1088/0022-3727/42/9/095407

[55] Qi, X., Liu, C. M., and Kuo, C. C. (2010). Pr3+ doped LaTiNbO6 as a single phosphor for white LEDs. *Journal of Alloys and Compounds*, 492(1–2), L61–L63. 10.1016/j.jallcom.2009.11.188

[56] Ju, G., Hu, Y., Wu, H., Yang, Z., Fu, C., Mu, Z., and Kang, F. (2011). A red-emitting heavy doped phosphor Li6Y (BO3) 3: Eu3+ for white light-emitting diodes. *Optical Materials*, 33(8), 1297–1301. 10.1016/j.optmat.2011.03.002

[57] Sakirzanovas, S., Dutczak, D., Kareiva, A., and Jüstel, T. (2012). Concentration influence on temperature-dependent luminescence properties of samarium substituted strontium tetraborate. *Journal of Luminescence*, 132(1), 141–146. 10.1016/j.jlumin.2011.08.011

[58] Chikte, D., Omanwar, S. K., and Moharil, S. V. (2014). Blue emitting KSCN: xCe phosphor for solid state lighting. *Journal of Luminescence*, 145, 729–732. 10.1016/j.jlumin.2013.08.057

[59] Chikte, D., and Omanwar, S. K. (2019). Synthesis and luminescence properties of novel NaSCN: xCe3+ phosphor. *Journal of Asian Ceramic Societies*, 7(3), 350–354. 10.1080/21870764.2019.1641885

[60] Liu, W. R., Chiu, Y. C., Yeh, Y. T., Jang, S. M., and Chen, T. M. (2009). Luminescence and energy transfer mechanism in Ca10K (PO4) 7: Eu2+, Mn2+ phosphor. *Journal of The Electrochemical Society*, 156(7), J165–J169. 10.1149/1.3121531

[61] Huang, C. H., Chen, T. M., Liu, W. R., Chiu, Y. C., Yeh, Y. T., and Jang, S. M. (2010). A single-phased emission-tunable phosphor Ca9Y (PO4) 7: Eu2+, Mn2+ with efficient energy transfer for white-light-emitting diodes. *ACS Applied Materials & Interfaces*, 2(1), 259–264. 10.1021/am900668r

[62] Kim, T. G., Kim, Y. S., and Im, S. J. (2009). Energy transfer and brightness saturation in (Sr, Ca) 2P2O7: Eu2+, Mn2+ phosphor for UV-LED lighting. *Journal of the Electrochemical Society*, 156(7), J203. 10.1149/1.3131324

[63] Guo, N., Zheng, Y., Jia, Y., Qiao, H., and You, H. (2012). A tunable warm-white-light Sr 3 Gd (PO 4) 3: Eu 2+, Mn 2+ phosphor system for LED-based solid-state lighting. *New Journal of Chemistry*, 36(1), 168–172. 10.1039/C1NJ20532C

[64] Lü, W., Luo, Y., Hao, Z., Zhang, X., Wang, X., and Zhang, J. (2012). A new dual-emission phosphor Ca4Si2O7F2: Ce3+, Mn2+ with energy transfer for near-UV LEDs. *Materials Letters*, 77, 45–47. 10.1016/j.matlet.2012.02.095

[65] Smet, P. F., Parmentier, A. B., and Poelman, D. (2011). Selecting conversion phosphors for white light-emitting diodes. *Journal of the Electrochemical Society*, 158(6), R37–R54 10.1149/1.3568524

[66] Amachraa, M., Wang, Z, Chen, C., Hariyani, S., Tang, H., Brgoch, J., and Ong, S. P. (2020 Jul 6). Predicting thermal quenching in inorganic phosphors. *Chemistry of Materials*, 32(14), 6256–6265. 10.1021/acs.chemmater.0c02231

[67] https://en.wikibooks.org/wiki/File:Human_photopic_response.jpg#metadata

[68] Chhajed, S., Xi, Y., Li, Y.-L., Gessmann, Th., and Schubert, E. F. (2005). Influence of junction temperature on chromaticity and colour-rendering properties of trichromatic white-light sources based on light-emitting diodes. *Journal of Applied Physics*, 97. 10.1063/1.1852073

[69] Mirhosseini, R., Schubert, M. F., Chhajed, S., Cho, J., Kim, J. K., and Schubert, E. F. (2009). Improved colour rendering and luminous efficacy in phosphor-converted white light-emitting diodes by use of dual-blue emitting active regions. *Optics Express*, 17 (13) 10805–10813. 10.1364/OE.17.010806

[70] https://upload.wikimedia.org/wikipedia/commons/thumb/8/8f/CIE_<u>1931</u>_XYZ_Color_Matching_Functions.svg/325px-CIE_1931_XYZ_Color_Matching_Functions.svg.png

[71] Ohno, Y. (2004, October). Color rendering and luminous efficacy of white LED spectra. In Fourth International Conference on Solid State Lighting (Vol. 5530, pp. 88-98). International Society for Optics and Photonics. 10.1117/12.565757

[72] Tian Y. (2014 Dec). Development of phosphors with high thermal stability and efficiency for phosphor-converted LEDs. *Journal of Solid State Lighting*, 1(1), 1–5. 10.1186/s40539-014-0011-8

[73] Dexter, D. L., and Schulman, J. H. (1954 Jun). Theory of concentration quenching in inorganic phosphors. *The Journal of Chemical Physics*, 22(6), 1063–1070. 10.1063/1.1740265

[74] Park, H. K., Oh, J. H., Do, Y. R. (2012 Apr 23). Toward scatter-free phosphors in white phosphor-converted light-emitting diodes. *Optics Express*, 20(9), 10218–10228. 10.1364%2FOE.20.010218

[75] Chen, R., and Lockwood, D. J. (2002). Developments in luminescence and display materials over the last 100 years as reflected in electrochemical society publications. *Journal of the Electrochemical Society*, 149(9), S69. 10.1149/1.1502258

[76] Blasse, G. (1967). Concentration quenching of Eu3+ fluorescence. *The Journal of Chemical Physics*, 46(7), 2583–2585. 10.1063/1.1841087

[77] Velchuri, R., Kumar, B. V., Devi, V. R., Prasad, G., Prakash, D. J., and Vithal, M. (2011). Preparation and characterization of rare earth orthoborates, LnBO3 (Ln= Tb, La, Pr, Nd, Sm, Eu, Gd, Dy, Y) and LaBO3: Gd, Tb, Eu by metathesis reaction: ESR of LaBO3: Gd and luminescence of LaBO3: Tb, Eu. *Materials Research Bulletin*, 46(8), 1219–1226. 10.1016/j.materresbull.2011.04.006

[78] Kim, K., Moon, Y. M., Choi, S., Jung, H. K., and Nahm, S. (2008). Luminescent properties of a novel green-emitting gallium borate phosphor under vacuum ultraviolet excitation. *Materials Letters*, 62(24), 3925–3927. 10.1016/j.matlet.2008.04.085

[79] Shahare, D. I., Deshmukh, B. T., Moharil, S. V., Dhopte, S. M., Muthal, P. L., and Kondawar, V. K. (1994). Synthesis of Li2B4O7: Cu phosphor. *Physica Status Solidi (A)*, 141(2), 329–334. 10.1002/pssa.2211410210

[80] Bessière, A., Jacquart, S., Priolkar, K., Lecointre, A., Viana, B., and Gourier, D. (2011). ZnGa 2 O 4: Cr 3+: a new red long-lasting phosphor with high brightness. *Optics Express*, 19(11), 10131–10137. 10.1364/OE.19.010131

[81] Huang, Y., Nakai, Y., Tsuboi, T., and Seo, H. J. (2011). The new red-emitting phosphor of oxy-fluoride Ca 2 RF 4 PO 4: Eu 3+(R = Gd, Y) for solid state lighting applications. *Optics Express*, 19(7), 6303–6311. 10.1364/OE.19.006303

[82] Yuan, S., Yang, Y., Zhang, X., Tessier, F., Cheviré, F., Adam, J. L., ... and Chen, G. (2008). Eu²⁺ and Mn²⁺ codoped Ba₂Mg (BO3)₂—new red phosphor for white LEDs. *Optics Letters*, 33(23), 2865–2867. 10.1364/OL.33.002865

[83] Bandi, V. R., Nien, Y. T., Lu, T. H., and Chen, I. G. (2009). Effect of calcination temperature and concentration on luminescence properties of novel Ca3Y2Si3O12: Eu phosphors. *Journal of the American Ceramic Society*, 92(12), 2953–2956. 10.1111/j.1551-2916.2009.03308.x

[84] Huang, C. H., and Chen, T. M. (2010). Ca 9 La (PO 4) 7: Eu 2+, Mn 2+: an emission-tunable phosphor through efficient energy transfer for white light-emitting diodes. *Optics Express*, 18(5), 5089–5099. 10.1364/OE.18.005089

[85] Kodaira, C. A., Brito, H. F., and Felinto, M. C. F. (2003). Luminescence investigation of Eu3+ ion in the RE2 (WO4) 3 matrix (RE = La and Gd) produced using the Pechini method. *Journal of Solid State Chemistry*, 171(1–2), 401–407. 10.1016/S0022-4596(02)00221-9

[86] Kuo, T. W., Huang, C. H., and Chen, T. M. (2010). Novel yellowish-orange Sr8 Al12O24S2: Eu2+ phosphor for application in blue light-emitting diode based white LED. *Optics Express*, 18(102), A231–A236. 10.1364/OE.18.00A231

[87] Duan, C. J., Delsing, A. C. A., and Hintzen, H. T. (2009). Photoluminescence properties of novel red-emitting Mn2+-activated MZnOS (M-= Ca, Ba) phosphors. *Chemistry of Materials*, 21(6), 1010–1016. 10.1021/cm801990r

[88] Ando, M., and Ono, Y. A. (1990). Role of Eu2+ luminescent centers in the electro-optical

characteristics of red-emitting CaS: Eu thin-film electroluminescent devices with memory. *Journal of Applied Physics*, 68(7), 3578–3583. 10.1063/1.346317#

[89] Kumar, V., Pitale, S. S., Mishra, V., Nagpure, I. M., Biggs, M. M., Ntwaeaborwa, O. M., and Swart, H. C. (2010). Luminescence investigations of Ce^{3+} doped CaS nanophosphors. *Journal of Alloys and Compounds*, 492(1–2), L8–L12. 10.1016/j.jallcom.2009.11.076

[90] Xie, R. J., Hirosaki, N., Mitomo, M., Takahashi, K., and Sakuma, K. (2006). Highly efficient white-light-emitting diodes fabricated with short-wavelength yellow oxynitride phosphors. *Applied Physics Letters*, 88(10), 101104. 10.1063/1.2182067

[91] Liu, T. C., Cheng, B. M., Hu, S. F., and Liu, R. S. (2011). Highly stable red oxynitride β-SiAlON: Pr3+ phosphor for light-emitting diodes. *Chemistry of Materials*, 23(16), 3698–3705. 10.1021/cm201289s

[92] Hecht, C., Stadler, F., Schmidt, P. J., auf der Günne, J. S., Baumann, V., and Schnick, W. (2009). SrAlSi4N7: Eu2+– A nitridoalumosilicate phosphor for warm white light (pc) LEDs with edge-sharing tetrahedra. *Chemistry of Materials*, 21(8), 1595–1601. 10.1021/cm803231h

[93] Wu, L., Sun, J. C., Zhang, Y., Jin, S. F., Kong, Y. F., and Xu, J. J. (2010). Structure determination and relative properties of novel chiral orthoborate KMgBO3. *Inorganic Chemistry*, 49(6), 2715–2720. 10.1021/ic901963t

[94] Wu, L., Zhang, Y., Kong, Y. F., Sun, T. Q., Xu, J. J., and Chen, X. L. (2007). Structure determination of novel orthoborate NaMgBO3: a promising birefringent crystal. *Inorganic Chemistry*, 46(13), 5207–5211. 10.1021/ic062429i

[95] Wu, L., Zhang, Y., Chen, X. L., Kong, Y. F., Sun, T. Q., Xu, J. J., and Xu, Y. P. (2007). The Na2O–SrO–B2O3 diagram in the B-rich part and the crystal structure of NaSrB5O9. *Journal of Solid State Chemistry*, 180(4), 1470–1475. 10.1016/j.jssc.2007.02.014

[96] Wu, L., Chen, X. L., Zhang, Y., Kong, Y. F., Xu, J. J., and Xu, Y. P. (2006). Ab initio structure determination of novel borate NaSrBO3. *Journal of Solid State Chemistry*, 179(4), 1219–1224. 10.1016/j.jssc.2006.01.003

[97] Wu, L., Chen, X. L., Li, H., He, M., Xu, Y. P., and Li, X. Z. (2005). Structure Determination and Relative Properties of Novel Cubic Borates MM '4 (BO3) 3 (M = Li, M '= Sr; M = Na, M '= Sr, Ba). *Inorganic Chemistry*, 44(18), 6409–6414. 10.1021/ic050299s

[98] Yuan, S., Yang, Y., Zhang, X., Tessier, F., Cheviré, F., Adam, J. L., ... and Chen, G. (2008). Eu 2+ and Mn 2+ codoped Ba 2 Mg (BO 3) 2—new red phosphor for white LEDs. *Optics Letters*, 33(23), 2865–2867. 10.1364/OL.33.002865

[99] Liu, W. R., Lin, C. C., Chiu, Y. C., Yeh, Y. T., Jang, S. M., and Liu, R. S. (2010). ZnB 2 O 4: Bi 3+, Eu 3+: a highly efficient, red-emitting phosphor. *Optics Express*, 18(3), 2946–2951. 10.1364/OE.18.002946

[100] Zhang, Y., Wu, L., Ji, M., Wang, B., Kong, Y., and Xu, J. (2012). Structure and photoluminescence properties of KSr 4 (BO 3) 3: Eu 3+ red-emitting phosphor. *Optical Materials Express*, 2(1), 92–102. 10.1364/OME.2.000092

[101] Yang, F., Liang, Y., Liu, M., Li, X., Zhang, M., and Wang, N. (2013). Photoluminescence properties of novel red-emitting NaSrBO3: Eu3+ phosphor for near-UV light-emitting diodes. *Optics & Laser Technology*, 46, 14–19. 10.1016/j.optlastec.2012.04.015

[102] Chikte, D., Omanwar, S. K., and Moharil, S. V. (2013). Luminescence properties of red emitting phosphor NaSrBO3: Eu3+ prepared with novel combustion synthesis method. *Journal of Luminescence*, 142, 180–183. 10.1016/j.jlumin.2013.03.045

[103] Li, J., Yan, H., and Yan, F. (2016). A novel orange-red emitting borate based phosphor for NUV pumped LEDs. *Optik*, 127(15), 5984–5989. 10.1016/j.ijleo.2016.04.022

[104] Bajaj, N. S., Koparkar, K. A., Nagpure, P. A., and Omanwar, S. K. (2017). Red and blue emitting borate phosphor excited by near Ultraviolet Light. *Journal of Optics*, 46(2), 91–94. 10.1007/s12596-016-0344-3

[105] Yuan, S., Yang, Y., Zhang, X., Tessier, F., Cheviré, F., Adam, J. L., ... and Chen, G. (2008). Eu 2+ and Mn 2+ codoped Ba 2 Mg (BO 3) 2—new red phosphor for white LEDs. *Optics Letters*, 33(23), 2865–2867. 10.1364/OL.33.002865

[106] Majling, J., Figusch, V., Čorba, J., and Hanic, F. (1974). Crystal data on calcium borate chloride, Ca2BO3Cl. *Journal of Applied Crystallography*, 7(3), 402-402. 10.1107/S0021889874009927

[107] Chikte, D., Omanwar, S. K., and Moharil, S. V. (2014). Yellow emitting phosphor for solid state lighting Ca2BO3Cl:Eu2+. *International Journal of Basic and Applied Research-Special issue*, 4161–4164.

[108] Song, W. S., Kim, Y. S., and Yang, H. (2009). Yellow-emitting phosphor of Sr3B2O6: Eu2+ for application to white light-emitting diodes. *Materials Chemistry and Physics*, 117(2-3), 500–503. 10.1016/j.matchemphys.2009.06.042

[109] Xue, Y., Xu, X., Hu, L., Fan, Y., Li, X., Li, J., ... and Tang, C. (2011). Synthesis and photo-luminescence characteristics of (Sr, Ca) 3B2O6: Eu for application in white light-emitting diodes. *Journal of Luminescence*, 131(9), 2016–2020. 10.1016/j.jlumin.2011.04.032

[110] Wang, R., Xu, J., and Chen, C. (2012). Luminescent characteristics of Sr2B2O5: Tb3+, Li+ green phosphor. *Materials Letters*, 68, 307–309. 10.1016/j.matlet.2011.10.005

[111] Zhijun, W. A. N. G., Zhiping, Y., Panlai, L. I., Qinglin, G., and Yanmin, Y. A. N. G. (2010). Luminescence characteristics of LiCaBO3: Tb3+ phosphor for white LEDs. *Journal of Rare Earths*, 28(1), 30–33. 10.1016/S1002-0721(09)60044-2

[112] Cai, G. M., Sun, Y., Li, H. K., Fan, X., Chen, X. L., Zheng, F., and Jin, Z. P. (2011). Crystal structure and photoluminescence of Tb3+-activated Ba3InB3O9. *Materials Chemistry and Physics*, 129(3), 761–768. 10.1016/j.matchemphys.2011.04.071

[113] Geng, W., Zhou, X., Ding, J., and Wang, Y. (2018). NaBaY (BO3) 2: Ce3+, Tb3+: A novel sharp green-emitting phosphor used for WLED and FEDs. *Journal of the American Ceramic Society*, 101(10), 4560–4571. 10.1111/jace.15693

[114] Guo, C., Jing, H., and Li, T. (2012). Green-emitting phosphor Na 2 Gd 2 B 2 O 7: Ce 3+, Tb 3+ for near-UV LEDs. *RSC Advances*, 2(5), 2119–2122. 10.1039/C2RA00808D

[115] Panse, V. R., Kokode, N. S., Yerpude, A. N., and Dhoble, S. J. (2016). Combustion synthesis of Sr2B2O5: Tb3+ green emitting phosphor for solid state lighting. *Optik*, 127(4), 1603–1606. 10.101 6/j.ijleo.2015.11.030

[116] Singh, V., Shinde, K. N., Singh, N., Pathak, M. S., Singh, P. K., and Dubey, V. (2018). Green emitting Tb doped LiBaB9O15 phosphors. *Optik*, 156, 677–683. 10.1016/j.ijleo.2017.11.145

[117] Zhuo, Y., Zhong, J., and Brgoch, J. (2019). Controlling Eu2+ substitution towards a narrow-band green-emitting borate phosphor NaBaB9O15: Eu2+. 10.26434/chemrxiv.10110773.v1 1-13

[118] Zhen, L. I., Zheng, S., Xing, Y. A. N. G., Bin, L. U. O., Shuai, L. I., Yong, S. U. N., and Cheng, Q. (2017). Luminescence properties of Sr2MgB2O6: Tb3+, Li+ green-emitting phosphor. *Journal of Rare Earths*, 35(3), 211–216. 10.1016/S1002-0721(17)60901-3

[119] Liu, W. R., Huang, C. H., Wu, C. P., Chiu, Y. C., Yeh, Y. T., and Chen, T. M. (2011). High efficiency and high colour purity blue-emitting NaSrBO 3: Ce 3+ phosphor for near-UV light-emitting diodes. *Journal of Materials Chemistry*, 21(19), 6869–6874. 10.1039/C1JM10765H

[120] Li, X., Liu, C., Guan, L., Wei, W., Yang, Z., Guo, Q., and Fu, G. (2012). An ideal blue Sr3B2O6: Ce3+ phosphor prepared by sol-combustion method. *Materials Letters*, 87, 121–123. 10.1016/ j.matlet.2012.07.094

[121] Reddy, A. A., Das, S., Ahmad, S., Babu, S. S., Ferreira, J. M., and Prakash, G. V. (2012). Influence of the annealing temperatures on the photoluminescence of KCaBO3: Eu3+ phosphor. *RSC Advances*, 2(23), 8768–8776. 10.1039/C2RA20866K

[122] Chikte, D., Omanwar, S. K., and Moharil, S. V.(2014) Blue Emitting Phosphor for Solid State Lighting Application by combustion synthesis: KCaBO3:Ce3+. *International Journal of Researches in Biosciences, Agriculture & Technology*, 1(2), 173–183.

[123] Nakamura, S., Inabe, K., and Takeuchi, N. (1992). Luminescence of sintered Ca2B5O9Cl: Eu2+ excited by UV-light or X-rays. *Japanese Journal of Applied Physics*, 31(6R), 1823. 10.1143/jjap.31.1823

[124] Guo, C., Xu, Y., Ding, X., Li, M., Yu, J., Ren, Z., and Bai, J. (2011). Blue-emitting phosphor M2B5O9Cl: Eu2+ (MSr, Ca) for white LEDs. *Journal of Alloys and Compounds*, 509(4), L38–L41. 10.1016/j.jallcom.2010.10.032

[125] Chikte, D., Omanwar, S. K., and Moharil, S. V. (2018) Synthesis and photoluminescence properties of Blue emitting Europium doped calcium chloro-borate phosphor suitable for solid state lighting application. *International Journal of Review of Research*, 10, 10–14.

[126] Wang, Q., Deng, D., Hua, Y., Huang, L., Wang, H., Zhao, S., ... and Xu, S. (2012). Potential tunable white-emitting phosphor LiSr4 (BO3) 3: Ce3+, Eu2+ for ultraviolet light-emitting diodes. *Journal of Luminescence*, 132(2), 434–438. 10.1016/j.jlumin.2011.09.003

[127] Panlai, L., Zhijun, W., Zhiping, Y., and Qinglin, G. (2010). A potential single-phased white-emitting LiBaBO3: Ce3+, Eu2+ phosphor for white LEDs. *Journal of Rare Earths*, 28(4), 523–525. 10.1016/ S1002-0721(09)60145-9

[128] Das, S., Reddy, A. A., Babu, S. S., and Prakash, G. V. (2011). Controllable white light emission from Dy 3+–Eu 3+ co-doped KCaBO 3 phosphor. *Journal of Materials Science*, 46(24), 7770–7775. 10.1007/s10853-011-5756-5

[129] Li, P., Yang, Z., Wang, Z., and Guo, Q. (2008). White-light-emitting diodes of UV-based Sr3Y2 (BO3) 4: Dy3+ and luminescent properties. *Materials Letters*, 62(10-11), 1455–1457. 10.1016/ j.matlet.2007.08.085

[130] Zhang, Q. Y., Yang, C. H., and Pan, Y. X. (2007). Enhanced white light emission from GdAl3 (BO3) 4: Dy3+, Ce3+ nanorods. *Nanotechnology*, 18(14), 145602. 10.1088/0957-4484/18/14/145602

[131] Chang, C. K., and Chen, T. M. (2007). Sr 3 B 2 O 6: Ce 3+, Eu 2+: a potential single-phased white-emitting borate phosphor for ultraviolet light-emitting diodes. *Applied Physics Letters*, 91(8), 081902. 10.1063/1.2772195

[132] Guo, C., Luan, L., Xu, Y., Gao, F., and Liang, L. (2008). White light–generation phosphor Ba2Ca (BO3) 2: Ce3+, Mn2+ for light-emitting diodes. *Journal of the Electrochemical Society*, 155(11), J310. 10.1149/1.2976215

[133] Xiao, F., Xue, Y. N., Ma, Y. Y., and Zhang, Q. Y. (2010). Ba2Ca (B3O6) 2: Eu2+, Mn2+: A potential tunable blue–white–red phosphors for white light-emitting diodes. *Physica B: Condensed Matter*, 405(3), 891–895. 10.1016/j.physb.2009.10.009

[134] Guo, C., Luan, L., Shi, F. G., and Ding, X. (2009). White-emitting phosphor Ca2BO3Cl: Ce3+, Eu2+ for UV light-emitting diodes. *Journal of the Electrochemical Society*, 156(6), J125. 10.1149/1.3106039

[135] Xiao, F., Xue, Y. N., and Zhang, Q. Y. (2009). Ca2BO3Cl: Ce3+, Eu2+: a potential tunable yellow–white–blue-emitting phosphors for white light-emitting diodes. *Physica B: Condensed Matter*, 404(20), 3743–3747. 10.1016/j.physb.2009.06.122

[136] Li, B., Huang, X., and Lin, J. (2018). Single-phased white-emitting Ca3Y (GaO) 3 (BO3) 4: Ce3+, Tb3+, Sm3+ phosphors with high-efficiency: photoluminescence, energy transfer and application in near-UV-pumped white LEDs. *Journal of Luminescence*, 204, 410–418. 10.1016/j.jlumin.2018.08.044

[137] Cai, G. M., Yang, N., Liu, H. X., Si, J. Y., and Zhang, Y. Q. (2017). Single-phased and colour tunable LiSrBO3: Dy3+, Tm3+, Eu3+ phosphors for white-light-emitting application. *Journal of Luminescence*, 187, 211–220. 10.1016/j.jlumin.2017.03.017

[138] Gaffuri, P., Salaün, M., Gautier-Luneau, I., Chadeyron, G., Potdevin, A., Rapenne, L., ... and Ibanez, A. (2020). Rare-earth-free zinc aluminium borate white phosphors for LED lighting. *Journal of Materials Chemistry C*, 8(34), 11839–11849. 10.1039/D0TC02196B

[139] https://www.gtvinc.com/latest-led-technology-trends-in-led-industry/

[140] https://www.earthled.com/blogs/light-2-0-the-earthled-blog-led-lighting-news-tips-reviews/ 37176324-how-to-choose-the-best-led-light-bulb-for-any-room-in-your-home

[141] https://www.standardpro.com/3-basic-types-of-lighting/

[142] https://www.hgtv.com/design/remodel/mechanical-systems/lighting-tips-for-every-room

[143] https://www.havells.com/en/consumer/lighting.html#gref

[144] https://5.imimg.com/data5/SELLER/Default/2020/11/FI/EN/ON/8313991/wipro-wifi-enabled-smart-led-bulb-b22-9-watt-125x125.jpg

[145] https://www.electrical4u.com/road-lighting-design/

[146] https://www.wiprolighting.com/products/outdoor/streetlights

[147] https://www.ledil.com/product-landing/

[148] https://innovationtoronto.com/<u>2017/03</u>/flexible-led-displays-containing-printed-electronics-ready-for-roll-to-roll-manufacturing/

[149] https://www.persistencemarketresearch.com/market-research/led-lighting-market.asp#:~:text=At %20present%2C%20LED%20based%20automotive,LEDs%20nearly%20achieve%20daylight %20quality

[150] https://www.ledsmagazine.com/manufacturing-services-testing/assembly-contract-manufacturing/ article/16695370/implementing-led-flash-in-camera-phones

[151] https://www.osram.com/am/light-for/led-lighting/index.jsp

[152] https://soundcertified.com/how-to-install-led-light-strips-car/

[153] Graydon, O., Jenkins, A., Pei Chin Won, R., and Gevaux, D. (2007). Haitz's law. *Nature Photonics*, 1(1). 10.1038/nphoton.2006.78

[154] https://www.lifitn.com/im-new

[155] https://www.techadvisor.com/feature/small-business/what-is-li-fi-everything-you-need-know-3788560/#:~:text=With%20Li%2DFi%2C%20your%20light,a%20TED%20Talk%20in%202011

[156] https://www.communicationstoday.co.in/two-gujarat-villages-first-in-india-to-get-lifi/

[157] https://www.theglobeandmail.com/drive/culture/article-how-to-automatic-high-beam-systems-work/

[158] https://medium.com/acmvit/li-fi-the-future-of-internet-e573eab6bd0d

[159] https://www.ledsmagazine.com/architectural-lighting/retail-hospitality/article/16696265/benefits-and-drawbacks-of-leds

5 Borate Phosphor for Phototherapy Application

A. O. Chauhan

CONTENTS

DOI: 10.1201/9781003207757-5

5.1   ULTRAVIOLET RADIATIONS

Sunlight is composed of a continuous spectrum of electromagnetic radiation that is divided into three main regions of wavelengths: ultraviolet, visible and infrared, which is diagrammatically represented in Figure 5.1. Out of the total electromagnetic spectrum, ultraviolet radiation covers a small part, nearly about 10%. The word 'ultraviolet' means "beyond violet". The violet color light has the highest frequencies in the visible light spectrum. Ultraviolet radiations possess a higher frequency than violet light. It is shorter than that of visible light but longer than X-rays [1].

5.1.1   Types of UV Radiation

According to biological and physical aspects, the Commission Internationale de l'Eclairage (CIE) has divided the ultraviolet radiation into three sections:

- UVA – (320–400 nm) or long wave UV is responsible for skin tanning effects,
- UVB – (280–320 nm) is known as the sunburn (erythemal) region,
- UVC – (200–280 nm) is the germicidal region

The UVA region is further subdivided into UVA1 (340–400 nm) and UVA2 (320–340 nm). Among these wavelengths, UVA is less strongly absorbed in skin than is UVB. UVC is potentially carcinogenic because DNA absorbs it, while narrow-band UVB (NB-UVB) (311–315 nm) was demonstrated to be effective in the treatment of skin diseases and disorders. The selection of the proper wavelength emitting phosphor is of prime importance, which is based on" action spectra and absorption spectra".

Ultraviolet (UV) light is employed in a wide range of applications as diverse as biochemical analysis in laboratories, curing of epoxies in industry, detection of watermarks used for countering money forging, medical treatment of some skin conditions, water purification and generation of white light by phosphor excitation, as in fluorescent tubes.

Our prime natural source for UV radiation is sun. Artificially it is produced by electric arcs and specialized lights such as mercury-vapor lamps, germicidal lamps, tanning lamps, black lights and some types of lasers (nitrogen lasers, excimer lasers and third harmonic Nd:YAG lasers). Unique hazards apply to the different sources depending on the wavelength range of the emitted UV radiation [2,3].

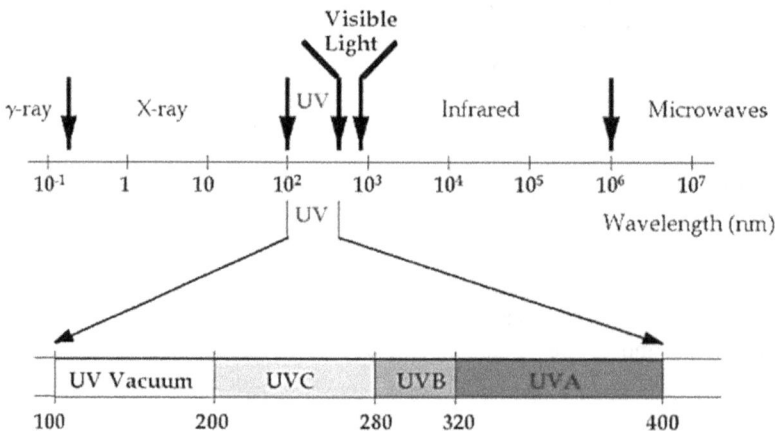

FIGURE 5.1   Ultraviolet component of the electromagnetic spectrum.

5.1.2 Quantities and Units

Quantities of UV radiation are uttered using radiometric terminology (Table 5.1). The terms relating to a beam of radiation passing through space are radiant energy and radiant flux. Terms relating to a source of radiation are radiation intensity and radiance. The term 'irradiance', which is the most commonly used term in photobiology, relates to the object (e.g. patient) struck by the radiation. The time integral of the irradiance is strictly termed the 'radiant exposure', but is sometimes expressed as exposure dose. The term 'dose' in photobiology is analogous to the term 'energy influence' in radiobiology and not to absorbed dose. Although radiometric terminology is widely used in photobiology, the units chosen vary throughout the literature. For example, exposure doses may be quoted in $mJ\text{-}cm^{-2}$ or $kJ\text{-}m^{-2}$ [3].

5.1.3 Radiometric Calculations

The most frequent radiometric calculation is to determine the time for which a patient (or other object), who is prescribed a certain dose (in $J\ cm^{-2}$), should be exposed when the radiometer indicates irradiance in $mW\ cm^{-2}$. The relationship between these three quantities (time, dose and irradiance) is simply exposure time (min):

$$= \frac{1000 \times \text{prescribed dose (J cm}^{-2})}{60 \times \text{measured irradiance (mW cm}^{-2})}$$

5.1.4 Standard Erythema Dose (SED)

The problem of dosimetry in photodermatology lies in the fact that the ability of UV radiation to elicit erythema in human skin depends strongly on wavelength, encompassing a range of four orders of magnitude between 250 and 400 nm. Thus, a statement that a subject received an exposure dose of $1\ J\ cm^{-2}$ of UV radiation conveys nothing about the consequences of that exposure in terms of erythema. If the radiation source was a UVA fluorescent lamp, no erythemal response would be seen apart from people exhibiting severe, abnormal pathological photosensitivity. The same dose delivered from an unfiltered mercury arc lamp or fluorescent sun-lamp would result in marked violaceous erythema in most white-skinned individuals.

Consequently, photobiologists have long recognized the need to express the exposure as an erythemally weighted quantity [4]. A CIE erythema action spectrum [5] was proposed in 1987 but no erythemal quantity and radiometric equivalence was agreed till up to now. It has been common practice for many years to use the term 'minimal erythema dose (MED)' as a "measure" of erythemal radiation. MED is not a standard measure of anything but, encompasses the variable nature of

TABLE 5.1
Radiometric terms and units

Term	Unit	Symbol
Wavelength	nm	λ
Radiant energy	J	Q
Radiant flux	W	φ
Radiant intensity	$W\ sr^{-1}$	I
Radiance	$W\ m^{-2}\ sr^{-1}$	L
Irradiance	$W\ m^{-2}$	E
Radiant exposure	$J\ m^{-2}$	H

individual sensitivity to UV radiation. To avoid further confusing abuse of the term 'MED', it has been proposed that this term be reserved solely for observational studies in humans and other animals [6]. The term 'standard erythema dose (SED)' is used to refer to erythemal effective radiant exposures from natural and artificial sources of UV radiation. One SED is equivalent to an erythemal effective radiant exposure of 100 J m^{-2} [7]. There is an international recommendation to use the so-called CIE action spectrum, McKinlay and Diffey (1987), to mimic the erythemal effect of UV radiation. The unit of SED sometimes referred to as dose, Whm^{-2} or J m^{-2} [3].

5.1.5   ERYTHEMAL EFFECTIVE IRRADIANCE

The erythemal effective irradiance (E_{er}) from a source of ultraviolet radiation is obtained by weighting the spectral irradiance of the radiation at wavelength λ in nm by the effectiveness of radiation of this wavelength to cause a minimal erythema and summing over all wavelengths present in the source spectrum. This can be expressed mathematically as

$$E_{er} = \int E_\lambda S_{er}(\lambda) d\lambda$$

E_λ is the spectral irradiance in Wm^{-2}-nm^{-1} at wavelength λ in nm and $d\lambda$ is the wavelength interval used in the summation. $S_{er}(\lambda)$ is a measure of the effectiveness of radiation of wavelength λ (in nm) relative to some reference wavelength in producing a minimal erythema. Integration has to be carried out in the wavelength range where neither E_λ nor $S_{er}(\lambda)$ equal zero. The effective irradiance is equivalent to a hypothetical irradiance of monochromatic radiation having a wavelength at which $S_{er}(\lambda)$ is equal to unity. The time integral of effective irradiance is the erythemal effective radiant exposure (also called the effective dose or erythemal dose). The dose (expressed as an erythemal quantity) received after an exposure period of t seconds is

$$H_{er} = E_{er} \cdot t/\phi$$

where ϕ is the numerical value in J m^{-2} equivalent to one erythemal quantity [3,7].

5.1.6   DEFINED ACTION SPECTRA IN THE SKIN

Action spectra within the skin will vary according to the absorption spectra of the chromophores they target. If action spectra are similar for different variables, e.g. erythema and DNA damage, this would imply common chromophores. The degree of erythema induced by UV is affected by individual susceptibility, site of irradiation, skin thickness, photo-adaptation of the skin due to previous UV exposure and dose of UV administered [8]. A number of erythema action spectra have been published, but only one has been endorsed as internationally recognized by the CIE, this was originally described by McKinlay and Diffey in 1987 and later updated (for a review of the erythema action spectra see Webb et al., 2011) [9,10]. This action spectrum also helps to calculate the UV index for public health information in weather forecasting, and to calibrate instruments to measure erythemally weighted UV which in turn will determine the accuracy of the dose of UV administered to patients. Figure 5.2 shows the erythema action spectrum peaks at 290–300 nm, falls off quickly between 300 nm and 325 nm, and thereafter declines steadily. On the other hand, it is unclear which layers of the epidermis are responsible for the release of erythema inducing cytokines, although shorter wavelengths of UV (e.g. 290 nm) can clearly penetrate the epidermis sufficiently deeply to induce these biological changes. However, although wavelengths between 250 nm and 300 nm have a similar erythemal response at low doses, there appears to be very different erythemal effect at higher doses, with wavelengths around 300 nm having a much steeper erythemal response than wavelengths of 254 nm and 280 nm, suggesting that different cytokine mediators may be involved [3,11,12].

Moreover, to optimize the efficacy of phototherapy for psoriasis it is important to know which wavelengths are the most effective in plaque clearance. An action spectrum for psoriasis was established by Parrish, as shown in Figure 5.3. This curve for psoriasis was accepted as having its peak at 311 nm (solid line) with the dotted line showing a steep decline in therapeutic effective at wavelengths $\leq$290 nm; the erythema action spectrum extends to 250 nm. This clearly indicates the utility of a light source in the NB-UVB range, paving way for phototherapy.

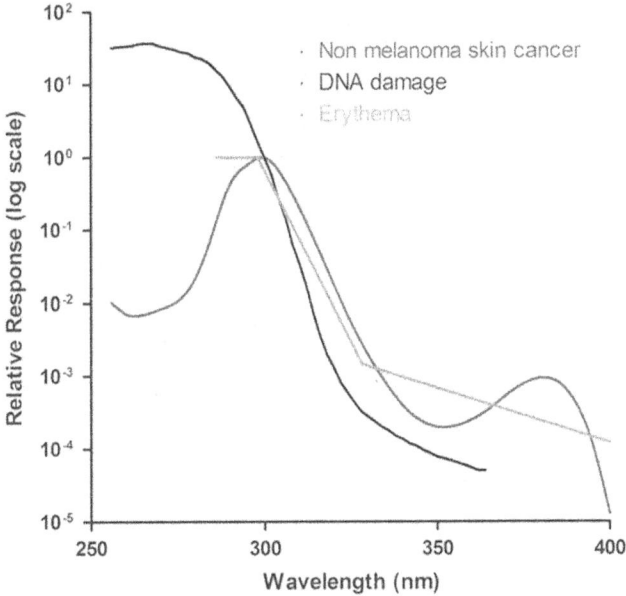

FIGURE 5.2   Recognized action spectra within the skin. Three action spectra are shown: non-melanoma skin cancer, DNA damage and erythema.

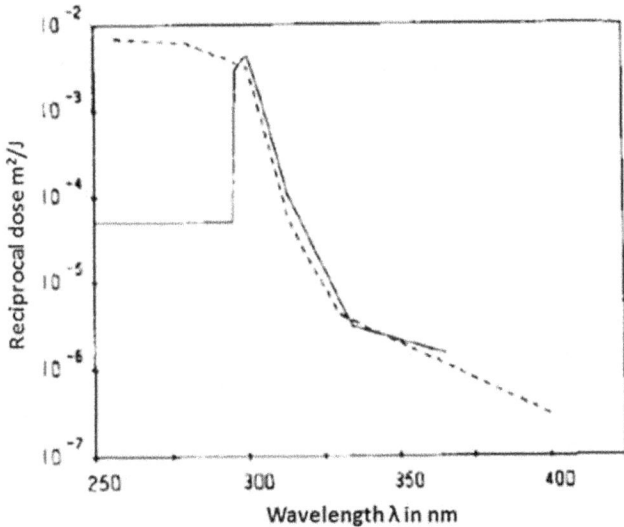

FIGURE 5.3   Action spectra for the lowest effective daily dose for clearing psoriasis (Parrish, 1982) and for minimal erythema at 24 h [8].

5.2   PHOTOTHERAPY AND ITS MECHANISM

The phototherapy (light therapy) is nothing but the direct exposure of light (UV and visible light) of specific frequencies with tissues to cause a change in immune response.

The phototherapy is the most effective and easiest route for the treatment of many skin disorders such as psoriasis, vitiligo, lupus erythematosus, etc [13–17]. This treatment is decided on the basis of physical, photo-biological and photochemical activities carried out at a micro-scale. Obviously, optical radiation can only be effective if absorbed by so called chromophores (*biological molecules that absorb UV radiation*).

In phototherapy the use of artificial UVB delivered by fluorescent lamps without the addition of exogenous photosensitizers. The given radiation is absorbed by this endogenous chromophore. After the appropriate wavelength's irradiance, the chromophore is transformed into a more energetic electronic state, either the unstable singlet or the more persistent triplet state. Although the duration of the singlet state is extremely small, the triplet state may persist for a few seconds, allowing for photochemical reactions to take place. Photochemical reactions between these absorbing biomolecules result in alternations in skin biology and subsequent thermal chemistry with the production of mediators and observable biological responses [18]. According to literature, nuclear DNA is the most excellent characterized chromophore for UVB. Absorption of UV by nucleotides leads to the formation of DNA photoproducts, mainly pyrimidine dimers, which are biologically more significant. It is well known that the psoriatic epidermal cells can synthesized after reducing the DNA using UVB exposure. UVB exposure up-regulates the tumor suppressor gene product P53, which is involved in control of the cell-cycle gene and thus important for reversal of the shortened cell cycle in keratinocytes in psoriatic epidermis. In psoriasis, both epidermal keratinocytes and cutaneous lymphocytes may be targeted by UVB. Immune suppression, alteration of cytokine expression and cell-cycle arrest may contribute to the suppression of disease activity in psoriatic lesions [3,19].

5.2.1   History of Phototherapy

Phototherapy (*light therapy*) is also referred as heliotherapy (*using sunlight rich in UVB at the Dead Sea*) was popular and widely practiced in ancient times. Indian medical literature dating to 1500 BCE describes a treatment combining herbs with natural sunlight to treat non-pigmented skin areas [20]. The similar references were built in Buddhist literature from about 200 CE and 10th-century Chinese documents. The sacred book Atharva-Veda (2000–1400 BC) describes the treatment of leukoderma (vitiligo) with the plants Vasuchika or Bavachee (Psoraliacotylifolia), and the need for exposure to solar radiation in the form of sun-worshipping prayer is stressed. Psoraliacotylifolia has been shown to contain Psoralen and several other furocoumarins [19].

In 1901, the Danish physician Niels Finsen (the father of modern phototherapy) developed the first artificial light source for this purpose of phototherapy. He used filtered carbon arc source for the UVR treatment of skin tuberculosis, erysipelas and rickets. Finsen received the Noble Prize in 1903 in recognition of his treatment of lupus vulgaris using a chemical ray lamp [21].

In 1923, William Henry Gorckerman established the use of artificial sourceof broad band UVB from a high-pressure mercury lamp plus topical coal tar for the treatment of psoriasis [21]. In 1947, in Egypt, E. L. Mofty et al. observed regimentation over several months in vitiligo sufferers exposed to sunshine after eating seeds of the Nile plant Ammimajus [19,22], which contains 8-methoxypsoralen (8-MOP). El Mofty also successfully treated vitiligo using sunlight and 8-MOP or 5-MOP as a sensitizer.

Subsequently, Parrish et al. introduced the use of oral psoralen with artificially source of UVR in the treatment of psoriasis, which is well known as photochemotherapy [23], later abbreviated to PUVA. PUVA is the combined treatment of skin disorders with a photosensitizing drug (psoralen) and UVA.

At the beginning the results were not especially encouraging due to the relatively low output of the UVA lamps available at that time. In 1960s, at Hamburg, Germany, Wiskemann manufactured the phototherapy unit using Osram ULTRA-VITALUX lamps with fluorescent tubes.

The first use of oral 8-MOP was reported in 1967 for the treatment of psoriasis and a new era of 8-MOP use in combination with UVA in the treatment of psoriasis was reported by Pathak, Fitzpatrick, Tanenbaum and Parrish. In 1974, a group from Harvard Medical School published a seminal study [23] that changed the practice of dermatology worldwide and was made possible by the availability of fluorescent lamps emitting UVA radiation at sufficient intensity to allow effective treatment over practicable irradiation times.

5.2.2 Need of Phototherapy

Worldwide millions of people suffer from the symptoms of skin diseases like psoriasis, neurodermitis, vitiligo, palmoplantarpustolosis, mycosis fungoides, atopic dermatitis, generalised lichen planus, urticariapigmentosa, alpoeciaareata, pityriasisrubrapilaris, prurigonodularis, scleromyxedema, etc.

There are wide ranges of options available for the treatment, such as skin creams or ointments, corticosteroids, antibiotics, antihistamines, drugs that suppress the immune system, protection from allergens, light therapy, a mix of light therapy and a drug called psoralen. All these treatments have either the side effects of medicine or adverse results in some cases. Light phototherapy is advocated to be beneficial and also found to be more effective with simplest form of the treatment for the skin diseases. Also, it is inexpensive and easy to use [3].

In case of neonatal hyperbilirubinemia, the phototherapy with white (day-light) or monochromatic (blue light) lamps is widely used for the prevention and treatment. The breakdown products of bilirubin are not toxic for the CNS and are rapidly eliminated through the kidneys and liver (Table 5.2).

5.2.3 Types of Phototherapy

The ability of UVR to affect the skin's immune system was first recognized in the early 1970s. It is now accepted that UVA, UVB and PUVA therapy exert a variety of immunomodulatory effects on human skin, and that this is of critical importance for the therapeutic efficacy of phototherapy. The actual therapeutic relevance of these effects is mainly determined by the depth of penetration of the type of UV radiation employed. Figure 5.4 shows the skin penetration of UV radiations. Most of the UVC rays are absorbed by the ozone layer and therefore they do not reach to the earth's surface. However, UVA and UVB rays pass through the ozone layer and come at the earth surface. UVB mainly affects epidermal keratinocytes and Langerhans cells, whereas UVA penetrates more deeply into the dermis, thereby affecting dermal fibroblasts, dermal dendritic cells, endothelial cells and skin – infiltrating inflammatory cells such as T lymphocytes, mast cells and granulocytes. Many of these effects have been identified using animal models or through in vitro studies employing cultured human skin cells [24].

There are many systemic and topical therapies are available for the treatment of skin disorders, as detailed below.

5.2.3.1 PUVA (Psoralen and Ultraviolet Light – A) Phototherapy

Photochemotherapy (PUVA) is a kind of ultraviolet radiation treatment used for severe skin diseases. The skin disorders such as psoriasis, vitiligo, polymorphic light eruption, cutaneous T-cell lymphoma and eczema (dermatitis) can be treated by using PUVA therapy. In this treatment combination of photosensitizing chemical compound (*Psoralen*) and long-wave ultraviolet radiation (UVA) are used to exposing the skin [13]. The spectral distribution of commercial Phillips TL/09 lamps is shown in Figure 5.5.

Psoralens are photosensitizing compounds found in five major plant families used in India and Egypt for thousands of years for vitiligo [25]. Bergapten (5-methoxypsoralen, 5-MOP), independently isolated from bergamot oil in 1834, was also recognized as a photosensitizer in 1916. Photochemotherapy is most widely under taken with 8-MOP, although 5-MOP or trimethyl psoralen (TMP), a synthetic compound, are occasionally preferred. Psoralens are usually administered orally, although topical application is occasionally preferred.

TABLE 5.2
Phototherapy by disease type [24]

Sr. no.	Disease	UV therapy
1	Psoriasis	PUVA, bath-PUVA, NB-UVB (311 nm), combined bath-PUVA and NB-UVB (311 nm), 308 nm excimer laser (targeted phototherapy)
2	Plaque parapsoriasis	PUVA (balneophototherapy), climatotherapy
3	Vitiligo	PUVA, PUVB (broad band), NB-UVB (311 nm), 308 nm excimer laser, punch grafting with NB-UVB (311 nm), climatotherapy
4	Morphea (scleroderma)	Bath-PUVA, UVA1
5	Lichen sclerosus	Selective UVB (SUP), bath-PUVA
6	Polymorphic light eruption	PUVA, NB-UVB (311 nm) (for testing)
7	Pruritus in polycythemia vera	NB-UVB
8	Actinic prurigo	NB-UVB, PUVA and broad band UVB
9	Solar urticaria	Phototesting with NB-UVB, BB-UVB, UVA alone, combination of UVA, UVB and visiblelight, photopheresis
10	Seborrheic dermatitis	SUP, NB-UVB (311 nm)
11	Graft versus host disease	Bath-PUVA
12	Atopic dermatitis	UVA1, NB-UVB (311 nm), UVB phototherapy and PUVA
13	Contact dermatitis	Bath - PUVA
14	Chronic actinic dermatitis	UVA1 (340–400 nm), NB-UVB (311 nm)
15	Mycosis fungoides	PUVA, bexarotene and NB-UVB
16	Necrobiosislipoidica	PUVA, UVA1 (340–400 nm)
17	Erythropoietin protoporphyria	Phototesting using UVB and UVA, NB-UVB (311 nm)
18	Eosinophilicpustular folliculitis (Ofuji's disease)	UVB
19	Lupuserythematous	UVA, UVB
20	Skin conditioning	Erb-YAG laser
21	Acne vulgaris	UVA, blue light spectrum (400–420 nm), blue and red spectrum (660 nm), yellow and green spectrum (500–600 nm), diode laser (1,450 nm)
22	Hyperbilirubinemia	Philips TL20W03T, Westinghouse F12 T20/BB, Philips TL20W/52, Sylvania F20F12/G, Sylvania F20F12/DL

NB = narrow band; PUVA = Psoralen combined with UVA; SUP = selective ultraviolet phototherapy; UV = ultraviolet.

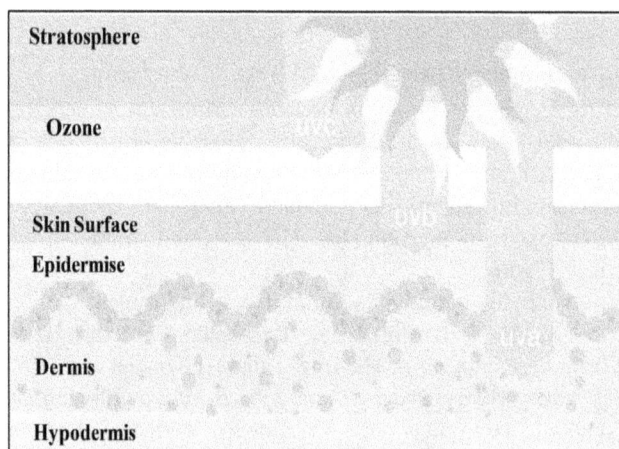

FIGURE 5.4 Skin penetration depth of optical radiation.

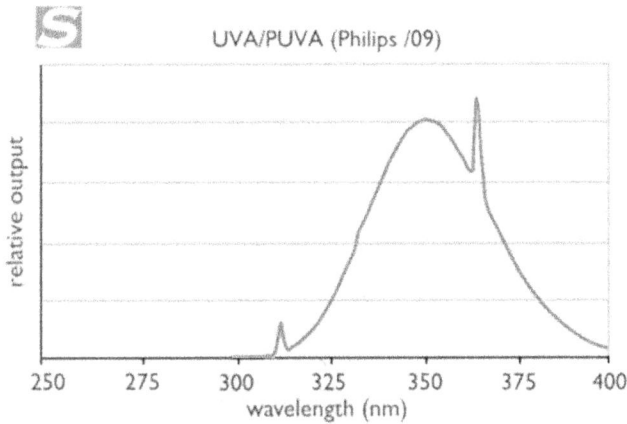

FIGURE 5.5   Spectral power distribution of UVA radiation emitted by PUVA lamp (Philips TL/09 tube).

5.2.3.2   UVB Phototherapy

At the early stage phototherapy and photochemotherapy with UV lamps as the only effective treatment accepted to cure the diseases like psoriases and vitiligo. Though, the eruption of erythema and other undesired side effects of PUVA therapy have led experts working in this field to discard its widespread use.

Moreover, in 1953, Ingram makes the first move toward treatment of psoriasis by the combination of UVB radiation, dithranol and tar-bathing. Subsequently, facts from Fisher et al. and Parrish et al. showed that wavelengths around 311 nm provoke fewest erythema while being most effective for clearing psoriasis [26,27].

5.2.3.3   308 nm Excimer Laser Therapy (Targeted Phototherapy)

One of the category of phototherapy, known as targeted phototherapy, using excimer (308 nm) lasers and lamps have become more and more popular in the last decade, as they permit for targeted phototherapy treatment with high doses of 308 nm light. Most targeted phototherapy devices emit radiation in the UVB range (308–311 nm), although some light-based non-laser machines emit UVA radiation also. Hence, the modes of action of targeted phototherapy are similar to those of UVB/UVA therapy [28]. The 308 nm laser emits wavelengths adjacent to those of NB-UVB 311 nm, and presumably may have similar biological and clinical effects. This laser has been found to be effective in the treatment of localized psoriasis. By allowing for effective and selective targeting (hence the name 'targeted phototherapy'), this device reduces the overall cutaneous cumulative burden and cuts down on the total number of treatments needed to clear psoriasis [29].

5.2.3.4   Photodynamic Therapy (PDT)

Photodynamic therapy (PDT) utilizes photosensitizing agents, oxygen and light, to create a photochemical reaction that selectively destroys cancer cells. PDT is also known as photochemotherapy or light-activated chemotherapy and is a highly selective form of cancer therapy when compared with radiation therapy or systemic chemotherapy. Photosensitizing agents used in PDT are drugs that are administered into the body through topical, oral or intravenous methods. In the body, they concentrate in cancer cells and only become active when light of a certain wavelength is directed onto the area where the cancer is. If not activated, the photosensitizing agent is not harmful to tissue [30]. The photodynamic reaction between the photosensitizing agent, light and oxygen kills the cancer cells [31].

Photodynamic therapy (PDT) commonly uses aminoleveulinic acid (ALA) or methyl aminolevulinate (MAL) as a topical agent for photosensitization, and blue light, red light, intense pulsed light, or pulse dye lasers for treatment [32]. One study observed a response rate of 65.32% at

6-month follow-up for two treatment sessions with ALA-PDT [33]. Compared with 5% imiquimod (IMIQ), PDT had a similar overall response rate (55.65% for 5% imiquimod cream), a similar response rate for grade I lesions (71.64% for PDT and 72.13% for IMIQ), and significantly higher response rate of grade II lesions (57.89% for PDT and 37.03% for imiquimod).

Common adverse effects (AEs) of PDT include erythema, a burning sensation that remits within hours of treatment [34] and pain [35]. The cosmetic outcome of PDT is usually excellent and does not differ significantly from 5% IMIQ. PDT has been shown to give better cosmetic results and higher patient satisfaction than cryotherapy [36]. Impediments to PDT use include prolonged treatment time, multiple visits for complete response (CR) and access to a costly light source. PDT offers an advantage over other therapies in cases of subclinical disease, large AKs, or disseminated AKs, especially in areas where tissue sparing is desirable [37]. It also may be useful with patients who are not candidates for destructive treatment (e.g. cryotherapy) or are noncompliant with topical treatment. Cost comparison of PDT versus cryotherapy and 5-FU for Bowen's disease indicates that PDT is a viable option [38].

PDT is a non-surgical treatment widely used for actinic keratoses and some thin nonmelanoma skin cancers [39]. Photodynamic therapy with placement of a biliary stent may improve bileduct patency in patients with cholangiocarcinoma (CCA) [40]. Whereas in human medicine, photodynamic therapy represents a well known and recognized treatment option for diverse indications, it is still little known and unfortunately not yet established treatment option for pets [41].

5.3   ULTRAVIOLET B PHOTOTHERAPY

The ultraviolet radiation between range 280 nm and 320 nm is used for UVB phototherapy. In addition, the UVB radiation further sub divided into broad band UVB (290 nm–320 nm) and narrowband UVB (311 ± 2 nm). Both of these radiations are used for phototherapy applications.

5.3.1   BROADBAND UVB

For decades, broadband ultraviolet B (UVB) radiation (280 to 320 nm), with or without topical tar, has been used for the treatment of moderate to severe psoriasis. In case of conventional broad band UVB lamps, the emission of light is in a much broader spectrum that includes both the therapeutic wavelengths precise to the treatment of skin diseases as well as the shorter wavelengths having higher energy which can result into sunburn which has increases the possibility of skin cancer, and confines the amount of beneficial therapeutic UVB that is needed for the treatment. In the early 1980s, researchers observe that wavelengths around 311 nm were more effective than broad spectrum UVB in clearing psoriasis directed to a major advancement in phototherapy with the development of fluorescent lamps emitting selective UVB spectra in the range of 311 to 313 nm (i.e. narrowband UVB) [42,43].

5.3.2   NARROWBAND UVB

The narrowband UVB radiation possesses a precise and unique spectral energy distribution with an emission peak around 311 nm concentrated primarily in the therapeutic range. In Figure 5.7, the smaller grey-shaded area shows that narrowband UVB devices have considerably less erythemogenic output than conventional UVB broadband devices that minimizes the side effects such as redness and itching. As a result, narrowband UVB is safer and effective than conventional UVB broadband radiation [44–47]. This therapy is a broadly used modality in the treatment of psoriasis and is also accepted to be safe in pregnancy [48]. The fluorescent bulb was developed (TL-01), emitting a major peak at 311(±2 nm) and a minor peak at 305 nm. The spectral distribution of commercial Phillips TL/01 lamps is shown in Figure 5.6.

Figure 5.7(d) represents the action spectrum of vitamin D content of narrowband UVB. By using narrowband UVB radiations it is possible to produce vitamin D in human skin. Vitamin D plays a

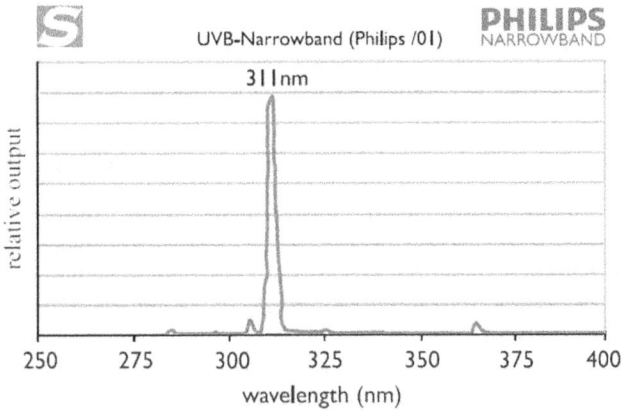

FIGURE 5.6 Spectral power distribution of NB-UVB radiation emitted by commercial Philips TL/09 tube.

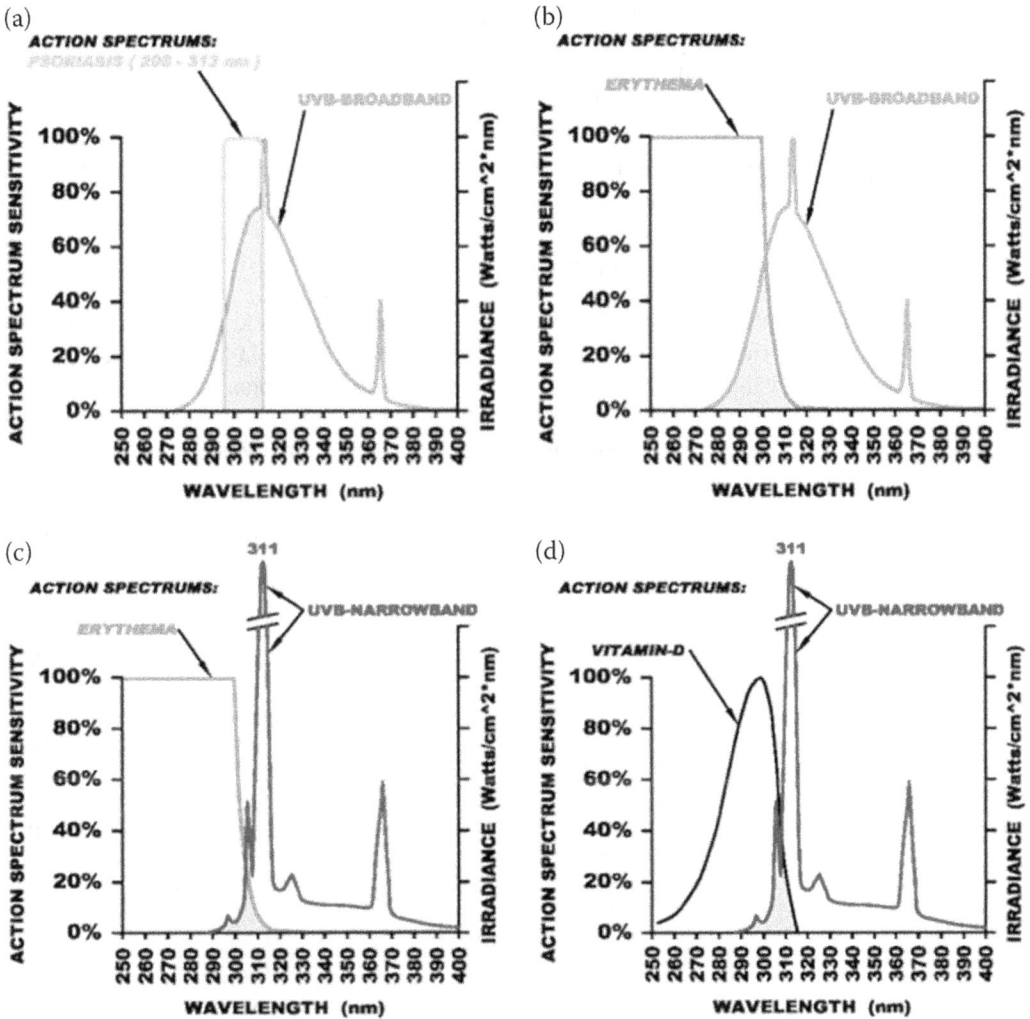

FIGURE 5.7 (a) Graphical representation of the action spectrum of psoriasis, (b) erythema content of broadband UVB, (c) action spectrum of erythema and (d) vitamin D content of narrowband UVB.

vital role as a catalyst for calcium absorption. The progress of many chronic diseases such as cancer, multiple sclerosis, cardiovascular disease, osteoporosis, osteomalacia, hypertension, type 1 diabetes mellitus, depression and rheumatoid arthritis can be reduced because of vitamin D [49].

5.4 ROLE OF PHOSPHOR AND SIGNIFICANCE OF SENSITIZER

The sharp and intense narrow band emission is more suitable for the treatment of skin disorder. Out of all the rare earths, some selected elements such as Ce^{3+}, Pr^{3+}, Gd^{3+} and Eu^{2+} have been reported as NB-UVB, UVA1 and UVA2 emitting rare earths. Generally, gadolinium oxide possesses the sharp and intense emission in the narrowband UVB region, which makes the Gd^{3+} ion ideal to use in a phototherapy lamp phosphor.

The optical property of the Gd^{3+} ion is very poor because of forbidden nature of the transitions within the $4f^7$ configuration of the Gd^{3+} ion. Therefore, it is very difficult to irradiate a large amount of excitation energy into the Gd^{3+} sublattice. These poor optical properties can affect the general performance and life span of phototherapy devices. To avoid these drawbacks, some sensitizing materials are incorporated to increase the intensity of phosphor materials. To make this possible, the sensitizer should have an allowed optical transition in the short-wavelength UV region. Its emission should overlap the Gd^{3+} absorption lines [50].

The energy level scheme of Gd^{3+} ion shows a bandgap of about 32×10^3 cm between the lowest excited 4f state $^6P_{7/2}$ and 4f ground state $^8S_{7/2}$. Therefore, sensitization of the Gd^{3+} ion is possible by a sensitizer with emission that overlaps the $^8S_{7/2} \rightarrow ^6P_{7/2}$ absorption of Gd^{3+} ion. The schematic energy level of Gd^{3+} ions is shown in Figure 5.8. In this regard, we have observed that the Sb^{3+}, Tm^{3+}, Pb^{2+}, Bi^{3+}, Pr^{3+} and Ce^{3+} could be promising sensitizer for the Gd^{3+} ion.

5.4.1 CHARACTERISTIC EMISSION OF PB^{2+} IONS

Lead compound is a relatively un-reactive post-transition metal. Its weak, metallic character is illustrated by its amphoteric nature. Compounds of lead are usually found in the 2+ oxidation state rather than the 4+ state common with lighter members of the carbon group.

The luminescence properties of Pb^{2+} ion is quite diverse and depends strongly upon the host lattice. The divalent lead (Pb^{2+}) is a promising UV luminescence centre, which typically shows a

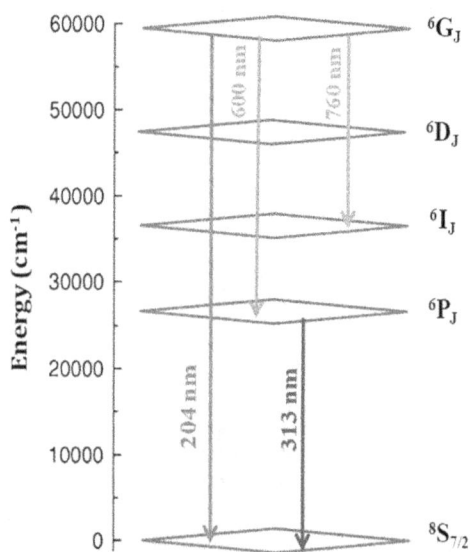

FIGURE 5.8 Schematic energy levels of Gd^{3+} ions.

broad UV emission band (220–400 nm; generally dominated by the $^3P_1 \rightarrow {}_1S^0$ transition) which is suitable for energy transfer in case of Gd^{3+} ion [51,52], but at room higher temperature, it is partially or totally quenched by the ligand to metal charge transfer state, whose emission shifts to the visible range.

The energy-level diagram of Pb^{2+} ion is shown in Figure 5.9.

Generally, the absorption and emission of Pb^{2+} ions usually appear in the UV region. The optical properties and transition levels shows that, Pb^{2+} ion can be act as good sensitizer for the Gd^{3+} ion.

5.4.2 Characteristics Emission of Pr³⁺ Ion

Praseodymium (Pr) is an element having atomic number 59. It is the third element of the lanthanide series and is usually considered to be one of the rare earth metals. Praseodymium is a soft, silvery, malleable and ductile metal possesses magnetic, electrical, chemical and optical properties. It is too reactive to be found in native form, and pure praseodymium metal slowly develops a green oxide coating. It occurs along with other lanthanide elements in a variety of minerals. The two primary sources are monazite and bastnaesite. It is extracted from these minerals by ion exchange and solvent extraction.

In praseodymium all five outer electrons can act as valence electrons, but the use of all five requires extreme conditions. Similarly, to most lanthanides, Pr usually uses three electrons as valence electrons, as afterwards the remaining 4f electrons are too strongly bound, this is because the 4f orbitals penetrate the most through the inert xenon core of electrons to the nucleus, followed by 5d and 6s, and this increases with higher ionic charge.

In 1968, the 4f energy level scheme of Pr^{3+} ion was proposed by Dieke [53]. There are two types of emission transitions from Pr^{3+}, namely inter-configurational 4f5d $\rightarrow$ 4f^2 or intra-configurational 4f^2 $\rightarrow$ 4f^2, could take place with a VUV excitation. The emission transition of Pr^{3+} depends on the host lattice and the energetic location of the 4f5d states relative to the $^1S_0$ state, which is located at about 47,000 cm^{-1}. If the lowest 4f5d state lies below the $^1S_0$ state, the high energetic excitation will stimulate the broadband emission from the lowest 4f5d state, which is parity allowed inter-configurational 4f5d $\rightarrow$ 4f^2 transition. The 4f5d $\rightarrow$ 4f^2 emission is usually situated at the UV region [54] (Figure 5.10).

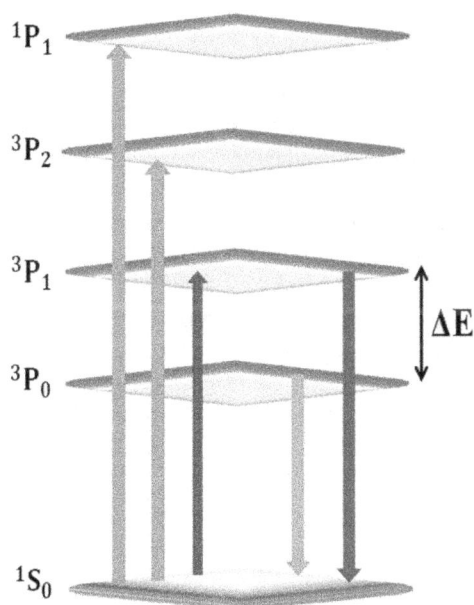

FIGURE 5.9 Energy level diagram of Pb^{2+} ion.

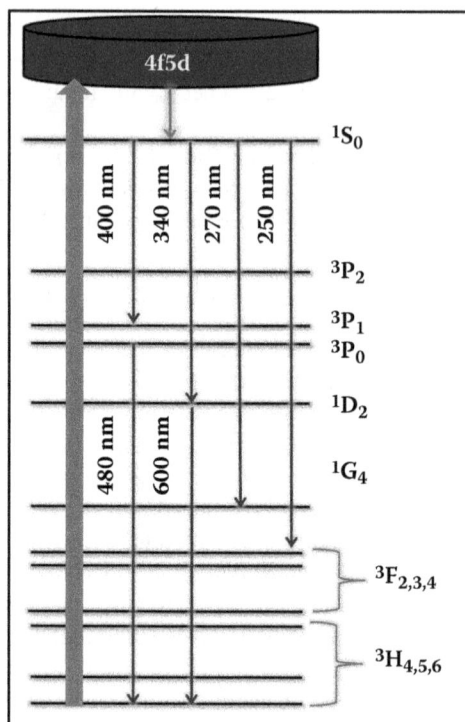

FIGURE 5.10 Schematic energy levels of Pr^{3+} ions.

5.4.3 Characteristic Emission of Bi^{3+} Ion

Bismuth, with $[Xe]4f^{14}5d^{10}6s^26p^3$ electronic configuration, is taking much attention due to its diverse oxidation states and multi-type electronic structures. It also been considered as 'the wonder metal'. The Bi^{3+} exhibits phenomenal properties in several host materials and it can be used as an activator or sensitizer. Bismuth shows various types of oxidation states, such as 0, +1, +2, +3 and +5, since that its outer 6s and 6p electron valence electrons can be adjusted by crystal field [55,56]. Generally, Bi^{3+} is the most common and stable ionic form compared to other valence states. Bismuth is considered to be a green element between toxic and heavy metals but it is less toxic than sodium chloride [57]. Hence, bismuth compounds have been successfully applied in a tremendous range of areas [58–61].

The phosphor doped with Bi^{3+} having different morphologies, such as spherical, rhombus, fibers and graininess, were synthesized through some simple methods, such as solid-state reaction technique, hydrothermal or solvothermal route, sol-gel synthesis, the co-precipitation method and combustion process.

Energy transfer between Bi^{3+} and other luminescent ions or hosts results to tunable luminescence of Bi^{3+} doped phosphors [62]. Early in 1967, Blasse and Bril [63] reported the energy transfer from Sb^{3+}, Bi^{3+} and Ce^{3+} sensitized ions to trivalent lanthanide ions, based on the theoretical Forster and Dexter models [64,65].

Moreover, the Bi^{3+} ion can efficiently absorb UV light and transfer energy to other luminescent centers, which significantly enhances the emission intensity of accept ions. Therefore, the Bi^{3+} ion can be used as an excellent sensitizer [66].

It has been well accepted that the Bi^{3+} ion transferred its energy to tetravalent ions (Mn^{4+}), trivalent ions (Gd^{3+}, Ce^{3+}, Sm^{3+}, Eu^{3+}, Tb^{3+}, Dy^{3+}, Ho^{3+}, Er^{3+}, Tm^{3+}, Yb^{3+}, Pr^{3+}, Nd^{3+} and Cr^{3+}) and divalent ions (Mn^{2+}, Eu^{2+} and Co^{2+}).

5.4.4 Characteristics Emission of Ce^{3+} Ion

Rare earth (RE) elements are well known due to their interesting energy level structures that result in light emission from ultraviolet (UV) to far infra-red regions, suitable for many applications [67].

Inorganic materials doped with Ce^{3+} have attracted the interest of many researchers due to applications of these materials as phosphors, scintillator for elementary particles, detectors for ionizing radiations, UV absorbing filters or UV emitters and activators for energy transfer [68].

The Ce^{3+} ion has a simple energy level structure compared to other lanthanide ions. Cerium ion can exist in the oxidation state +3 and +4 in the phosphors. The luminescence efficiency is greatly affected by the existence of Ce^{4+}, as it provides a non-radiative pathway and reduces the number of Ce^{3+}.

The ground state configuration of the Ce^{3+} ion has one 4f electron and excited state configuration has one 5d electron with empty 4f shell. The $4f^1$ ground state configuration yields two levels viz. $^2F_{5/2}$ and $^2F_{7/2}$ due to spin orbit coupling and the $5d^1$ excited state configuration is split by crystal field in 2 to 5 components. The Ce^{3+} emission occurs from the lowest crystal field component of $5d^1$ configuration to the 4f ground state levels. Since 4f $\rightarrow$ 5d transition is parity allowed and spin selection is not appropriate, the emission transition is fully allowed one [69]. Due to this, the excitation spectrum and emission spectrum of Ce^{3+} ions in inorganic solids consists of broad bands. The Ce^{3+} ion shows allowed optical transitions in absorption and emission, which are of the f $\rightarrow$ d type. The Ce^{3+} ion shows an efficient broadband luminescence in inorganic solids due to its 4f $\rightarrow$ 5d parity allowed electric dipole transitions [70]. In Ce^{3+} ions due to the crystalline environment, the 5d configuration may split into at most five 5d states. The average energy of the five 5d levels may shift downwards towards the 4f ground state. This downward shift is defined as the centroid shift ε_c of the 5d configuration, although it is not solely caused by the nephelauxetic effect [71].

The energy-level diagram of Ce^{3+} ion is shown in Figure 5.11.

Generally, the absorption and emission of Ce^{3+} ions usually appear in the UV region. Depending on the position of the emission, Ce^{3+} can act as a good sensitizer for Gd^{3+} ions.

5.5 BORATE-BASED PHOTOTHERAPY PHOSPHORS

Basically, phosphors exhibit characteristic emissions of an activator (rare earth) when excited. This is achieved by the technique down-conversion. When excitation is by high-energy photons (high frequency/lower wavelength), the phenomenon is known as down conversion (widely observed). The phosphors with UV emitting radiation are useful for phototherapy application.

From literature survey, it is observed that most of host lattice used for the UV emitting phosphors belong to silicates, fluorides, aluminates or phosphates families. Very few attempts were done to explore the borate-based UV emitting phosphor. Borate compounds in luminescent materials has attracted much attention due to their excellent thermal stability, variety of structure type, transparency to a wide range of wavelengths, high optical damage threshold and high optical quality, potential low-cost synthesis and inherent large bandgap [72–75].

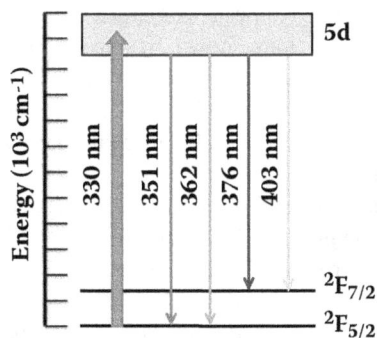

FIGURE 5.11 Energy level diagram of Ce^{3+} ion.

5.5.1 UVB-EMITTING PHOSPHOR

Commercial NB-UVB emitting phosphor (i.e. LaB3O6:Gd^{3+}, Bi^{3+}) is used in phototherapy lamps for the treatment of psoriasis [76]. Eu^{2+}-activated SrB4O7 and Ba2B5O9Cl phosphors are used in the treatment of hyperbilirubinemia [77].

S. Tamboli et al. synthesized LiCaBO3:Gd^{3+} phosphor via the solid-state diffusion method and studied the NB-UVB emission (315 nm) under 274 nm excitation. Phosphors with various concentrations of dopant ion were studied and optimum PL emission intensity was found to be at 1mol% concentration of dopant. The critical energy transfer distance was found to be 22.77 Å. The EPR spectrum of the LiCaBO3:Gd^{3+} phosphor exhibited signals with effective g = 6.11, 3.94, 2.65, 1.99 [78].

V. Singh et al. synthesized BaAl2O4:Gd^{3+} phosphor successfully using a combustion technique and reported that under-excitation of 273 nm, the main emission peak of the phosphor is located at 314 nm, and this is attributed to the 6P7/2→8S7/2 transition of Gd^{3+}. EPR spectrum appears to be U-shaped, and lines are evident at geff ~ 2.14, 4.56 and 6.75. The PL and EPR analyses indicate the presence of Gd^{3+} in this sample [79].

P. A. Nagpure et al. synthesized the SrZr(BO3)2: Gd^{3+}, Pr^{3+}, Al4B2O9:Ce^{3+}, LaMgB5O10:Ce^{3+} phosphor using Combustion method and proposed that the SrZr(BO3)2: Gd^{3+}, Pr^{3+} phosphor show narrow band UVB emission which is useful for the treatment of psoriasis. It also reported that Al4B2O9:Ce^{3+} and LaMgB5O10:Ce^{3+} phosphors possess the emission at 356 nm and 440 nm, respectively. The UVA spectrum (315–400 nm) is used in combination with photosensitizer in PUVA therapy and the band just above 400 nm without the UV component is desired for the treatment of hyperbilirubinemia [80].

D. Taoli et al. synthesized the YPO4 phosphors single-doped with Sb^{3+} or Gd^{3+} and co-doped with Sb^{3+} and Gd^{3+} were prepared by a solid-state reaction method. In codoped phosphor Sb^{3+} is act as a sensitizer which made the phosphors excitable by short-wave ultraviolet (UV) at a wavelength between 220 and 260 nm. The phosphor was studied with different doping concentrations of Gd^{3+} and Sb^{3+}. The phosphor shows strong emission at a wavelength of 312 nm under 253.7 nm excitation [81].

R. G. Kunghatkar et al. synthesized Gadolinium doped LiSrBO3 by the sol-gel technique. It reported the UVB emission at 316 nm under the excitation of 274 nm. The second-order emission is also observed in PL emission spectra at 612 nm and 627 nm. The diffuse reflectance spectrum of LiSrBO3 was studied and energy bandgap were calculated using the Kubelka-Munk function. The bandgap was found to be 5.81 eV [82].

S. P. Bhagat et al. reported that the novel UV emitting phosphor Sr2Mg(BO3)2:Pb^{2+},Gd^{3+} with varying concentration of Pb^{2+} and Gd^{3+} were synthesized by combustion synthesis method. It was reported that the phosphor possesses two emission spectra, one line emission at 313 nm due to 6Pj–8S7/2 transition of Gd^{3+} ions and one band emission at 330 nm due to 3P1–1S0 transition of Pb^{2+} ions. The emission is in UVA and UVB region of electromagnetic spectrum therefore it was suggested that phosphor material could be a potential candidate as a phototherapy lamp phosphor for combination therapy of narrowband UVB plus PUVA [83].

A. B. Gawande et al. studied UVB emission in Pr^{3+} - Gd^{3+} co-doped Sr2Mg(BO3)2 phosphor with varying concentration of Pr^{3+} and Gd^{3+} by solution combustion synthesis technique. The co-operative energy transfer mechanism between Pr^{3+} ions and Gd^{3+} ions shown by the overlap between the excitation spectrum and emission spectrum. The charge compensations by Li^+, Na^+ and K^+ anions all increased the luminescent intensity but all of that Sr2Mg(BO3)2:Pr^{3+}. Na^+ was observed to be the optimal charge compensator. It is suggested that there is an efficient energy transferred from Pr^{3+} → Gd^{3+} ions in Sr2Mg(BO3)2 phosphor [84].

D. D. Ramteke et al. have studied the effect of Gd2O3 addition on optical properties of 27.5 Li2O-72.5 B2O3 glass systems. In this work, Gd^{3+}-activated lithium borate glasses with general formula 27.5 Li2O-(72.5-X)B2O3-XGd2O3 were prepared by melt quenching process. The PL emission spectra showed intense band at 311.5 nm and weak band and at 305 nm corresponded to transition 6P7/2→8S7/2 and 6P5/2→8S7/2, respectively. From the result it is concluded that Gd^{3+}-doped glasses could be useful as source of narrowband UV light for the treatment of skin diseases [85].

5.5.2 UVA-Emitting Phosphors

The use of UVA (320–400 nm) emission along with a sensitizing agent in the treatment of skin diseases is well established. Some of the phosphors, such as SrB4O7: Eu, SrLiB4O15: Eu, CeMgB5O10: Gd, GdBO3: Pr, LaB3O6: Ce, Bi, LaB3O6: Ce, BaAlBO3F2: Eu^{2+} [86] are attempted by various researchers in the context of the above-mentioned applications.

Moreover, the Pb^{2+} ion doped phosphors like SrB2O4:Pb2+ [87], Sr6YAl(BO3)6:Pb^{2+}, Sr5La2Mg(BO3)6:Pb^{2+} [88], SrAl2B2O7:Pb^{2+} [89], BaAl2B2O7:Pb^{2+} [90], CaAl2B2O7:Pb^{2+} [91], CaZr(BO3)2:Pb^{2+} [92], Ca3B2O6:Pb $^{2+}$ [93] also shows emission in UVA region and it could be the potential candidates for phototherapy application.

Ilhan Pekgozlu et al. synthesized the pure Li6CaB3O8.5 and different concentration of Pb^{2+} ion doped Li6CaB3O8.5 materials by a solution combustion method. In study the emission band observed at 307 nm from the 3P1 excited state level to the 1S0 ground state under excitation of 268 nm. The Stokes shift of the synthesized phosphor was calculated to be 4,740 cm^{-1} [94].

R. G. Kunghatkar et al. synthesized new phosphor material, BaAlBO3F2 doped with Eu^{2+} ions via the simple wet chemical method. In the PL study, the emission of phosphor before reduction, is in the blue region at 431 nm and after reduction, excitation is at 258 nm and emission is at 361 nm, which is in the UVA region. The TL study also been carried out after exposed to γ-radiation from a 60Co source at room temperature. The TL intensity increases randomly as the concentration of Eu^{2+} increases. For calculation of the trapping parameters 0.1 mol% concentration of phosphor is used. The TL glow curve has two prominent peaks, one at a low temperature of 137°C and another at a higher temperature of 250°C [95].

5.6 APPLICATIONS

The list of viable UV applications continues to grow. The UV radiation is widely used in industrial processes and in medical and dental practices for a variety of purposes, such as killing bacteria, creating fluorescent effects, curing inks and resins, phototherapy and suntanning. Different UV wavelengths and intensities are used for different purposes. Some of the applications are given below.

5.6.1 Germicidal Lamps

Ultraviolet germicidal irradiation (UVGI) is an established means of disinfection and can be used to prevent the spread of certain infectious diseases. For the purpose of germicidal irradiation low-pressure mercury (Hg) discharge lamps are commonly used, the shortwave ultraviolet-C (100–280 nm) radiation, primarily at 254 nm is used for germicidal application. The UVC radiation kills or inactivates microbes by damaging their deoxyribonucleic acid (DNA). The principal mode of inactivation occurs when the absorption of a photon forms pyrimidine dimers between adjacent thymine bases and renders the microbe incapable of replicating [96].

As we know that, a strong germicidal effect is provided by the light in the short-wave UVC band. In addition, erythema and conjunctivitis can also be caused by this form of light. Therefore, when germicidal UV-light lamps are used, it is important to design systems to exclude UVC leakage and so avoid these effects.

5.6.2 Water Applications

The world faces an invisible crisis of water quality and on the other hand clean water is a key factor for economic growth. Deteriorating water quality is stalling economic growth, worsening health conditions, reducing food production and exacerbating poverty in many countries. Due to the presence of a wide variety of microorganisms in the water can cause disease, especially for young and senior people, who may have weaker immune systems.

The purification of water can be achieved by means of UV light without the addition of che-micals that can produce harmful by-products and add an unpleasant taste to water.

The UV radiation has the capability to inactivate bacteria, viruses and protozoa. The dose of radiation is set as per the type of organism for inactivation such as viruses require higher doses than bacteria and protozoa. Additional benefits include easy installation, low maintenance and minimal space requirements. Some of the water purification examples are as follows:

- Municipal drinking water
- Residential drinking water
- Water coolers dispensers
- Semiconductors process water
- Spas and swimming pools
- Fish ponds and aquariums
- Municipal wastewater

5.6.3 Surface and Air Disinfection

The UVC radiation is also used for surface and air disinfection. Air disinfection can play an important role in a broad range of applications such as hospitals, schools, offices, industry, (food) retail and museums. UVC radiation has the ability to destroy the outer protein coating of the SARS-Coronavirus, which is a different virus from the current SARS-CoV-2 virus. The destruction ultimately leads to inactivation of the virus. The air disinfection is achieved via several methods given as follows:

- Irradiating the upper-room air only
- Irradiating the full room (when the room is not occupied)
- Irradiating air as it passes through enclosed air circulation and heating, ventilation and air conditioning (HVAC) systems

The upper air UV-C disinfection luminaires can be installed to continuously disinfect air where people are present. These systems are usually installed at a height of above 2,4 m. By making use of parabolic reflectors the UVGI can be concentrated in a zone. In this zone a high intensity UV-C is realized without having too high intensities below 2,0 m. Because of the presence of natural or mechanical ventilation in the room, the contaminated room-air will pass through the UV-C zone and thus will be disinfected [97].

The entire volume of air is directly exposed to high intensities of UV-C light by installing open UV-C disinfection fixtures on walls or ceilings. Therefore, strong reduction in the number of microorganisms is achieved very quickly and additionally directly exposed surfaces are also disinfected. This form of air disinfection is used when no people or animals can be present in the room or without protective measures.

5.6.4 UV Curing

UV lamps are useful tools for actinic purposes such as producing chemical reactions in industrial applications such as curing or drying chemical compounds, inks and coatings. Common substrates for MPUV curing have been wood, glass, metal, labelling and plastic.

Common applications for the use of UV light in curing or drying include:

- Adhesive bonding
- Coatings
- Decorative adhesive transfers
- Engraving and plating resist

- Inks/varnishes/lacquers
- Decorative glazes
- Packaging
- Printed circuit boards
- Printing/label printing
- Graphic arts
- Sheet decorating

5.7 LIMITATIONS AND CHALLENGES

Presently, devices used for phototherapy are still based on the old technology of plasma emission mercury-based lamps that suffers from many drawbacks namely low lifetime and plasma phosphor interaction resulting in severe glass solarization, phosphor degradation, plasma efficiency loss, high energy consumption, high temperature operation, fixed discharge wavelengths and in addition toxicity of mercury harmful to the environment. On the other hand, we have witnessed the revolution in lighting devices which are now based on LEDs. The above drawbacks can be overcome by designing LED-based phototherapy devices. The important role played in such devices is of phosphor materials.

The prepare phosphors which could be coated on UV-LED so that low-cost LED lamps could be designed for the phototherapy treatment. As the phosphor-coated UV-LEDs will be able to overcome the drawback like high energy consumption, high temperature operation and toxicity.

5.7.1 UV-LED

In the past two decades, group III-nitride-based ultraviolet light-emitting diodes (UV-LEDs) and their applications have undergone a progressively accelerating development. This can be demonstrated by many metrics.

It is well known that, the bandgap of LED is dependent on the ration of indium (In) or aluminium (Al) to the gallium nitride (GaN). By adding aluminium nitride (AlN) to the GaN alloy system, the emission wavelength of AlGaN-based LEDs can be tuned over almost the entire UV spectra. Therefore, the aim is to prepare the phosphors having emission in narrowband UVB region which could be coated on UV-LED so that low-cost LED lamps could be designed for the phototherapy treatment. And the drawback such as high energy consumption, high temperature operation and toxicity can be overcome by the phosphor coated UV-LEDs [98].

5.7.2 BENEFITS OF UV-LED-BASED DEVICES

Following are the some other benefits of UV-LED-based devices.

- **Sturdy devices** – LEDs are solid-state devices that utilize semiconductor material instead of a filament or Hg-discharge gas. An LED light is a tiny chip encapsulated in an epoxy resin enclosure, which makes LEDs far sturdier than traditional incandescent lamps or fluorescent tubes. Since LEDs don't use fragile components such as glass and filaments, LEDs are able to withstand shock, vibration and extreme temperature.
- **Low-voltage operation** – The devices could be operated at very low voltages (DC operation) as compared to existing devices.
- **Optimal wavelength** – The LED-based lamps can focus tightly on the optimal wavelength (450–465 nm) for maximizing bilirubin metabolism in case of UV-A phototherapy or UV-B skin treatment (313 nm).
- **Ideal irradiance levels** – The LED-based lamps allows treatment tailored to each patient's needs by using sophisticated electronic technology (with a high irradiance of 45 $\mu w/cm^2/nm$ and a low irradiance at 22 $\mu w/cm^2/nm$).

- **Uniform light distribution** – The optical design using LED-based lamps can be made to ensure uniform light distribution to the patient's exposed surface area.
- **Whisper-quiet operation** – The LED-based lamps can be designed with no fan and no other mechanical moving parts, so that it will operates significantly quieter than pre-scribed IEC guidelines of 60 dB(A).
- **No spill, no glare** – The optical design using LED-based lamps can be made so that it minimizes spill light outside the bed so neither caregivers nor other patients receive undesired exposure to light.
- **Temperature** – LED lights generate virtually no heat therefore they are cool to the touch and can be left on for hours without incident or consequence if touched. LEDs produce 3.4 btus/hour, compared to 85 btus/hour for incandescent bulbs. In comparison, in-candescent lighting expels 90% of the energy it consumes via heat, making the bulbs hot to the touch. LEDs reduce the potential for safety risks such as burns and fires.
- **Environment friendly** – LEDs are made from non-toxic materials, unlike fluorescent lighting that uses mercury that may pose a danger to the environment. LEDs are also recyclable and considered "green" or earth-friendly.

5.8 FUTURE PROSPECTS AND SCOPE

From a literature survey, it is clearly understood that the phosphors possessing UV emission in the range 200 nm–400 nm significantly used for phototherapy applications. Hence, many such devices like fluorescent lamps for medical use, black-light fluorescent lamps, medical and scientific UV lamps are developed. The efficiency of such devices is strongly depended on luminescence properties of phosphors used in them.

It has been observed that obtained surface morphology of phosphor possesses irregular grain size and rough nature of phosphor restrict the uniformity/thickness of coating over UV-LED chip and it also affects the conversion efficiency of a phosphor. These hurdles can be overcome by the preparation of agglomeration free, particles with pre-designed morphology. For that synthesis method can be optimized for getting better performance of phosphors.

It is also important to investigate the effect on the photoluminescent properties of the phosphors by co-doping with suitable sensitizers to obtain efficient and stabilized phosphor materials for various applications in medical field.

Generally, it is well known that the cost of LEDs chips reduces as output wavelength increase towards the higher value of wavelength. Therefore, it is important to focus on the preparation of phosphors (up-conversion) to be used with near-UV LEDs (blue chips) for designing the solid-state phototherapy illuminators (lamps). This strategy will help to reduce the cost of phototherapy lamps.

REFERENCES

[1] Soehnge, H., Ouhtit, A., and Ananthaswamy, H. N. (1997). *Front. Biosci.*, 2, 1, d538.
[2] Hardy K., Meltz M., and Glickman R. (1997). *Medical Physics Publishing*. Madison.
[3] Gawande, A. B. (2015). *Ultra Violet light Emitting Borate Phosphors*. ISBN-978-3-659-71157-2.
[4] Diffey, B. L. (1984). *Photodermatology*, 1, 103.
[5] McKinlay, A. F., and Diffey, B. L. (1987). *CIE-Journal*, 6, 17.
[6] Diffey, B. L., Janseen, C. T., Urbach, F., and Wulf, H. C. (1997). *Photodermatol. Photoimmunol. Photomed.*, 13, 64.
[7] CIE Standard. (1998). Commission Internationale de l" Eclairage, Vienna.
[8] Gordon, P. M., Saunders, P. J., Diffey, B. L., and Farr, P. M. (1998). *Br. J. Dermatol.*, 139, 811.
[9] Diffey, B. L. (1994). *Photochem. Photobiol.*, 60, 380.
[10] Webb, A. R., Slaper, H., Koepke, P., and Schmalwieser, A. W. (2011). *Photochem. Photobiol.*, 87, 483.
[11] Farr, P. M., and Diffey, B. L. (1985). *Br. J. Dermatol.*, 113, 65.
[12] Parrish., J. A., and Jaenicke, K. F. (1981). *J. Invest. Dermatol.*, 76, 359.

[13] Dare, R. A. S., Goodfield, M. J., and Rowell, N. R. (1989). *Br. J. Dermatol*, 65, 121.
[14] Krutmann, J., Czech, W., and Diepgen, T. (1992). *J. Am. Acad. Dermatol.*, 225, 26.
[15] Kroft, E. B., Berkhof, N. J., and Van de Kerkhof, P. C. (2008). *J. Am. Acad. Dermatol.*, 59, 1017.
[16] Cates, E. L. H. (May, 2013). Ph.D. Thesis. Georgia.
[17] Berneburg, M., Cken, M. R., and Benedix, F. (2005). *Acta. Derm. Venereol.*, 85, 98.
[18] Honigsmann, H. (2001). *Clin. Exp. Dermatol.*, 26, 343.
[19] Hemne, P. S., Kunghatkar, R. G., Dhoble, S. J., Moharil, S. V., and Singh, V. (2017). *Luminescence*, 32, 260.
[20] Fitzpatric, T. B., and Pathak, M. A. (1959). *J. Invest. Dermatol.*, 32, 229.
[21] Honigsmann, H. (2013). *Photochem. Photobiol. Sci.*, 12, 16.
[22] Mofty A. M. E. (1968). *Vitiligo and Psoralens*, Pergamon Press, Oxford, 1.
[23] Parrish, J. A., Fitzpatrick, T. B., Tanenbaum, L., and Pathak, M. A. (1974). *New Engl. J. Med.*, 291, 1207.
[24] Hemne, P. S., Kunghatkar, R. G., Dhoble, S. J., Moharil, S. V., and Singh, V. (2017). *Luminescence*, 32, 260–270.
[25] Pathak, M. A., and Daniels, F. (1962). *J. lnvssr. Dermnrol*, 39, 225.
[26] Berneburg, M., Rocken, M., and Benedix, F. (2005). *Acta Derm Venereol*, 85, 98.
[27] Green, C., Ferguson, J., Lakshmipathi, T., and Johnson, B. E. (1988). *Br. J. Dermatol.*, 119, 691.
[28] Mysore, V. (2009). *Indian J. Dermatol. Venereol. Leprol.*, 75, 119.
[29] Fritz, K. (2008). *Med. Laser. Appl.*, 23, 87.
[30] Lucroy, M. D. (2002). *Vet. Clin. Small. Anim.*, 32, 693.
[31] Wiegell, S. R., Wulf, H. C., Szeimies, R. M., Seguin, N. B., Bissonnette, R., Gerritsen, M. J., Gilaberte, Y., Pinton, P. C., Morton, C. A., Sidoroff, A., and Braathen, L. R. (2012). *J. Eur. Acad. Dermatol. Venereol.*, 26, 673.
[32] Kalisiak, M. S., and Rao, J. (2007). *P. Dermatol. Clin.*, 25, 15.
[33] Sotiriou, E., Apalla, Z., Maliamani, F., Zaparas, N., Panagiotidou, D., and Ioannides, D. (2009). *J. Eur. Acad. Dermatol. Venereol.*, 23, 1061.
[34] Chamberlain, A. J., and Kurwa, H. A. (2003). *Am. J. Clin. Dermatol.*, 4, 149.
[35] Langan, S. M., and Collins, P. (2006). *Br. J. Dermatol.*, 154, 146.
[36] Szeimies, R. M., Karrer, S., and Fijan, S. R. (2002). *J. Am. Acad. Dermatol.*, 47, 258.
[37] Carlson S., and Bordeaux J. (2011). *Evidence Based Dermatology*. 2nd ed. People's Medical Publishing House, Shelton, CT.
[38] Adam, R., Schmitt, B. A., and Jeremy, S. (2013). *Clin. Dermatol.*, 31, 712.
[39] Jennifer, D., Bahner, M. D., and Jeremy, S. (2013). *Clin. Dermatol.*, 31, 792.
[40] Leggett, C. L., Gorospe, E. C., Murad, M. H., Montori, V. M., Barona, T. H., and Wang, K. K. (2012). *Photodiag. Photod. Ther.*, 9, 189.
[41] Julia Buchholz, D. V. M., and Walt, H. (2013). *Photodiag. Photod. Ther.*, 10, 342.
[42] Fischer, T. (1976) UV-light treatment of psoriasis. *Acta Derm Venereol*, 56, 473.
[43] Parrish, J. A., and Jaenicke, K. F. (1981) Action spectrum for phototherapy of psoriasis. *J. Invest. Dermatol.*, 76, 359.
[44] Chauhan, A. O., Gawande, A. B., and Omanwar, S. K. (2016). *J. Inorg. Organ. Poly. Mater.*, 26, 1023.
[45] Weelden, H. V., Faille, H. B., Young, E., and Leun, J. C. V. (1988). *Br. J. Dermatol*, 119, 119.
[46] Karvonen, J., Kokkonen, E., and Ruotsalainen, E. (1989). *Acta. Derm. Venereol.*, 69.
[47] Johnson, B., Green, C., and Lakshmipathi, T. (1988). J. Ferguson Proc. 2nd Eur. Photobiol. Congr., Padua, Italy.
[48] Rose, R. F., Batchelor, R. J., Turner, D., and Goulden, V. (2009). *J. Am. Acad. Dermatol.*, 61, 259.
[49] Haykal, K. A., Groseilliers, J. P. D. (2006). *J. Cutan. Med. Surg.*, 10, 234.
[50] de Vries, A. J., and Blasse, G. (1986). *Mat. Res. Bull.*, 21, 683.
[51] Folkerts, H. F., Ghianni, F., Blasse, G. (1996). *J. Phys. Chem. Solids*, 57, 1659.
[52] Dijken, V., Folkerts, H. F., and Blasse, G. (1997). *J. Lumin.*, 72, 660.
[53] Dieke G. H. (1968). *Spectra and Energy Levels of Rare Earth Ions in Crystals*. Interscience, New York.
[54] Zhang, Q. Y., and Huang, X. Y. (2010). *Prog. Mater. Sci.*, 55, 353.
[55] Sun, H. T., Zhou, J., and Qiu, J. (2014). *Prog. Mater. Sci.*, 64, 1.
[56] De Jong, M., Meijerink, A., Barandiaran, Z., and Seijo, L. (2014). *J. Phys. Chem. C*, 118, 17932.
[57] Mohan, R. (2010) Green bismuth. *Nat. Chem.*, 2, 336.
[58] Xia, Z., Chen, D., Yang, M., and Ying, T. (2010). *J. Phys. Chem. Solids*, 71, 175.
[59] Kang, F., Peng, M., Xu, S., Ma, Z., Dong, G., and Qiu, J. (2014). *Eur. J. Inorg. Chem.*, 2014, 1373.
[60] Zeng, X., Im, S. J., Jang, S. H., Kim, Y. M., Bin Park, H., Son, S. H., Hatanaka, H., Kim, G. Y., and Kim, S. G. (2006). *J. Lumin.*, 121, 1.

[61] Chen, L., Jiang, Y., Zhang, G., Wu, C., Yang, G., Wang, C., and Li, G. (2008). *Chinese Physic Lett.*, 25, 1884.
[62] Ju, G., Hu, Y., Chen, L., Wang, X., Mu, Z., Wu, H., and Kang, F. (2012). *J. Lumin.*, 132, 1853.
[63] Blasse, G., and Bril, A. (1967). *J. Chem. Phys.*, 47, 5139.
[64] Forster, V. T. (1948). *Ann. Phys.*, 437, 55.
[65] Dexter, D. L. (1953). *J. Chem. Phys.*, 21, 836.
[66] Xu, C., Sheng, Y., Yuan, B., Guan, H., Ma, P., Song, Y., Zou, H., and Zheng, K. (2016). *RSC Adv.*, 6, 89984.
[67] Zhu, J., Cheng, W. D., Wu, D. S., Zhang, H., Gong, Y. J., Tong, H. N., and Zhao, D. (2008). *J. Alloys Compd.*, 454, 419.
[68] Ternane, R., Adad, M. T. C., Panczer, G., Goutaudier, C., Dujardin, C., Boulon, G., Ariguib, N. K., and Ayedi, M. T. (2002). *Solid State Sci.*, 4, 53.
[69] Blasse, G., Grabmaier, B. C. (1994). *Lumin. Mater.* Springer, Berlin, Heidelberg.
[70] Caldino, U., Speghini, A., and Bettinelli, M. (2006). *J. Phys. Condens. Matter*, 18, 3499.
[71] Wang, T., Xia, Z., Xiang, Q., and Qin, S. (2015). *Q. Liu. J. Lumin.*, 166, 106.
[72] Wu, L., Chen, X. L., Tu, Q. Y., He, M., Zhang, Y., and Xu, Y. P. (2003). *J. Alloys Compd.*, 358, 23.
[73] Wu, L., Chen, X. L., Zhang, Y., Kong, Y. F., Xu, J. J., and Xu, Y. P. (2006). *J. Solid State Chem.*, 179, 1219.
[74] Wu, L., Chen, X. L., Li, H., He, M., Xu, Y. P., and Li, X. Z. (2005). *Inorg. Chem.*, 44, 6409.
[75] Gravereau, P., Chaminade, J. P., Pechev, S., Nikolov, V., Ivanova, D., and Peshev, P. (2002). *Solid State Sci.*, 4, 993.
[76] Nagpure, P. A., Bajaj, N. S., Sonekar, R. P., and Omanwar, S. K. (2011). *Indian J. Pure. Appl. Phys.*, 49, 799.
[77] Nagpure, P. A., and Omanwar, S. K. (2012). *J. Lumin.*, 132, 2088.
[78] Tamboli, S., Rajeswari, B., and Dhoble, S. J. (2016). *Luminescence*, 31, 551–556.
[79] Singh, V., Sivaramaiah, G., Singh, N., Pathak, M. S., Rao, J. L., Singh, Pramod K, and Nagpure, A. S. (2019). *Bull. Mater. Sci.*, 42, 19.
[80] Nagpure, P. A., and Omanwar, S. K. (2015). *Indian J. Pure. Appl. Phys.*, 53, 77.
[81] Deng, T., Yan, S., and Hu, J. (Feb 2016). *J. Rare Earths*, 34, 2, 137.
[82] Kunghatkar, R. G., Hemne, P. S., and Dhoble, S. J. (2018). *AIP Conf. Proc.*, 1953, 080038, doi: 10.1063/1.5032844
[83] Bhagat, S. P., Gawande, A. B., and Omanwar, S. K., 10.1016/j.optmat.2014.11.043
[84] Gawande, A. B., Sonekar, R. P., and Omanwar, S. K. (2014). *Mater. Res. Bull.*, 60, e285–e291.
[85] Ramteke, D. D., and Gedam, R. S. (May 2014). *J. Rare Earths*, 32, 5, 389.
[86] Kunghatkar, R. G., Dhoble, S. J., and Hemne, P. S. (2016). *Luminescence*, 31, 1503, doi: 10.1002/bio.3136
[87] Pekgozlu, I., and Karabulut, H. (2009). *Inorg. Mater.*, 45, 1, 61.
[88] Sankar, R., and Rao, G. V. S. (1998). *J. Alloys Compd.*, 281, 126.
[89] Tascıoglu, S., Pekgozlu, I., and Mergen, A. (2008). *Mater. Chem. Phys.*, 112, 78.
[90] Pekgozlu, I., et al. (2008). *J. Lumin.*, 128, 9, 1541.
[91] Pekgozlu, I., Tascıoglu, S., and Menger, A. (2008). *Inorg. Mater.*, 44, 10, 1151.
[92] Blasse, G., Sas, S. J. M., Smit, W. M. A., and Konijnendijk, W. L. (1986). *Mater. Chem. Phys.*, 14, 253.
[93] Gawande, A. B., Sonekar, R. P., and Omanwar, S. K. (2013). *AIP Conf. Proc.*, 1536, 601.
[94] Pekgozlu, I., Erdogmus, E., Demirel, B., Gok, M. S., Karabulut, H., and Basak, A. S. (2011). *J. Lumin.*, 131, 2290–2293.
[95] Kunghatkar, R. G., Dhoble, S. J., and Hemne, P. S. (2016). *Luminescence.* 10.1002/bio.3136
[96] Nicholas, G. (Jan-Feb 2010). *Reed, Public Health Rep.*, 125, 1, 15–27.
[97] https://www.lighting.philips.com/main/products/uv-disinfection/air
[98] Kneiss, M. III-Nitride Ultraviolet Emitters Technology and Applications. *Springer Series in Materials Science* , ISBN 978-3-319-24098-5 10.1007/978-3-319-24100-5

6 Borate Phosphor
PDP Phosphor

R. G. Korpe

CONTENTS

DOI: 10.1201/9781003207757-6

6.1 LIGHTING TECHNOLOGIES

Light sources play an important role in our daily life, as it is difficult to imagine the world without different lamps used today. Different lamps are used for different purposes: Most of them exist in a multitude of wattages and voltages and with different lamp caps. The lamps can be as small as 0.1 watts or as large as 20 kilowatts and their physical appearance can be clear, frosted, white, coloured or mirrored.

The development of electric lamps dates from the second half of the 19th century. The first commercial electric lamps were incandescent light sources, introduced after Thomas Edison in the USA and Joseph Swan in England separately developed a lamp in 1878 in which electric current was passed through a filament made from carbon thread. At that time, there was also considerable interest in the use of electrical discharges through mixtures of gaseous vapours for lighting. It was in the 1930s that fluorescent lamps and high-pressure mercury lamps achieved commercial success.

6.2 CLASSIFICATION OF LAMPS

With regard to the principles of light generation and their relevant characteristics lamps are classified into filament lamps, gas discharge lamps and solid-state lamps (Figure 6.1).

6.2.1 FILAMENT LAMPS

The filament lamps are of two types: incandescent lamps and tungsten halogen lamps.

6.2.1.1 Incandescent Lamps

In incandescent lamps, light is produced by leading current through a tungsten wire. The working temperature of tungsten filament is about 2,700 K. Therefore, the main emission occurs in the

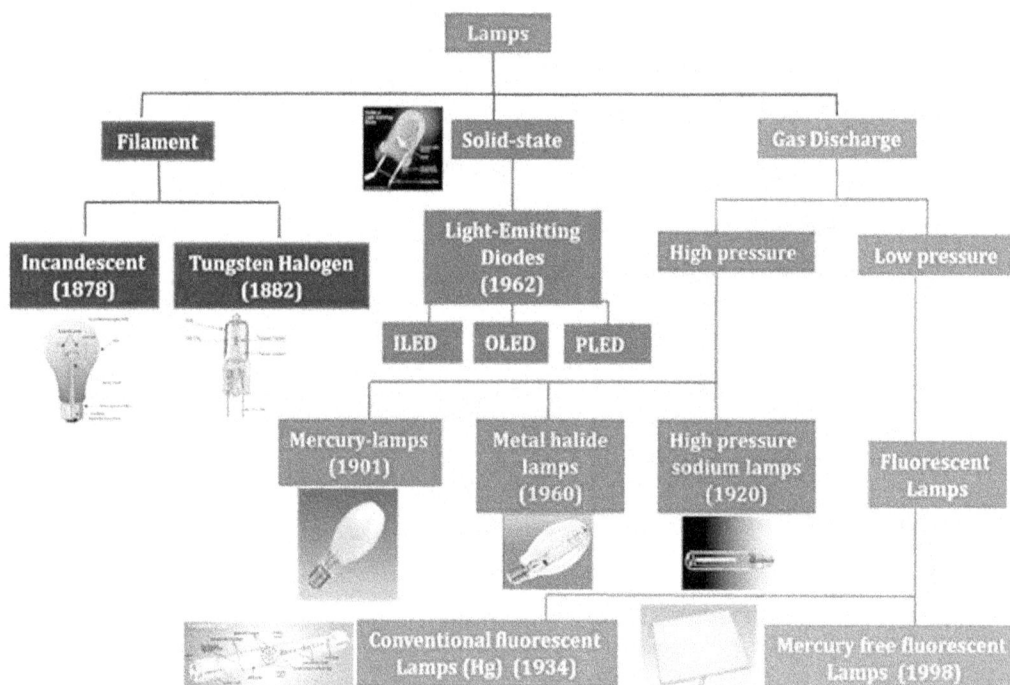

FIGURE 6.1 Classification of lamps on the principle of light generation.

TABLE 6.1
Advantages and disadvantages of incandescent lamps

Advantages	Disadvantages
Inexpensive	Short lamp life (1,000 hours)
Easy to use, small and does not need auxiliary equipment	Low luminous efficacy
Easy to dim by changing the voltage	Heat generation is very high
Excellent colour rendering properties	Lamp life and other characteristics are strongly dependent on the supply voltage
Directly work at power supplies with fixed voltage	The total costs are high due to high operation costs.
Free of toxic components	

infrared region. The typical luminous efficacy of different types of incandescent lamps is in the range between 5 and 15 lm/W. Advantages and disadvantages of incandescent lamps are listed in Table 6.1.

6.2.1.2 Tungsten Halogen Lamps

A tungsten halogen lamp is derived from an incandescent lamp. Inside the bulb, halogen gas limits the evapouration of the filament and redeposits the evapourated tungsten back to the filament through the so-called halogen cycle. Compared to incandescent lamp the operating temperature is higher, and consequently the colour temperature is also higher, which means that it is whiter. The lifetime spans from 2,000 to 4,000 hours and luminous efficacy is 12–35 lm/W. Advantages and disadvantages of tungsten halogen lamps are listed in Table 6.2.

6.2.2 SOLID-STATE LIGHTING

Solid-state lighting (SSL) refers to a type of lighting that utilizes light-emitting diodes (LEDs), organic light-emitting diodes (OLEDs) or polymer light-emitting diodes as a source of illumination rather than electrical filaments or gas discharge. The term "solid state" refers to the fact that light in an LED is emitted from a solid object – a block of semiconductor rather than form a vacuum or gas tube, as is the case in traditional incandescent lightbulbs and fluorescent lamps. It is the newest lighting technology to emerge in over 40 years and, with its energy efficiencies and cost savings, is one of the promising lighting technologies.

6.2.2.1 Light-Emitting Diodes (LEDs)

Electroluminescence was first discovered by Round in a crystal of silicon carbide and nearly 60 years later the first practical p-n junction emitter of red light was developed by Holonyak and Bevacqua [1].

TABLE 6.2
Advantages and disadvantages of tungsten halogen lamps

Advantages	Disadvantages
Small size	Low luminous efficacy
Directional light (narrow beams)	Surface temperature is high
Low-voltage alternatives	Lamp life and other characteristics are strongly dependent on the supply voltage
Easy to dim	High power consumption
Excellent colour rendering properties	Limited design variation due to filament restrictions

The development of light emitters from gallium nitride and related materials by Nakamura et al. in 1994 resulted in a new generation of white light emitters [2,3].

LEDs are the fastest-growing light technology for general lighting and in other lighting applications. A light-emitting diode (LED) is a solid-state semiconductor device that is closer to a semiconductor chip than a lighting device. It is a semiconductor device with a junction formed by joining p-type and n-type semiconductors together. Like a regular diode, electrical current flows across the junction in only one direction. When current is flowing, the LED is forward-biased, and the LED emits light. Typically, GaAlAs (red), AlGaInP (orange-yellowvgreen) and AlGaInN (green-blue), which all emit monochromatic light at a frequency corresponding to the bandgap of their semiconductor compounds, are used. Depending on the chemical composition of the semiconductor layers, the colour of light emission varies. The LED may be of different types, such as organic light-emitting diodes (OLEDs) and light-emitting polymers (LEPs) depending on the semiconductor layer organic or polymers. Advantages and disadvantages of LEDs are discussed in Table 6.3.

6.2.3 GAS DISCHARGE LAMPS

Gas discharge lamps are a light source that generates light by sending and electrical discharge through an ionized gas, plasma. The character of the gas discharge depends on the pressure of the gas as well as the frequency of the current. Typically, such lamps used a noble gas (argon, neon, krypton and xenon) or a mixture of these gases. Most lamps are filled with additional materials, like mercury, sodium and metal halides. In operation, the gas is ionized and free electrons are accelerated by the electrical field in the tube collide with gas and metal atoms. Some electrons in the atomic orbitals of these atoms are excited by these collisions to a higher energy state. When the excited atom falls back to a lower energy state, it emits a photon of a characteristic energy, resulting in infrared, visible light or ultraviolet radiation. Some lamps convert the ultraviolet radiation to visible light with the help of a phosphor coating on the inside of a lamp's glass surface. The gas discharge lamps are mainly classified into high-pressure and low-pressure discharge lamps.

6.2.3.1 High-Pressure Discharge Lamps

Without any temperature limitation (e.g. melting point of tungsten) it is possible to use gas discharge (plasmas) to generate optical radiation. Unlike thermal solid sources with continuous spectral emission, radiation from the gas discharge occurs predominantly in the form of single spectral lines. These lines may be used directly or after spectral conversion by phosphors for emission of light. A discharge lamp generates light of different colour quality, according to how the spectral lines are distributed in the visible range. To prevent runway current and ensure stable operation from a constant voltage supply, the negative current- voltage characteristics of a gas discharge lamp must be counterbalanced by a circuit element such as conventional magnetic or electronic ballasts. In all cases, higher voltages are needed for igniting the discharge.

TABLE 6.3

Advantages and disadvantages of LED lamps

Advantages	Disadvantages
Contains no mercury	High price
No optical heat on radiation	Low luminous flux/package
Long lifetime expectancy (with proper thermal management)	CRI can be low
Excellent low ambient temperature operation	Need for thermal management

---

6.2.4 Types of High-Pressure Discharge Lamps

6.2.4.1 Mercury Lamps

In mercury lamps, light is produced with electric current passing through mercury vapour. An arc discharge in mercury vapour at a pressure of about 2 bars emits five strong spectral lines in the visible wavelengths at 404.7 nm, 435.8 nm, 546.1 nm, 577 nm and 579 nm. The red-gap is filled up by a phosphor layer at the outer bulb. Typical values of these lamps are luminous efficacy 40–60 lm/W, CRI between 40 and 60 and CCT 4,000 K. The lamp life is 12,000 hours. Advantages and disadvantages of a mercury lamp are discussed in Table 6.4.

6.2.4.2 Metal Halide Lamps

To increase the luminous efficacy and CRI of mercury high-pressure lamps, it is useful to add a mixture of metal components to the filling of the discharge tube. These additives emit their own line spectra in the arc discharge, leading to an enormous diversity of light colour. For sufficient vapour pressure, it is better to use metal halides (compounds with iodine or bromine) instead of elemental metals. When the vapour enters the high-temperature region of the discharge, molecules dissociate and metal atoms are excited and radiation is emitted.

The applications of metal halide lamps reach from electric torches (10 W miniature variants) to diverse purposes in indoor lighting (wattages up to 20 kW). The lamps are available with luminous efficacy typically from 50 to 100 lm/W, CCT value from 3,000 to 6,000 K and CRI from 70 to over 90. The lamp life is typically from 6,000 to 12,000 hours. Advantages and disadvantages of metal halide lamps are listed in Table 6.5.

6.2.4.3 High-Pressure Sodium Lamps

In high-pressure sodium lamps light is produced by sodium vapour, the gas pressure being about 15 kPa. The golden-yellowish emission spectrum applies to wide parts of the visible area. The CRI is low (≈ 20), but the luminous efficacy is high. The most common application today is in street and road lighting. Luminous efficacy of the lamps is 80–100 lm/W and lamp life is 12,000 to 16,000 hours. The CCT is 2,000 K.

TABLE 6.4
Advantages and disadvantages of a mercury-lamp

Advantages	Disadvantages
• Extremely good alternative to incandescent lamps	Use of mercury
• Longer service life than incandescent lamps	Limited to general lighting purposes
• No ballast or igniter necessary	Considerable amount of time to warm up

TABLE 6.5
Advantages and disadvantages of metal halide lamps

Advantages	Disadvantages
Good luminous efficacy	Expensive
Alternatives with good colour rendering available	Starting and re-starting time 2–minutes
Different colour temperatures available	Lamp life strongly dependent on supply voltage

TABLE 6.6

Advantages and disadvantages of high-pressure sodium lamps

Advantages	Disadvantages
Very good luminous efficacy	Low CCT, about 2,200 K
Long lamp life 12,000 to 16,000 hours	Low CRI about 20 (colour improved 65, white 80)
High luminous flux for street lighting	Starting and re-starting time 2–5 minutes

An improvement of the CRI is possible by pulse operation or elevated pressure, but this reduces the luminous efficacy. Colour improved high-pressure sodium lamps have CRI of about 65 and white high-pressure sodium lamps of more than 80. There CCT is 2,200 and 2,700, respectively. Advantages and disadvantages of metal halide lamps are listed in Table 6.6.

6.2.4.4 Low-Pressure Discharge Lamps

These lamps have gas inside the tube, with lower pressure than the atmospheric pressure. The classic fluorescent lamp is the best example of low-pressure fluorescent lamps.

6.2.4.5 Fluorescent Lamps

A fluorescent lamp is a low-pressure gas discharge light source, in which light is produced predominantly by fluorescent powders activated by ultraviolet radiation generated by discharge in mercury. The lamp, usually in the form of a long tubular bulb with an electrode at each end, contains mercury vapour at low pressure with a small amount of inert gas for starting. Figure 6.2 demonstrates operation principle of a fluorescent lamp [4]. The majority of the emission (95%) takes place in the ultraviolet (UV) region and the wavelengths of the main emission peaks are 254 nm and 185 nm. Hence, the UV radiation is converted into light by a phosphor layer on the inside of the tube. Since one UV-photon generates only one visible photon, 65% of the initial photon energy is lost as dissipation heat. On the other hand, the final spectral distribution of emitted light can be varied by different combinations of phosphors. Correlated colour temperatures (CCT) vary from 2,700 K (warm white) and 6,500 K (daylight) up to 17,000 K and colour rendering indices (CRI) from 50 to 95 are available. The luminous efficacy of the latest T5 fluorescent lamp is up to

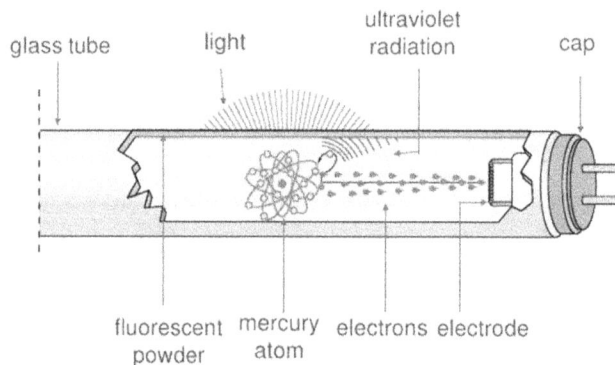

FIGURE 6.2   Operation principle of a fluorescent lamp.

Source: For details, see: Bommel, W. V. 2015. Interior Lighting Fundamentals, Technology and Application. Springer, Cham.

100 lm/W (without ballast losses). Dimming is possible down to 1% of the normal luminous flux, and with special high-voltage pulse circuits down to 0.01%.

Fluorescent lamps display negative voltage-current characteristics, requiring a device to limit the lamp current. Otherwise, the ever-increasing current would destroy the lamp. Pure magnetic (inductive) ballast needs an additional starting element such as a glow switch. Electronic control gear incorporates all the equipment necessary for starting and operating a fluorescent lamp. Compared to conventional magnetic ballasts that operate lamps at a line frequency of 50 Hz (or 60 Hz), electronic ballasts generate high-frequency currents, most commonly in the range of 25–50 kHz. High-frequency operation reduces the ballast losses and also makes the discharge itself more effective. Other advantages of the electronic ballasts are that the light is flicker-free and there is the opportunity of using dimming devices.

Fluorescent lamps are generally categorized by the electrode type. There are many electrode types with different shapes and materials but for fluorescent lamps there are mainly two types of operational modes. Fluorescent lamps that used coiled tungsten can be categorized as hot cathode lamps; these operate at high temperatures to liberate electrons and are the most common type. The other type is cold cathode fluorescent lamps (CCFL) which operates at high voltage to liberate electrons. Unlike conventional fluorescent lamps or compact fluorescent lamps (CFLs), the cathodes used in CCFLs are metals that are constructed to maximize surface area. Nickel is generally selected as the cathode material because it results in easier manufacturing, better electrical properties and lower material cost [5–7]. In addition to the different operational mode, CCFLs have the advantages of smaller size, longer lifetime and higher efficiency [8]. Therefore, they have been used as light sources for LCD panels. There are other types of cold cathode lamps including external electrode fluorescent lamps (EEFL) and flat fluorescent lamps (FFL). EEFLs have other benefits over conventional CCFLs, the most attractive of which is the absence of electrodes in the discharge tube, which is the major factor that limits the lifetime of the lamp [9–13]. An FFL is a flat shaped CCFL. The operational theory is the same as for the CCFL and the electrodes can be both internal and external. It has better light uniformity when it is applied as the light source for LCD panels [14–17]. Although cold cathode lamps offer better performance than conventional FLs and CFLs, they require a discharge tube and a tube that contains at least a milligram of mercury.

Despite the many advantages that CCFLs have over conventional FLs, the operation principles and required materials remain the same as FLs. In one of the most efficient FLs, only 63% of energy can be converted to UV radiation. After the UV radiation is converted to visible light by the phosphor, the overall efficiency of the fluorescent lamp cannot exceed 28% [18]. All fluorescent lamps contaminate the environment with mercury when broken or disposed of after use. Because of their shapes and the structure of the materials, collection and recycling of the lamps is not applicable to every situation. Therefore, CCFL for LCD backlights are fast disappearing due to the development of more efficient, more environmentally friendly and brighter sources of light like mercury-free fluorescent lamp (MFFL) and LEDs. Advantages and disadvantages of LED lamps are listed in Table 6.7.

TABLE 6.7
Advantages and disadvantages of fluorescent lamps

Advantages	Disadvantages
Inexpensive	Ambient temperature affects the switch-on and output
Good luminous efficacy	Need of auxiliary ballast and starter or electronic ballast
Long lamp life, 10,000–16,000 h	Light output depreciates with age (shorten lamp life)
Large variety of CCT and CRI	Contain mercury

6.2.4.6   Mercury-Free Fluorescent Lamp (MFFL) Technology

The mercury-free fluorescent lamp (MFFL) technology is a modern and distinguished technique to fabricate and develop a lamp with environmentally friendly technology (Hg free), fast switching cycles, temperature independent, dimmable, high lifetime and discharge efficiency ~65%. Due to the merits over present lighting technology, it has attracted the attention of researchers for the development of next-generation lamps with various applications such as general lighting, LCD backlight, scanning, photo copy lamp, lighting in automobile industry, PDP, etc. In MFFL, phosphors generate light when excited by plasma discharge between two glass plates. The gas discharge contains no mercury (contrary to the conventional fluorescent lamps); a mixture of noble gas (neon and xenon) is used instead. This gas mixture is inert and entirely non-harmful [19–22]. The phenomenon of generation of visible light in MFFL is shown in Figure 6.3. In the development of MFFL, the main ingredient, which is legitimately responsible for producing visible light, is phosphor. Phosphors in the MFFLs are mainly excited by vacuum-ultraviolet (VUV) radiation lines of Xe atoms at 147 nm and Xe_2 molecular line at 172 nm wavelengths.

6.3   PLASMA DISPLAY PANELS (PDP) TECHNOLOGY

Plasma display panels (PDPs) can be defined as flat-panel information display devices in which light is created by phosphors excited by a plasma discharge between two flat panels of glass. The gas discharge contains no mercury (contrary to the backlights of an AMLCD), a mixture of noble gases (neon and xenon) is used instead. This gas mixture is inert and entirely harmless [23].

6.3.1   History and Development of PDP

The plasma display panel was invented at the University of Illinois at Urbana Champaign by Donald L. Bitzer and H. Gene Slottow in 1964 for the PLATO computer system. The original monochrome (usually orange or green) panels enjoyed a surge of popularity in the early 1970s because the displays were rugged and neither needed memory nor refresh circuitry. There followed a long period of sales decline in the late 1970s as semiconductor memory made CRT displays incredibly cheap. On the other hand, plasma's relatively large screen size and thin profile made the displays attractive for high-profile placement, such as lobbies and stock exchanges. In 1983, IBM introduced a 19-inch orange on black monochrome display (model 3290 – information panel) which was able to show four simultaneous 3270 virtual machine (VM) terminal sessions. In 1992, Fujitsu introduced the world's first 21-inch full colour display. It was a hybrid based on the plasma display created at the University of Illinois at Urbana-Champaign and NHK STRL, achieving

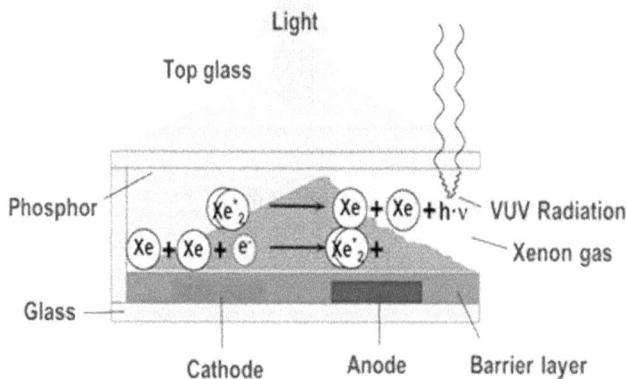

FIGURE 6.3   Generation of visible light in MFFL.

superior brightness. In 1997, Pioneer started selling the first plasma TV to the public. Screen sizes have increased since the 21-inch display in 1992. The largest plasma display in the world was shown at the CES (Consumer Electronics Show) in Las Vegas in 2006. It measured 103 inches and was made by Matsushita Electrical Industries (Panasonic).

Until quite recently (2004), the superior brightness and viewing angle of colour plasma panels when compared to LCD, made them one of the most popular forms of display for HDTV. However, since that time, improvements in LCD technology have closed the gap dramatically. The much lower weight, price and power consumption of LCDs have seen them make large inroads into the former plasma market. Sony now only sells a very limited range of plasma screens and appears set to quit this market altogether [24,25].

6.3.2 ELEMENTS OF PDP TECHNOLOGY

A plasma display panel (PDP) is essentially a collection of very small fluorescent type lamps, each a few tenths of a millimetre in size. If we look closely, it is easy to distinguish the individual PDP cells the tiny colour elements of red, green and blue light that together form what is called a pixel (Figure 6.4). In reality, the front and back electrodes run perpendicular to each other rather than parallel [26]. As in a fluorescent lamp, the light we see does not come from the plasma directly but rather from the phosphor coatings on the inside walls of the cells when they are exposed to vacuum ultraviolet (VUV) radiation emitted by the plasma. Because each cell emits its own light, a plasma display panel is called an *'emissive display'*. This contrasts with the familiar liquid crystal display (LCD), a type of flat display in which the light comes from a lamp (actually a plasma lamp) behind the liquid crystal, which has arrays of small switches controlling where light is allowed to pass through.

All plasmas require a source of energy. As in fluorescent lamps, the plasma in a PDP is produced by applying a voltage across a gap that contains gas (mix of neon and xenon gases). The plasmas used in PDPs are considered *'cold'* plasmas in the sense that the background gas stays relatively cold while the electrons (and ions) in the plasma are heated by the applied voltage. When the hot electrons collide with the background gas atoms and transfer energy to them many of those atoms respond by emitting VUV radiation (150–180 nm). The operating conditions of the display (gas composition, pressure, voltage, geometry, etc.) represent a compromise taking into account performance requirements such as low voltage operation, long life, high brightness and high contrast.

The plasma display itself is a simple device consisting of two parallel glass plates separated by a precise spacing of some tenths of a millimetre and sealed around the edges. The space between the plates is filled with a mixture of rare gases at a pressure somewhat less than one atmosphere. Parallel stripes of transparent conducting material with a width of about a tenth of a millimetre are deposited on each plate, with the stripes on one plate perpendicular to those on the other. These stripes are the

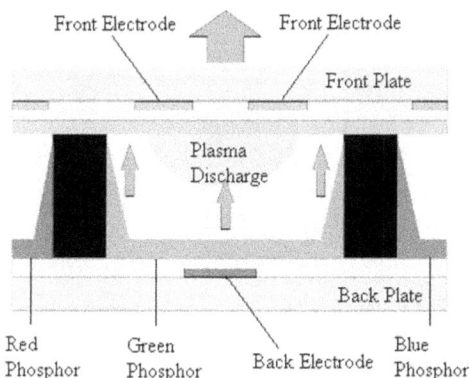

FIGURE 6.4 Schematic of a plasma display pixel. In reality, the front and back electrodes run perpendicular to each other rather than parallel.

Source: For details see: Nagpure, P. A. 2012. Thesis submitted to Sant Gadgw Baba Amravati University, Amravati.

'*electrodes*' to which voltages are applied. The intersections of the rows of electrodes on one side and the columns of electrodes on the opposite glass plate define the individual colour elements (or cells) of a PDP. For high-quality colour images, it is important to keep the VUV radiation from passing between cells. To isolate the individual cells barriers are created on the inside surface of one of the plates before sealing. Troughs, honeycomb-like structures and other shapes have been used (Figure 6.5). The red, green and blue phosphors are deposited inside these structures.

An important feature of PDPs is that the plasma in each individual cell can be turned on and off rapidly enough to produce a high-quality moving picture. (To help turn the individual cells on and off, there are actually two electrodes on one side and a third electrode on the opposite side of each cell.) Switching the cells on and off cheaply and efficiently is now possible because of advances over the past 20 years in the miniaturization and efficiency of electronics.

A commercial panel consists of several million cells which have to be switched at a rate that will create 60 TV picture frames per second. A computer translates an image into a sequence of on and off voltage pulses which are applied to the electrode arrays line by line and row by row to select individual cells. Such control is possible because the plasma is fast and can respond to the voltage pulses in a millionth of a second. The complexity increases significantly when we consider that each small picture element or pixel consists of three colour cells and each colour cell can display 256 intensity levels. Thus, each pixel can display over 16.7 million (or more exactly, 256 x 256 x 256) colours. Variation in light intensity from a particular cell is not accomplished by changing the voltage or the current through the cell. Rather it is achieved by changing the length of time that the cell is *on* during one TV frame. Since the eye response is slower than the TV picture frequency, it perceives different colours depending on how long each cell is *on*. Each company has made its own contribution to the switching systems to improve efficiency, speed and performance.

Many years of research and development as well as major advances in electronics and manufacturing techniques have led to the plasma display panels, we see in the market today. The plasma display panel itself was invented in 1964 by researchers at the University of Illinois, with the first PDPs being single colour (or monochrome) displays. Research on multi-colour PDPs was going strong in the 1980s and the first commercially available colour displays appeared in the late 1990s. It is now possible to manufacture PDPs with diagonals as large as 80 inches and with a thickness of only 3 to 4 inches. Considerable progress has also been made recently to reduce the power consumption and increase the efficiency and lifetime of PDPs.

Large screens, excellent image quality, brightness and greater than 160° viewing angle characterise today's plasma panels, which are perfectly flat and perform well even in bright environments.

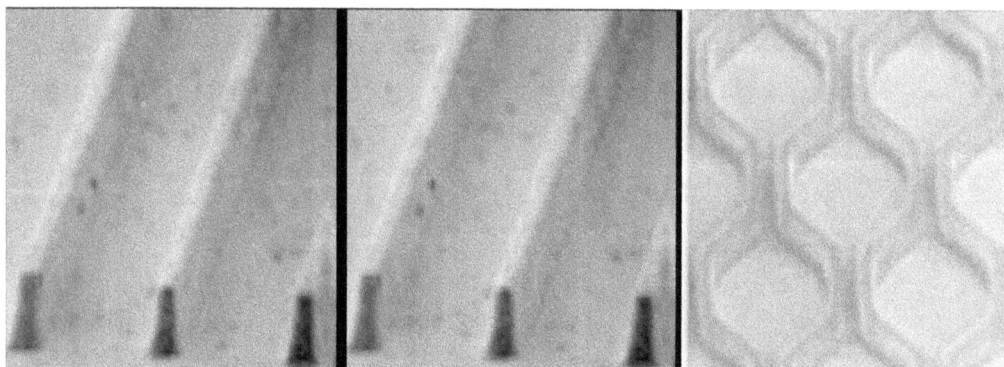

FIGURE 6.5 Examples of the barriers that isolate each cell. The distance between the walls of each cell is a couple hundred micrometers or about ten times the diameter of a human hair.

Source: For details see: Nagpure, P. A. 2012. Thesis submitted to Sant Gadgw Baba Amravati University, Amravati.

Long-term commercial success of the PDP now lies in the manufacturers' ability to produce low-cost displays. Significant reductions in cost have been realized over the last few years and new PDP designs and processes are being introduced all the time, continuing to reduce production cost. The PDP is sure to be one of the predominant large format displays of the future [27–29].

6.3.3 Mechanism of VUV Radiation Generation

The mercury-free fluorescent lamp is fundamentally a Xe discharge lamp with a phosphor layer inside the flat glass [30]. This emission mechanism, based on Xe-gas discharge, has already been utilized in plasma display panels (PDPs). The phosphors of the mercury-free fluorescent lamp and PDP are almost identical compounds. A difference between the present mercury-based fluorescent lamp and the mercury-free fluorescent lamp is the radiation wavelength of the gas discharge. The mercury-free fluorescent lamp contains a rare gas mixture of Xe-Ne. Xe is used as VUV emitter and Ne is used as a buffer gas to lower the breakdown voltage. Potential discharge gases must have high radiation intensity characteristics, be chemically stable and exhibit a weak visible emission in order to prevent degradation of colour purity character. When the xenon molecules are electrically excited, Xe excimer state (excimer stands for excited dimmer) is formed. When it returns to the original state, intense VUV light is produced. The range of VUV radiation in electromagnetic spectrum is from 100 to 200 nm.

The generation of VUV radiation line by bombardment of electron with atoms or molecules of Xe gas is shown in Figure 6.6 and described below.

Xe discharge has a sharp VUV emission at 147 nm (Xe resonance line) and a broad emission around 172 nm (Xe excimer band). Their ratio depends on gas pressure in the discharge cell. The VUV radiation output of the plasma source depends on the gas component, composition and pressure. The energy conversion efficiency of the VUV in the range of 7–12 eV into the visible is generally higher at lower energies. Thus, xenon gas is preferred constituent for the MFFL. In the PDP and the mercury-free fluorescent lamp, the resonance line (147 nm) provides the main radiation. Vacuum ultraviolet (VUV) irradiation from the Xe gas is converted into visible light by various VUV phosphors. Highly efficient VUV phosphors are required for mercury-free fluorescent lamps because the main emission peak of this discharge is located at a shorter wavelength (147 nm) than that of the conventional mercury discharges (254 nm).

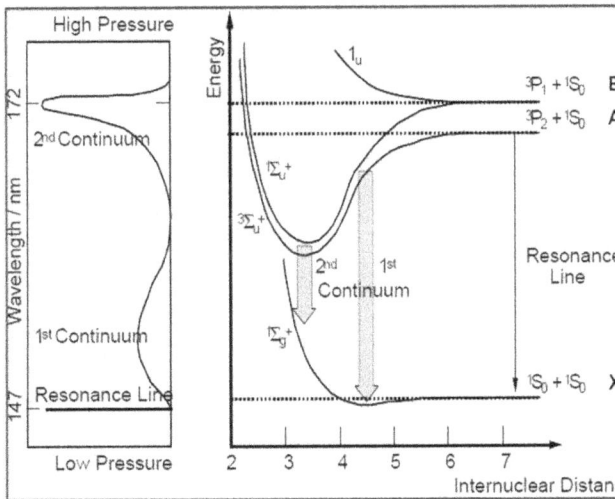

FIGURE 6.6 Schematic diagram of generation of VUV radiation lines.

$$Xe + e^- \rightarrow Xe\left(^3P_1\right) + e^-$$

$$\rightarrow Xe\left(^3P_2\right) + e^-$$

$$\rightarrow Xe**$$

$$Xe** \rightarrow Xe\left(^3P_1\right) + h\nu \ (828 \ nm)$$

$$\rightarrow Xe\left(^3P_2\right) + h\nu(823 \ nm)$$

$$Xe\left(^3P_1\right) \rightarrow Xe + h\nu(\mathbf{147 \ nm}) \ – \ ––––\textbf{Resonance radiation of Xe atoms}$$

$$Xe\left(^3P_1\right) + Xe + M \rightarrow Xe_2 * + M$$

$$Xe_2 * \rightarrow 2Xe + h\nu(\mathbf{150 \ and \ 172 \ nm}) \ \ – \ –––– \ \textbf{Excited state of Xe excimer}$$

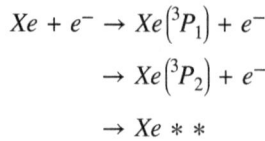

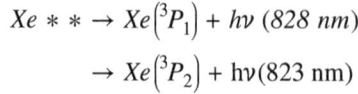

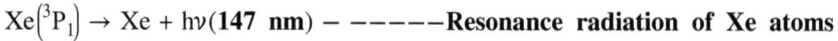

6.3.4 Requirements for VUV phosphors in a PDP

An understanding of the role that phosphors play in colour display technology requires an examination of the response of the human eye to colour. Visible light constitutes only a small part of the electromagnetic spectrum and the portion perceived by the human eye covers a fairly narrow band from 380 nm to 780 nm. Not all human eyes react the same, so a quantitative method for describing colour is necessary and can be achieved by testing the perception of many individuals and averaging the results.

The main ingredient in PDP, which is legitimately responsible to produce visible light, is phosphor. In PDPs and mercury-free fluorescent lamps are excited by VUV emission of 147 and 172 nm from a discharge in a gas containing Xe. For both applications phosphors are needed that efficiently convert VUV into visible light. Phosphors applied in PDP are excited by host-lattice adsorption, contrary to the phosphors in fluorescent lamp (FL), in which excitation mostly takes place on optical centres Figure 6.7(a,b).

As a consequence, the penetration depth for VUV radiation is very small (about 0.1–1 um). This requires rather perfect VUV phosphor surfaces to minimize surface losses. Crystal defect concentration within the first atomic layers involved in VUV interaction is to be tackled. High-efficiency lamp phosphors require intense and wide absorption bands (via host, f-d or CT) and efficient energy transfer from the excited levels to the emission centre (sensitized luminescence).

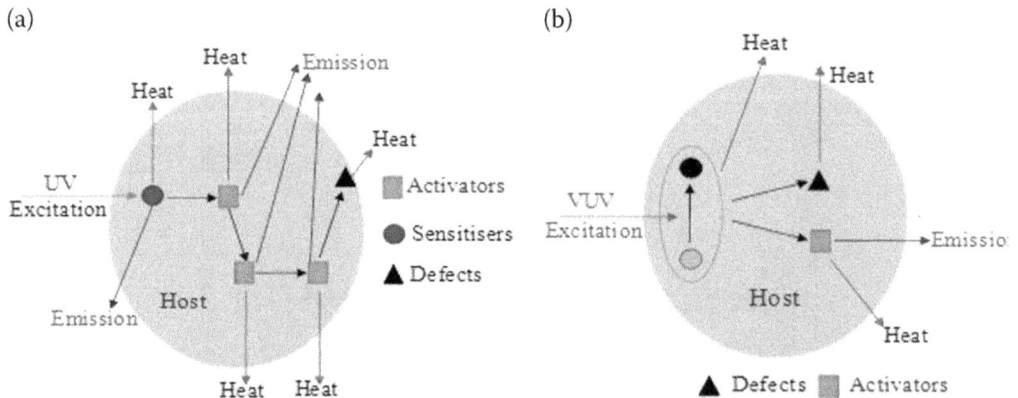

FIGURE 6.7 (a) and (b) Demonstrate excitation and emission phenomenon in FL and MFFL phosphors.

On the contrary, high-efficiency VUV phosphors require efficient energy transfer from host lattice to activator. This means host lattice (exciton like) emission following VUV absorption should overlap well with the excitation spectra of activator ion [31]. Rare-earth-ion doped compounds have received attention for their high luminescence efficiency [32,33]. However, many problems remain, including deterioration of phosphors, surface damage and poor excitation efficiency because of their high-energy VUV excitation. For VUV phosphors, the host employed must have high chemical stability in the high-energy plasma.

6.3.5 POSSIBLE CAUSES OF PHOSPHOR DAMAGE IN PDP

Phosphors based on sulphides (emission of S or SO_2 gas) and fluorides (rapid discolouration) have been found to be unstable in VUV discharge environment. Hence, host lattice based on oxides, aluminates or borates are generally preferred for VUV applications. The possible causes of phosphor damage in VUV applications are as follows:

- Ionic bombardment
- Oxygen depletion and formation of oxygen vacancies
- Surface amorphization
- Trapping of electron-hole pairs and colour center production
- Changes in host lattice
- Thermal damage in air
- Activator valence conversion (eg., $Eu^{2+} \rightarrow Eu^{3+}$)
- Binder burn-off during panel baking (in case of flat panel MFFL lamp)

6.3.6 THE QUANTIFICATION OF COLOUR

Colours could be duplicated by mixing the three primaries of red, green, and blue to obtain various hues, including shades of white. Colour perceived from a self-emitting object is typically described by its hue, brightness and saturation. Hue is the attribute that we denote by red, yellow or green and is determined by the dominant wavelength. The degree to which a colour differs from white describes its saturation and provides a description of the purity of chromaticity. Finally, brightness or luminance is the attribute used to describe the perceived intensity of light [34]. These attributes describe colour in a qualitative manner. For scientific purposes, a quantitative method of describing colour is necessary. To do this, the CIE system developed by the Commission Internationale de l'Eclairage (International Commission on Illumination) is used.

It is based on the tri-chromatic theory and deals with three fundamental aspects of colour: the object, the light source and the observer [35]. A set of tri-stimulus values (X, Y, Z), are used to denote the intensities of the colours in any arbitrary sample spectra. Although it is a simplified way of thinking of the tri-stimulus values, one may view these as the amount of red, green and blue in a sample. The tri-stimulus values X, Y, Z can be calculated by using the colour matching functions given in Figure 6.8 [36].

As noted previously, these functions have been derived by determining the average response of the human eye. Each tri-stimulus value for a sample of interest is calculated by integrating the product of the power output of the sample and the colour matching function:

$$X = \int_- S(\lambda).\, x(\lambda).\, d\lambda \qquad\qquad (1.3,\ a)$$

$$Y = \int_- S(\lambda).\, y(\lambda).\, d\lambda \qquad\qquad (1.3,\ b)$$

$$Z = \int_- S(\lambda).\, z(\lambda).\, d\lambda \qquad\qquad (1.3,\ c)$$

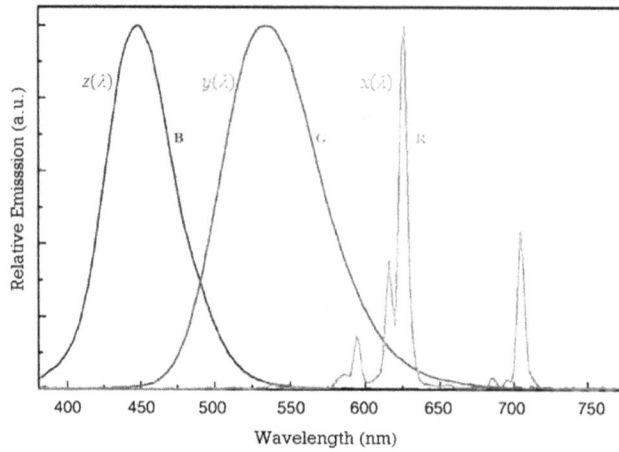

FIGURE 6.8 Colour matching functions for the CIE standard observer.

Source: For details see: Ingle, J. T. 2015. Thesis submitted to Sant Gadge Baba Amravati University, Amravati.

Chromaticity coordinates x, y and z are then derived by normalizing each of the tri-stimulus values as follows. Since the sum of x, y and z must equal 1, we need only to consider the pair (x, y) which is commonly plotted on a CIE chromaticity diagram that gives a quantitative value of colour (Figure 6.9). Colours with chromaticity coordinates near the perimeter of the diagram are regarded as being saturated, whereas true white light has equivalent chromaticity coordinates, i.e. x = y = 0.33 [37].

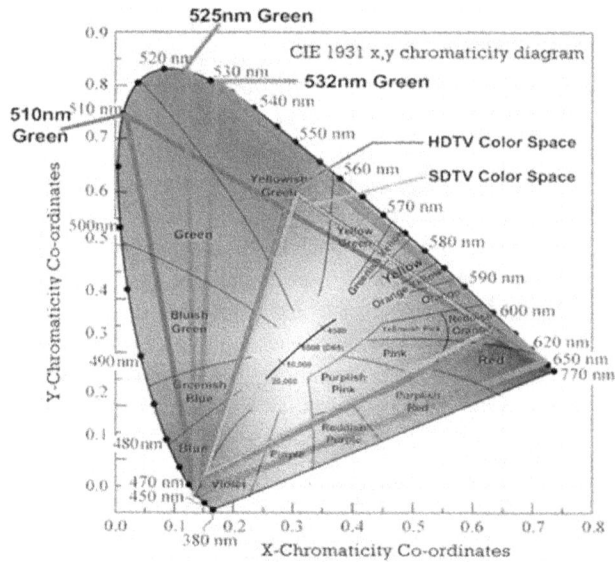

FIGURE 6.9 CIE chromatic colour rendering diagram.

Source: For details see: Ingle, J. T. 2015. Thesis submitted to Sant Gadge Baba Amravati University, Amravati.

6.4 PRESENTLY USED BORATE-BASED PDP PHOSPHORS

The luminescence efficiency of a PDP depends upon various components such as phosphors, gas mixture, dielectric layer, reflective layer, black matrix, etc. The phosphor for PDP applications should have good luminescent characteristics under vacuum ultraviolet (VUV) light consisting of the resonance radiation of Xe atoms (147 nm) and the excited state of molecular Xe (172 nm). Among various host materials, borates and oxyborates have been extensively investigated as host lattices for luminescent materials because of their large bandgap, strong nonlinear optical properties, chemical and environmental stabilities and mechanical robustness. Most borate hosts are transparent up to 140–180 nm, so that the VUV light can directly excite impurity activator in these hosts [38]. The strong absorption due to the impurity ions would give efficient conversion of the VUV light. Several Inorganic Borate host materials have been reported as PDP phosphors considering their potential application in VUV region [39].

6.4.1 Red/Orange-Emitting Phosphors

In most of the PDPs devices well known and conventional phosphors, like Y_2O_3: Eu and (Y, Gd) BO_3:Eu as red, which were already known as lamp, CTV phosphors are used as PDP materials. However, quite consistent problems are associated with them; for example, Y_2O_3: Eu phosphor has less luminescence efficiency [40]. YBO_3:Eu^{3+} while $(Y,Gd)BO_3$: Eu has poor colour purity as it generates orange-red emission instead of red [41–43]. Number of inorganic borate materials doped with Eu^{3+} which were used in conventional devices are now used in commercial applications. $BaGdB_9O_{16}$: Eu^{3+} [44], is preferred as a red VUV phosphor for PDP. $BaZr(BO_3)_2$:Eu^{3+} [45] and $YCa_4O(BO_3)_3$: Eu^{3+} [46] phosphors are said to have higher efficiency as compared to a widely used $(Y,Gd)BO_3$: Eu^{3+} phosphor and proves to be a potential candidate as a red phosphor in PDP and Xe lamps. In the recent years several researchers have contributed to enhance the luminescence performance, improvement in the colour purity and increase the quantum yield of red PDP phosphors. Several important materials reported with achieved target and properties for red PDP phosphors have been listed in the following Table 6.8. Several PDP nano-phosphors have also been reported [47]. (Several borates of Y, Gd, Sc and Lu are acknowledged to be good host materials at VUV excitation so that they can be used for PDPs. When activated with Eu^{3+} ions they emit a considerably bright red light at the VUV excitation [48]. Gadolinium aluminium borate crystallizes in hexagonal form where the Gd^{3+} ions are separated from each other by BO_3 groups without Gd_{3+} ions sharing the same oxygen ion. The colour coordinates of this emission is close to that required by the NTSC colour standard. In case of $(Y,Gd)Al_3(BO_3)_4$:Eu, the emission spectrum is dominated by 617 nm peak more favourable as a TV phosphor. Eu^{3+} activated yttrium lithium borate,

---

TABLE 6.8

The samples of different rare earth–doped concentration

Sr. no.	Molecular formula	$Y(NO_3)_3 . 6H_2O$	$Gd(NO_3)_3 . 6H_2O$	$Eu(NO_3)_3 . 6H_2O$ m/g	$H_2C_2O_4$	H_3BO_3
1	$(Y_{0.85}Gd_{0.1})BO_3$:$Eu_{0.05}$	13.0224	1.8051	0.8920	5.0065	2.4730
2	$(Y_{0.83}Gd_{0.1})BO_3$:$Eu_{0.07}$	12.7149	1.8048	1.2487	5.0074	2.4731
3	$(Y_{0.8}Gd_{0.1})BO_3$:$Eu_{0.1}$	12.2552	1.8056	1.7836	5.0079	2.4733
4	$(Y_{0.7}Gd_{0.1})BO_3$:$Eu_{0.2}$	10.7261	1.8049	3.5679	5.0078	2.4732
5	$(Y_{0.9}Gd_{0.05})BO_3$:$Eu_{0.05}$	13.7874	0.9027	0.8925	5.0076	2.4734
6	$(Y_{0.8}Gd_{0.15})BO_3$:$Eu_{0.05}$	12.2553	2.7081	0.8929	5.0070	2.4731
7	$(Y_{0.75}Gd_{0.2})BO_3$:$Eu_{0.05}$	11.4902	3.6179	0.8927	5.0073	2.4730

$Li_6(BO)_3$ and lithium lanthanum borate, $Li3La_2(BO_3)_3$ and similar oxyborate phosphors are being considered for PDP applications due their colour point in the deep red. It has been reported that in lanthanum-based phosphor Gd^{3+} plays an important role in acting as a sensitizer. The CTS excitation band of Eu^{3+} is enhanced by co-doping Al^{3+} into the $BaZr(BO_3)_2$ lattice. The stability has been improved by replacing a small portion of Y_2O_3 with Gd_2O_3. The complex of $(Y,Gd)_2O_3$ phosphor exhibits the same level of brightness and persistence with better stability. By blending $(Y,Gd)BO_3$ with $(Y,Gd)_2O_3$, colour purity of the phosphor can be improved with a little loss in brightness [49]. It is reported that the brightness of these oxides can also be improved by co-doping with L^+. Li^+ substitution in the lattice leads to a decrease in interstitial oxygen an increase in the quantum yield. Currently, PDP manufacturers are not willing to replace $(Y,Gd)BO_3:Eu^{3+}$ as a red-emitting phosphor [50].

The $(Y, Gd)BO_3:Eu^{3+}$ phosphors were prepared by the co-precipitation method. Table 6.8 shows the different rare earth–doped concentrations in the preparation of $(Y, Gd)BO_3:Eu^{3+}$ phosphors. The samples were characterized by thermo-gravimetric analyzer, X-ray powder diffraction and vacuum ultraviolet (VUV) test system for the plasma display panel.

The $(Y, Gd)BO_3:Eu^{3+}$ phosphors are found to have the best luminous performance at 611 nm with respect to intensity and colour purity under VUV excitation. At the same time, the $(Y,Gd)BO_3:Eu^{3+}$ phosphors prepared by co-precipitation method show high crystallinity, high distribution and good luminescence intensity. The excitation spectra of different prepared Eu-doped $(Y_{0.9}-xGd_{0.1})BO_3:Eu_x$ phosphors are shown in Figure 6.10. The excitation spectra of Eu-doped $(Y_{0.9}-xGd_{0.1})BO_3:Eu_x$ phosphors, it is noted that the location and shape of the excitation spectra of $(Y_{0.9}-xGd_{0.1})BO_3:Eu_x$ phosphors change with the change of Eu doping concentration.

With the increase of concentration of Eu, the relative intensity decreases and the excitation peak of 214 nm is disappeared. When the concentration of Eu-doped is relatively low (x = 0.05), the intensity of excitation spectra shows the maximum value.

Figure 6.11 displays the excitation spectra of different Gd-doped$(Y_{0.95}-yGdy)BO_3:Eu_{0.05}$ phosphors, it can be seen that the location and shape of the excitation spectra of $(Y_{0.95}-yGdy)BO_3:Eu_{0.05}$ phosphors unchanged with the change of Eu doping concentration. However, the relative intensity increases with an increase of Gd concentration when the concentration of Gd-doped

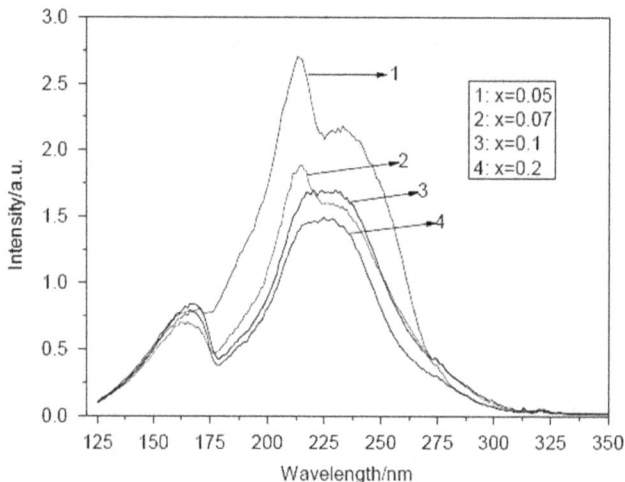

FIGURE 6.10 The excitation spectra of Eu-doped $(Y_{0.9}-xGd_{0.1})BO_3:Eu_x$ phosphors.

Source: For details see: "Luminescent properties of (Y, Gd) BO$_3$: Eu^{3+} under VUV excitation for PDP prepared by co-precipitation method" by Hu, Y., Tao, Y., Huang, Y., Yu, X., Zhang, C., Liang,T., Yu, J., (2011). Optoelectronics and Advanced Materials – Rapid Communications, 5(4), 348–352.

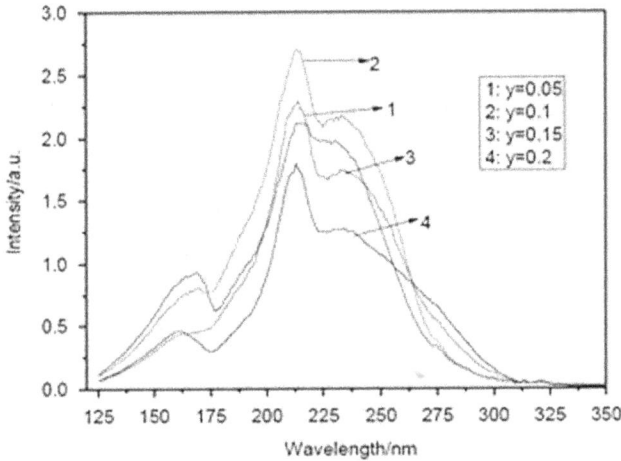

FIGURE 6.11 The excitation spectra of Gd-doped $(Y_{0.95}\text{-}yGd_y)BO_3$: $Eu_{0.05}$ phosphors.

Source: For details see: "Luminescent properties of (Y, Gd) BO_3: Eu^{3+} under VUV excitation for PDP prepared by co-precipitation method" by Hu, Y., Tao, Y., Huang, Y., Yu, X., Zhang, C., Liang,T., Yu, J., (2011). Optoelectronics and Advanced Materials – Rapid Communications, 5(4), 348–352.

is relatively low (y < 0.1). By contrast, when y > 0.1, the relative intensity decreases and the maximum value appears in y = 0.1, which shows the best excitation properties.

As is shown in Figure 6.11, the excitation spectra have different excitation intensities with different concentration of gadolinium under VUV excitation. There are two wide excitation bands at 165 nm and 214 nm, which can be attributed to the absorption due to charge transfer transition from the $2p^6$ orbit of ligand O^{2-} to the $4f^6$ of Eu^{3+} empty orbit.

Figure 6.12 shows emission spectra of different Eu-doped $(Y_{0.9}\text{-}xGd_{0.1})BO_3:Eu_x$ phosphors. For $(Y_{0.9}\text{-}xGd_{0.1})BO_3:Eu_x$ phosphors, there are three emission peaks seen at 593 nm, 611 nm and 627 nm, respectively. The intensity of emission peak of 593 nm exhibits the biggest change in comparison with those of other two peaks with a decrease of the Eu doping concentration. However, the intensity of three emission peaks decreases with the increase of the Eu doping concentration. The Eu doping concentration is the most important factor for luminescent properties of (Y, Gd) $BO_3:Eu^{3+}$ phosphors. The three sharp emissions peaked at 593 nm, 611 nm and 627 nm originate from transitions of $^5D_0 \rightarrow{}^7F_1$, $^5D_0 \rightarrow{}^7F_2$and $^5D_0 \rightarrow{}^7F_3$, respectively. When the doping concentration of Eu is 0.05, the emission peak at 611 nm ($^5D_0 \rightarrow{}^7F_2$) is the main emission peak.

When increasing Eu, intensity of peak at 593 nm ($^5D_0 \rightarrow{}^7F_1$) increase and becomes dominant at the Eu concentration of 0.2. The peak at 593 nm shows the orange colour. Because the orange colour influences the excitation purity of red emission, the application of (Y, Gd)$BO_3:Eu^{3+}$ phosphors is limited in excitation of VUV light. So, when the doping concentration of Eu is 0.05, the optimum excitation purity of red emission was obtained. Figure 6.13 shows the emission spectra of different Gd-doped $(Y_{0.95}\text{-}yGd_y)BO_3:Eu_{0.05}$ phosphors prepared by the co-precipitation method.

As is seen in Figure 6.13, the three emission peaks for $(Y_{0.9}\text{-}xGd_{0.1})BO_3$: Eu_x phosphors were obtained, which belong to 593 nm, 611 nm and 627 nm, respectively. Among four emission spectra, the emission peak at 611 nm was always the highest emission peak. However, the intensity of three emission peaks reduces with the increase of Gd-doping concentration. Three sharp emission peaks at 593 nm, 611 nm and 627 nm are due to the transitions of $^5D_0 \rightarrow{}^7F_1$, $^5D_0 \rightarrow{}^7F_2$, and $^5D_0 \rightarrow{}^7F_3$, respectively. When the Gd doping concentration is 0.1, the emission peak at 611 nm ($^5D_0 \rightarrow{}^7F_2$) is the main emission peak. With increasing Gd-doping concentration, the intensity of emission peak at 593 nm increases relatively. This phenomenon has disadvantages for the excitation purity of red emission.

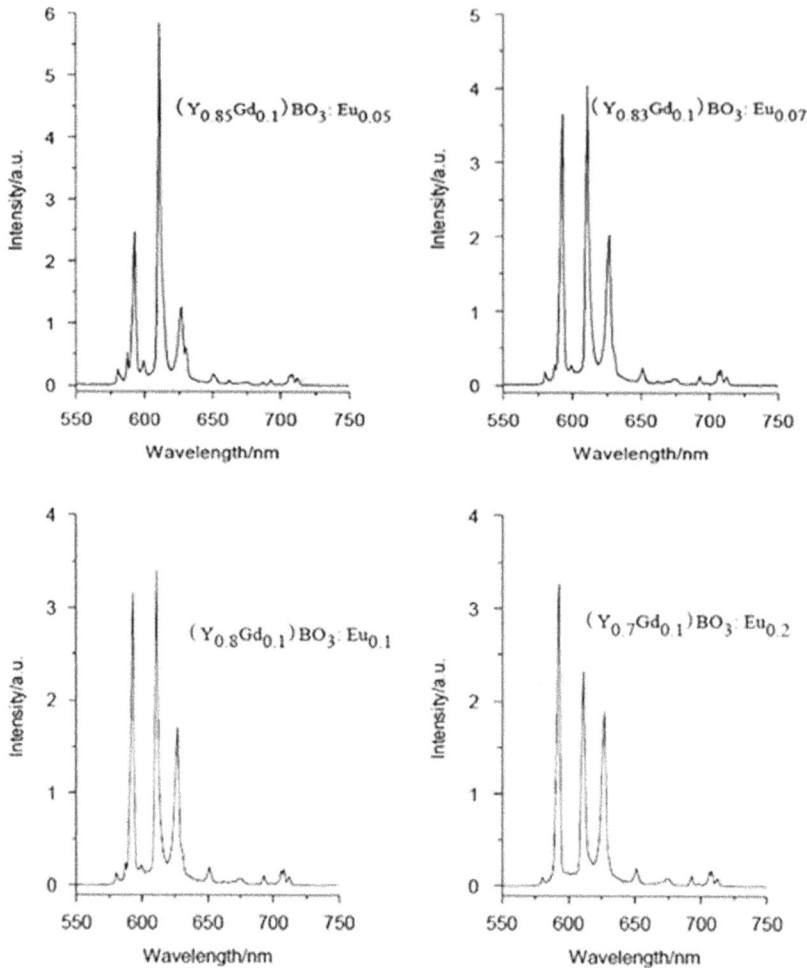

FIGURE 6.12 The emission spectra of different Eu-doped $(Y_{0.9}\text{-}xGd_{0.1})BO_3$: Eu_x phosphors.

Source: For details see: "Luminescent properties of (Y, Gd) BO_3: Eu^{3+} under VUV excitation for PDP prepared by co-precipitation method" by Hu, Y., Tao, Y., Huang, Y., Yu, X., Zhang, C., Liang, T., Yu, J., (2011). **Optoelectronics and Advanced Materials – Rapid Communications**, 5(4), 348–352.

Therefore, the $(Y_{0.9}Gd_{0.05})BO_3$:$Eu_{0.05}$ phosphor has better colour purity assumed as a red PDP phosphors candidate, though its intensity is lower than the $(Y_{0.85}Gd_{0.1})BO_3$:$Eu_{0.05}$ phosphors. Thus, it has been seen that phosphors (Y, Gd)BO_3:Eu^{3+} prepared by the co-precipitation method combined with ultrasonic-disperse and characterized by TG-DSC, XRD and VUV. The phosphor prepared showed many desirable characteristics such as high crystallinity, high distribution and good photoluminescence intensity required for PDP application. It is also seen that luminescence of red phosphor (Y, Gd)BO_3:Eu^{3+} is influenced by the mole ratio of Y, Gd and Eu. With respect to luminescence intensity and colour purity under VUV excitation, $(Y_{0.9}Gd_{0.05})BO_3$:$Eu_{0.05}$ phosphor showed the best luminous performance at 611 nm [51] (Table 6.9).

6.4.2 Green-Emitting Phosphors

Green component phosphors used in PDP panels are mostly obtained by doping terbium (III) Tb^{3+} or Mn^{2+} impurity ions in suitable composition of host material, whereas in a few crystal fields

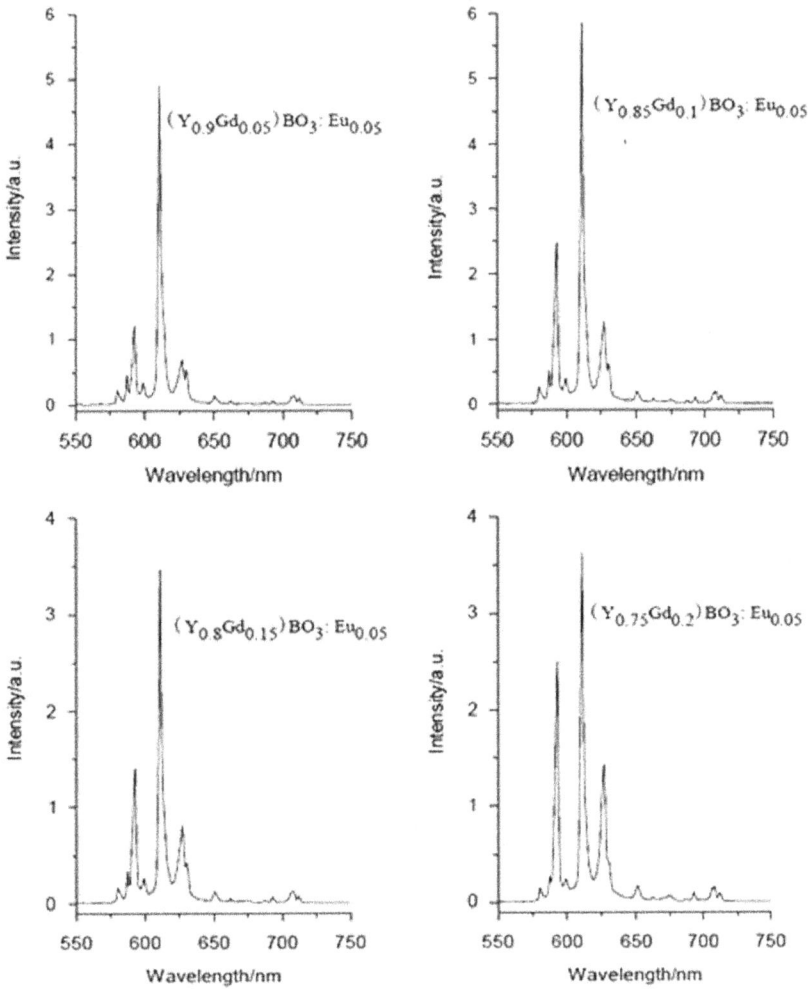

FIGURE 6.13   The emission spectra of Gd-doped $(Y_{0.95}-yGdy)BO_3$: $Eu_{0.05}$ phosphors.

Source: For details see: "Luminescent properties of (Y, Gd) BO_3: Eu^{3+} under VUV excitation for PDP prepared by co-precipitation method" by Hu, Y., Tao, Y., Huang, Y., Yu, X., Zhang, C., Liang, T., Yu, J., (2011). Optoelectronics and Advanced Materials – Rapid Communications, 5(4), 348–352.

sensitive cases, Eu^{2+} have also been incorporated. Transitions of electrons in levels of Tb are responsible for the emission of green colour. Several researchers reported borates, aluminates, silicates and phosphates as hosts for green-emitting phosphors with detailed crystal structure and PL properties. X. Huibing et. al. reported the effect of Li^+ ions doping on the structure and luminescence of $(Y,Gd)BO_3:Tb^{3+}$ [79]. Enhancement of green emission has been observed in the phosphor in $YBO_3:Tb^{3+}$ prepared by solid-state method when co-doped by Al^{3+} [80]. RE^{3+}-activated phosphors $CaAl_{12}B_2O_7$ (RE = Tb, Ce) [21] were reported as potential green-emitting phosphors for PDP applications. Tb^{3+}-activated $BaGdB_9O_{16}$ reported by SSD as a potential green phosphor for PDP applications [81]. Decay times of $Y_3Al_5O_{12}:Ce^{3+}$, $(Y,Gd)Al_3(BO_3)_4:Tb^{3+}$, and $(Y,Gd)_2O_3:Eu^{3+}$ of new phosphors for 3D PDP application was studied. The results showed that the decay time of these phosphors was significantly shorter than conventional PDP phosphors [82]. In an effort to develop an intense green emitting, fast decaying and degradation resistant Xe/Xe_2 plasma excitable green phosphor, an improved yttrium terbium ortho-borate phosphor has been developed. As degradation of luminescence intensity under thermal baking process is a major

TABLE 6.9
Literature survey of some borate-based red phosphor for PDP application

Sr. no.	Material	Method of synthesis	Reference
1	$Sr_3Y_2(BO_3)_4: Eu^{3+}$	Solid-state diffusion method	52
2	$Na_3Y_9O_3(BO_3)_8:Eu^{3+}$	Solid-state diffusion method	53
3	$(Gd_{1-x}Eu_x)Ba_3B_9O_{18}$	Solid-state diffusion method	54
4	$Al_3GdB_4O_{12}:Eu$	Solid-state diffusion method	55
5	$La_2SrB_{10}O_{19}:Eu^{3+}$	Solid-state diffusion method	56
6	$Ca_4GdO(BO_3)_3:Eu^{3+}$	Solid-state diffusion method	57
7	$LnCa_4O(BO_3)_3:RE^{3+}$ (Ln = Y, La, Gd; Re = Eu, Tb, Dy, Ce)	Solid-state diffusion method	58
8	$YAl_3(BO_3)_4:Eu^{3+}$	Combinatorial approach	59
9	$YAl_3(BO_3)_4:Eu^{3+}$	Combustion synthesis	60
10	$(Y_{0.65}Gd_{0.35})BO_3:Eu_{0.05}^{3+}$	Combinatorial approach	61
11	$YBO_3:Eu^{3+}$	Solid-state diffusion method	62–65
12	$YBO_3:Eu^{3+}$	Co-precipitation method	66
13	$YBO_3:Eu^{3+}, Tb^{3+}$	Sol-gel method	67
14	$YBO_3:Eu^{3+}$	Hydrothermal method	68
15	$YBO_3:Eu^{3+}$, (Y, Gd)BO_3: Eu^{3+} or Tb^{3+}	Spray pyrolysis	69
16	$(Y,Gd)Al_3(BO_3)_4:Eu^{3+}$	Combustion synthesis	70
17	$YBO_3:RE^{3+}$ (RE = Eu^{3+}, Tb^{3+})	Combustion synthesis	71
18	$YCaBO_4: Eu^{3+}$	Solution combustion synthesis	72
19	$YCaBO_4:Eu^{3+}$	Combustion synthesis	60
20	$(Y_{1-x-y}, Gd_x) BaB_9O_{16}:Eu^{3+}_y$	Solution combustion synthesis	73
21	$M_3Y_2 (BO_3)_4: Eu^{3+}$ (M = Ba, Sr)	Combustion synthesis	74
22	$(Y,Gd)Al_3(BO_3)_4:Eu^{3+}$	Solution combustion synthesis	75
23	$YCa_4O(BO_3)_3:RE^{3+}$ (RE = Eu^{3+}, Tb^{3+})	Combustion synthesis	76
24	$LaBaB_9O_{16}:Eu^{3+}$	Combustion synthesis	77
25	$BaZr(BO_3)_2: Eu^{3+}$	Combustion synthesis	60
26	$Ba_6Gd_9B_{79}O_{138}:Eu^{3+}$	Solid-state diffusion method	78

problem, individual phosphor grains have been coated with nanometre thick layer of silica nanoparticles for creating an inert interface layer [83]. Comparing the Tb-activated Y borates with Gd borates, it was noted that they all show a wide range of strong excitation spectra below 160 nm (see Figures 6.14 and 6.15) with some differences in the peak shape and intensity (Figure 6.14). But GdB_3O_6: Tb shows a strong band with a peak at around 155 nm. The Tb-activated Gd borates, in particular, show high quantum efficiencies. Figure 6.15 suggests there are some differences in the host bandgap between the borates and these gaps match well with the excitation energy of activators (i.e. Tb^{3+}) except GdB_3O_6: Tb.

As peaks near 160 nm are present in several Tb-activated borates, the absorption in this region is most probably due to the host lattice, i.e. in the BO3 groups. Excitation spectra of $Tb_{0.05}La_{0.95}$ MgB_5O_{10} and $BaLa_{0.9}Tb_{0.1}B_9O_{16}$ show also the weak host sensitization band of BO_3 groups at around 150 nm along with a strong Tb excitation band, as in Figure 6.16 [84]. Rao reported that the Tb-activated borate phosphor is stable and also improved the uniformity of discharge characteristics in PDPs [85]. To overcome the blue peak and take advantage of a higher lifetime and lower persistence of yttrium borate, he suggested the blends of $Zn_2SiO_4:Mn^{2+}$ (ZSM) and yttrium borate or barium aluminate and yttrium borate phosphors for the practical use in PDPs. Due to their

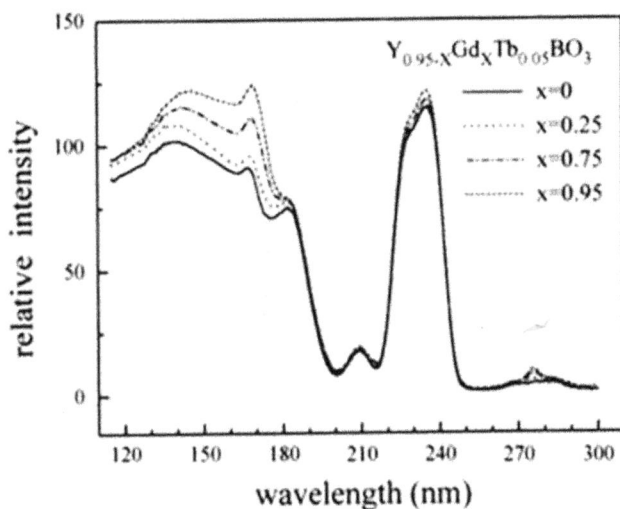

FIGURE 6.14 Excitation spectrum of $Y_{0.95-x}$ Gd x $Tb_{0.05}$ BO_3.

Source: For details see: "Phosphors for plasma display panels. Journal of Alloys and Compounds" by Kim, C., Kwon, I., Park, C., Hwang, Y., Bae, H., Yu, B., Pyun, C., Hong, G., (2000), 311(1), 33–39. https://doi.org/10.1016/S0925-8388(00)00856-2.

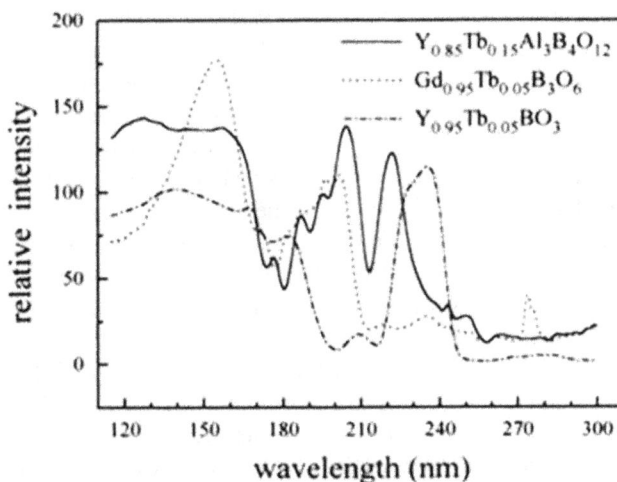

FIGURE 6.15 Excitation spectra of $Y_{0.85}Tb_{0.15}Al_3B_4O_{12}$, $Gd_{0.95}Tb_{0.05}B_3O_6$ and $Y_{0.95}Tb_{0.05}BO_3$.

Source: For details see: "Phosphors for plasma display panels. Journal of Alloys and Compounds" by Kim, C., Kwon, I., Park, C., Hwang, Y., Bae, H., Yu, B., Pyun, C., Hong, G., (2000), 311(1), 33–39. https://doi.org/10.1016/S0925-8388(00)00856-2.

quantum efficiency and stability at high temperatures, Tb-activated green-emitting lanthanum borate phosphors have been well studied and are widely used in VUV applications.

Luminescence properties of YBO_3-$GdBO_3$ prepared by solid-state diffusion method and doped with Tb^{3+} have been investigated in order to understand the energy transfer processes between Gd^{3+} and Tb^{3+} in these phosphors. It was found that YBO_3 - $GdBO_3$: Tb solid solutions, (Gd^{3+} - BO_3^{+3}) absorbs excitation energy more efficiently than (Y^{3+} - BO_3^{+3}) in the VUV region [86]. The borate

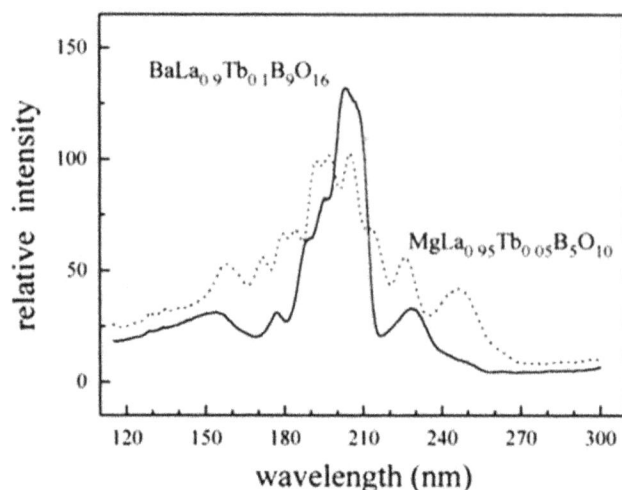

FIGURE 6.16   Excitation spectra of $BaLa_{0.9}Tb_{0.1}B_9O_{16}$ and $MgLa_{0.95}Tb_{0.05}B_5O_{10}$.

Source: For details see: "Phosphors for plasma display panels. Journal of Alloys and Compounds" by Kim, C., Kwon, I., Park, C., Hwang, Y., Bae, H., Yu, B., Pyun, C., Hong, G., (2000), 311(1), 33–39. https://doi.org/10.1016/S0925-8388(00)00856-2.

phosphors of type $LaBaB_9O_{16}$ doped with Tb and Ce-Tb were prepared by combustion synthesis technique using urea as fuel an ammonium nitrate as oxidizer the synthesised material was annealed in reducing atmosphere. The $LaBaB_9O_{16}$: Ce^{3+}, Tb^{3+} exhibited intense green emission at 543 nm when excited at 173 nm. Thus, $LaBaB_9O_{16}$: Ce^{3+}, Tb^{3+} proved to be a potential green borate phosphor for PDP and mercury-free fluorescent lamp applications [87]. Optimization of phosphors already applied as conventional lamp phosphors have resulted in improvement in their luminescent properties and match the requirements as PDP phosphors. The success and failure of phosphors for PDP application depends on quality and longevity. As quality and longevity of a phosphor depends mainly on their luminous efficiency, colour rendering and durability. All of these properties depend on the nature and quality of phosphors. Furthermore, it is still possible to improve the colour rendering by using new luminescent materials able to have better colour coordinates, i.e. a colour more saturated, especially for green component. Another vital characteristic of phosphors for PDP is the fluorescence lifetime: the shorter, the better in order to produce the maximum number of grey levels. Optimisation of Tb-doped borate phosphors like [$YAl_3(BO_3)_4$: Tb^{3+}, $LaBaB_9O_{16}$: Tb^{3+}, $LaBaB_9O_{16}$: (Ce^{3+}, Tb^{3+}), $Na_3La_2(BO_3)_3$: Tb^{3+}] have been reported by synthesising the material by a low-temperature combustion method. The emission under VUV excitation (173 nm) clearly proves them to be promising candidates for PDP and Hg-free lamp applications. The photoluminescence properties of the optimized green borate phosphors are summarized in Table 6.10.

TABLE 6.10

Literature survey of some borate-based green phosphor for PDP application

Sr. No.	Material	Method of synthesis	Reference
1	$Na_3La_2(BO_3)_3$: Tb^{3+}	Solution combustion technique	75
2	$YCa_4O(BO_3)_3$: Tb^{3+}	Solid-state reaction	92
3	$YCa_4O(BO_3)_3$: Tb^{3+}	Solution combustion technique	76
4	$LaBaB_9O_{16}$:Tb^{3+} & $LaBaB_9O_{16}$:Ce^{3+}, Tb^{3+}	Solution combustion technique	77

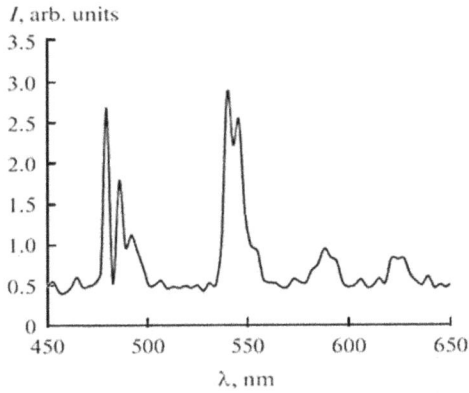

FIGURE 6.17 PL spectrum of YAl$_3$(BO$_3$)$_4$: Tb^{3+} at 173 nm excitation.

Source: For details see: "International Journal of Self - Propagating High - Temperature Synthesis" by Nagpure, P. A., Omanwar, S. K., (2013). Combustion synthesis of borate phosphors for use in plasma display panels and mercury-free fluorescent lamps. 22(1), 32–36. https://doi.org/10.3103/S106138621301007X.

6.4.2.1 Photoluminescence of YAl$_3$(BO$_3$)$_4$:Tb^{3+}

The $^5D_4 \rightarrow ^7F_J$ transitions of Tb^{3+} in YAl$_3$(BO$_3$)$_4$ powder are at 480 nm (J = 6), 540 nm (J = 5), 591 nm (J = 4) and 621 nm (J = 3) (Figure 6.17). The green emission of $^5D_4 \rightarrow ^7F_5$ transition is obviously predominant. The characteristic emission lines split into a fine structure of the Stark levels in the crystal field [60].

6.4.2.2 Photoluminescence of LaBaB$_9$O$_{16}$: (Ce^{3+}, Tb^{3+})

The emission spectrum of LaBaB$_9$O$_{16}$: Tb^{3+} in the visible (450–650 nm) is shown in Figure 6.18. Under excitation at 173 nm, the emission peaks around 486, 543, 585 and 627 nm are clearly observed and can be assigned to the $^5D_4 \rightarrow ^7F_J$(J = 6, 5, 4, 3) transition of Tb^{3+}. Materials co-doped with Ce^{3+} and Tb^{3+} were extensively used in fluorescent lamps, CRT and PDP applications because of high efficiency energy transfer between Ce^{3+} and Tb^{3+} ions [88–90,91]. The comparison of the photoluminescence parameters of LaBaB$_9$O$_{16}$:Tb^{3+} and LaBaB$_9$O$_{16}$: (Ce^{3+}, Tb^{3+}) gave a convincing evidence for efficient energy transfer from Ce^{3+} to Tb^{3+} ions. It has been found that the green emission (543 nm), corresponding to the transition $^5D_4 \rightarrow ^7F_5$ of Tb^{3+} in LaBaB$_9$O$_{16}$, in LaBaB$_9$O$_{16}$:(Ce^{3+}, Tb^{3+}) enhanced by a factor of about four (Figure 6.18, curve b) [55].

6.4.2.3 Photoluminescence of Na$_3$La$_2$(BO$_3$)$_3$: Tb^{3+}

Under excitation at 173 nm, the Na3La$_2$(BO$_3$)$_3$: Tb^{3+} phosphor emits intense green light (Figure 6.19) due to characteristic f–f transitions of Tb^{3+} ion. The emission peaks around 480, 540,

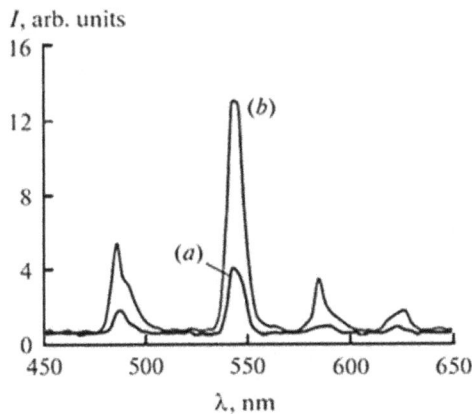

FIGURE 6.18 PL spectra of (a) LaBaB$_9$O$_{16}$: Tb^{3+} and (b) LaBaB$_9$O$_{16}$: (Ce^{3+}, Tb^{3+}) at 173 nm excitation.

Source: For details see: "International Journal of Self - Propagating High - Temperature Synthesis" by Nagpure, P. A., Omanwar, S. K., (2013). Combustion synthesis of borate phosphors for use in plasma display panels and mercury-free fluorescent lamps. 22(1), 32–36. https://doi.org/10.3103/S106138621301007X.

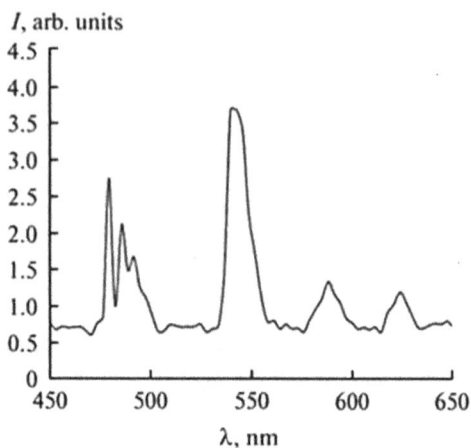

FIGURE 6.19 PL spectrum of Na$_3$La$_2$(BO$_3$)$_3$:Tb^{3+} at 173 nm excitation.

Source: For details see: "International Journal of Self - Propagating High - Temperature Synthesis" by Nagpure, P. A., Omanwar, S. K., (2013). Combustion synthesis of borate phosphors for use in plasma display panels and mercury-free fluorescent lamps. 22(1), 32–36. https://doi.org/10.3103/S106138621301007X.

588 and 624 nm are clearly observed and assigned to the $^5D_4 \rightarrow {}^7F_J$ (J = 6, 5, 4, 3) transition of Tb^{3+} [60]. The inorganic borate host phosphor YBO$_3$:Tb^{3+} is synthesized by a simple, time saving, low temperature and inexpensive novel solution combustion technique. Under 147 nm VUV excitation, YBO$_3$:Tb^{3+} exhibits strong absorption and efficient emission at 546 nm in the intense green region. Thus, a phosphor could be potential candidates for PDP applications [71].

6.4.3 BLUE-EMITTING PHOSPHORS

Blue component phosphors used in PDP panels are mostly obtained by doping Eu^{2+} ions in suitable crystal field sensitive composition of host material. Transitions of electrons in level of Eu^{2+} are responsible for the emission of blue colour. Several researchers reported Eu^{2+}-activated borates, aluminates, silicates, vanadates and phosphates host for blue-emitting phosphors with detailed crystal structure and photoluminescence properties. As it is known in PDPs, the phosphors coated inside the cell are excited by VUV (147 and 172 nm). The phosphors than convert the VUV into the visible light, which includes blue, green and red emissions. As the phosphors of PDP are directly exposed to high-energy VUV light, the stability of the applied phosphors is very important for the durability of the PDP devices [84,92,93]. In present commercial PDPs [84,94] mostly BaMgAl$_{10}$O$_{17}$:Eu^{2+} (BAM) is used as blue phosphor. As it is reported by several workers that with its efficient properties it has some drawbacks, such as it degrades under high-energy VUV excitation, this leads to a sharp decrease in the luminous efficiency [95]. This instability is caused by the degradation of the host structure and the oxidation of the dopant Eu^{2+} during the annealing process in the panel manufacture and the operation of the panels under VUV bombardment [95]. This is a serious problem that affects the luminance of the PDPs. Several efforts have been made to overcome the degradation of BAM, like the compositional variations and the introduction of trivalent ions into BAM [96]. However, no notable breakthrough has been achieved up till now. In addition to this, new phosphor materials for substituting BAM are widely being investigated. As discussed earlier, the requirement for efficient blue PDP phosphors is that the phosphor host lattice must absorb efficiently in the wavelength range between 140 and 180 nm, which can then transfer the energy to the dopant ion having a strong emission in the visible region [97–101]. It has been reported that the borates have special potential for VUV excitation since the host absorption band is located in the VUV region. In the ongoing efforts to prepare competent PDP blue phosphor, synthesis and luminescent properties of a new blue phosphor Ba$_2$CaB$_6$O$_{12}$:Eu^{2+} were studied under VUV and ultraviolet (UV) excitation. The material was synthesised by a high-temperature solid-state method. The phosphors dibarium magnesium orthoborate of type Ba$_2$Mg (BO$_3$)$_2$ doped with Ce^{3+} and Eu^{2+} ions were prepared by high-temperature, solid-state reaction technique. The luminescent properties of Ba$_2$Mg (BO$_3$)$_2$: Ce^{3+} and Ba$_2$Mg (BO$_3$)$_2$: Eu^{2+} in VUV-vis range were investigated. It was found that the host-related absorptions with the maxima at about 172 and 146 nm. The 5d crystal field splitting components were found to have the energies of about 34,400 cm^{-1}, 36,600 cm^{-1}, 39,200 cm^{-1}, 45,000 cm^{-1} and 46,900 cm^{-1}, the barycentre of 40,400 cm^{-1}, and Stokes shifts ~9,000 cm^{-1} for Ce^{3+} in Ba$_2$Mg (BO$_3$)$_2$. Eu^{2+}-activated phosphors show red emission with maximum at about 594 nm, FWHM is about 2,800 cm^{-1}, Stokes shift is 7,950 cm^{-1}, CIE chromaticity coordinates are (0.51, 0.47) and the optimal dopant concentration is x = 3 mol%. The lowest 4f-5d absorption of Eu^{2+} was found to be approximately 403 nm in the host [102]. Eu^{2+} concentration variation has been carried out in Sr$_3$Al$_{10}$SiO$_{20}$ and the effect of boron substitution in Sr$_3$Al$_{10}$SiO$_{20}$:Eu^{2+} phosphor has also been investigated to improve the blue emission intensity of Eu^{2+}. Powder samples of Sr$_3$Al$_{10}$SiO$_{20}$:Eu^{2+} and Sr$_3$Al$_{10-x}$BxSiO$_{20}$:Eu^{2+} were prepared by a high-temperature solid-state diffusion method. The compositions were checked for phase formation by X-ray powder diffraction (XRD) using Cu-Ka1 radiation (Rigaku, D/max-IIIc (3 kW) at 40 kV and 45 mA). The emission spectra were recorded using a standard spectrofluorometer setup from DARSA PRO PL System (Professional Scientific Instrument Co., Korea), which utilizes a deuterium lamp as the excitation source. The sample chamber was maintained at about 5 × 10^{-5} Torr using a turbo pumping system. PL spectra were obtained by scanning a wavelength region from 300 to 700 nm under an excitation of 147 nm radiation from the deuterium lamp at room temperature. The photoluminescence excitation and emission spectrum of Sr$_{2.979}$Eu$_{0.021}$Al$_{10}$SiO$_{20}$ are shown in Figure 6.20. The excitation and emission spectra of Sr$_{2.979}$Eu$_{0.021}$Al$_{10}$BxSiO$_{20}$. The broad absorption bands were due to the host lattice absorption as well as the Eu^{2+} ion present in the Sr^{2+} site. The host lattice absorption in Sr$_3$Al$_{10}$SiO$_{20}$ mainly occurred in the VUV region (electronic transition from valence band to conduction band), due to insulating nature of the compound. The Eu^{2+} absorption is due to the electronic transition of ^{4}f$_7$ $-^4$f$_6$^{5}d$_1$. In the study, two absorption bands in the VUV and UV region were observed. The corresponding emission spectrum showed intense and broad blue emission in the region of 400–500 nm with peak maxima at around 460 and 466 nm [103].

The emission spectra of Sr$_3$Al$_{10-x}$BxSiO$_{20}$ (x = 0, 0.25, 0.5, 0.75 and 1) is shown in Figure 6.21. Emission spectra of Sr$_{2.979}$Eu$_{0.021}$Al$_{10-x}$BxSiO$_{20}$ (x = 0, 0.25, 0.5, 0.75 and 1). All the compositions show blue emission, however, the PL intensity of the compositions found to be different for each composition.

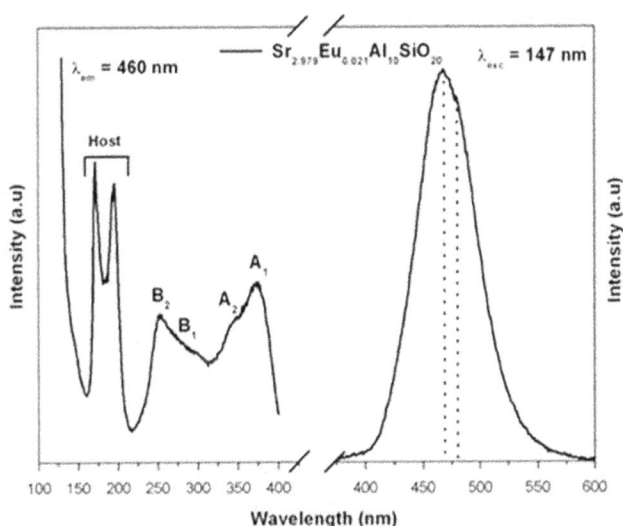

FIGURE 6.20 The excitation and emission spectra of Sr$_{2.979}$Eu$_{0.021}$Al$_{10}$BxSiO$_{20}$.

Source: For details see: "Novel blue-emitting phosphor Sr$_{2.979}$Eu$_{0.021}$Al$_{10-x}$B$_x$SiO$_{20}$ (x=0, 0.25, 0.5, 0.75 and 1) for PDP applications by Sivakumar, V., Hong, G. Y., Kim, J. S., Jeon, D.Y., (2009). Journal of Luminescence, 29 (12), 1632–1636. https://doi.org/10.1016/j.jlumin.2008.12.021.**

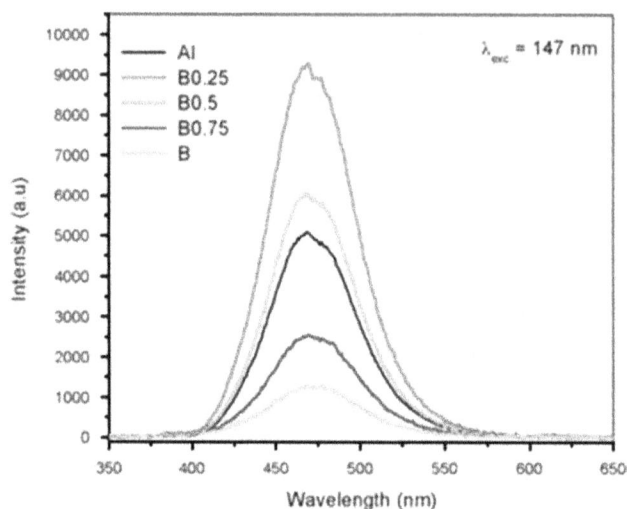

FIGURE 6.21 Emission spectra of $Sr_{2.979}Eu_{0.021}Al_{10-x}B_xSiO_{20}$ (x = 0, 0.25, 0.5, 0.75 and 1).

Source: For details see: "Novel blue-emitting phosphor $Sr_{2.979}Eu_{0.021}Al_{10-x}B_xSiO_{20}$ (x=0, 0.25, 0.5, 0.75 and 1) for PDP applications by Sivakumar, V., Hong, G. Y., Kim, J. S., Jeon, D.Y., (2009). Journal of Luminescence, 29(12), 1632–1636. https://doi.org/10.1016/j.jlumin.2008.12.021.

The $B_{0.25}$-substituted compositions show maximum blue emission and it is almost two times higher than that of x = 0 composition (PL intensity 62% vs. commercial BAM). The boron-substituted compositions show maximum intensity, due to the formation anionic tetrahedra BO_4^{5-} network, which facile the electron transfer process ($Eu^{3+} \rightarrow Eu^{2+}$ reduction). Due to its excellent colour purity and emission intensity, this phosphor can be considered for a promising candidate for PDP applications [103].

The phosphor $BaB_8O_{13}:Eu^{3+}$ were synthesized by solid-state reaction, and their luminescent properties were studied under 254 and 147 nm excitation. The excitation spectrum showed two broadbands in the range of 100–300 nm: one was the host lattice absorption with the maxima at 160 nm and the other was Ba-O absorption overlapped with the CT band of Eu^{3+}, which indicated that the energy of the host lattice absorption could be efficiently transferred to the Eu^{3+}. The overlapped bands were tended to separate when monitored by different wavelength, which in-dicated that at least two Ba^{2+} sites were available in BaB_8O_{13}. The emissions of Eu^{3+} (612 nm) and Eu^{2+} (405 nm) were both observed in the emission spectra of $BaB_8O_{13}:Eu^{3+}$ under the excitation of either 254 or 147 nm. With the doping concentration of Eu^{3+} increasing, the 612 nm emission was enhanced while 405 nm emission was decreased under 254 nm excitation, which was due to the persistent energy transfer from Eu^{2+} to Eu^{3+}. While under 147 nm excitation, the 612 nm emission was quenched and the 405 nm emission was enhanced. It was concluded that the preferential excitation of Eu^{2+} under 147 nm excitation was one of the reasons for these facts [104].

Luminescent properties under VUV (147 nm) excitation were investigated for a new Eu^{2+}-doped borophosphate phosphor. The new phosphor has a relatively strong absorption band in a short wa-velength region because of the strong covalent characteristics of B O and P O bonds. Powder samples of the new borophosphate, $Sr_6BP_5O_{20}:Eu^{2+}$ (SBP), were synthesized by a conventional solid-state reaction. The X-ray diffraction patterns revels that SBP contains small amounts of the impurity $Sr_2P_2O_7$. The SBP compound is an isomorph of $Pb_6BAs_5O_{20}$ [105]. The anion framework in this compound is complex: it resembles a propeller, with a central $[BO_4]$ tetrahedron surrounded by an array of four $[PO_4]$ tetrahedrons. The excitation energy is transferred from the surrounding propeller to the luminescence centre Eu^{2+} ion. This mechanism is similar to the antenna effect in luminescent complex between a rare-earth and an organic ligand [106]. The VUV excitation spectrum and the

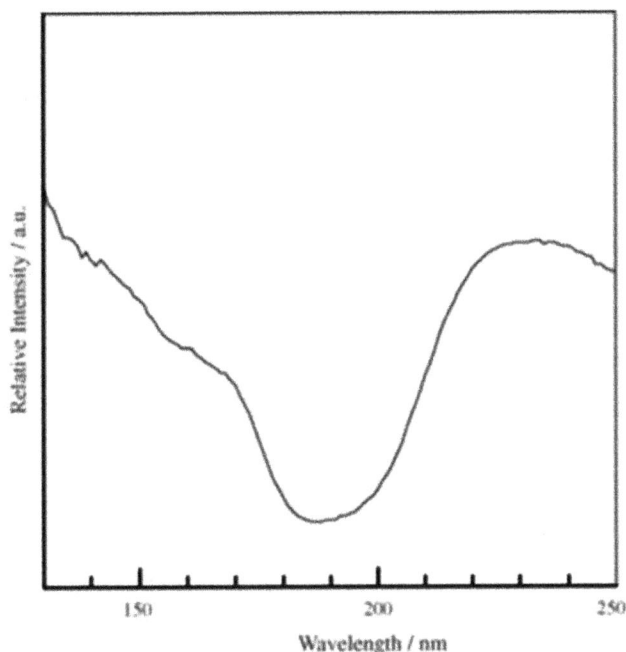

FIGURE 6.22 VUV excitation spectrum of $Sr_6BP_5O_{20}$:Eu^{2+}.

Source: For details see: "Recent research and development of VUV phosphors for a mercury-free lamp" Toda, K., (2006). Journal of Alloys and Compounds, 408–412, 665–668. https://doi.org/10.1016/j. jallcom.2005.01.080.

emission spectrum of the SBP are shown in Figure 6.22 and Figure 6.23. The broad excitation absorption band around 147 nm and broad cyan emission (476 nm) were confirmed. The fundamental absorption band edge of the SBP is present around 180 nm.

Therefore, the SBP efficiently absorbs the VUV photons. The composition $(Sr_{0.99}Eu_{0.01})_6BP_5O_{20}$ shows its highest luminance under VUV excitation (147 nm), which corresponds to 216% that of the commercial BAM phosphor as a reference. Therefore, this borophosphate phosphor is an attractive candidate for a new VUV phosphor for a mercury-free fluorescent lamp [107].

6.5 CURRENT TRENDS AND INNOVATIONS

Luminescent materials with lanthanides are found in fluorescent tubes, colour televisions, X-ray photography, lasers, infrared (IR) to visible light up conversion materials and fibre amplifiers [108–110]. Such applications depend on the luminescence properties of lanthanide ions (sharp lines and high efficiency). In fluorescent lamps, phosphors on the inside wall of the glass tube convert the ultraviolet (UV) radiation (mainly with a wavelength l of 254 nm) that is generated in the Hg discharge to blue, green and red light, yielding white light. The quantum efficiency of the lanthanide-based lamp phosphors is high (90%) (1): For every 100 UV photons that are absorbed, 90 photons in the visible spectral region are emitted.

A challenge in the field of luminescence of lanthanide ions is the research in the vacuum ultraviolet (VUV) (1,200 nm) spectral region. Recently, this field has become important because of the need for phosphors for VUV excitation [108,110–112]. For example, Hg-free fluorescent lamps can be made by replacing Hg with a noble gas, such as Xe, as the discharge medium. A Xe discharge offers the advantage of immediate start-up: There is no delay in the emission of light (as is the case in

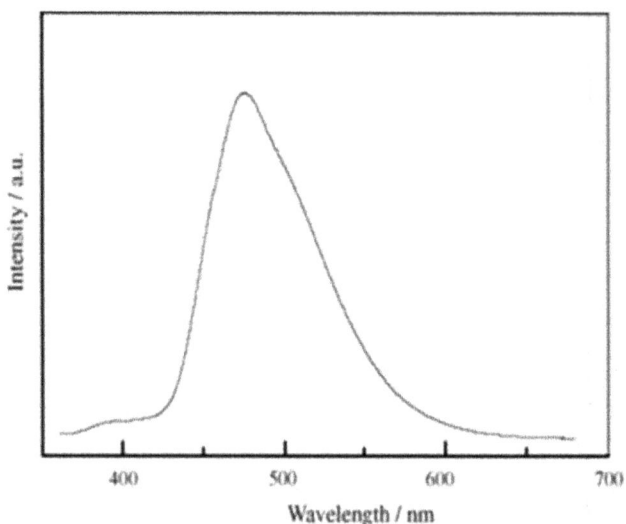

FIGURE 6.23   Emission spectrum of $Sr_6BP_5O_{20}:Eu^{2+}$.

Source: For details see: "Recent research and development of VUV phosphors for a mercury-free lamp"
Toda, K., (2006). *Journal of Alloys and Compounds*, 408–412, 665–668. https://doi.org/10.1016/j.
jallcom.2005.01.080.

conventional fluorescent lamps, where Hg must first evapourate). Immediate start-up is important in special applications (for example, for lamps in facsimile and copying machines and for car brake lights).

In plasma display panels, a Xe discharge is also used to generate VUV radiation. In each pixel, the VUV radiation is converted to blue, green or red light by a phosphor [112]. For the conversion of the VUV radiation of a Xe discharge ($\lambda \approx 172$ nm) to visible light, alternative phosphors are needed. Three aspects are important in the development of new VUV phosphors: a higher efficiency, a higher stability and a higher VUV absorption [112]. Although the generation efficiency of VUV radiation is high in a Xe discharge (higher than in other noble gas discharges), the light output (in lumens per watt) of Xe-discharge lamps is lower than that of Hg-based fluorescent lamps. An important limitation in obtaining high energy efficiencies is inherent to the conversion of one VUV photon ($\lambda \approx 172$ nm) to one visible photon ($\lambda = 400$ to 700 nm). Even if the quantum efficiency of a phosphor is 100%, in this process, ~65% of the energy from the VUV photon is lost by nonradiative relaxation processes. For a Hg discharge (($\lambda = 254$ nm), the losses are lower (~50%). To reduce the higher energy losses, a quantum efficiency of more than 100% is required for phosphors in Xe-discharge fluorescent lamps. In theory, this is possible as the high energy of VUV photons that are generated in a noble gas discharge allows the emission of two photons in the visible spectral region per each VUV photon absorbed. This phenomenon has been studied for single lanthanide ions like Pr^{3+} and Tm^{3+} [113,114], but for Tm^{3+}, a substantial amount of light is lost in the IR or UV spectral region, whereas for Pr^{3+}, the main emission occurs in the deep violet region (~405 nm), where eye sensitivity is very low. An alternative concept for obtaining quantum efficiencies of more than 100% is based on a combination of two lanthanide ions. Through partial energy transfer between the ions, a high quantum efficiency (close to 200%) can be obtained. This process is the opposite of the process that is known as "Addition de Photons par Transfertd' Energie" (APTE) [which was discovered by Auzel in 1966 [115] and also known as up-conversion [109,110]. Because of this analogy, we call the quantum cutting through energy transfer "down-conversion." The concept is illustrated (Figure 6.24) with two types of ions, I and II, with hypothetical energy level schemes. Efficient visible quantum cutting by a two-photon emission from a high energy level for a single lanthanide ion is theoretically possible (Figure 6.24A, red lines).

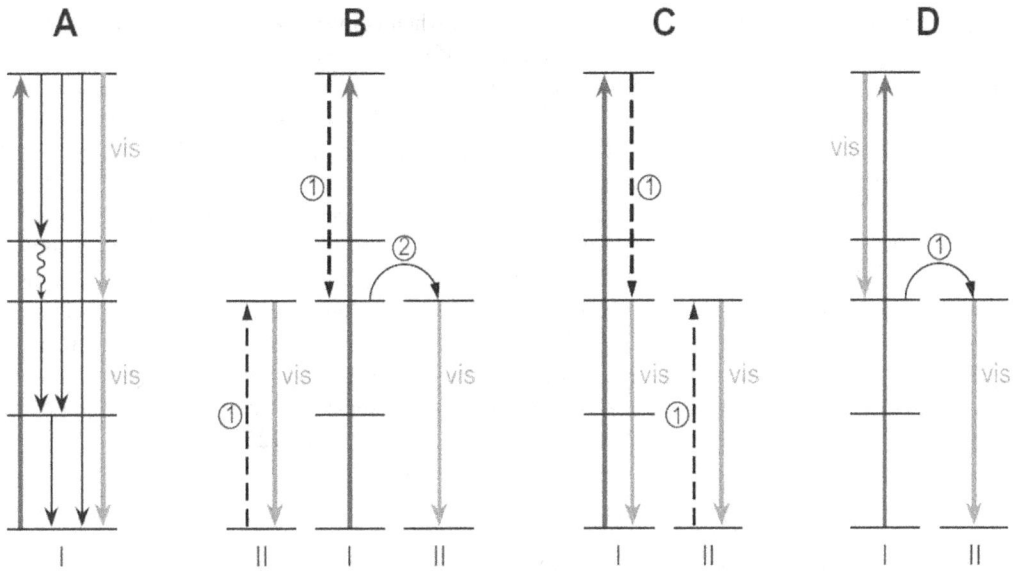

FIGURE 6.24 Energy level diagrams for two (hypothetical) types of lanthanide ions (I and II), showing the concept of down conversion.

Source: For details see: "Quantum cutting through down conversion in rare-earth compounds. Journal of Luminescence" by Wegh, R.T., Donker, H., Loef, E. V. Dvan., Oskam, K. D., Meijerink, A., (2000), 87–89, 1017–1019. https://doi.org/10.1016/S0022-2313(99)00514-1.

The observation of photon cascade emission in Pr^{3+}-activated $LaMgB_5O_{10}$ under 185 nm vacuum ultraviolet excitation is reported and discussed. In this host lattice, the lowest energy state of the Pr^{3+} 4f5d excited configuration is located above the $^1S_0$ state. The excitation of the 4f5d state by vacuum ultraviolet photons results in radiative emission originating from the $^1S_0$, level state. The absence of emission from the $^3P_0$ state is attributed to the efficient multiphonon $^3P_0 \rightarrow {}^1D_2$ relaxation in this lattice [116]. The oxide materials are usually considered better choices than fluoride materials for VUV phosphor because of its relatively easier preparation method and higher VUV absorption efficiency. Among oxide materials, SrB_4O_7 is a host in which the lowest 4f5d level of Pr^{3+} is located above its $^1S_0(4f^2)$ level. So SrB_4O_7 was selected as a host to incorporate Pr^{3+} and Mn^{2+} to investigate the possibility of the energy transfer between Pr^{3+} and Mn^{2+}. The samples were prepared by high-temperature solid state reaction method, using $SrCO_3$, HBO_3, Pr_6O_{11}, and $MnCO_3$ as raw materials. The Pr^{3+} concentrations of 0.05, 0.1, 0.5 and 1 mol% along with Mn^{2+} concentrations of 1, 2, 5 and 8 mol% were used in the preparation process. The comparison of the overall emission intensities of samples with different concentrations disclosed that the optimal concentration was 0.1 mol% and 5 mol% for Pr^{3+} and Mn^{2+}, respectively. The observation exhibited the feasibility of achieving VUV phosphor with quantum efficiency greater than 1 based on Pr^{3+} and Mn^{2+} combination in oxide material [117–119].

As the investigations of rare earth (RE) ions doped phosphors in the vacuum ultraviolet (VUV) region have gained much attention for the increasing need of theory and their potential applications in plasma display panels (PDPs), mercury-free lamps, etc [115,116]. Especially, the interest in three-dimensional (3D) displays has steadily grown due to increasing needs for advanced image devices, for which the 3D PDPs operate using the images that are separated to the left and right eyes by dividing the scanning period [120–122]. This 3D device requires higher brightness and shorter response time of phosphors than the conventional PDPs. However, the efficiency constraint with regard to the quantum efficiency (QE) in relation to the current phosphors is still one major problem due to too much energy is lost associated with nonradiative relaxation processes when converting one VUV photon into one visible photon [123]. Considering this, quantum cutting (QC) via down-conversion as a phenomenon occurring in materials that emit more than one photon per phonon absorbed has been extensively studied. Recently, more attention has been paid to QC in oxide hosts, such as SrB_4O_7:Pr^{3+}, Mn^{2+}, $GdPO_4$:Tb^{3+}, $Sr_3Gd (PO_4)_3$: Tb^{3+}, $K_2GdZr (PO_4)_3$: Eu^{3+}, and so on [117,123,124].

Visible quantum cutting (QC) through down-conversion was observed upon 4f8-4f75d1 excitation of Tb^{3+} in both $BaGdB_9O_{16}$ and $Ca_8MgGd (PO_4)_7$ compounds. The QC involves a cross relaxation process between two neighbouring Tb^{3+}. Modified and new QC models are proposed based on the QC mechanisms in Gd^{3+}-Tb^{3+} system. Two calculation equations for the cross-relaxation efficiency are suggested according to the energy transfer theory and spectral results. By studying the spectral characteristics of Tb^{3+} in hosts from fluoride to oxide, it indicates that the visible QC in Gd^{3+}-Tb^{3+} system may occur mainly depending on the structural features of phosphors, rather than the phonon energies of matrixes. All of the above conclusions have meaning of guidance for investigating other phosphors with QC phenomenon [125].

6.6   LIMITATIONS AND CHALLENGES

The VUV phosphors engender numerous problems including phosphor deterioration, surface damage and poor excitation efficiency. These problems should be resolved by a new design concept such as a new excitation mechanism "quantum cutting" [117,126–128] and new materials offering new crystal structures. "Research and development of phosphors for PDP and mercury-free lamp is underway. Performances of the PDP mainly depends upon luminescence performance of phosphors used in it, thus selection of efficient phosphors with desired luminescence characteristics is of utmost important. We need to continue research in this field because it is known that good phosphor for electronic or ultraviolet excitation is not necessarily a good choice for excitation in vacuum ultraviolet (VUV). Therefore, the fluorescence properties of phosphors induced by such VUV photons, has to be studied and improved if efficacy decreases.

In the PDP technology, the key factors between success and failure are going to be quality and longevity; both factors depend directly on phosphors: luminous efficiency, colour rendering and longevity are all properties which depend on the nature and the quality of phosphors. Furthermore, it is still possible to improve the colour rendering by using new luminescent materials able to produce better colour coordinates, i.e. a colour more saturated, especially for red and green component. Another crucial characteristic of phosphors for PDP is the fluorescence lifetime. Highly efficient phosphors under vacuum ultraviolet (VUV) excitation are still demanded for the development of plasma display panels (PDP) and Hg-free fluorescent lamps [107,128].

6.7   FUTURE PROSPECTUS AND SCOPE

6.7.1   IN TERMS OF SYNTHESIS

The thoughtful interest in combustion synthesis is due to the ease and cost efficiency of the process followed by the advanced nature of the VUV properties of the products because of narrow distribution of particles and homogeneity in morphology.

For further/future work in synthesis, special attention has to be focused towards the preparation of agglomeration free, nanosize particles with predesigned morphology. It is therefore very important to carry out the combustion synthesis of these phosphors by trying the different fuels such as carbohydrazide (CH), oxalyl dihydrazide (ODH), diformyl hydrazine (DFH), HMT ($C_6H_{12}N_4$), etc. and using various fuels to oxidizer ratios in each case and investigate the effect on VUV properties.

It may also be a theme of interest to prepare the materials via different synthesis techniques as suggested in chapter 2 and compare the results obtained.

6.7.2   IN TERMS OF QUANTUM CUTTING PHOSPHORS

The promising technique of plasma display panels (PDP) and mercury-free fluorescent tubes demands new phosphors to efficiently convert the vacuum ultraviolet (VUV) radiation generated by noble gas discharges into visible light. To convert the disadvantage of low discharge efficiency of noble gas relative to mercury and make the mercury-free fluorescent lamp more competitive, new VUV-excited phosphors with quantum efficiency higher than 100% are expected because the phosphors used in a mercury lamp have quantum efficiency close to 100%. So, for every one photon absorbed by the new VUV-excited phosphors, two or more visible photons should be emitted out. This is the so-called quantum cutting, or two-photon emission. Theoretically, it is possible for VUV excitation to produce two visible photons because a VUV photon has enough energy. Thus, it would be exciting to successfully develop phosphors with quantum cutting properties in near future. Major advances have been made by many research groups around the world recently in demonstrating QC, fabricating new luminescent materials, developing high efficiency QC phosphors and the mechanism involved. There are still many areas that need additional work, including:

1. Development of new QC materials, to realize efficient visible-QC. This will require a better understanding of ET and DC of the visible- and NIR-QC mechanism, and also a better understanding of materials properties and new materials fabrication process.
2. Realization of higher VUV absorption and effectively ET in QC phosphors with good characteristics. This will require suitable sensitizer ions.
3. Improved QE in dual/ternary ions activated QC phosphors. This will need a better understanding of mechanism of DC, ET, CR and non-radiative relaxation in ion pairs such as Pr^{3+}–Mn^{2+}, Gd^{3+}–Eu^{3+}, Gd^{3+}–Tb^{3+}, etc. It is also necessary to understand that nature of the role of sensitizer ions, new QC materials and its characteristics: synthesis, characterization and optical properties.
4. Clear and direct demonstrations of ET and DC between ions. It is necessary to develop new methods for the measurements of QC luminescent properties including QE.

It is likely that many of these areas could be possible in attaining more significant progress in the near future concerning the QC phosphors for their use in VUV-excited PDPs, MFFLs and even in the development of efficient solar cells converter becomes more widely known.

6.7.3   IN TERMS OF ACTIVATORS

It would be important to investigate the difference and effect on VUV properties of the phosphors by co-doping with the suitable sensitizers, in the search for efficient stabilizing associates.

6.7.4   IN TERMS OF NEW HOST MATERIALS

Along with the study of different activators and their combinations it will be challenging to study new host materials for VUV applications as the host is equally vital as of activators. With the invention of new display technologies it was thought that PDP would become the invention of the last decade and will not compete with the future technologies. But this has proved wrong as PDP has advantages, such as a high-speed response with emissive, an extensive viewing angle, high luminous efficacy, low cost, an easy manufacturing for making large screen over LCD in large-screen field. PDPs will continue raising its advantages in providing a new applications and market for screens with wide size. Phosphors are considered to be basic and important component in lighting and display technology.

As per conclusive remarks, in this chapter we have seen a demonstration of efficient VUV-excited borate phosphors that have an important role in the development of superior luminescent materials and devices. The progress of efficient red, green and blue borate phosphors is discussed herewith. It is well known that the VUV phosphors have numerous problems including phosphor deterioration, surface damage and poor excitation efficiency. These problems can be resolved by a new concept, such as a new excitation mechanism "quantum cutting" and new materials with novel crystal structures. Continuous research in the area of PDP, especially in the development of a new, efficient VUV material, and mainly borate as a host will provide a desirable solution to various problems faced. A great number of fresh concepts and novel materials are still in the research stage. Some of them may lead to much higher efficiency and lower cost in the coming decades, although there is a long way ahead in achieving the goal from efficient phosphors. However, the prospects of reaching this goal have been found to be satisfactory and more interesting with the emergence of different and promising materials and concepts.

REFERENCES

[1] Holonyak, N., and Bevacqua, S. F. (1962). Coherent (Visible) light emission from Ga $(As_{1-x}P_x)$. *Junctions*, 4 (1), 82–83. 10.1063/1.1753706

[2] Nakamura, S., Senoh, N., Iwasa, N., and Nagahama, S. (1995). High-brightness ingan blue, green and yellow light-emitting diodes with quantum well structures. *Japanese Journal of Applied Physics*, 34 (2), L797–L799. 10.1143/JJAP.34.L797

[3] Nakamura, S. (1997). III–V nitride-based light-emitting devices. *Solid State Communications*, 102 (2–3), 237–248. 10.1016/S0038-1098(96)00722-3

[4] Bommel, W. V. (2015). Interior lighting fundamentals. *Technology and Application*. Springer, Cham

[5] Lin, M., Ho, W., Shih, F., Chen, D., and Wu, Y. (1998). A cold-cathode fluorescent lamp driver circuit with synchronous primary-side dimming control. *IEEE Transactions on Industrial Electronics*, 45 (2), 249–255. 10.1109/41.681223

[6] Lin, D., and Yan, W. (2010). Modeling of cold cathode fluorescent lamps (CCFLs) with realistic electrode profile. *IEEE Transactions on Power Electronics*, 25(3), 699–709. 10.1109/TPEL.2009.2035359

[7] Alberts, I., Barratt, D., Ray, A., Disp, J. (2010). Hollow cathode effect in cold cathode fluorescent lamps: A review. *Journal of Display Technology*, 6(2), 52–59. 10.1109/JDT.2009.2031924

[8] LEECL, U.S. patent No. 2005057143 (2005) 11 08.

[9] Guangsup, C., Lee, J. Y., Lee, D. H., Kim, S. B., Song, H. S., Jehuan, K., Kim, B. S., Kang, J. G., Choi, E. H., Lee, U. W., Yang, S. C., and Verboncoeur, J. P. (2005). Glow discharge in the external electrode fluorescent lamp. *IEEE Transactions on Plasma Science*, 33 (4), 1410–1415. 10.1109/TPS.2005.852434

[10] Cho, K., Oh, W., Moon, G., Park, M., and Lee, S. (2007). A study on the equivalent model of an external electrode fluorescent lamp based on equivalent resistance and capacitance variation. *Journal of Power Electronics*, 7(1), 38–43. 10.6113/JPE.2007.7.1.38

[11] D. S. Lim, U. S. Patent No. 2006126332 (2006) 6 15.

[12] I. Hironori, Japanese Patent No. 2004079270 (2004) 0311.

[13] Jinno, M., Okamoto, M., Takeda, M., and Motomura, H. (2007). Luminance and efficacy improvement of low-pressure xenon pulsed fluorescent lamps by using an auxiliary external electrode. *Journal of Physics D: Applied Physics*, 40 (13), 3889–3895. 10.1088/0022-3727/40/13/S08

[14] Park, J., and Lim, S. (2007). LCD backlights, light sources, and flat fluorescent lamps. *Journal of the Society for Information Display*, 15(12), 1109–1114. 10.1889/1.2825100

[15] Hu, W., Liu, Z., and Yang, M. (2010). Luminescence characteristics of mercury-free flat fluorescent lamp with arc-shape anodes. *IEEE Transactions on Consumer Electronics*, 56(4), 2631–2635. 10.1109/TCE.2010.5681150

[16] Jung, J. C., Lee, J. K., Seo, I. W., Oh, B. J., and Whang, K. W. (2009). Electro-optic characteristics and areal selective dimming method for a new highly efficient mercury-free flat fluorescent lamp (MFFL). *Journal of Physics D: Applied Physics*, 42(12), 125205. 10.1088/0022-3727/42/12/125205

[17] Winsor, M., and Flynn, J. (2007). *SID Symposium Digest of Technical Paper*, 38, 979–982.

[18] Shur, M., and Zukauskas, A. (2005). Solid-state lighting: Toward superior illumination. *Proceedings of the IEEE*, 93(10), 1691–1703. 10.1109/JPROC.2005.853537

[19] Lin, H., Liang, H., Han, B., Zhong, J., Su, Q., Dorenbos, P., Birowosuto, M. D., Zhang, G., Fu, Y., and Wu, W. (2007). Luminescence and site occupancy of Ce^{3+} in $Ba_2Ca(BO_3)_2$. *Physical Review B*, 76(3), 035117. 10.1103/PhysRevB.76.035117

[20] Komeno, A., Uematsu, K., Toda, K., and Sato, M. (2006). VUV properties of Eu-doped alkaline earth magnesium silicate. *Journal of Alloys and Compounds*, 408–412, 871–874. 10.1016/j.jallcom.2005. 01.070

[21] Zhang, Z., Yuan, J., Chen, S., Chen, H., Yang, X., Zhao, J., Zhang, G., and Shi, C. (2008). Investigation on the luminescence of RE^{3+} (RE = Ce, Tb, Eu and Tm) in $KMGd(PO_4)_2$ (M = Ca, Sr) phosphates. *Optical Materials*, 30(12), 1848 - 1853. 10.1016/j.optmat.2007.12.002

[22] Lee, K., Yu, B., Pyun, C., and Mho, S. (2002). Vacuum ultraviolet excitation and photoluminescence characteristics of $(Y,Gd)Al_3(BO_3)_4/Eu^{3+}$. *Solid State Communications*, 122 (9), 485–488. 10.1016/ S0038-1098(02)00195-3

[23] Shinoya, S., and Yen, W. M. (1998). *Phosphor Handbook*. CRC Press, Boca Raton, FL, 623.

[24] Weston, G. F. (1975). Plasma panel displays. *Journal of Physics E: Scientific Instruments*, 8(12), 981. 10.1088/0022-3735/8/12/001

[25] Boeuf, J. P. (2003). Plasma display panels: Physics, recent developments and key issues. *Journal of Physics D: Applied Physics*, 36(6), R53. 10.1088/0022-3727/36/6/201

[26] Nagpure, P. A. (2012). Thesis submitted to Sant Gadge Baba Amravati University, Amravati.

[27] http://www.plasmacoalition.org © 2004 Coalition for Plasma Science.

[28] Jüstel, T., Nikol, H., (2000). Optimization of luminescent materials for plasma display panels. *Advanced Materials*, 12(7), 527–530. 10.1002/(SICI)1521-4095(200004)12:7<527::AID-ADMA52 7>3.0.CO;2-8

[29] Weber L.F. (1997). *"Color Plasma Displays," The Electrical Engineering Handbook*, edited by Richard C. Dorf, CRC Press, Boca Raton, Florida.

[30] Ishii, M., Takeda, Y., Shiga, T., Igarashi, K., and Mikoshiba, S. (2000). Data-pulse-voltage reduction of AC-PDPs by using metastable-particle priming.*Journal of the Society for Information Display*, 8(3), 217–221.

[31] Lakshamanan, A. (2008). *Luminescence and Display Phosphors Phenomena and Applications*. Nova Science Publishers, Inc, New York, NY, USA.

[32] Ronda, C. R. (1995). Phosphors for lamps and displays: An applicational view. *Journal of Alloys and Compounds*, (1995) 534–538. 10.1016/0925-8388(94)07065-2

[33] Justel, T., Krupa, J.-C., and Wiechert, D. U. (2001). VUV spectroscopy of luminescent materials for plasma display panels and Xe discharge lamps. *Journal of Luminescence*, 93(3), 179–189. 10.1016/ S0022-2313(01)00199-5

[34] McCauley, R. A., Hummel, F. A., and Hoffman, M. V. (1971). Phase equilibria and Eu^{2+}-, Tb^{3+}-, and Mn^{2+}-activated luminescent phases in the $CaO-MgO-P_2O_5$ system. *Journal of the Electrochemical Society*, 118 (5), 755–759. 10.1149/1.2408158

[35] B. L. Clark (2001). , Ph.D. Thesis, Oregon State, University.

[36] Ingle, J. T. (2015). Thesis submitted to Sant Gadge Baba Amravati University, Amravati.

[37] F. Meyer, H. Spanner, and E. Germet, US 2 182(1939) 732.

[38] Jung, K.Y., Lee, H. W., Kang, Y. C., Park, S. B., and Yang, Y. S. (2005). Luminescent properties of $(Ba,Sr)MgAl_{10}O_{17}$:Mn,Eu green phosphor prepared by spray pyrolysis under VUV excitation. *Chemistry of Materials*, 17(10), 2729–2734. 10.1021/cm050074f

[39] Ingle, J. T., Sonekar, R. P., Nagpure, P. A., and Omanwar, S. K. (2013). Synthesis and UV, VUV photoluminescence of red emitting borate host PDP phosphors $YCaBO_4$: Eu^{3+} and YBO_3:Eu^{3+}. *International Journal of Current Research*, 5(03), 529–531. http://www.journalcra.com

[40] Kim, C. H., Kwon, I. E., Park, C. H., Wang, Y. J., Bae, H., Yu, B.Y., Pyun, C. H., and Hong, G.Y. (2000). Phosphors for plasma display panels. *Journal of Alloys and Compounds*, 311(1), 33–39. 10.1 016/S0925-8388(00)00856-2

[41] Wang, L., and Wang, Y. (2007). Enhanced photoluminescence of YBO_3:Eu^{3+} with the incorporation of Sc^{3+} Bi^{3+} and La^{3+} for plasma display panel application. *Journal of Luminescence*, 122–123, 921–923. 10.1016/j.jlumin.2006.01.327

[42] Tian, L. H., Yu, B.Y., Pyun, C. H., Park, H. L., and Mho, S. I. (2004). New red phosphors BaZr $(BO_3)_2$ and $SrAl_2B_2O_7$ doped with Eu^{3+} for PDP applications. *Solid State Communications*, 129(1), 43–46. 10.1016/j.ssc.2003.09.012

[43] Tian, L.H., Kim, S.J., Park, H.L., and Mho, S.I., (2006). Variation of the photoluminescence and vacuum ultraviolet excitation characteristics of $BaZr(BO_3)_2$:Eu^{3+} by the incorporation of Al^{3+}, La^{3+}, or Y^{3+} into the lattice. *Materials Research Bulletin*, 41(1), 29–37. 10.1016/j.materresbull.2005.07.039

[44] Park, W., Summers, C. J., Do, Y. R., and Yang, H.G. (2002). Photoluminescence properties of red emitting $BaGdB_9O_{16}$:Eu phosphor. *Journal of Materials Science*, 37, (2002), 4041–4045. 10.1023/ A:1019652832744

[45] He, L., and Wang, Y. (2007). Synthesis of $Sr_3Y_2(BO_3)_4:Eu^{3+}$ and its photoluminescence under UV and VUV excitation. *Journal of Alloys and Compounds*, 431(1–2), 226–229. 10.1016/j.jallcom.2006.05.047

[46] Tian, L., and Mho, S. I. (2005). Luminescence and VUV excitation characteristics of Eu^{3+}- or Tb^{3+}-activated $Ca_4YO(BO_3)_3$. *Journal of the Korean Physical Society*, 47(6), 1070–1073. https://www.researchgate.net/profile/Sun-il_Mho/publication/282368805__

[47] Chen, Z., and Yan, Y., (2006). Nano-sized PDP phosphors prepared by solution combustion method. *Journal of Materials Science*, 41, 5793–5796. 10.1007/s10853-006-0129-1

[48] Sohn, K. S., Kim, C. H., Park, J. T., and Park, H. D. (2002). Optimization of red phosphor for plasma display panel by the combinatorial chemistry method. *Journal of Materials Research*, 17, 3201–3205. 10.1557/JMR.2002.0463

[49] Bechtel, H., Justel, T., Glaser, H., and Wiechert, D. U., (2012). Phosphors for plasma-display panels: Demands and achieved performance. *Journal of the Society for Information Display*, 10(1), 63–67. 10.1889/1.1827845

[50] Recent Developments in Display Phosphors, Ravi P Rao, Proc. of ASID '06, 8-12 Oct, New Delhi.

[51] Hu, Y., Tao, Y., Huang, Y., Yu, X., Zhang, C., Liang,T., and Yu, J. (2011). Luminescent properties of (Y, Gd) BO_3: Eu^{3+} under VUV excitation for PDP prepared by co-precipitation method. *Optoelectronics and Advanced Materials – Rapid Communications*, 5(4), 348–352. https://oam-rc.inoe.ro/articles/luminescent-properties-of-y-gdbo3eu3-under-vuv-excitation-for-pdp-prepared-by-co-precipitation-method/

[52] He, L., and Wang, Y. (2007). Synthesis of $Sr_3Y_2(BO_3)_4$: Eu^{3+} and its photoluminescence under UV and VUV excitation. *Journal of Alloys and Compounds*, 431(1-2), 226–229. 10.1016/j.jallcom.2006.05.047

[53] Ingle, J. T., Sonekar, R. P., Omanwar, S. K., Wang, Y., and Zhao, L. (2014). Solution combustion synthesis and optimization of phosphors for plasma display panels. *Optical Materials*, 36(8), 1299–1304. 10.1016/j.optmat.2014.03.015

[54] Xinmin, Z., Hong, C., and Jongsu, K. (2009). Photoluminescence properties of $(Gd_{1-x}Eu_x)Ba_3B_9O_{18}$ red emitting phosphors. *Journal of Rare Earths*, 27(1), 50–53. 10.1016/S1002-0721(08)60189-1

[55] Park, W., Lee, R. Y., Summers, C. J., Do, Y. R., Yang, H. G. (2000). Photoluminescence properties of $Al_3GdB_4O_{12}:Eu$ phosphors. *Materials Science and Engineering: B*, 78(1), 28–31. 10.1016/S0921-5107(00)00509-2

[56] Guo, R., Tang, S., Cheng, B., and Tan, D. (2013). A new red emitting phosphor: $La_2SrB_{10}O_{19}:Eu^{3+}$. *Journal of Luminescence*, 138, 170–173. 10.1016/j.jlumin.2013.02.008

[57] Wang, Y., Endo, T., Xie, E., He, D., and Liu, B. (2004). Luminescence properties of $Ca_4GdO(BO_3)_3:Eu$ in ultraviolet and vacuum ultraviolet regions. *Microelectronics Journal*, 35(4), 357–361. 10.1016/S0026-2692(03)00245-3

[58] Yanga, H. C., Lia, C. Y., Hea, H., Tao, Y., Xu, J. H., and Sua, Q. (2006). VUV-UV excited luminescent properties of $LnCa_4O(BO_3)_3:RE^{3+}$ (Ln = Y, La, Gd; Re = Eu, Tb, Dy, Ce). *Journal of Luminescence*, 118(1), 61–69. 10.1016/j.jlumin.2005.06.007

[59] Guifang, L., Quanxi, C., Zhimin, L., Yunxia, H., and Jiang, H. (2010). Solution combustion synthesis and luminescence properties of $(Y,Gd)Al_3(BO_3)_4:Eu^{3+}$ phosphors. *Journal of Rare Earths*, 28 (5), 709–712. 10.1016/S1002-0721(09)60185-X

[60] Nagpure, P. A., and Omanwar, S. K. (2013). Combustion synthesis of borate phosphors for use in plasma display panels and mercury-free fluorescent lamps. *International Journal of Self-Propagating High-Temperature Synthesis*, 22(1), 32–36. 10.3103/S106138621301007X

[61] Chen, L., Fu, Y., Zhang, G., Bao, J., and Gao, C. (2008). Optimization of Pr^{3+}, Tb^{3+}, and Sm^{3+} co-doped $(Y_{0.65}Gd_{0.35})BO_3:Eu_{0.05}^{3+}$ VUV phosphors through combinatorial approach. *Journal of Combinational Chemistry*, 10 (3), 401–404. 10.1021/cc700172e

[62] Wei, Z., Sun, L., Liao, C., Yan, C., and Huang, S. (2002). Fluorescence intensity and colour purity improvement in nanosized $YBO_3:Eu$. *Applied Physics Letters*, 80(8), 1447–1449. 10.1063/1.1452787

[63] Wanga, L., and Wang, Y. (2007). Enhanced photoluminescence of $YBO_3:Eu^{3+}$ with the in-corporation of Sc^{3+} Bi^{3+} and La^{3+} for plasma display panel application. *Journal of Luminescence*, 122–123, 921–923. 10.1016/j.jlumin.2006.01.327

[64] Dubey, V., Kaur, J., Agrawal, S., and Suryanarayana, N. S. (2013). Synthesis and characterization of Eu3+ doped YBO3 phosphor. *International Journal of Luminescence and Applications*, 3, 98–101.

[65] Dubey, V., Kaur, J., Agrawal, S., Suryanarayana, N.S. (2014). Effect of Eu^{3+} concentration on photoluminescence and thermoluminescence behavior of $YBO_3:Eu^{3+}$ phosphor. *Superlattices and Microstructures*, 67, 156–171. 10.1016/j.spmi.2013.12.026

[66] Yadav, S. R., Dutta, R. K., Kumar, M., Pandey, A. C. (2009). Improved color purity in nano-size Eu3+-doped YBO3 red phosphor. *Journal of Luminescence*, 129, 1078–1082.

[67] Zhu, H., Zhang, L., Zuo, T., Gu, X., Wang, Z., Zhu, L., and Yao, K. (2008). Sol–gel preparation and photoluminescence property of $YBO_3:Eu^{3+}/Tb^{3+}$ nanocrystalline thin films. *Applied Surface Science*, 254(20), 6362–6365. 10.1016/j.apsusc.2008.03.183

[68] Zhang, X., Marathe, A., Sohal, S., Holtz, M., Davis, M., Hope-Weeks, L. J., and Chaudhuri, J. (2012). Synthesis and photoluminescence properties of hierarchical architectures of $YBO_3:Eu^{3+}$. *Journal of Materials Chemistry*, 22(13), 6485–6490. 10.1039/C2JM30255A

[69] Dexpert-Ghys, J., Mauricot, R., Caillier, B., Guillot, P., Beaudette, T., Jia, G., Tanner P. A., and Cheng, B. M. (2010). VUV excitation of YBO_3 and $(Y,Gd)BO_3$ phosphors doped with Eu^{3+} or Tb^{3+}: Comparison of efficiencies and effect of site-selectivity. *Journal of Physical Chemistry C*, 114(14), 6681–6689. 10.1021/jp909197t

[70] Ingle, J. T., Sonekar, R. P., Omanwar, S. K., and Wang, Y. (2014). Modified route of combustion synthesis and photoluminescence of $(Y,Gd)Al_3(BO_3)_4:Eu^{3+}$ phosphors for PDPs applications. *International Journal of Chemical and Physical Sciences*, 3, 37–42. https://citeseerx.ist.psu.edu/viewdoc/download?doi=10.1.1.679.8744&rep=rep1&type=pdf

[71] Ingle, J. T., Sonekar, R. P., Omanwar, S. K., Wang, Y., and Zhao, L. (2014). Combustion synthesis and VUV photoluminescence studies of borate host phosphors $YBO_3:RE^{3+}$ (RE = Eu^{3+}, Tb^{3+}) for PDPs applications. *Combustion Science and Technology*, 186(1), 83–89. 10.1080/00102202.2013.846332

[72] Ingle, J. T., Gawande, A. B., Sonekar, R. P., Nagpure, P. A., and Omanwar, S. K. (2013). Synthesis and photoluminescence of inorganic borate host red emitting VUV phosphor $YCaBO_4$: Eu^{3+}. *Proceeding of International Conference on Recent Trends in Applied Physics and Material Science AIP Conference Proceedings* 1536, 895–896. 10.1063/1.4810517

[73] Ingle, J. T., Sonekar, R. P., Omanwar, S. K., Wang, Y., and Zhao, L. (2014). Combustion synthesis and photoluminescence study of novel red phosphor $(Y1xy, Gdx)BaB_9O_{16}:Eu^{3+}y$ for display and lighting. *Journal of Alloys and Compounds*, 608, 235–240. 10.1016/j.jallcom.2014.04.079

[74] Ingle, J. T., Sonekar, R. P., Omanwar, S. K., Wang, Y., and Zhao, L. (2014). Combustion synthesis and luminescent properties of metal yttrium borates $M_3Y_2 (BO_3)_4$: Eu^{3+} (M = Ba, Sr) for PDPs applications. *Solid State Sciences*, 33, 19–24. 10.1016/j.solidstatesciences.2014.04.007

[75] Ingle, J. T., Sonekar, R. P., Omanwar, S. K., Wang, Y., and Zhao, L. (2014). Solution combustion synthesis and optimization of phosphors for plasma display panels. *Optical Materials*, 36(8), 1299–1304. 10.1016/j.optmat.2014.03.015

[76] Ingle, J. T., Gawande, A. B., Sonekar, R. P., Omanwar, S. K. Wang, Y., and Zhao, L. (2014). Combustion synthesis and optical properties of Oxy-borate phosphors $YCa_4O(BO_3)_3:RE^{3+}$ (RE = Eu^{3+}, Tb^{3+}) under UV, VUV excitation. *Journal of Alloys and Compounds*, 585 (5), 633–636. 10.1 016/j.jallcom.2013.09.178

[77] Ingle, J. T., Sonekar, R. P., and Omanwar, S. K. (2016). Combustion synthesis and superior photo-luminescence from rare earth doped (Eu, Tb) lanthanum borates phosphors for display. *Journal of Materials Science: Materials in Electronics*, 27, 10735–10741. 10.1007/s10854-016-5175-0

[78] Zhang, Z., Zhang, S., Zhang, W., and Yang, W. (2016). VUV spectroscopic properties of rare-earth (RE3+, Sm3+, Eu3+, Tb3+, Dy3+) -Activated Layered Borate Ba6Gd9B79O138. *Solid State Sciences*, 64, 69–75. 10.1016/j.solidstatesciences.2016.12.014

[79] Huibing, X., Weidong, Z., Xiaofan, Ronghui, W., Yunsheng, L. H., and Tian. X., (2010). Effect of Li^+ ions doping on structure and luminescence of $(Y,Gd)BO_3:Tb^{3+}$. *Journal of Rare Earths*, 28 (5), 701–704. 10.1016/S1002-0721(09)60183-6

[80] Park, K., Kim, J., and Kim, K. Y. (2012). Enhancement of green emission for Al^{3+}-doped $YBO_3:Tb^{3+}$. *Materials Chemistry and Physics*, 136 (1), 264–267. 10.1016/j.matchemphys.2012.06.067

[81] Li, X., and Wang, Y. (2010). VUV-excited luminescence properties of Tb^{3+} activated $BaGdB_9O_{16}$. *Journal of Rare Earths*, 28 (3), 361–364. 10.1016/S1002-0721(09)60112-5

[82] Zang, D. S., Song, J. H., Park, D. H., Kim, Y. C., and Yoon, D. H. (2009). New fast-decaying green and red phosphors for 3D application of plasma display panels. *Journal of Luminescence*, 129 (9), 1088–1093. 10.1016/j.jlumin.2009.05.004

[83] Chawla, S., Ravishanker, Khan, A. F., Yadav, A., Chander, H., and Shanker, V. (2011). Enhanced luminescence and degradation resistance in Tb modified Yttrium Borate core–nano silica shell phosphor under UV and VUV excitation. *Applied Surface Science*, 257 (16), 7167–7171. 10.1016/j.apsusc.2011.03.082

[84] Kim, C., Kwon, I., Park, C., Hwang, Y., Bae, H., Yu, B., Pyun, C., and Hong, G. (2000). Phosphors for plasma display panels. *Journal of Alloys and Compounds*, 311 (1), 33–39. 10.1016/S0925-8388(00)00856-2

[85] Rao, R. P. (2003). Tb^{3+} activated green phosphors for plasma display panel applications. *Journal of The Electrochemical Society*, 150 (8), H165–H171. 10.1149/1.1583718

[86] Kwon, E., Yu, B. Y., Bae, H., Hwang, Y., Kwon, T., Kim, C., Pyun, C., and Kim, S. (2000). Luminescence properties of borate phosphors in the UV/VUV region. *Journal of Luminescence*, 87–89, 1039–1041. 10.1016/S0022-2313(99)00532-3

[87] Nagpure, P. A., Sonekar, R. P., Omanwar, S. K., Pande, A. C. (2011). Photoluminescence study of green PDP phosphor LaBaB$_9$O$_{16}$:Tb/CeTb. *International Journal of Advanced Engineering Sciences And Technologies*, 8 (2), 311–315.

[88] Yang, Y., Bao, A., Lai, H., Tao, Y., and Yang, H. (2009). Luminescent properties of SrAl$_2$B$_2$O$_7$: Ce^{3+}, Tb^{3+}. *Journal of Physics and Chemistry of Solids*, 70 (10), 1317–1321. 10.1016/j.jpcs.2009.06.012

[89] Cheng, S. D., Kam, C. H., and Buddhudu, S. (2001). Enhancement of green emission from Tb^{3+}:GdOBr phosphors with Ce^{3+} ion co-doping. *Materials Research Bulletin*, 36 (5–6), 1131–1137. 10.1016/S0025-5408(01)00587-6

[90] Henderson, E.W., and Meehan, J. P. (1974). Optical properties of divalent rare earth ions in SrAlF$_5$. *Journal of Luminescence*, 8 (5), 415–427. 10.1016/0022-2313(74)90082-9

[91] Tian, L., Mho, S., Yu, B., and Park, H. (2005). Luminescence and VUV excitation characteristics of Eu3+- or Tb3+- activated Ca4YO(BO3)3. *Journal of the Korean Physical Society*, 47 (6), 1070 –1073.

[92] Justel, T., and Nikol, H. (2000). Optimization of luminescent materials for plasma display panels. *Advanced Materials*, 12 (7), 527–530. 10.1002/(SICI)1521-4095(200004)12:7<527::AID-ADMA527>3.0.CO;2-8

[93] Dawson, B., Ferguson, M., Marking, G., and Diaz, A. (2004). Mechanisms of VUV damage in BaMgAl$_{10}$O$_{17}$:Eu^{2+}. *Chemistry of Materials*, 16 (25), 5311–5317. 10.1021/cm0489284

[94] Rao, R., and Devine, D. (2000). RE-activated lanthanide phosphate phosphors for PDP applications. *Journal of Luminescence*, 87–89, 1260–1263. 10.1016/S0022-2313(99)00551-7

[95] Kim, K., Koo, K., Cho, T., and Chun, H. (2003). Effect of heat treatment on photoluminescence behavior of BaMgAl$_{10}$O$_{17}$:Eu phosphors. *Materials Chemistry and Physics*, 80 (3), 682–689. 10.1016/S0254-0584(03)00110-X

[96] Zhang, S., Kono, T., Ito. A., Yasaka, T., and Uchiike, H. (2004). Degradation mechanisms of the blue-emitting phosphor BaMgAl10O17:Eu2+ under baking and VUV-irradiating treatments. *Journal of Luminescence*, 106 (1), 39–46. 10.1016/S0022-2313(03)00132-7

[97] Hata, E., Kunimoto, T., Tanaka, M., Yamaguchi, S., Ohmi, K., Tanaka, S., and Kobayashi, H. (2003). Proceedings of the IDW'03 (Fukuoka, Japan), pp. 853.

[98] Tanamachi, N., Egoshi, K., Tanno, H., Zeng, Q., and Zhang, S. (2004). Proceedings of the IDW'04 (Niigata, Japan), pp. 1085.

[99] Liang, H., Zeng, Q., Tao, Y., Wang, S., and Su, Q. (2003). VUV-UV excited luminescent properties of calcium borophosphate doped with rare earth ions. *Materials Science and Engineering: B*, 98 (3), 213–219. 10.1016/S0921-5107(03)00034-5

[100] Justel, T., Bechtel, H., Mayr, W., and Wiechert, D. (2003). Blue emitting BaMgAl$_{10}$O$_{17}$:Eu with a blue body colour. *Journal of Luminescence*, 104 (1–2), 137–143. 10.1016/S0022-2313(03)00010-3

[101] Saubat, B., Fouassier, C., Hagenmuller, P., and Bourcet, J. C. (1981). Luminescent efficiency of Eu^{3+} and Tb^{3+} in LaMgB$_5$O$_{10}$ – Type borates under excitation from 100 to 400 nm. *Materials Research Bulletin*, 16(2), 193–198. 10.1016/0025-5408(81)90081-7

[102] Lin, H., Liang, H., Tian, Z., Su, Q., Xie, H. and Ding, J. (2006). Vacuum-ultraviolet–vis luminescence of dibarium magnesium orthoborate Ba$_2$Mg(BO$_3$)$_2$ doped with Ce^{3+} and Eu^{2+} ions. *Journal of Materials Research*, 21, 864–869. 10.1557/jmr.2006.0123

[103] Sivakumar, V., Hong, G. Y., Kim, J. S., and Jeon, D. Y. (2009). Novel blue-emitting phosphor Sr$_{2.979}$Eu$_{0.021}$Al$_{10-x}$B$_x$SiO$_{20}$ (x = 0, 0.25, 0.5, 0.75 and 1) for PDP applications. *Journal of Luminescence*, 29 (12), 1632–1636. 10.1016/j.jlumin.2008.12.021

[104] Ling, H. E., Yuhua, W., and Weimin, S. (2009). Luminescence properties of BaB$_8$O$_{13}$:Eu under UV and VUV excitation. *Journal of Rare Earths*, 27 (3), 385–389. 10.1016/S1002-0721(08)60256-2

[105] Park, C. H., and Bluhm, K. (1996). Pb$_6$(AsO$_4$)[B(AsO$_4$)$_4$] – A new crystal structure type in the system PbO/B$_2$O$_3$/As$_2$O$_3$ with remarks about Pb(BAsO$_5$). *Z. Naturforsch*, 51b, 313–318. 10.1515/znb-1996-0303

[106] Hasegawa, Y., Murakoshi, K., Wada, Y., Yanagida, S., Kim, J. H., Nakashima, N., and Yamanaka, T. (1996). Enhancement of luminescence of Nd^{3+} complexes with deuterated hexafluoroacetylacetonato ligands in organic solvent. *Chemical Physics Letters*, 248 (1–2), 8–12. 10.1016/0009-2614(95)01279-6

[107] Toda, K. (2006). Recent research and development of VUV phosphors for a mercury-free lamp. *Journal of Alloys and Compounds*, 408–412, 665–668. 10.1016/j.jallcom.2005.01.080

[108] Blasse, G., and Grabmaier, B. C. (1994). *Luminescent Materials*. Springer-Verlag, Berlin. 10.1007/978-3-642-79017-1

[109] Henderson, B., and Imbusch, G. F. (1989). *Optical Spectroscopy in Inorganic Solids*. Clarendon, Oxford.

[110] Ronda, C. R. (1995). Phosphors for lamps and displays: An applicational view. *Journal of Alloys and Compounds*, 225 (1–2), 534–538. 10.1016/0925-8388(94)07065-2

[111] Depp, S. W., and Howard, W. E. (1993). Flat-Panel Displays. *Scientific American*, 268 (3), 90–97.

[112] Vollkommer, F., and Hitzschke, L. (2013). Beratender Ausschuß der Industriephysiker: Durchbruch bei der effizienten Erzeugung von Excimer-Strahlung: Hohe Erwartungen an die Leuchtstofforschung. *Physikalische Blätter*, 53 (9), 887–889. 10.1002/phbl.19970530912

[113] Sommerdijk, J. L., Bril, A., and Jager, A. W. (1974). Two photon luminescence with ultraviolet excitation of trivalent praseodymium. *Journal of Luminescence*, 8 (4), 341–343. 10.1016/0022-2313(74)90006-4

[114] Pappalardo, R. (1976). Calculated quantum yields for photon-cascade emission (PCE) for Pr^{3+} and Tm^{3+} in fluoride hosts. *Journal of Luminescence*, 14 (3), 159–193. 10.1016/S0022-2313(76)90592-5

[115] Auzel, F., and Acad, C. R. (1966). *Science (Paris)*, 262, 1016. (In French).

[116] Srivastava, A. M., and Doughty, D. A. (1996). Photon cascade luminescence of Pr^{3+} in $LaMgB_5O_{10}$. *Journal of The Electrochemical Society*, 143 (12), 4113–4116. 10.1149/1.1837346

[117] Chen, Y., Shi, C., Yan, W., Qi, Z., and Fu, Y. (2006). Energy transfer between Pr^{3+} and Mn^{2+} in SrB_4O_7: Pr, Mn. *Applied Physics Letters*, 88 (6), 061906-1-061906-3. 10.1063/1.2172731

[118] Han, B., Liang, H., Huang, Y. Tao, Y., and Su, Q. (2010). Vacuum ultraviolet–visible spectroscopic properties of Tb^{3+} in Li(Y, Gd)$(PO_3)_4$: Tunable emission, quantum cutting, and energy transfer. *Journal of Physical Chemistry*, 114 (14), 6770–6777. 10.1021/jp100755d

[119] Hou, D., Liang, H., Xie, M., Ding, X., Zhong, J., Su, Q., Tao, Y., Huang, Y., and Gao, Z. (2011). Bright green-emitting, energy transfer and quantum cutting of $Ba_3Ln(PO_4)_3$: Tb^{3+} (Ln = La, Gd) under VUV-UV excitation. *Optics Express*, 19 (12), 11071–11083. 10.1364/OE.19.011071

[120] Song, J. H., Song, Y., Kim, J., Kim, M., Kwon, S., Park, D., Kim, Y., and Zang, D., (2008). *IMID/IDMC/ASIA Display'08 Digest*, 34 (3), 1255.

[121] Moon, T., Hong, G. Y., Lee, H. C., Moon, E. A., Jeoung, B. W., Hwang, S. T., Kim, J. S., and Ryu, B. G. (2009). Effects of Eu^{2+} co-doping on VUV photoluminescence properties of $BaMgAl_{10}O_{17}$:Mn^{2+} phosphors for plasma display panels. *Electrochemical and Solid-State Letters*, 12 (7), J61–J63. 10.1149/1.3126528

[122] Oskam, K. D., Wegh, R. T., Donker, H., Dvan, Loef, E.V. D van, and Meijerink, A. (2000). Downconversion: A new route to visible quantum cutting. *Journal of Alloys and Compounds*, 300–301, 421–425. 10.1016/S0925-8388(99)00755-0

[123] Wang, D., and Kodama, N. (2009). Visible quantum cutting through downconversion in $GdPO_4$:Tb^{3+} and $Sr_3Gd(PO_4)_3$:Tb^{3+}. *Journal of Solid-State Chemistry*, 182 (8), 2219–2224. 10.1016/j.jssc.2009.05.026

[124] Liang, W., and Wang, Y. (2010). Visible quantum cutting through downconversion in Eu^{3+}-doped $K_2GdZr(PO_4)_3$ phosphor. *Materials Chemistry and Physics*, 119 (1–2), 214–217. 10.1016/j.matchemphys.2009.08.058

[125] Zhang, J., Wang, Y., Chen, G., and Huang, Y. (2014). Investigation on visible quantum cutting of Tb^{3+} in oxide hosts. *Journal of Applied Physics*, 115 (9), 093108-1-093108-7. 10.1063/1.4867612

[126] Maarten, L., Heerdt, H., Kolk, Evan der, Yen, W. M., Srivastava, A. M. (2002). Vacuum ultraviolet spectroscopy of Pr^{3+} in $CaAl_4O_7$, $LaMgAl_{11}O_{19}$ and $SrLaAlO_4$. *Journal of Luminescence*, 100 (1–4), 107–113. 10.1016/S0022-2313(02)00454-4

[127] Wegh, R. T., Donker, H., Loef, E. V. D., Oskam, K. D., and Meijerink, A. (2000). Quantum cutting through downconversion in rare-earth compounds. *Journal of Luminescence*, 87–89, 1017–1019. 10.1016/S0022-2313(99)00514-1

[128] Moine, B., and Bizarri, G. (2003). Rare-earth doped phosphors: Oldies or goldies? *Materials Science and Engineering: B*, 105 (1–3), 2–7. 10.1016/j.mseb.2003.08.004

7 Borate Phosphors for Radiation Dosimetery

C. B. Palan and Y. K. More

CONTENTS

DOI: 10.1201/9781003207757-7

7.1 INTRODUCTION

7.1.1 RADIATION

Radiation is one of the modes of transfer of energy. It is the emission or transmission of energy in the form of electromagnetic waves (X-ray and gamma radiation, etc.) or streams of particles (beta, alpha, neutron, etc). In electromagnetic waves, there is transmission of only energy without any mass, whereas in particle radiation it includes fast-moving tiny particles that have both energy and mass. These are termed two physical forms of the radiation. The electromagnetic spectrum is the range of frequencies and photon energies of radiation. Figure 7.1 presents an overview of the electromagnetic spectrum. Table 7.1 shows the Frequency range of various types of electromagnetic radiation.

7.1.2 RADIATION SOURCES

Radiation can be categorized into two categories on the basis of their ability to ionize the ordinary chemical matter as follows:

- Ionizing radiation
- Non-ionizing radiation

7.1.2.1 Ionizing Radiation

Ionizing radiation is capable of knocking electrons out of their orbits around atoms, disturbing the electron/proton balance and converting the atom into its respective ion. Ionizing radiations can be classified into two categories: directly ionizing and indirectly ionizing radiations. Based on their electrical

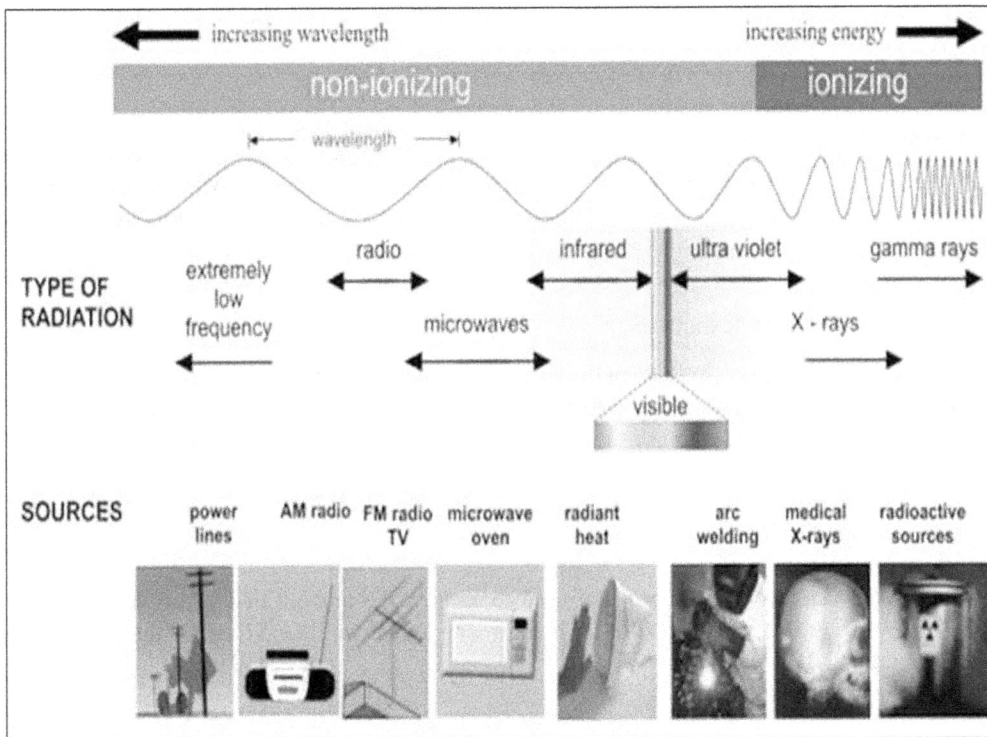

FIGURE 7.1 Overview of the electromagnetic spectrum.

TABLE 7.1

Frequency range of various types of electromagnetic radiation

Type of radiation	Frequency range (Hz)	Wavelength range
Gamma rays	10^{20}–10^{24}	$<10^{-12}$ m
X-rays	10^{17}–10^{20}	1 nm–1 pm
Ultraviolet	10^{15}–10^{17}	400 nm–1 nm
Visible	4–$7.5*10^{14}$	750 nm–400 nm
Near-infrared	1×10^{14}–4×10^{14}	2.5 μm–750 nm
Infrared	10^{13}–10^{14}	25 μm–2.5 μm
Microwaves	3×10^{11}–10^{13}	1 mm–25 μm
Radio waves	$< 3 \times 10^{11}$	>1 mm

properties, the ionizing radiation can be classified into charged radiation, such as alpha particle, beta particle and uncharged radiations such as gamma rays and neutrons. These are harmful, since ions travelling through the living body can break DNA and cells in the body, potentially leading to mutated cells, which can lead to cancer. Also, according to their penetration power, radiation is classified into soft radiation and hard radiation. Ionizing radiation includes the radiation that comes from both natural and man-made radioactive materials.

The International Commission on Radiological Protection (ICRP) is an independently registered charity established to advance, for the public benefit, the science of radiological protection, in particular by providing recommendations and guidance on all aspects of protection against ionizing radiation. Table 7.2 representsthe main sources of radiation.

7.1.2.2 Non-Ionizing Radiation
Non-ionizing radiation has less energy than ionizing radiation; it does not possess enough energy to produce ions. It therefore will not produce breaks in body cells and DNA and does not lead to cancer related to radiation exposure. Examples of non-ionizing radiation are visible light, infrared, radio waves, microwaves and sunlight.

7.1.3 BASIC UNITS OF RADIATION
There are many different quantities and units used to quantify the radiation, which are discussed below. Depending on the number of different aspects of radiation that can be used to express the amount of radiation, the selection of the most appropriate quantity in the specific unit depends on specific application.

Table 7.3 is a list of most of the physical quantities and their units encountered in radiology.

7.1.3.1 Exposure
Exposure is the quantity most commonly used to express the amount of radiation delivered to a point. The conventional unit for exposure is the roentgen (R) and the SI unit is the coulomb per kilogram of air (C/kg):

$$1\ R = 2.\ 58\ \times\ 10^{-4} C/Kg$$

TABLE 7.2
Main sources of radiation

Category	Source/Machine	Radiation
Environmental	Cosmic rays	Neutrons, protons, electrons and photons
	Radioactivity	α, β, γ particles and neutrons
Artificial	Orthovoltage X-rays	X-rays
	Linac/betatron	MV X-rays, electrons and radioactivity
	van de Graff generator and cyclotron	Protons, neutrons and radioactivity
	Synchrotron	Electrons, protons, X-ray and UV photons
	Nuclear reactor	Neutrons, γ-rays and residual radioactivity

TABLE 7.3
Physical quantities and units

Exposure	Conventional unit	SI unit	Conversions
Exposure	roentgen (R)	coulomb/kg of air (C/kg)	1 C/kg = 3,876 R
Dose	rad (R)	gray (Gy)	1 Gy = 100 rad
Dose equivalent	rem	sievert (Sv)	1 Sv = 100 rem
Activity	curie (Ci)	becquerel (Bq)	1 mCi = 37 mBq

7.1.3.2 Absorbed Dose

Absorbed dose is the quantity that expresses the concentration of radiation energy absorbed at a specific point within the body tissue. Since an X-ray beam is attenuated by absorption as it passes through the body, all tissues within the beam will not absorb the same dose. The absorbed dose will be much greater for the tissues near the entrance surface than for those deeper within the body. **Absorbed dose is defined as the quantity of radiation energy absorbed per unit mass of tissue.**

The conventional unit for absorbed dose is the rad. The SI unit is the gray (Gy), the **gray (Gy)** is a derived unit of ionizing radiation dose in the International System of Units (SI).

The **rad**: A basic unit of an absorbed dose of radiation, equal to the absorption of 100 ergs of energy per gram of material or 0.01 joule per kilogram of material (0.01 gray).

The **gray (Gy)**: It is defined as the absorption of one joule of radiation energy per kilogram of matter (1 Gy–100 rad).

The **sievert** represents the equivalent biological effect of the deposit of a joule of radiation energy in a kilogram of human tissue.

7.1.3.3 Dose Equivalent

Dose equivalent (some refer to this as equivalent dose) is a quantity which takes into account 'radiation quality', which relates to the degree in which a type of ionizing radiation will produce biological damage. The dose equivalent is obtained by a product of absorbed dose and a quality factor. The resulting quantity can then be expressed numerically in rem (old units) or more commonly in Sieverts (Sv). It is worth emphasizing that the quantity is independent of the absorbing material (i.e. tissue, water, air).

7.1.3.4 Integral Dose

Integral dose is the total amount of energy absorbed in the body. It is determined by the absorbed dose values as well as by the total mass of tissue exposed.

The conventional unit for integral dose is the gram-rad, which is equivalent to 100 ergs of absorbed energy. The concept behind the use of this unit is that if we add the absorbed doses (rads) for each gram of tissue in the body, we will have an indication of total absorbed energy. Since integral dose is a quantity of energy, the SI unit used is the joule. The relationship between the two units is

$$1J = 1000 \text{ gram} - \text{rad}$$

7.1.3.5 Dose Rate

'Dose rate' is a term applied for the rate at which ionizing radiation is being absorbed by a medium (e.g. tissue). Whilst it is accurate to apply a prefix or postfix (e.g. 'absorbed dose rate'), the term is probably most often applied to effective dose, which is a useful quality in expressing harm or overall risk of harm from whole body irradiation.

7.1.3.6 Effective Dose

The effective dose is obtained by taking the equivalent dose (dose equivalent) and multiplying it by a tissue weighting factor that relates to the organs/tissues under consideration.

$$\textbf{Effective Dose (Gy)} = \textbf{absorbed Dose (Gy)} \times \mathbf{W_T}$$

where
W_T is tissue weighting factor value.

7.1.3.7 KERMA

KERMA is a term is short for kinetic energy released per unit mass (although many refer to the term 'kinetic energy released in material'). Either term can mean the same thing, although by definition the former term is more accurate since the quantity is based on functions of energy and mass. KERMA is a quantity that expresses the initial kinetic energy of all charged ionising radiations (e.g. photo-electrons) that are produced as a result of indirectly ionising radiations (e.g. photons) interaction with material. The quantity is material specific (e.g. air-KERMA or water-KERMA) and has the units J/kg or gray. KERMA is related to absorbed dose but is not the same quantity.

7.1.4 Radiation Dosimetry

Radiation dosimetry is a well-established technique of calculation and assessment of the ionizing radiation dose received by the human body due to both external irradiation and the ingestion or inhalation of radioactive materials. It becomes a scientific sub-specialty in the fields of health and medical physics [1]. The internal dose can be calculated from a variety of physiological techniques, while external dose can be measured with a dosimeter or inferred from other radiological protection instruments.

Dosimetry technique is extensively used for radiation protection and routinely applied for occupational radiation workers where radiation dose is expected, without exceeding regulatory level. It is also used where radiation is unexpected, such as in the outcome of Three Mile Island, Chernobyl or Fukushima radiological release incidents, where the public dose is measured and calculated from a variety of indicators such as ambient dosimetry of radiation and radioactive contamination.

Other significant areas of radiation dosimetry are in the field of medicine, where the required treatment dose and any collateral dose are monitored. Also, it is used in the environmental dosimetry such as radon monitoring in buildings.

7.1.5 Need of Radiation Dosimetry

The radiation monitoring could be effective only when the radiation dosimetry is accurate. Radiation processing using high-doses presents significant advantages in industry (water purification, organic polymer materials crosslinking and pasteurization), medicine (radio-sterilization) and agriculture (disinfections, inhibition of sprouting) [2].

The nuclear energy is a clean and efficient alternative over traditional methods of power generation worldwide. The number of nuclear plants is increasing every year. Also radioactive and radiological techniques are widely used for treatment of food, seed, edible items and deceases like cancer (radiation therapy). Apart from nuclear installations, there is a need of regular radiation monitoring in cancer hospitals and preservation of agricultural products. In fact, there is a separate division working for monitoring the radiation exposures received by the workers in such installations.

7.1.6 Radiation Detection and Measurements

The radiation detection system is composed of a detector, signal processor electronics and data output display device such as counter or multichannel analyzer. A radiation detector is the backbone component of any radiation detection system. The physical properties and characteristics of the detector control the features of the detection system. A radiation detector is composed of three main components: a detector sensitive volume where the radiation interactions occur, structural components that enclose the sensitive volume to maintain the proper conditions for its optimum operation and a signal output display device that extracts the information from the sensitive volume and transfer it to the signal processing devices.

The main types of radiation detectors are as follows:

1. Gas-filled detectors
2. Scintillation detectors
3. Semiconductor detectors
4. **Thermoluminescence detectors (TLDs)**

5. **Optically stimulated luminescence detector (OSLDs)**
6. Cerenkov counter
7. Nuclear track detectors
8. Neutron detectors

Out of the above, **TLDs** and **OSLDs** follow the luminescence phenomenon.

7.1.7 LUMINESCENCE TECHNIQUES IN RADIATION DOSIMETRY

Radiation dosimetric methods are used for the estimation of dose absorbed by radiation in a detector material. These methods are required for estimation of absorbed dose in various applications of radiation, such as personnel and environmental dosimetry, retrospective/accidental, dosimetry and medical applications of radiation. Solid-state luminescent dosimetry is based on radiation energy storage in dosimetric material in the form of lattice defects and captured charge carriers. The stored energy can be released as light at luminescence centres. Energy release is stimulated either by heating (thermally stimulated luminescence, TSL, TL) or by irradiating with light quanta of proper energy (optically stimulated luminescence, OSL). TL-based dosimeters are widely used in radiation dose monitoring. Both detectors are equipped with TL dosimeters that are very compact and portable, but still require a readout system with a heating element, which is less compact compared with a system based on the OSL readout. OSL dosimetry systems can be really portable due to the application of light-emitting diodes.

The use of thermoluminescence (TL) as a method for radiation dosimetry of ionizing radiation has been established for many decades and has found many useful applications in various fields, such as personnel, environmental, medical, archaeological, geological dating and space dosimetry. Several highly sensitive TL phosphor materials and thermoluminescent dosimeters (TLDs) are now commercially available in different physical forms. There are many commercial TLD systems that are being used for various dosimetric applications and even presently, TL is a popular technique in the field of radiation dosimetry, particularly in personnel monitoring [3–6].

7.1.7.1 The Advantages of the Technique

- No electronic gadgets are needed during radiation monitoring.
- The size of the detector is very tiny and could be used as badges, I-cards, jewellery like a ring, necklace, earring, etc.
- The instrumentation for taking readouts is very simple.
- The detectors (TLD materials) are generally nontoxic and easy to use.
- The detectors are reusable and cost effective.
- The detectors could be coded and a large number of them could be processed simultaneously and the records could be maintained easily.
- Some of them are tissue equivalent and could be used in mixed files also.
- Dosimetry of swift heavy ions, neutrons, alpha/beta/gamma rays is possible.

In the last two decades, an alternative technique, optically stimulated luminescence (OSL), has been developed and widely used. The OSL technique is the best of all the known techniques for measurement of radiation exposures since the out process does not involve problems of blackbody radiation and thermal quenching as in TL [7–10]. The technique is simple, easy and best for monitoring. The technique uses the OSL material in tablet or pellet form. These are the materials that are sensitive to the radiation exposures. When these materials are exposed to radiation, there is creation of localized disorders in the lattice that leads to the energy traps where the free electrons could sit for considerably longer time. Upon stimulation by optical treatment, the trapped electrons are set free and recombine with hole (hole trapped local disorders), giving emission of the photons in a visible region. The OSL emission in the near-UV (360–400 nm) region is preferred for accurate measurement of exposure. The intensity of the emitted light should be directly proportional to the absorbed radiations; this is known as linearity of the OSL phosphor. The OSL signals are often accompanied by photoconductivity phenomena.

Radiophotoluminescence (RPL) refers to the property of certain substances to become luminescent after irradiation to ionizing radiations in a spectral region where no luminescence exists prior to irradiation. The RPL substances (luminescent glasses or crystals) when exposed to radiation develop colour centres within themselves. When stimulated by light belonging to the absorption zone of the colour centres, they emit fluorescence. Such centres are called radiophotoluminescent centres and the phenomenon as radiophotoluminescence. Silver-activated phosphate glass possesses RPL. When it is exposed to ionizing radiation, a stable luminescent centre is formed. When excited by ultraviolet rays (~365 nm), it emits an orange luminescence. The intensity of this fluorescence is proportional to the radiation dose. RPL dosimeters, due to their high range and stability over a wide range of ambient conditions, are particularly suited for personal monitoring of a rather high dose (i.e. accident dosimetry) of gamma rays. Table 7.4 gives comparative advantages of TL, OSL and RPL dosimeters.

7.2 OPTICALLY STIMULATED LUMINESCENCE

7.2.1 BASIC PHENOMENON OF OSL

Optically stimulated luminescence (OSL) is the emission of light from an irradiated semi-conductor (wide bandgap) or insulator material during optical stimulation. Irradiation with ionizing radiation creates free electrons and holes in the material. Some of these electron-hole pairs can be captured by defects of the material, which are called **traps**. These traps are deep enough to accumulate charges. After the storage period, stimulation with light gives rise to the transition of the trapped electrons to the conduction band. As a result, emission of luminescence occurs after recombination of these electrons with optically active trapped holes that are also called recombination centres. Figure 7.2 represents the OSL process. In this process, optical energy required to release the charges from the traps depends on the depth of the trap. Thus, depending on the depth of the trap, the optical stimulation energy may lie anywhere in the range of UV to infra. In this process, OSL emission is plotted as a function of time, as shown in Figure 7.3. The OSL emission monotonically decreases as a function of time because of a decrease in electron population in the traps.

TABLE 7.4
Comparative advantages of TL, OSL and RPL dosimeters [11]

TL dosimeters	OSL dosimeters	RPL dosimeters
High sensitivity of materials:LiF:Mg,Cu,P, LiF:Mg,Cu,Si,CaSO$_4$:Dy	High sensitivity due to absence of thermal quenching and IR background: Al$_2$O$_3$:C and BeO	High sensitivity of materials such as silver-activated phosphate glass
Easy handling using LiF:Mg,Cu,P, LiF:Mg,Ti	Highly light sensitive, but can be managed with appropriate precautions during storage and use	No light sensitivity for visible light (λ>366 nm), as RPL emission is stimulated by UV light
Dose re-estimation possible	Multiple dose re-estimations possible	Dosimeter can be read as many times as necessary without depleting the signal
Flat photon energy response for LiF:Mg, Cu,P (100 keV–3 MeV)	Flat photon energy response for BeO OSL dosimeters (100 keV–3 MeV)	Use of appropriate filters gives nearly flat photon energy response:10 keV to 10 MeV
Dose imaging difficult	Dose imaging possible	Dose imaging possible
Dose range: 1 µGy to 100 Gy	10 µGy to 10 Gy	14 µGy–10 Gy
Post-irradiation fading: <2% in 6 months	<5% per year	1% per month

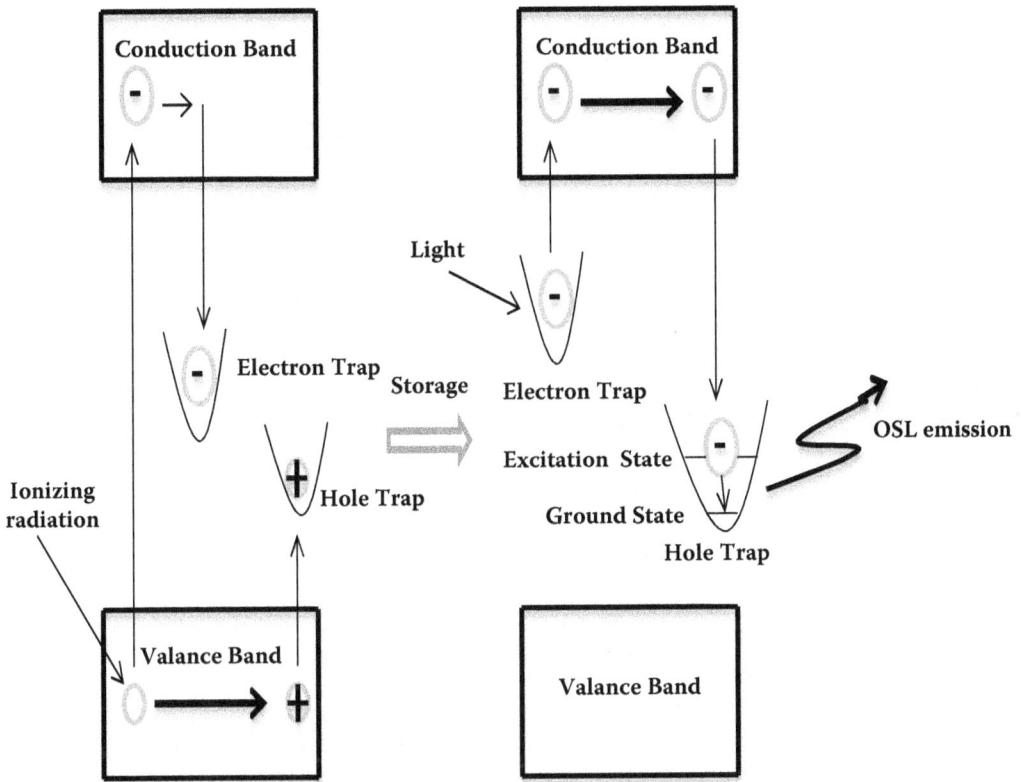

FIGURE 7.2 Schematic representations of optically stimulated luminescence phenomena.

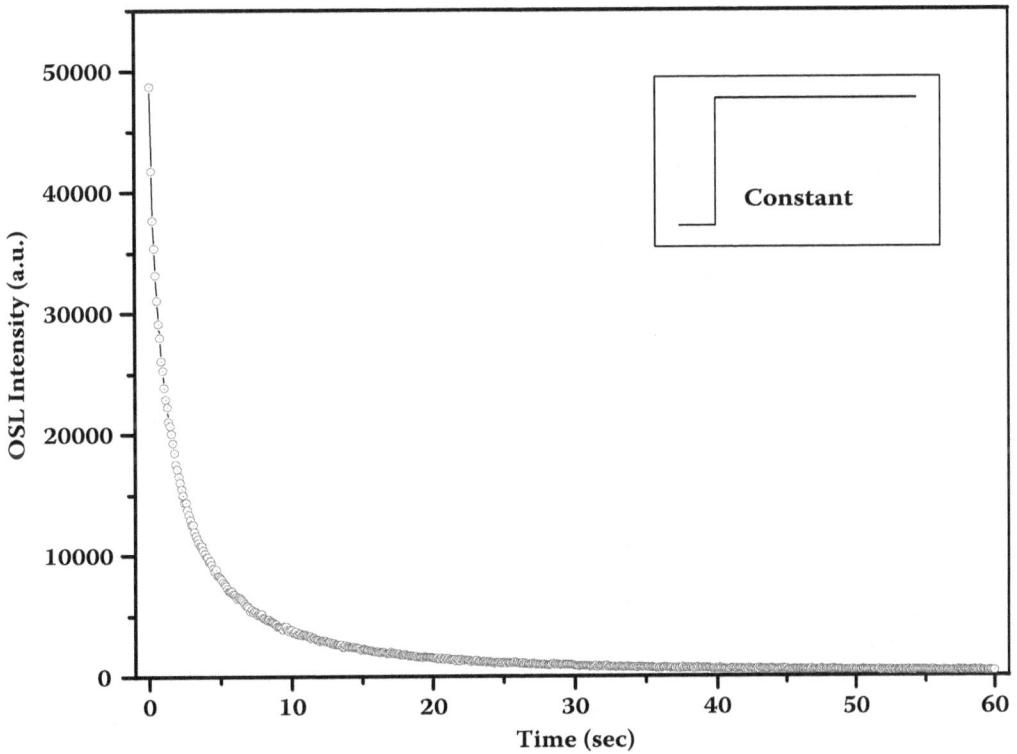

FIGURE 7.3 Typical OSL decay curve. Inset to graph represents the stimulation light intensity against time.

7.2.2 Different Readout Modes for TL/OSL

The basis of OSL measurement is to stimulate an irradiated sample with the light of a selected wavelength and to monitor the emission from the sample at a different wavelength. Different modes of stimulation that are widely used are continuous wave-OSL (CW-OSL), linearly modulated (LM-OSL) and pulsed OSL (P-OSL).

7.2.2.1 CW-OSL

In this technique, a laser or a broadband source and monochromator (or filter) is used to select a particular stimulation wavelength. The luminescence emission is monitored continuously while the stimulation beam is on and narrowband filters are used to discriminate between the excitation light and emission light and to prevent scattered stimulation light from entering the detector. The OSL is monitored from the instant that the stimulation source is switched on and is usually of the form of the exponential-like decay until all the traps are emptied and the luminescence ceases. The integrated emission (i.e. the area under the decay curve, less background) is recorded and is used to determine the dose of absorbed radiation.In CW-OSL, the recorded luminescence is very fast because most of the OSL emission occurs immediately. In this mode, the OSL pattern looks like a decay curve, as shown in Figure 7.2 [12].

7.2.2.2 LM-OSL

In this technique, OSL output is observed to increase, initially linearly as the stimulation power increases until the traps become depleted, after which the OSL intensity decreases non-linearly to zero. Thus, the OSL signal is in the shape of a peak, the position of which depends upon the rate of linear increase in intensity of the stimulating light and the photo-ionization cross section of the trap being emptied. For a given ramp rate and stimulation wavelength, traps with different photo-ionization cross section values thus appear at different times and the method is thus able to distinguish between OSL originating from different traps. Figure 7.4 shows the schematic representation of the three main OSL stimulation modes, namely: CW-OSL, LM-OSL and P-OSL.

7.2.2.3 P-OSL

Pulsed OSL results when the stimulation source is pulsed at a particular modulation frequency and with a particular pulse width appropriate to the lifetime of the luminescence being observed. In this mode of excitation, one measures the OSL emission only between pulses. In this way, discrimination between the excitation light and the emission light is achieved by time resolution, rather than wavelength resolution.

7.2.3 Advantages of OSL over TL [13]

The OSL is a comparatively recent technique as an alternative to the TL technique. The physical principles of OSL are closely related to those of the TL technique:

   a. OSL is normally measured at room temperature. (However, some additional advantages may be gained by performing the OSL at slightly elevated temperature for the academic interest.) The OSL dosimetry is becoming a serious contender to TL dosimetry due to its operational ease.
   b. The OSL readout method is fully optical in nature.
   c. The avoidance of heating during OSL removes the problems of thermal quenching from phosphors and thus a significant increase in sensitivity is achieved.
   d. Another important advantage of OSL is that trapped charges in the irradiated phosphor can be emptied by simply exposing it to the daylight for some hours, enabling its reset as well as reuse.
   e. In the OSL technique, one has a fine control over the degree to which the traps can be emptied by varying the intensity and wavelength of stimulation light. Because it is not necessary to stimulate all the trapped charges in order to obtain sufficient luminescence

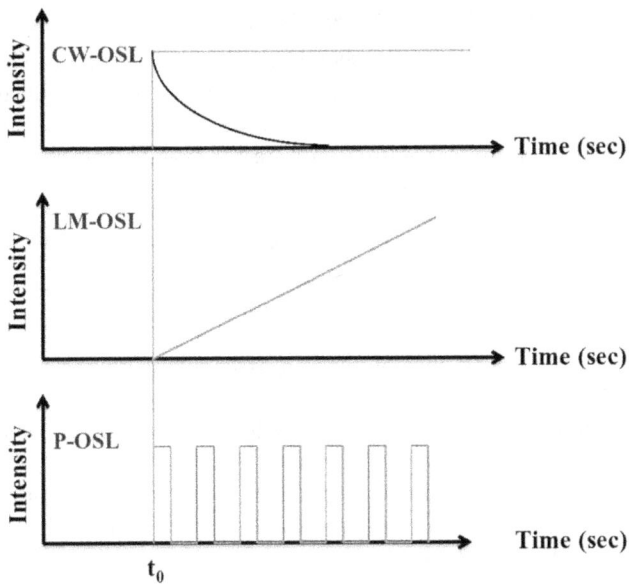

FIGURE 7.4 Schematic representation of the three main OSL stimulation modes: CW-OSL, LM-OSL and P-OSL.

signal, with an appropriate choice of stimulation wavelength and power, the dose can be reanalyzed multiple times subject to material sensitivity. The residual information, in the form of trapped charges, can be stimulated later to verify the dose.

In short, the lower detection limits (nGy), convenient optical stimulation, short readout time (~ms), optical erasing of the dose, high dynamic dose range (7 order), high sensitivity, dose imaging and optical fibre–based online remote dosimetry are some of the advantages of OSL over the TL technique. However, the OSL technique is relatively more expensive than the TL as it requires a darkroom facility during the complete readout process. Another disadvantage associated with the OSL technique is that the optical stimulation itself introduces background counts. This is because at a given stimulating wavelength (λ), the optical filter never attenuates the stimulating photon intensity perfectly i.e. 100%. It is also worth mentioning that the stimulating light is ~10^{18} orders of magnitude stronger than the emitted luminescence [14]. Additional details about advantages and disadvantages of the dosimetry carried out using OSL and TL processes can be seen from the debate of McKeever and Moscovitch [15].

7.2.4 Requirements of Good OSL Materials

The important requirements of a material to be a good OSL phosphor are listed below:

- Tissue equivalence (effective atomic number; Z_{eff} = 7.4).
- Sufficiently high sensitivity and accuracy.
- Linearity of dose response over a wide range (µGy–few Gy).
- Dose rate independence of the response.
- Negligible fading.
- Emission spectra preferably in the range 300–425 nm.
- Minimum detectable dose (MDD) in the (µGy) range.
- Low cost of preparation.

7.2.4.1   Tissue Equivalence (Effective Atomic Number (Z_{eff}))

The effective atomic number of the OSL phosphor is an important parameter to be considered for the personnel dosimetry application. Hence, it is necessary to determine the effective atomic number (Z_{eff}) of the phosphor, prior to knowing the response of the phosphor to radiations of different energies. One method to determine the Z_{eff} is to obtain the fractional number of electrons of each element constituting the phosphor. Equation (7.1) is used for the calculation of Z_{eff} of phosphor.

$$Z_{eff} = \sqrt[2.94]{f_1 \times (Z_1)^{2.94} + f_2 \times (Z_2)^{2.94} + f_3 \times (Z_3)^{2.94} + \ldots\ldots\ldots + f_n \times (Z_n)^{2.94}} \tag{7.1}$$

where

f_n is the fraction of the total number of electron associated with each element and Z_n is the atomic number of each element.

For calculation of the effective atomic number of human tissue, Z_{eff}, (water (H_2O) used as a reference) [16].

Water (H_2O) is made up of two hydrogen atoms ($Z = 1$) and one oxygen atom ($Z = 8$). The total number of electrons is $1 + 1 + 8 = 10$, so the fraction of electrons for two hydrogens is (2/10) and for one oxygen is (8/10). So the Z_{eff} for water is:

$$Z_{eff} = \sqrt[2.94]{0.2 \times 1^{2.94} + 0.8 \times 8^{2.94}} \tag{7.2}$$

$$Z_{eff} = 7.42$$

7.2.4.2   Sufficiently High Sensitivity and Accuracy

Sensitivity of any OSL phosphor in general is the OSL signal strength per unit absorbed dose. This absolute definition of sensitivity is practically inadequate because the OSL signal depends on many instrumental parameters used in the OSL readout system, such as PMT voltage, stimulating sources used, method of measurement and so on. (The OSL sensitivities are compared with two different methods. In the first method, the OSL during the first second was compared. In the second method, the total area under OSL curve was used.) Hence, in order to overcome these experimental difficulties, one normally defines what is called the relative sensitivity by comparing the OSL signal of the phosphor of interest with that of the standard phosphor (normally α-Al_2O_3:C or $LiMgPO_4$:Tb^{3+}, B (BARC) are used as standard) given by

$$R(D) = \frac{S(D)_{tested}}{S(D)_{s\,tan\,dard}} \tag{7.3}$$

where $S(D)_{tested}$ phosphor and $S(D)_{standard}$ phosphor are the strength of OSL signals at dose D of the phosphor which is under test and the standard phosphor measured using the same measurement system. This relative sensitivity also depends on dose, dose rate, OSL readout system, PMT voltage and stimulated sources.

The accuracy of dosimetry measurements is the proximity of their expectation value to the 'true value' of the measured quantity. Results of the measurement cannot be absolutely accurate and the inaccuracy of a measurement result is characterized as uncertainty.

7.2.4.3   Linearity of Dose Response over a Wide Range (µGy–few Gy)

This is one of the most desirable properties for a phosphor to qualify as a OSL dosimeter. That is, the phosphor should exhibit a linear relationship between OSL output (either in terms of the area under the curve or intensity height) and absorbed dose (D). OSL output as a function of absorbed dose is known as a dose response curve, which is linear only over a range of doses. R^2 value is often taken as a goodness of fit. R^2 values range from 0 to 1 with 1 representing a perfect fit between the data and the line drawn through them and 0 representing no statistical correlation between the data and a line.

7.2.4.4 Dose Rate Independence of the Response

An integrating system measures the integrated response of a dosimetry system. For such a system, the measured dosimetric quantity should be independent of the rate of that quantity. Ideally, the response of a dosimetry system is M/Q at two different dose rates: $\left(\frac{dQ}{dt}\right)_1$ and $\left(\frac{dQ}{dt}\right)_2$ and should remain constant.

7.2.4.5 Negligible Fading

Fading is a loss of OSL signal with time. It may be caused either due to thermally or optically stimulated release of the electrons or a combination of both. High thermal fading is observed when the glow curve contains low temperature peaks. If the trap depth E is too small, then severe fading of the signal will occur, both during irradiation and between irradiation and readout. Phosphors, which are to be used for dosimetry purposes, should possess a glow curve with a TL dosimetric peak in the range 200–250°C. Usually in this temperature range, the trap depth is larger and no appreciable trap emptying can take place.

7.2.4.6 Emission Spectra Preferably in the Range 300–425 nm [17]

The general requirement for material to be a good OSL phosphor is that the emission should be in between 350 and 425 nm and the defects should have a high photo-ionization cross section in the blue-green region (450–550 nm) or IR region (650–800 nm). This limit on wavelength is due to availability of suitable filters, stimulation sources as well as sensitive PM tubes in this range and most importantly the requirement of separation of stimulating wavelength from the emission wavelength, which ensures better signal-to-noise ratio [18].

7.2.4.7 Reusability

This is an important property for a phosphor to qualify as a good OSL dosimeter. The OSL phosphor should be reused several times for dosimetry without losing its original OSL response.

7.2.4.8 Minimum Detectable Dose (MDD) in the (μGy) Range

The minimum detectable dose of a phosphor depends on the standard deviation of the background signal, which affects the signal-to-noise ratio. Minimum detectable dose in CW-OSL mode can be expressed as

$$MDD\,(\mu Gy) = \frac{3\sigma}{\text{Integrated CW} - \text{OSL signal per unit dose at given } \phi} \qquad (7.4)$$

where σ is the standard deviation in background counts integrated for time t and 3σ implies MDD evaluated with 99.9% confidence level. Integrated CW-OSL signal for time t is obtained by subtracting the integrated background counts for the same time from the area under the CW-OSL curve. ϕ is stimulation intensity in mW/cm^2.

MDD is a function of both phosphor sensitivity and instrumentation. Higher OSL counts per unit absorbed dose (higher the sensitivity) imply lower dose detection threshold or MDD. As far as instrumentation is concerned, MDD is lower for higher collection efficiency of a detector photo-multiplier tube (PMT). Another criterion for choice of detector is that it should possess high quantum efficiency in the wavelength region of OSL emission. Despite best instrumentation and highly sensitive phosphor, the MDD evaluated using the CW-OSL technique can be substantially improved. To achieve lower detection thresholds (better MDD) from a phosphor on a given instrument, the technique of CW-OSL is addressed quantitatively in terms of optimizing the OSL parameters.

7.2.4.9   Low Cost of Preparation

Synthesis of phosphor materials should be easy, simple, inexpensive and scalable to bulk production and involve no chemicals that are toxic.

7.2.5   GENERALIZED MATHEMATICAL DESCRIPTION OF OSL

The total concentration of occupied metastable states in the system at time t may be represented by

$$\mu(t) = \int_{\gamma_1} \int_{\gamma_2} \cdots \int_{\gamma_m} n\,(\gamma_1,\, \gamma_2,\, \dots \gamma_m,\, t)\,d\gamma_1 d\gamma_2 \dots d\gamma_m \qquad (7.5)$$

where $n\,(\gamma_1,\, \gamma_2\, \dots,\, \gamma_m,\, t)$ is the concentration of occupied states $1 \rightarrow$ m, described by state parameters $\gamma_1,\, \gamma_2\, \dots,\, \gamma_m$ and t. In general, $n(\gamma,\, t) = N\,(\gamma)\,f\,(\gamma,\, t)$. Here, $n\,(\gamma)$ is the concentration of occupied states, $N\,(\gamma)$ the concentration of available states and $f\,(\gamma)$ the occupancy of the state. (f = 1 when a state is full and f = 0 when it is empty.) Both $n\,(\gamma)$ and $f\,(\gamma)$ are time-dependent functions. The state parameters $\gamma_1,\, \gamma_2\, \dots,\, \gamma_m$ indicate the stability of the metastable state under the prevailing conditions of temperature and illumination intensity; that is, they govern the probability per unit time that the system will return to equilibrium. $n\,(\gamma_1,\, \gamma_2\, \dots,\, \gamma_m,\, t)$ is a weighting function, or distribution, expressing the concentration of occupied states possessing the parameters $\gamma_1,\, \gamma_2\, \dots,\, \gamma_m$. Equation (7.5) is a time- and dose-dependent function since it increases during irradiation and decreases during stimulation. In stimulated luminescence measurements, the emitted luminescence intensity is monitored during the return of the system to equilibrium. The luminescence intensity 'I' is proportional to the rate, at which the metastable states decay, such that:

$$I\,(t) = \left| \frac{d\mu(t)}{dt} \right| \qquad (7.6)$$

Thus, the stimulated luminescence is time-dependent through the functional dependence of equation (7.6) upon $n\,(\gamma_1,\, \gamma_2\dots,\, \gamma_m,\, t)$. In order to evaluate $I(t)$, we can write an expression of the form:

$$\frac{d\mu(t)}{dt} = -\mu'(t)P\,(t) \qquad (7.7)$$

where $P(t)$ is the probability per unit time of the decay of the metastable states $\mu(t)$.
For optical stimulation, we have

$$P\,(E_0) = \Phi\sigma\,(E_0) \qquad (7.8)$$

where Φ is the optical stimulation intensity, $\sigma(E_0)$ is the photo-ionization cross section for interaction of the metastable state with an incident photon and E_0 is the threshold optical stimulation energy required for charge release and a return of the system to equilibrium. For optical stimulation, when the traps are emptied using a fixed wavelength and a steady illumination intensity, the luminescence recorded is known as continuous wave OSL, or CW-OSL. However, time dependence to P can be introduced by scanning the above terms with time i.e. $\Phi\,(t)$ or $\lambda(t)$. Thus, for a linear increase in the intensity of optical stimulation at a fixed wavelength:

$$\Phi(t) = \Phi_0 + \beta_\Phi t \qquad (7.9)$$

with $\beta_\Phi = d\Phi/dt$. OSL recorded in this manner is known as LM-OSL. Other schemes can be imagined in which the intensity is modulated in nonlinear manner ways. For example, one can imagine an exponentially increasing stimulation intensity [$\Phi(t) = \Phi_0 exp\,(t)$], which is a scheme that can have advantages when emptying a range of traps with photo-ionization cross sections that differ by orders of magnitude. Alternatively, the stimulation may be pulsed, such that $\Phi(t) = \Phi_0$ for $t < t_0 + \Delta t$ and $\Phi(t) = 0$ for $(t_0 + \Delta t) \leq t < (t_0 + \tau)$, where Δt is the pulse width and τ is the period. Such a scheme is known as P-OSL.

7.2.6 Models and Rate Equations

7.2.6.1 CW-OSL

The transitions of charge between energy levels during irradiation and subsequent optical stimulation of a dosimeter can be described by a series of non-linear, coupled equations. The equations themselves are intractable and several simplifying assumptions have to be introduced in order to arrive at analytical expressions for the evolution of OSL intensity with time during optical stimulation and ultimately, the dependence of the OSL signal on the absorbed dose. Taking into consideration the quasi-equilibrium conditions, the rate equations describing the flow of charge during CW-OSL production are

$$\frac{dn_c}{dt} = -\frac{dn}{dt} + \frac{dm}{dt} \qquad (7.10\text{a})$$

where n_c and n are the concentrations of electrons in the conduction band and OSL active traps, respectively; m is the concentration of holes in recombination traps. Therefore, OSL intensity (I_{OSL}) is proportional to the rate of depleting OSL active electron traps:

$$I_{OSL} = -\frac{dm}{dt} = -\frac{dn}{dt} = np \qquad (7.10\text{b})$$

Charge neutrality for this system can be written as $n_c + n = m$.

Under this assumption, the solution of the equation is given as

$$I_{OSL} = n_0 p \exp(-tp) = I_0 \exp(-t/\tau) \qquad (7.11)$$

where I_0 is the initial intensity when $t = 0$ and $\tau = 1/p$ is the decay constant. Thus, this first-order model leads to an exponentially decaying OSL intensity as the stimulation light intensity is applied to the sample. Eventually, all traps are depleted and the OSL becomes zero. If additional traps introduced into the energy band model are not sensitive to the optical stimulation, they can still act as traps for released charge and affect the OSL decay kinetics (Figure 7.5b). In this model, the deep traps may be viewed as competitors to the recombination centres. The OSL intensity is given as

$$I_{OSL} = n_{l0} p \exp(-t/\tau) - n_c (N_2 - n_2) A_2 \qquad (7.12)$$

where N_2, n_2, A_2 are the concentrations of the available traps, concentrations of filled traps and trapping probability, respectively, for the deep trap. Thus, the OSL intensity is reduced by an amount indicated by the relative rates of recombination and trapping into the deep competing traps. The decay is now no longer exponential since the second term in equation (7.12) is time dependent and approaches zero as t → ∞. If the competing traps are shallow (Figure 7.5c), equation (7.12) becomes

$$I_{OSL} = n_{l0} p \exp(-t/\tau) + n_2 s \exp(-E/KT) - n_c (N_2 - n_2) A_2 \qquad (7.13)$$

(a)   (b)

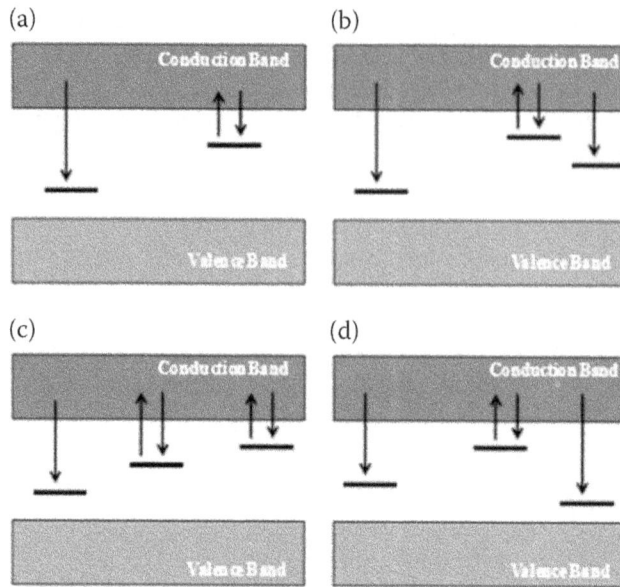

(c)   (d)

FIGURE 7.5 Simple models for OSL: (a) simplest model involving one trap and one radiative re-combination centre. (b) Model containing an additional deep, competing trap. (c) Model containing a shallow, competing trap. (d) Model containing a competing non-radiative recombination centre.

The last two terms in equation (7.13) combine to produce a long-lived, temperature-dependent component to the OSL decay curve. The form of this component is an initial increase, followed by a decrease at longer times. Considering further the final model (Figure 7.5d) that deals with one trap and two recombination centres (of concentrations m_1 and m_2, respectively) one of which is non-radiative, under these conditions we have

$$I_{OSL} = np\exp(-t/\tau) - \frac{dm_2}{dt} \tag{7.14}$$

where we can observe that the OSL intensity is reduced by the presence of a non-radiative pathway for charge interaction. The relative size of the two recombination centres is time dependent. Thus,

$$\frac{m_1}{m_2} \approx \frac{m_{10}}{m_{20}} \exp[-tn_c(A_{m_1} - A_{m_2})] \tag{7.15}$$

However, if $A_{m1} \approx A_{m2}$ the ratio remains approximately constant, in this case, the CW-OSL decay curve remains approximately exponential according to

$$I_{OSL} = \frac{1}{K}n_0 p\exp(-tp) \tag{7.16}$$

where K is a constant given by $K = (m_1 + m_2)/m_1$.

7.2.6.2 LM-OSL

Bulur [19] introduced an alternative technique in which the intensity of the stimulation source is ramped linearly and the OSL is monitored throughout the ramp. By adopting this stimulation mode, the OSL is seen as a series of peaks, with each peak corresponding to the release of charge from different traps. Thus, the traps for which the cross section is large at the particular wavelength used are emptied first and are shown as a peak in a plot of OSL vs. time. Traps with smaller photo-ionization cross sections empty more slowly and give rise to OSL peaks that appear at later times. Thus, traps with fast, slow and medium rates of de-trapping may be more easily resolved using LM-OSL compared with CW-OSL.

Consider a one-trap/one-centre model in which electrons of concentration n are trapped at a localized state until stimulated into the conduction band by absorption of a photon. For first-order kinetics (negligible re-trapping), the rate of de-trapping is given by equation (7.10b) and the corresponding luminescence intensity is given by equation (7.11), where $\tau = 1/\sigma\Phi$. If the intensity is linearly ramped from zero to a maximum value Φ_m according to $\Phi(t) = \gamma t$, then equation (7.10b) becomes

$$\frac{dn}{dt} = -n\sigma\gamma t \tag{7.17}$$

From which a Gaussian function can be obtained as

$$n = n_0 \exp\left(-\frac{\sigma\gamma t^2}{2}\right) \tag{7.18}$$

The luminescence intensity is then given by

$$I_{OSL} = n_0\sigma\gamma t \exp\left(-\frac{\sigma\gamma t^2}{2}\right) \tag{7.19}$$

The shape of the LM-OSL curve for a single trap is that of a linearly increasing function followed by a Gaussian decrease in OSL intensity as the traps depleted. The time at which the maximum is achieved is given by

$$t_m = \sqrt{1/\sigma\gamma} \tag{7.20}$$

And the LM-OSL maximum intensity is

$$I_{OSL}^{max} = \frac{n_0}{t_{max}} \exp\left(-\frac{1}{2}\right) \tag{7.21}$$

Thus, the ionization cross section at the wavelength used can be determined from the known value of γ and the observed value of t_{max}. Adopting the general order kinetic model in which the rate of re-trapping of the released charge is significant compared to the rate of recombination:

$$\frac{dn}{dt} = -\frac{n^b\sigma\gamma t}{n_0^{b-1}} \tag{7.22}$$

where b is a dimensionless positive number; b > 0, b ≠ 1. The solution is

$$I_{OSL} = n_0\sigma\gamma t \left[(b-1)\frac{\sigma\gamma t^2}{2} + 1\right]^{b/(1-b)} \tag{7.23}$$

the maximum of general order LM-OSL peak

$t_{\max} = \sqrt{\frac{2}{\sigma\gamma(b+1)}}$ is achieved at time at which the maximum intensity is

$$I_{OSL}^{\max} = \left(\frac{2n_0}{b+1}\right)\left(\frac{1}{t_{\max}}\right)\left(\frac{2b}{b+1}\right)^{b/(1-b)} \tag{7.24}$$

7.2.6.3 P-OSL

The third major stimulation mode, as shown in Figure 7.4, is pulsed OSL. To describe the principle behind the measurement of P-OSL, we begin by considering several stimulation pulses, of different intensities $\Phi_i(i = 1, 2 ...)$ and durations (pulse widths, T_i) such that $\Phi_i T_i$ is kept constant. The stimulation rate is proportional to the stimulation power absorbed by the sample and thus, by decreasing the pulse width in proportion to an increase in the stimulation power, the absorbed energy per pulse may be maintained fixed. Furthermore, for first-order kinetics we have (from equation (7.10b))

$$\Delta n = \int_0^T n\sigma\Phi.\,dt \tag{7.25}$$

and therefore for weak stimulation (i.e. $\Delta n << n$) we see that $\Delta n \propto \Phi T$ and thus, the total charge released from the traps is approximately equal in all cases.

Normally, when writing rate equations we consider that each recombination event leads instantly to a photon emission event, and thus we normally write $I_{OSL} \propto dm/dt$. However, actually there is a built-in delay between recombination and photon emission due to the fact that each recombination event leads to the excitation of the recombination centre (luminescence centre) into an excited state, where it remains for a characteristic lifetime τ before relaxation occurs with the emission of a luminescence photon. Thus, if the concentration of excited states is n_e, then we may write:

$$\frac{dn_e}{dt} = \frac{dm}{dt} - \frac{n_e}{\tau} \tag{7.26}$$

A photon is only emitted when the excited state relaxes to the ground state and thus $I_{OSL} = n_e/\tau$. If the relaxation time is extremely fast (i.e. τ_e is very small compared with the pulse width T which is the usual case and is certainly true for CW-OSL), then quasi-equilibrium conditions hold $(dn_e/dt \sim 0)$ and we have $I_{OSL} \sim dm/dt$, as usual. However, if τ is comparable to or larger than T, the later relationship is not true. The key to understanding P-OSL is to consider how many of the excited states relax during the excitation pulse, versus how many relax after the excitation pulse. If one chooses the pulse width such that $T < \tau$, then a larger concentration of centres exist in the excited state after a short pulse, compared with those in the excited state after a long pulse ($T \geq \tau$), for the same energy input (ΦT). At the end of the stimulation pulse, those centres in the excited state relax with a time constant τ. The net effect is that the ratio of the photons emitted after the pulse to those emitted during the pulse, increases as the pulse width decreases, for constant stimulation energy. For $T << \tau$, most of the photons emerge after the pulse.

7.2.6.4 Relationship between LM-OSL and CW-OSL

If the stimulation ramp in an LM-OSL experiment is arranged so that it reaches a final stimulation power Φ_f in time t, such that Φ_f is equal to the fixed stimulation power used in a CW-OSL experiment [20], then the observed CW-OSL decay rate will be related to the observed maximum LM-OSL by:

$$\tau_d = \frac{1}{\sigma\gamma t_f} = \frac{t_{\max}^2}{t_f} \tag{7.27}$$

Bulur [21] describes a simple mathematical transformation that allows one to convert CW-OSL curves into LM-OSL curves. First, define a variable u, thus:

$$u = \sqrt{2tp} \text{ or } t = \frac{u^2}{2p} \tag{7.28}$$

where p is the total measurement period in an LM-OSL experiment and u has the dimensions of time. Substituting equation (7.28) in the expression for CW-OSL (equation (7.11)) and multiplying by u/p yields:

$$I_{OSL} = \frac{n_0 \sigma \Phi u}{p} \exp\left(-\frac{\sigma \Phi}{2p} u^2\right) \tag{7.29}$$

which is of the same form as the expression for LM-OSL (equation (7.18)) where u maps with t and $\Phi/p = \gamma$. Note that if P is made equal to the observation time for the CW-OSL experiment, then the scaling factor is $\sqrt{2}$.

7.2.7 Order Kinetics

The simplest, first-order model for TL and OSL processes assumes the absence of re-trapping, i.e. all charges escaping from the traps recombine immediately producing luminescence. Therefore, the luminescence intensity is simply proportional to the variation in the trapped charge concentration given by equation (7.10b). If the OSL readout is done by simply illuminating the dosimeter continuously with light of appropriate wavelength [22], the shape of the curve of OSL intensity versus stimulation time can be obtained by solving equation (7.10b) using the probability (equation (7.8)) and a constant stimulation intensity. As the escape probability is constant, the OSL intensity is an exponential decay following the trapped charge concentration. In a real dosimeter, the contribution from different trapping levels appears as a superposition of exponential decays. Figure 7.6 illustrates the OSL curves for different stimulation intensities. For a photon flux (stimulation intensity) Φ such that the probability $p = \sigma\Phi = 0.05$ s^{-1}, the initial OSL intensity is higher and decays faster than an OSL curve obtained with a lower stimulation intensity ($p = \sigma\Phi = 0.01$ s^{-1}). Integration of equation (7.11) leads to $\int I_{OSL}(t)dt \propto n_0$, from which one can see that the integrated OSL signal is

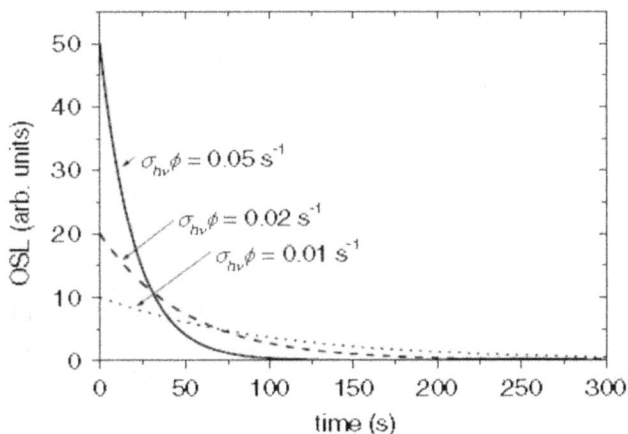

FIGURE 7.6 OSL intensity versus stimulation time for various light intensities, i.e., various values of $\sigma\Phi$, where σ is the photoionization cross section and Φ is the stimulation photon flux.

proportional to the initial trapped charge concentration n_0 (i.e. to the absorbed dose of radiation) and is independent of the stimulation power. The initial OSL intensity is proportional to both n_0 (i.e. to the absorbed dose of radiation) and the photon flux Φ.

Factors that can affect the OSL curve shape and/or the total OSL signal include the following:

a. trapping at competing traps (deep and shallow);
b. thermal stimulation of trapped charges from shallow traps;
c. re-trapping at the dosimetric trap;
d. simultaneous stimulation from multiple trapping levels;
e. recombination at multiple recombination centres;
f. photo-transfer of charges from deep to the dosimetric and shallow energy levels.

In fact, the OSL decay curve is the result of a dynamic process that can be affected by the concentration of competitors represented by shallow and deep traps, the extent to which they are occupied and the level of the absorbed dose. These factors in turn result in a dependence on the ionization density created by the radiation field in the crystal [23,24]. Some of the processes and events relevant to the final character and properties of the OSL signal are outlined as follows.

7.2.7.1 Charge Capture by Shallow Traps

Shallow traps are localized energy levels close to the edge of the conduction band in the case of electron traps or close to the valence band in the case of hole traps. 'Close' in this context means that the energy difference between the trap and the delocalized band is such that the escape probability for charges trapped in shallow traps is significant even at room temperature; consequently, the charge concentration in these energy levels can decrease over time scales ranging from minutes to days. Shallow traps are generally associated with two important phenomena viz phosphorescence observed immediately after irradiation and an initial increase in the intensity of the OSL signal upon stimulation. The phosphorescence is caused by the recombination of charges escaping from the shallow traps following the end of the irradiation period. To avoid its influence in the OSL readout, it may be necessary to introduce a delay between dosimeter irradiation and readout to allow for the decay of the phosphorescence signal. Phosphorescence can also be observed immediately after optical stimulation due to charges stimulated from the dosimetric trap and captured by the shallow trap, resulting in a long 'tail' in the OSL signal. The initial increase in the OSL curve is related to competition by shallow traps. At the beginning of the OSL readout, the shallow traps are empty and can capture charges escaping from the main dosimetric trap that would otherwise recombine. As a result, the overall OSL intensity is smaller than would be expected. However, as the shallow traps are filled and the number of charges captured by the shallow traps equals the number of charges escaping from them, the competition process becomes less important and the OSL intensity increases. Figure 7.7 illustrates the effect of shallow and deep traps on the OSL curves.

7.2.7.2 Charge Capture by Deep Traps

Deep electron trap and deep hole traps can act as competitors during the irradiation and readout stages, capturing charges released to the conduction and valence bands. As deep traps are filled, the sensitivity of the crystal can increase or decrease, depending on the nature of the deep trap [25,26], but the overall effect is to introduce an undesirable dependence of the OSLD sensitivity on the irradiation history. For example, the deep electron trap can capture charges released from the dosimetric trap, therefore decreasing the overall intensity of the OSL signal. This effect is illustrated in Figure 7.7. However, as the deep traps are filled, becoming less competitive for the capture of charges, more electrons are available for recombination. This results in an effective increase in the sensitivity of the dosimeter. In some cases, it is possible to reset the sensitivity by annealing the dosimeter at an appropriate temperature to empty the deep traps. In the case of Al_2O_3:C, an annealing at 900°C for 15 min is recommended for both TL and OSL [27].

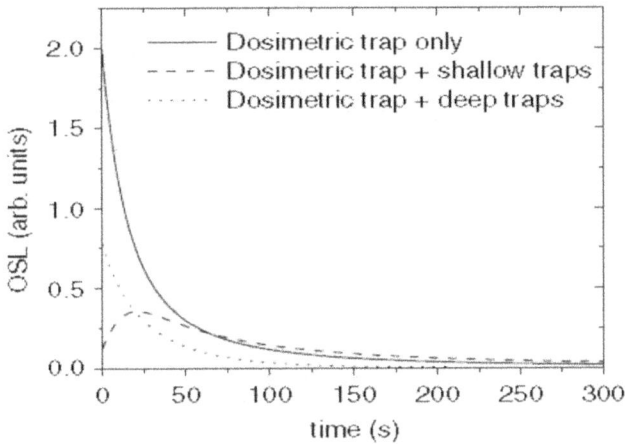

FIGURE 7.7 Theoretical OSL curves for a model containing only a dosimetric trap and recombination centre or containing in addition shallow traps or deep traps.

7.2.7.3 Prompt Recombination

Radioluminescence (RL) is produced by prompt electron-hole pair recombination during irradiation. In principle, the RL intensity is proportional to the rate of creation of electron–hole pairs and, therefore, it can be used to determine the dose rate. However, the RL intensity is also affected by the complex dynamic of trap filling in the crystal, since trapping is a process competing with recombination. As an example of the dynamic interplay between these competing processes, the reader is referred to the numerical calculations by Polf et al. [28], explaining the RL signal of Al_2O_3:C optical fibers subjected to simultaneous irradiation and stimulation. Therefore, the RL sensitivity may not be constant with the dose and corrections have to be implemented to take this effect into account.

7.2.7.4 Induction of Signal by Light

OSL can be induced by UV light by two mechanisms direct ionization of defects and photo-transfer of charges from deep traps. In a wide bandgap semiconductor or insulator, the ionization of atoms of the crystal requires photon energies higher than the bandgap energy and, therefore, this process can usually be neglected. However, the ionization of defects can occur at photon energies lower than the bandgap energy, as in the case of the ionization of F-centres in Al_2O_3, which requires photon energies of the order of 6 eV, as compared to a bandgap of 9 eV [29]. The photo-ionization of defects creates free charges that can be trapped, producing TL and OSL in a similar way as ionizing radiation. In the case of dosimeters previously irradiated, UV can also transfer charges from deep traps to the dosimetric trap, resulting in the so-called photo-transferred TL or OSL. For these reasons, it is important to protect the dosimeters from light even before irradiation, depending on the dose level to be measured.

7.2.8 DOSIMETERS

A radiation dosimeter is a device or system that measures either directly or indirectly, the quantity of radiation exposure, KERMA, absorbed dose or equivalent dose, or their time derivatives (rates), or related quantities of ionizing radiation. A dosimeter along with its reader is referred to as a dosimetry system.

With the increasing use of nuclear energy, there is a need for a wider range of efficient dosimeters for radiation detection and assessment. There has been a tremendous growth in the development of radiation detectors and devices in the past few decades. In recent years, the development of new materials for radiation dosimetry has progressed significantly. Alkaline earth sulfides (AES) have been known for a long time as excellent and versatile phosphor materials. Now, different dosimeters are available commercially. Radiation dosimetry is now used in medical and securities purposes. In this field, demand for accurate radiation detection and dose assessment has been higher. Dosimeters have been extremely valuable in monitoring the safety of radiation workers and performing environmental dose control. Generally dosimeters are classified into two broad categories.

a Passive Dosimeter

A passive dosimeter produces a radiation-induced signal, which is stored in the device. The dosimeter is then processed and the output is analysed. The OSLDs and TLDs are the examples of passive dosimeters.

b Active Dosimeter

An active dosimeter produces a radiation-induced signal and displays a direct reading of the detected dose or dose rate in real time. The example of active dosimeter is a direct reading dosimeter (DRD).

7.2.8.1 Properties of Dosimeters

7.2.8.1.1 Acuracy and Precision

In radiation dosimetry, the uncertainty associated with the measurement is often expressed in terms of accuracy and precision. The precision of dosimetry measurements specifies the reproducibility of the measurements under similar conditions and can be estimated from the data obtained in repeated measurements. High precision is associated with a small standard deviation of the distribution of the measurement results. TL/OSL dosimeters should be individually calibrated, for calibrating dosimeters the amount of signal response to a known dose must be measured before use. The less sensitivity variation and small standard deviation helps to maintain the accuracy and precision in dosimeters batch.

7.2.8.1.2 Linearity

Ideally, the dosimeter reading (absorbed dose) should be linearly proportional to the dosimetric quantity (dose). However, beyond a certain dose range, non-linearity sets in. The linearity range and the non-linearity behaviour depend on the type of dosimeter and its physical characteristics. Any dosimeter may show the linearity, supra-linearity and saturation behaviour within certain dose range. Linearity of the OSL/TL dosimeter helps to choose a dosimeter in particular application. A dosimeter and its reader may both exhibit non-linear characteristics, but their combined effect could produce linearity over a wider range.

7.2.8.1.3 Dose Rate Dependence

Integrating system measures the integrated response of a dosimetric system. The response of a dosimetric system M/Q at two different dose rates $((dQ/dt)_1$ and $(dQ/dt)_2)$ should remain constant. Where M is absorbed dose, Q is dosimetric quantity (exposure dose). In reality, the dose rate may influence the dosimeter readings and appropriate corrections are necessary.

7.2.8.1.4 Energy Dependence

The response of a dosimetry system M/Q is usually a function of radiation beam quality (energy). Since the dosimetry systems are calibrated at a precise radiation beam quality (or qualities) and used over a much wider energy range, the variation of the response of a dosimetry system with radiation quality (called energy dependence) requires correction. Ideally, the energy response should be flat (i.e. the system calibration should be independent of energy over a certain range of radiation qualities). In reality, the energy correction has to be included in the determination of the quantity Q for most measurement situations. In radiotherapy, the quantity of interest is the dose to water (or to tissue). As no dosimeter is water or tissue equivalent for all radiation beam qualities, the energy dependence is an important characteristic of a dosimetric system.

7.2.8.1.5   Directional Dependence

The deviation in response of a dosimeter with the angle of incidence of radiation is known as the directional, or angular dependence of the dosimeter. Dosimeters usually exhibit directional dependence, due to their constructional details, physical size and the energy of the incident radiation. Directional dependence is an important property in certain applications; for example, in in-vivo dosimetry while using semiconductor dosimeters. Therapy dosimeters are generally used in the same geometry as that in which they are calibrated.

7.2.8.1.6   Spatial Resolution and Physical Size

Since the dose is a point quantity, the dosimeter should allow the determination of the dose from a very small volume (i.e. one needs a 'point dosimeter' to characterize the dose at a point). The position of the point where the dose is determined (i.e. its spatial location) should be well defined in a reference coordinate system. TL and OSL dosimeters come in very small dimensions and their use to a great extent approximates a point measurement. Film dosimeters have excellent 2-D and gels 3-D resolution, where the point measurement is limited only by the resolution of the evaluation system.

7.2.8.1.7   Readout Convenience

Direct reading dosimeters (e.g. ionization chambers) are generally more convenient than passive dosimeters (i.e. those that are read after processing following the exposure, for example TLDs and films). While some dosimeters are inherently of the integrating type (e.g. TLDs and gels), others can measure in both integral and differential modes (ionization chambers).

7.2.8.1.8   Convenience of Use

Ionization chambers are reusable in which sensitivity is almost invariable within their life span. Semiconductor dosimeters are reusable, but after some re-uses gradual loss of sensitivity occurs within their life span; however, some dosimeters are not reusable (e.g. films, gels and alanine). Some dosimeters measure dose distribution in a single exposure (e.g. films and gels) and some dosimeters do not change their sensitivity upon rough handling (e.g. ionization chambers).

Dosimeters should have the ability to withstand severe environmental conditions with smaller, lighter sizes and portability.

While all dosimeters have advantages and disadvantages, an ideal instrument would have the following qualities:

- low energy dependence and angle dependence
- the ability to detect several radiation types
- high resistance to fading (stability when exposed to high temperatures and humidity levels)
- a linear response to dose (response does not change with increasing dose)
- a low minimum measurable dose

No single dosimeter has all of these characteristics in it. Therefore, a dosimeter user must understand the environment where the instrument will be used, as well as the shortcomings of different dosimeter types, in order to select a dosimeter that is most appropriate to the projected use.

7.2.8.2   Passive Optically Stimulated Luminescence Dosimeters

There are currently approximately 5 million personnel dosimetry badges being used by radiation workers around the world (2002 data), including TLDs, film badges, OSLDs and others. Although it may be argued that the use of OSL in personnel dosimetry is not yet widespread in terms of the number of different laboratories running OSL instruments, it has to be recognized nevertheless that OSL is indeed a major player in the personnel dosimetry arena since approximately 25% of the 5 million badges are in fact OSLDs, primarily based on P-OSL from Al_2O_3:C.

As already described earlier, OSL has been suggested as a personnel dosimetry method for several decades, with various forms of the OSL method being suggested at one point or another in the published literature. None of these early pioneering studies, however, led to a commercial-scale personnel dosimetry system until the advent of P-OSL of Al_2O_3:C [30], and the subsequent development of Luxel^{TM} personnel dosimeters by Landauer Inc. in the late 1990s. A possible exception to this statement is the use of radiophotoluminescence (RPL) of phosphate glasses in personnel dosimetry throughout the 1980s [31–33]. However, as discussed already, RPL and OSL are different physical phenomena and we limit the discussion here to a description of OSL personnel dosimetry. Following the success of P-OSL in this application, newly developed commercial OSL personnel dosimetry systems are now available using CW-OSL as a readout mode. Descriptions of some commercial OSL personnel dosimetry systems are given in the following sections.

In general, the key features of OSL that make it so attractive for personnel dosimetry include the ability to tune the performance across a wide dynamic dose range by being able to vary the stimulation power. This allows optimum sensitivity at both the low-dose and the high-dose ends of the dose response. Furthermore, by adjusting the stimulation power appropriate to the dose range, one has the ability to re-read the OSL signal since the high sensitivity allows dose assessment even though only a fraction of the stored charge is depleted during the measurement. Thus, a complete re-analysis is achieved and true, independent dose readings are possible. This is to be distinguished from film badge dosimetry in which the developed image can be read a second time, but the film can only be developed once; thus the dose readings are not independent in this sense. With TL, of course, the entire signal is destroyed (100% depletion of the trapped charge) during one reading and re-reading is not possible. Both fast readout and fast analysis, important for processing large numbers of dosimeters, are made possible with the speed of P-OSL, although at the cost of expensive equipment. Multiple analyses are possible with cheaper LEDs as the stimulation source, but at a cost of slower analysis time. In addition, the optical nature of the OSL process removes the need for heating and allows for imaging capabilities normally only expected with film badges.

7.2.8.2.1   Landauer's Luxel^{TM} Personnel Dosimetry System

The Luxel^{TM} badge is illustrated in Figure 7.8a–7.8c. The heart of the badge, i.e. the detector itself, is shown in Figure 7.8a and is a thin layer of Al_2O_3:C powder deposited onto a clear polyester film base.

(a)   (b)   (c)

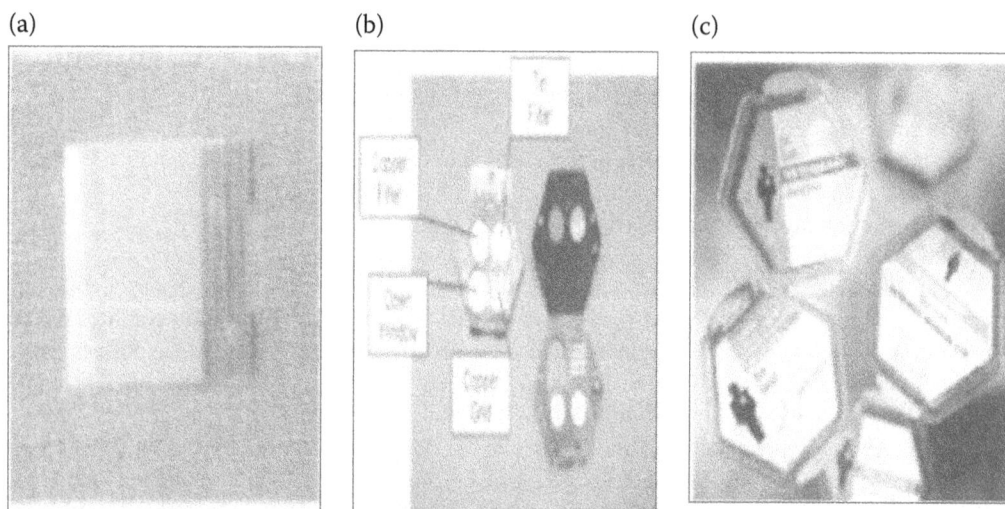

FIGURE 7.8   The Landauer Luxel^{TM} personal dosimetry badge. (a) The Al_2O_3:C detector. (b) The filter pack, showing the open window, the Cu filter, the Sn filter and the Cu-grid filter. (c) The complete Luxel TM badge (courtesy of Landauer Inc., USA).

The size of the active area of the detector is 16.5 mm by 18.5 mm and the grain size of the Al_2O_3:C powder is in the range of 20–90 µm. The powder layer is protected by a thin, clear polyester tape. The detector is located inside a filter pack (Figure 7.8b) consisting of an open window, a copper filter (250 µm thick), a tin filter (500 µm thick) and a 500 µm-thick copper grid consisting of a 5 × 5 array of 1.0 mm diameter holes. When the filter pack is folded over the detector, each of the filters sits over its own quadrant, covering part of the Al_2O_3:C powder film. In normal analysis mode, a POSL reading is taken from those three parts of the Al_2O_3:C film that sit under the three circular filters (open, Cu and Sn). From these three independent POSL readings deep dose, shallow dose and eye dose may be evaluated through use of a suitable algorithm of the form:

$$D = AI_{OW} + BI_{Cu} + CI_{Sn} \qquad (7.30)$$

where $I_i (i = OW, Cu, Sn)$ is the POSL intensity measured under the open window, copper or tin filters, respectively, and A, B and C are the coefficients unique to each of the different filters.

7.2.8.2.2 Landauer's InLight^{TM} Personnel Dosimetry System

A new development in OSL personnel dosimetry is the InLight^{TM} system, also by Landauer. This is a bench-top, CW-OSL system specifically designed for personnel dosimetry and uses bright LEDs (green) to stimulate the OSL. It is again based on Al_2O_3:C. The use of LEDs results in longer readout times compared with P-OSL, but enables dose readings for very little depletion of the signal. This in turn leads to a different paradigm for personnel dosimetry badge wearers, namely that the InLight^{TM} badges are normally distributed to individuals for a full year, with periodic readings taken twice every month during normal operation. In this way, not only are the doses corresponding to the individual wear periods evaluated, but the last reading gives the individual's total accumulated dose for the whole year. The stimulation power is such that the twice monthly readings deplete the signal by only approximately 10% over one year.

The InLight TM badge is illustrated in Figure 7.9. Four Al_2O_3:C powder films are used as dosimeters. The operational dose range is up to 10 Gy, for use with photons >5 keV, and with electrons >150 keV. As with all Al_2O_3:C dosimeters, the system is insensitive to neutrons. InLight^{TM} is designed for organisations that wish to perform their own dosimetry, whereas Luxel^{TM} is designed for use by radiation dosimetry service providers.

7.3 BORATE-BASED TL AND OSL PHOSPHORS

Borates are naturally occurring minerals containing boron, the fifth element on the periodic table. The element boron does not exist by itself in nature. Rather, boron combines with oxygen and

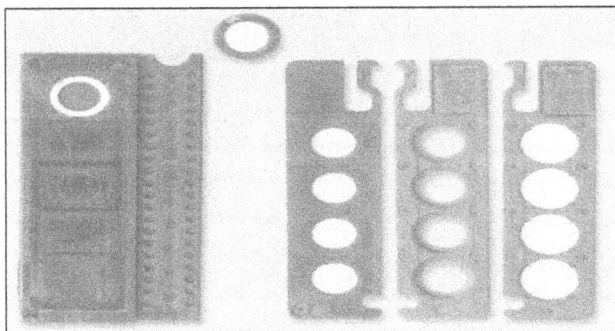

FIGURE 7.9 The Landauer InLight^{TM} personal dosimetry badge. Four Al_2O_3:C detectors, constructed in the same way as a Luxel^{TM} detector, fit within the badge holder (courtesy of Landauer Inc., USA).

other elements to form inorganic salts called borates. It is a semiconductor rather than a metallic conductor. Chemically it is closer to silicon than to aluminum, gallium, indium and thallium. Furthermore, it is an electron-deficient element, possessing a vacant p-orbital, which causes Lewis acidity. The reactions of boron are dominated by a requirement for electrons.

Borate compounds are known for their wide bandgap. These are best hosts for various activators. From the literature it is found that borate compounds find several interesting applications [34,35]. In borate compounds boron atom is coordinated by oxygen atoms to form a variety of atomic groups that affect the physical properties in general and optical properties in particular [36,37]. Borate-based materials are widely used for TLD/OSLD because of their tissue equivalent, absorption coefficient, low cost, thermal stability, neutron sensitivity, physical and optical properties [38]. Rare earth polyborates with high B_2O_3 contents are especially attractive because of their excellent, chemical, thermal stability, small alkaline earth content and the covalent boron oxygen network [39].

Also, borate compounds show a variety of applications due to their physical and optical properties. The borates with isolated planer $[BO_3]^{3-}$–group in their structure proven to be good birefringent materials. From the literature it is found that borate compounds find several interesting applications, e.g. $Li_2B_4O_7$;RE is dosimetric materials used for clinical and personnel dosimetry. In borate compounds, a boron atom is coordinated by oxygen atoms to form a variety of atomic groups that affect the physical properties in general and optical properties in particular [40].

In the last five decades, much research interest has been focused on the synthesis and characterization of inorganic borates for exploring nonlinear optical materials. Many borates, such as β-BaB_2O_4 [41], LiB_3O_5 [42] and $Sr_2Be_2B_2O_7$ [43], have been synthesised and structurally characterised and proposed to be new optical materials. Several borate compounds find important application as TLD phosphors. $Li_2B_4O_7$:Cu [44,45], $Li_2B_4O_7$:Cu-Ag [46] and MgB_4O_7:Dy [47] MgB_4O_7: Ag [48] are some of the low Z phosphors, used in personal dosimetry. CaB_4O_7 (Z_{eff} = 12.58) is an intermediate Z phosphor between tissue equivalent LiF:Mg,Ti (Z_{eff} = 8.4) phosphor, and high Z $CaSO_4$:Dy (Z_{eff} = 15.5) TLD phosphor [49,50]. Fakuda et al. [51,52] have studied the sintered CaB_4O_7 phosphors with the activators Pb, Cu, Eu and Dy. However, they have not observed promising TL sensitivity in the phosphor so as to propose the same for TLD applications. Recently, a series of attempts have been made [53–55] to investigate new borate materials as useful optical crystals in view of the demand for the nonlinear optical crystals in the deep UV band. The demand for such material is increasing consistently because of the development of optical communications and the semiconductor large-scale integrated circuit.

L. Wu et al.[56] has investigated the phase relations in Li_2O-CaO-B_2O_3 and found several new ternary compounds. They further identified that by using the flux method a high-quality single crystal of $LiCaBO_3$ (along with other three $LiMgBO_3$, $LiSrBO_3$ and $LiBaBO_3$) could be obtained. Its structure has also been refined from a powder X-ray diffraction data and found that it is very different from the other three alkaline-earth metal compounds, $LiMgBO_3$ [57] and $LiBaBO_3$ [58] though they have similar chemical formula. The other three compounds all belong to monoclinic crystal system. Specifically, $LiCaBO_3$ compound crystallizes in an orthorhombic unit cell (space group $Pbca$) with lattice parameters a = 13.227(13) Å, b = 6.1675(6) Å, and c = 6.0620(6) Å. There are eight formulas per unit cell and six unique atoms sites in the unit. Its structure is stacked alternately with [LiBO] and [CaO] layers along the [100] direction. In the [LiBO] layers, the isolated $[BO_3]^{3-}$ anionic groups, distributed along two directions, [011] and [0 1^- 1], are almost perpendicular to each other. Because of the low Z, $LiMgBO_3$:D & $LiCaBO_3$:D could be the potential candidates as TLD materials.

In 2006, L. Wu et al. reported the synthesis via a standard solid-state reaction of a series $MM'_4(BO_3)_3$ (M = Na, M' = Ca; M = K, M' = Ca, Sr). A comparison of the structures of these novel compounds and three other novel cubic compounds with the same formula, $MM'_4(BO_3)_3$ (M = Li, M = Sr; M = Na, M' = Sr, Ba), are discussed by them [59]. In their work, the crystallize are in the non-centro symmetric space group Ama2 with the following lattice parameters: a = 10.68004(11) Å, b = 11.28574(11) Å, c = 6.48521(6) Å for $NaCa_4(BO_3)_3$; a = 10.63455(10) Å, b = 11.51705(11) Å, c = 6.51942(6) Å for KCa$_4$(BO$_3$)$_3$; and a = 11.03843(8) Å, b = 11.98974(9) Å, c = 6.88446(5) Å for KSr4(BO3)3. Also, they predicted that the fundamental building units are isolated BO$_3$ anionic groups. Their second harmonic generation (SHG) coefficients were one-half (NaCa$_4$(BO$_3$)$_3$), one-third (KCa$_4$(BO$_3$)$_3$) and two-thirds (KSr$_4$(BO$_3$)$_3$) as large as that of KH$_2$PO$_4$ (KDP). Following the above reference, in 2007 Q. Su et al., from China synthesised LiSr$_4$(BO$_3$)$_3$ borates by a high-temperature solid-state reaction and discussed the thermoluminescence (TL) with their dosimetric characteristics when activated Ce3+ rare earth ions [60]. In their findings, the TL glow curve was composed of only one peak located at about 209°C and obtains the highest TL intensity for 1 mol% of Ce^{3+} concentration. The TL dose response shown by their material was linear in the protection dose ranging from 1 mGy to 1 Gy. Also, in the TSL spectra it shows a characteristic transition of Ce^{3+}. Again in 2009, Q. Su et al. investigated the thermoluminescence (TL) properties of Ce^{3+}-doped NaSr$_4$(BO$_3$)$_3$ material under the β-ray irradiation [61]. In their studied the same process of high temperature solid-state reaction was employed for synthesis. The TL glow curve of NaSr$_4$(BO$_3$)$_3$:Ce^{3+} material was composed of only one peak and found the activation energy (E) 0.590 eV and the frequency factor 1.008 × 106 s^{-1} deduced by the peak shape method. This material also shows linear dose response with characteristic emission of Ce^{3+} ions as per the TSL spectra recorded.

In same year, 2009, there was a report from Su et al. on the TL, photoluminescence (PL) and some dosimetric properties of KSr$_4$(BO$_3$)$_3$ material with different rare earth dopants, i.e. Tb^{3+}, Tm^{3+} and Ce^{3+} [62]. In their findings, they found Ce^{3+}-activated KSr$_4$(BO$_3$)$_3$ material shows good and comparable TL as compared to other dopants. The material prepared by them shows the optimum TL intensity for 0.2 mol% of Ce^{3+} ion. Again, the studies done by them followed the same pattern on finding dose response, effect of fading and recording TSL spectra to investigate the state of Ce ions.

Annalakshmi et al. have studied the kinetic parameter of lithium tetraborate using adaptive Simpson quadrature algorithm [63].

Rare earth activated Li$_2$B$_4$O$_7$ materials are attractive from the clinical and personnel dosimetry point of view because of their near tissue equivalence. The effective atomic number of lithium borate is 7.4, which is very close to human tissue. Materials with higher effective atomic number over-respond at low energies up to 100 keV because photoelectric absorption becomes predominant. The importance of lithium borate-based TL/OSL phosphors is its response to neutron radiation. This application is specific to nuclear industry. In lithium borate, if both boron and lithium are enriched, then the neutron response improves [64]. Li$_2$B$_4$O$_7$ is a relatively stable chemical compound. Normally, transition metal doped as impurity in Li$_2$B$_4$O$_7$ has found to have increased TL sensitivity.

In 1967, Schulman et al. reported Mn-activated Li$_2$B$_4$O$_7$ was the first material for TL dosimetry [65]. Takenaga et al. have reported a Cu-activated Li$_2$B$_4$O$_7$ phosphor, for radiation dosimetry and found that 20 times more sensitive to that of LiF phosphor with TL emission at 368 nm [66]. But Cu-activated Li$_2$B$_4$O$_7$ has many advantages over the Li$_2$B$_4$O$_7$:Mn phosphor. Moreover, different synthesis methods of Cu-activated Li$_2$B$_4$O$_7$ phosphor have shown different TL responses [67,68].

Rawat et al. have successfully synthesised Li$_2$B$_4$O$_7$: Cu-Ag phosphor by using a Czochralski method and proposed that this borate-based Li2B4O7: Cu-Ag Phosphor as a potential phosphor for remote optical fibre–based real-time OSL dosimetry for medical physics applications and neutron monitoring [69].

Kerikmäe et al. studied the trapped-hole centre with a high thermal stability in Li$_2$B$_4$O$_7$:Be phosphor and found that in EPR spectrum a hyperfine splitting due to interaction of the trapped hole with a ^{9}Be nucleus. [70]. Prokic et al. reported main dosimetric characteristics of sintered solid Li$_2$B$_4$O$_7$:Cu, Ag, P TL detectors for different dosimetry applications particularly in medical dosimetry and also for individual monitoring.

Ozdemir et al. reported the TL dosimetric characteristics of Mn-activated Li$_2$B$_4$O$_7$ phosphor synthesised by a solution combustion method. It is found that the LTB:Mn 0.04 M % phosphor is a promising material that can be used in dosimetry of ionizing radiations. Particularly, it will be used in clinical therapy dosimetry [71].

Brant et al. have reported the electron and hole trapping in Ag-doped single crystals of $Li_2B_4O_7$ using EPR and ENDOR techniques [72].

Danilkin et al. have studied the manganese clustering effects in $Li_2B_4O_7$ (LTB) ceramics and found that the radiation damage of LTB occurs due to breaking of chemical bonds in the boron-oxygen network [73]. Mn-doped LTB is a commercial thermolumin-escence dosimeter, denoted as TLD-800 [74]. The copper-activated LTB phosphors were investigated by many researchers and found the sensitivity higher or comparable to TLD-100 material (LTB:Mg,Ti) [75].

Drozdowski et al. have reported the luminescence properties of $Li_2B_4O_7$ (LTB) single crystals doped with 0.5 mol% Mn and 0.005 mol% Eu by using the Czochralski method and discuss the nature of the trap [76].

Holovey et al. have reported the influence of reducing annealing on the luminescent properties of $Li_2B_4O_7$:Cu single crystals [77]. Kar et al. have reported the TL properties of Mn-activated $Li_2B_4O_7$ under γ irradiation and studied the kinetic parameter of the TL glow curve [78]. Laxamanan et al. have reported the radiation dosimetric characteristic of $Li_2B_4O_7$:Cu TLD phosphor and studies batch to batch variation in TL sensitivity, spurious luminescence effect, effect of storage at ambient temperature and humidity and of pre-irradiation annealing temperature on TL sensitivity, dependence of TL sensitivity on the irradiation and fading of radiation induced TL on exposure to sunlight [79].

Patra et al. have reported the optically stimulated luminescence properties of Ag-activated $Li_2B_4O_7$ crystal under various nuclear radiations and found that high sensitivity of the LTB:Ag to thermal neutrons will be useful in a variety of applications including personal dosimetry in mixed fields and imaging devices for neutron radiography [80]. Ratas et al. have reported the thermoluminesce curve, kinetics and EPR (electro paramagnetic resonance) parameter of $Li_2B_4O_7$:Mn and $Li_2B_4O_7$:Mn,Be phosphors [81].

MgB_4O_7 is a known TL material. It is attractive due to an effective atomic number of $Z_{eff} = 8.2$, which is close to that of water ($Z_{eff} = 7.51$) [82,83]. This host has a suitable for TL dosimetry, particularly for Dy and Mn doping [84]. Because of its high sensitivity, tissue equivalency ($Z_{eff} = 8.4$) and interest for neutron dosimetry.

Yukihara et al. have reported thermoluminescence properties of $Li_2B_4O_7$:Cu, Ag (LBO), MgB_4O_7:Dy, Li (MBO) and $CaSO_4$:Ce phosphor for passive temperature sensors [85]. Gustafson et al. have developed a new OSL MgB_4O_7:Ce,Li material synthesised by using the solution combustion synthesis (SCS) method. The samples were characterised using photoluminescence, radioluminescence (RL), thermoluminescence (TL) and both blue- and green-stimulated luminescence (BSL and GSL). From these results, it is found that high sensitivity to ionizing radiation, low effective atomic number ($Z_{eff} = 8.2$) and fast luminescence when doped with cerium (Ce^{3+}) [86]. Yukihara et al. have reported crystallographic data of MgB4O7 phosphor and related crystallographic data are given in Table 7.5 [16].

TABLE 7.5

Crystallographic data of MgB_4O_7 phosphor

Chemical formula	MgB_4O_7
Crystal structure	Orthorhombic
Space group	Pbca
a (Å)	8.5960
b (Å)	13.7290
c (Å)	7.9560
α (°)	90
β (°)	90
γ (°)	90

The neutron-gamma discrimination can be achieved using detectors prepared with [10]B, which has a high thermal neutron capture cross section (3,800 barns), or [11]B, which has low thermal neutron capture cross section (<0.01 barn). The [10]B neutron capture reaction results in energetic recoil [4]He (0.84 MeV) and [7]Li nuclei (1.47 MeV), as well as gamma rays (0.48 MeV), which escape the detector. The ranges of the [4]He and [7]Li nuclei are small (<5 μm in water for the 0.84 MeV alpha particle, and <7 μm in water for 1.47 MeV [7]Li). Therefore, boron must be part of the detector host matrix to be used as an efficient neutron converter. Yukihara et al. have developed a new optically stimulated luminescence (OSL) material for dosimetry applications that is tissue equivalent and has high sensitivity to ionizing radiation, fast luminescence lifetime and intrinsic neutron sensitivity and found that dominant TL peak at 210°C with intensity comparable to LiF:Mg,Ti,. Also found is that sample crystallinity and luminescence intensity increases with annealing temperature at least up to 900°C [87].

First reports on the MgB_4O_7:Dy came from groups in Japan and Serbia and studied the TL glow curves, which are plots of TL light vs. temperature, were used to determine the trapping parameters. However, the TL response strongly depends on the heating rate, an important parameter in the TL measurements [88,89]. Magnesium tetraborate doped with dysprosium (MgB_4O_7:Dy) is also recognized as a good thermoluminophor for personal dosimetry of gamma rays and X-rays [90,91]. There are several methods used for synthesis of magnesium tetraborate; for example, sol-gel method, combustion, wet reaction synthesis, solid-state route and precipitation (crystal growth). The most commonly used synthesis methods are the wet reaction (precipitation) and solid-state synthesis [92]. M. Paluch-Ferszt et al. reported a comparison of the main characteristics of MgB_4O_7: Dy and LiF:Mg,Ti, and the main features of the detectors are summarised in Table 7.6 [93–95].

Palan et al. have reported polycrystalline MgB_4O_7:Tb^{3+} phosphor via the solid-state diffusion method for radiation dosimetry and found that the TL glow curve consists of a single peak in the temperature range 150–350°C and CW-OSL decay pattern of MgB_4O_7:Tb^{3+} phosphor is very similar to the decay pattern of commercially available α-Al_2O_3:C phosphor [96].

Souza et al. have reported dosimetric properties of MgB_4O_7:Dy,Li and MgB_4O_7:Ce,Li phosphor and found that MgB_4O_7:Ce,Li has a strong emission that peaked at 420 nm that is connected to the Ce^{3+} electronic transitions, while the emission of MgB_4O_7:Dy,Li has several peaks connected to the Dy^{3+} transitions. The OSL decay curves from both materials are composed by two components: a slow one and a fast one. MgB_4O_7:Ce,Li is 10 times more sensitive than MgB_4O_7:Dy, Li, specially due to the wavelengths of the emission peaks. The dose response for both materials were

TABLE 7.6

A comparison of the main characteristics of MgB_4O_7:Dy and LiF:Mg,Ti

Characteristics	Phosphor	
	MgB_4O_7:Dy (40% Tefl on)	LiF:Mg,Ti
Annealing procedure	Pre-irradiation: 450°C for 1 hPost-irradiation: 160°C for 5 s	Pre-irradiation: 400°C for 1 hfollowed by 100°C for 2 hPost-irradiation: 100°C for 10 min
Form	Solid disc 6 mm diameter	Solid disc 4.5 mm diameter
Effective atomic number Z	8.4	8.2
Main peak temperature [°C] for the heating rate of 5°C/s	230	230
Thermal fading [% at room temp.]	<20%/year	<5%/year
Detection threshold [mGy]	50	10

sublinear from 0.2 Gy to 100 Gy, for MgB_4O_7:Ce,Li, and from 0.2 Gy to 40 Gy, for the MgB_4O_7:Dy,Li. The OSL signal from MgB_4O_7:Ce,Li showed good stability over 40 days (with a fading <1%), while MgB_4O_7:Dy,Li presented a complete fading of the signal after 40 days. [97].

Gustafson et al. have developed a new optically stimulated luminescence (OSL) material based on the MgB_4O_7:Ce,Li phosphor via the solution combustion method and found that an increase in the OSL signal with an increase in the photon energy, as expected for a transition from the trapping centre to the delocalized band. TL sensitization was reduced from ~40% to ~10% using high-purity Mg precursor, but no significant improvement in the OSL sensitization (~30–40%) [98].

Kitagawa et al. have developed Ce^{3+} and Li^+ co-doped magnesium borate glass ceramics for optically stimulated luminescence dosimetry [99].

The MB_4O_7 (M = Sr, Ba and Ca) phosphor possesses a rigid three-dimensional (B_4O_7) network of corner-linked BO_4 tetrahedrons [100,101]. The SrB_4O_7 material belongs in the MB_4O_7 family. NUV phosphors currently known are SrB_4O_7:Eu^{2+} phosphors [102–104]. In recent years, the SrB_4O_7 (SBO) crystal has attracted researchers' attention due to its distinctive non-linear optical (NLO) properties and good mechanical properties [105]. The SrB_4O_7 crystal showed excellent piezoelectric, acoustic, elastic, piezooptic and acoustooptic properties and these studies showed SrB_4O_7 crystal may find potential applications in acoustoelectronic and acoustooptic areas. Sakirzanovas et al. reported the SrB_4O_7:Sm phosphor by using a solid-state reaction [106]. Iwamoto et al. reported optical properties of NUV-emitting SiO_2-SrB_4O_7:Eu^{2+} glass ceramic thin film [107]. Many researchers reported luminescence properties of rare earth–doped SrB_4O_7 phosphors like SrB_4O_7:Eu [108], SrB_4O_7:Pr [109], SrB_4O_7:Ag [110], SrB_4O_7:Bi [111], SrB_4O_7:Sm [112], SrB_4O_7:Eu [113], SrB_4O_7:Tm [114], SrB_4O_7:Dy [115], SrB_4O_7:Dy [116], SrB_4O_7:Cu [117], SrB_4O_7:Dy[118].

Also reported is crystallographic data of the SrB_4O_7 phosphor and related crystallographic data are given in Table 7.7.

In the recent past, SBO has attracted great attention of researchers aimed at the development of do-simetric materials. It is reported that in the temperature range 300–670°K, un-activated SBO crystals show thermally stimulated luminescence (TSL), whose intensity is almost comparable with that of TLD-700 solid-state dosimeter based on LiF (Mg, Ti). Due to the non-hygroscopic nature combined with a high TSL yield, SBO can be considered as a prospective material for TSL dosimetry [119,120].

Also, various alkaline earth (Li, Mg, Ca, Sr, Ba) based on tetraborate doped with different ions have been synthesised for radiation dosimetry. Because their atomic number is very close to that of soft tissue equivalent (Z_{eff} = 7.4) material, large linearity of dose-response, low fading, good reusability, MDD was found to be in the μGy range and good TL/OSL sensitivity to gamma, beta and neutron radiation makes tetraborates an interesting phosphor to develop and to investigate on its stimulated luminescent (TL and OSL) properties for radiation dosimetry [121–124].

TABLE 7.7

Crystallographic data of SrB_4O_7 phosphor

Chemical formula	SrB_4O_7
Crystal structure	Orthorhombic
Space group	Pnm21 (31)
a (Å)	4.4263
b (Å)	10.7074
c (Å)	4.2338
α (°)	90
β (°)	90
γ (°)	90

Bajaj et al. have prepared $LiCaBO_3$:D (D = Tb and Dy) phosphors by using a one-step, low-cost and low-temperature solution combustion method. This reaction produced very stable crystalline $LiCaBO_3$:D (D = Tb^{3+} and Dy^{3+}) phosphors. The TL glow curve of X-ray irradiated that $LiCaBO_3$:Tb^{3+} and $LiCaBO_3$:Dy^{3+} samples showed two major well-separated glow peaks. The TL sensitivity of these phosphors to X-ray radiation was comparable with that of TLD-100 (Harshaw). Photoluminescence spectra of $LiCaBO_3$:Tb^{3+} and $LiCaBO_3$:Dy^{3+} showed the characteristic Tb^{3+} and Dy^{3+} peaks. The TL response to X-ray radiation dose was linear up to 25 Gy [125].

In 2013, Bajaj et al. have reported TL dosimetric properties of $KSr_4(BO_3)_3$:Dy^{3+} phosphor under X-ray radiation and found that this phosphor shows 37% fading over the period of 20 days and shows linear behaviour on an increase of dose to which it exposed. This phosphor shows characteristic emissions of Dy^{3+} ions when excited with 350 nm, and this phosphor may also be used for solid-state lighting application.

Also reported is crystallographic data of the $KSr_4(BO_3)_3$ phosphor and related crystallographic data are given in Table 7.8.

Bajaj et al. have developed the low Z polycrystalline $LiMgBO_3$:Dy^{3+} phosphor via a novel solution combustion synthesis method and studied its luminescence characteristics like TL, PL etc. and found that the TL dosimetric glow peak at 154°C and TL sensitivity has half of the as TL sensitivity of available commercial TLD-100 phosphors. The $LiMgBO_3$ [LMB] material from the group has the lowest Z value and is suitable for a dosimetric application. However, very few reports are found on TL dosimetric applications of $LiMgBO_3$ [126–128].

Also reported is the crystallographic data of the $KSr_4(BO_3)_3$ phosphor and related crystallographic data are given in Table 7.9.

Bajaj et al. have reported the thermoluminescence and some of the dosimetric characteristics of Tb^{3+}-activated $NaSr_4(BO_3)_3$ under the X-ray irradiation and found that the TL glow curve consists of two well-separated glow peaks at 185°C and 318°C. The optimum concentration is 0.005 mol to obtain the highest intensity approximately comparable to commercial TLD-100. Also found is that the TL dose response is linear in the range of measurement. Also, photoluminescence spectra of $NaSr_4(BO_3)_3$:Tb^{3+} shows characteristic green emission when excited with a 236 nm UV source [129].

In recent years, the AB_4O_7 (A = monovalents or diavalent ion) borate hosts doped with RE get more concentration on its synthesis, PL, TL and OSL properties. The CaB_4O_7 is a typical example in this kind of compound and effective atomic number (Z_{eff}) of CaB_4O_7 phosphor (Z_{eff} = 12.5) is very nearer to Z_{eff} of α-Al_2O_3:C phosphor (Zeff = 11.28) and also this phosphor is an intermediate phosphor between LiF:Mg, Ti phosphor and $CaSO_4$:Dy phosphor.

Fukuda et al. have first reported TL properties of a Cu-doped CaB_4O_7 phosphor after Fukuda et al. have reported TL properties of doped and undoped CaB_4O_7 phosphors. Fukuda et al. also studied TL

TABLE 7.8

Crystallographic data of $KSr_4(BO_3)_3$ phosphor

Chemical formula	$KSr_4(BO_3)_3$
Crystal structure	**Orthorhombic**
Space group	**Ama2**
a (Å)	**11.03843**
b (Å)	**11.98974**
c (Å)	**6.88446**
α (°)	**90**
β (°)	**90**
γ (°)	**90**

TABLE 7.9

Crystallographic data of LiMgBO$_3$ phosphor

Chemical formula	LiMgBO$_3$
Crystal structure	**Monoclinic**
Space group	**P21/c(14)**
a (Å)	**9.9155(8)**
b (Å)	**8.8847(6)**
c (Å)	**5.1577(4)**
α (°)	**90.000**
β (°)	**91.2**
γ (°)	**90.000**

properties of Dy-, Eu- and Pb-doped CaB$_4$O$_7$ phosphors. After that, Tekin et al. have reported TL studies of thermally treated CaB$_4$O$_7$:Dy phosphors and various reports related [130–135].

In 2016, Palan et al. have reported thermoluminescence/optically stimulated luminescence properties of CaB$_4$O$_7$:Ce phosphors via the solid-state method and found that TL sensitivity of the CaB$_4$O$_7$:Ce phosphor was 0.8 times that of TLD-100 and OSL sensitivity about 0.46 times than that of α-Al$_2$O$_3$:C (BARC). Also found was that the decay pattern of CaB$_4$O$_7$:Ce phosphor was faster than the decay pattern of a-Al$_2$O$_3$:C phosphor. The effective atomic number (Z$_{eff}$) of a prepared phosphor is nearly similar to Z$_{eff}$ of Al$_2$O$_3$:C phosphor. The PL spectrum of CaB$_4$O$_7$:Ce showed emissions in the near blue region for the excitation of 290 nm under a UV source [136]. Also reported is thecrystallographic data of the CaB$_4$O$_7$ phosphor and related crystallographic data are given in Table 7.10.

The main characteristics of some borate-based TL/OSL materials are tabulated in Table 7.11.

7.4   ROLE AND SIGNIFICANCE OF SELECTED RARE EARTH IONS

From the literature survey, it has been understood that near-UV emitting phosphors are more preferable for TL/OSL. Thus, the phosphors with activators Tb^{3+}, Ce^{3+}, Cu, Ag, Cu-Ag and Eu^{2+} should have excellent TL/OSL properties and it has been advocated by many researchers. The emission of

TABLE 7.10

Crystallographic data of CaB$_4$O$_7$ phosphor

Chemical formula	CaB$_4$O$_7$
Crystal structure	Monoclinic
Space group	P21/n (14)
sa (Å)	12.264
b (Å)	9.895
c (Å)	7.7960
α (°)	90.000
β (°)	91.2
γ (°)	90.000
Volume	945.834
Z	8

TABLE 7.11
The main characteristics of some borate-based TL/OSL materials

Sr. no.	Material	Z_{eff}	Glow peak temp.	Linearity range	MDD	Fading	Synthesis method	Application	Ref.
1	$Li_2B_4O_7$:Cu	7.4	P_1 at 115°C and P_2 at 243 °C	For TL 10 mGy to 150.00 Gy ForOSL 10 mGy to 76.50 Gy	10 μGy	8–10% after 40 days	sintering technique	Personal dosimetry	[137]
2	$Li_2B_4O_7$: Cu,Ag	7.4	P_1 at 102 °C, P_2 at 168°C and $P_3$243°C	For TL 1 mGy to 1 kGy	10 mGy		Czochralski method	Medical physics application and neutron monitoring	[138]
3	SrB_4O_7:Eu		P1 at 305°C P_2 at 317°C	For TL 0.02 Gy To 2 Gy For OSL 0.02 Gy–20 Gy	1.26 mGy	-	Solid-state method	Radiation dosimetry	[139]
4	$Li_2B_4O_7$: Mn, Cu-Ag. In, P	7.4					Solid-state synthesis	Clinical and personnel dosimetry	[140]
5	CaB_4O_7:Ce	12.5	P_1 at 138°C To 172°C P_2 at 309°C	For TL 100–24,000 mGy	35.40 mGy	-	Solid-state method	Radiation dosimetry	[141]
6	$LiKB_4O_7$: Dy	8.07	P_1 at 171°C P_1 at 195°C	For 26 to 156 Gy			Czochralski technique	TL dosimetry	[142]
7	$LiKB_4O_7$:Eu	8.07	P_1 at 161°C P_2 at 201°C	For 26–156 Gy		7% for 60 days	Czochralski technique	TL dosimetry	[143]
8	$Li_2B_4O_7$: Cu-Ag	7.4	P_1 at 179 °C P_2 at 248 °C	For TL/OSL 20 mGy to 1,000 mGy			Solid-state method	Radiation dosimetry	[144]
9	MgB_4O_7:Ag	8.4	In bet^n 50°C to 300°C				Solid-state reaction	Radiation dosimetry	[145]
10	MgB_4O_7:Ce,Li	8.4	P_1 at 201°C	For OSL 0.5 Gy to 800 Gy	0.4 mGy	4% within 6 days	Solution combustion		[146]

(Continued)

TABLE 7.11 (Continued)
The main characteristics of some borate-based TL/OSL materials

Sr. no.	Material	Z_{eff}	Glow peak temp.	Linearity range	MDD	Fading	Synthesis method	Application	Ref.
11	MgB_4O_7:Ce,Li	8.4		For OSL 0.2 Gy to 100 Gy				2D dose mapping and neutron dosimetry	[147]
	MgB_4O_7:Dy,Li	8.4		For OSL 0.2 Gy to 40 Gy		1% for 40 days	Solid-state synthesis	Radiation dosimetry	
12	MgB_4O_7:Tb^{3+}	8.4	P$_1$ at 246°C				Solid-state diffusion method	Radiation dosimetry	[148]
13	MgB_4O_7:Tm,Ag	8.4	P$_1$ at 308°C				Solution combustion technique	Radiation dosimetry	[149]
14	$Li_2B_4O_7$:Cu	7.4	115°C and 243°C	For TL Up to 150.00 Gy and For OSL 76.50 Gy,		Low fading	Sintering technique	Clinical dosimetry	[150]
15	ZnB_2O_4:Dy	22.5	200°C	1 Gy to 100 Gy					[151]
	SrB_6O_{10}:Tb	31.8	210°C	50 Gy to 200 Gy					[152]

these rare earth ions (Tb^{3+}, Ce^{3+}, Cu, Ag, Cu-Ag and Eu^{2+}) doped phosphors are observed in the 330–425 nm region and these emissions are suitable for available PMT. The emission characteristics of Tb^{3+}, Ce^{3+} and Eu^{2+} are given below.

7.4.1 CHARACTERISTIC EMISSION OF Tb^{3+} IONS

It is well known that the activation of rare earth ions plays an important role in the development of the novel luminescence materials. Generally, luminescent materials consist of an activator and host. The host must have excellent physical and chemical stability. The selection of activator is dependent upon different applications. Trivalent terbium is a rare earth ion widely used as a dopant in a variety of host matrices and it is generally used as a blue-green-emitting centre in a variety of commercial phosphors. The Tb^{3+} ion exhibits a series of narrow line emissions located in the range of wavelength 370–650 nm. The blue light below 480 nm can be attributed to the $^5D_3$-^7F_J (J = 6, 5. 4, 3) transitions, while the green emission above 480 nm results from the $^5D_4$-^7F_J (J = 6, 5, 4, 3) transitions.

Figure 7.10 shows a partial energy level diagram of trivalent terbium with labeled transitions that correspond to the observed emission lines. The intensity of blue fluorescence from the $^5D_3$ level is highly dependent on the terbium concentration. Since the energy difference between $^5D_3$ and $^5D_4$ levels is approximately equal to the energy difference between $^7F_0$ and $^7F_6$ levels, the excited electrons in the $^5D_3$ level populate the $^5D_4$ level through cross-relaxation, which involves a transfer of energy from an ion in the $^5D_3$ state to a nearby ion in the ground state and can be represented as Tb^{3+} ($^5D_3$) + Tb^{3+} (^7F_j) $\rightarrow$ Tb^{3+} ($^5D_4$) + Tb^{3+} ($^7F_0$). At low Tb^{3+} concentration where cross-relaxation is improbable, both blue $^5D_3$-^7F_J and green $^5D_4$-^7F_J emissions are observed, while the blue $^5D_3$-^7F_J emission is generally not observed in samples containing higher Tb^{3+} concentrations.

This concentration quenching effect is well known and blue emission can be expected only if the cross-relaxation is inhibited [153]. The Tb^{3+}-activated phosphors show a series of sharp line emissions in the blue-green region and it is a general requirement for a material to be a good OSL phosphor.

7.4.2 CHARACTERISTIC EMISSION OF Ce^{3+} IONS

The rare earth ions doped inorganic compounds are widely used as lamp phosphors, cathode ray tube phosphors and scintillator phosphors because of their unique spectroscopic properties. New host materials doped with rare earths are getting much attention owing to their potential for applications. Inorganic materials doped with Ce^{3+} have attracted the interest of many researchers due to applications of these materials as phosphors, scintillator for elementary particles, detectors for ionizing radiations,

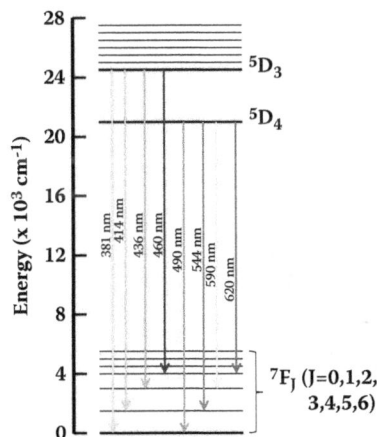

FIGURE 7.10   Energy level diagram of Tb^{3+} ion.

UV-absorbing filters or UV emitters and activators for energy transfer [154]. Cerium ion can exist in the oxidation state +3 and +4 in the phosphors. The luminescence efficiency is greatly affected by the existence of Ce^{4+}, as it provides a non-radiative pathway and reduces the number of Ce^{3+}.

The ground-state configuration of Ce^{3+} ions has one 4f electron and excited state configuration has one 5d electron with an empty 4f shell. The $4f^1$ ground-state configuration yields two levels viz. $^2F_{5/2}$ and $^2F_{7/2}$ due to spin orbit coupling and the $5d^1$ excited state configuration is split by a crystal field in 2 to 5 components. The Ce^{3+} emission occurs from the lowest crystal field component of $5d^1$ configuration to the 4f ground-state levels. Since the 4f $\rightarrow$ 5d transition is parity allowed and spin selection is not appropriate, the emission transition is fully allowed one [155].

Due to this, the excitation spectrum and emission spectrum of Ce^{3+} ions in inorganic solids consists of broadbands. The Ce^{3+} ion shows allowed optical transitions in absorption and emission, which are of the f $\rightarrow$ d type. A Ce^{3+} ion shows an efficient broadband luminescence in inorganic solids due to its 4f $\rightarrow$ 5d parity allowed electric dipole transitions [156]. The energy-level diagram of Ce^{3+} ion is shown in Figure 7.11.

7.4.3 Characteristic Emission of Eu²⁺ Ions

Luminescence properties of europium have been extensively studied. Eu^{3+} exhibits red or orange luminescence, while the luminescence of Eu^{2+} strongly depends on the host and it may lie anywhere from UV to the deep red region of the spectrum. Luminescence properties of Eu^{3+} have been used to obtain the red component of the full colour display devices [157,158]. Blue emission of Eu^{2+} has been extensively used in obtaining the blue component of lamp phosphors [159].

The luminescence of the Eu^{2+} ions in inorganic compounds has been widely investigated [160]. Eu^{2+} emission arises from the lowest band of the $4f^65d^1$ configuration to the $^8S_{7/2}$ state of the $4f^7$ configuration. The excitation arises from the transition from the $^8S_{7/2}$ state of the $4f^7$ configuration to the states belonging to the $4f^6 5d^1$ configuration. The ground-state electronic configuration of Eu^{2+} is $4f^7$. The emission and absorption spectra of Eu^{2+} ion usually consists of a broad band due to transitions between the $^8S_{7/2}$ ($4f^7$) ground state and the crystal field components of the $4f^65d^1$ excited states [161]. When the Eu^{2+} ion is incorporated in a crystalline host, the $^4f_6{}^5d_1$ levels experience much

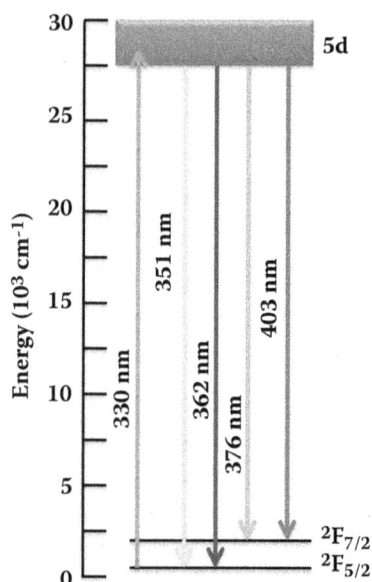

FIGURE 7.11 Energy level diagram of Ce^{3+} ion.

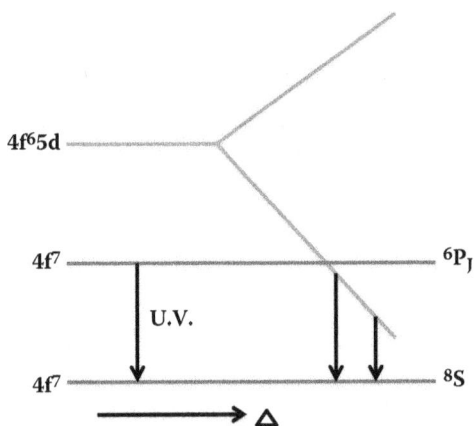

FIGURE 7.12 Schematic diagram of the energies of $4f^n$ and $4f^n$ 5d levels in Eu^{2+} influenced by crystal field.

more crystal field splitting than the $^4f_7$ levels due to the increased spatial extent of the 5d orbital and often are the metastable state, or the lowest excited state. The most commonly observed emission is the dipole and spin allowed d–f emission starting from the relaxed $^4f_6$ ($^7F_0$) $^5d_1$ level. However, due to the allowed nature of the transition, d–f emission is intense. This dependence is attributed to the crystal field splitting of the 5d level, as shown schematically in Figure 7.12.

The emission band shifts to a longer wavelength with increasing crystal field strength. The emission due to the 5d-4f transitions of Eu^{2+} are affected most by crystal parameters indicating electron-electron repulsion. The sharp lines corresponding to $^6P_j \rightarrow ^8S_{7/2}$ transitions are observed when $^4f_6$ 5d band is higher than ^6P_j states then the f-f transitions become parity allowed. This can happen when the crystal field is weak [162]. These transitions have a longer decay time of the order of several milliseconds. In many compounds, d-f emission is observed at room temperature, while at low enough temperatures sharp line f-f emission becomes dominant. A third type of emission involving the Eu^{2+} ions is often characterized by a very large Stokes (5,000–10,000 cm^{-1}) shift and very broad (44,000 cm^{-1}) emission band. The auto-ionization of the 5d electron to conduction band level causes this 'anomalous' emission of Eu^{2+} ions. An impurity trapped exciton state is created when the electron is localized on the cations around the hole that stays behind on Eu^{2+} ions. The 'anomalous' emission is the radiative transfer that causes the electron back to the ground state of Eu^{2+} ions. The auto-ionization can also be a cause for absence of Eu^{2+} luminescence.

7.5 APPLICATIONS

7.5.1 PERSONNEL DOSIMETRY

Personnel dosimetry is concerned with the evaluation of deep, shallow and eye radiation doses. The main objective of personnel dosimetry is the monitoring of radiation dose delivered to a person during routine occupational exposure, which includes workers in nuclear industry, medical physicist, radiotherapy technicians, workers in industrial radiography and naval personnel on nuclear-powered vessels. By means of such monitoring it is hoped to limit the exposure of such person to within prescribed safety limits, which are based on recommendations of the International Commission of Radiological Protection, publication 60 (ICRP 60). As per the United Nations Sources and Effects of Ionizing Radiation (UNSEAR) report, about 10 million workers are exposed to radiation from artificial sources. External monitoring of these radiation workers is carried out using personal dosimetry badges, which include TLDs, OSLDs, RPLDs and film badges. These personal dosimeters should comply with IEC requirements for "passive integrating dosimetry systems for personnel and environmental monitoring". Due to non-reliability of photographic film for long-term stability of stored information particularly in tropical climates, this method of personnel dosimetry has nearly phased out and is being replaced by TL, OSL and RPL methods [163–166]. A major

TABLE 7.12

General characteristics of borate phosphor for personnel dosimetry

Sr. no.	Phosphor	Z_{eff}	Synthesis method	Reference
	$Li_2B_4O_7$:Ag/Cu	7.4	Czochralski method	[167]

requirement of OSLDs in these applications is that they are approximately tissue equivalent. Thus, materials with effective atomic numbers (Z_{eff}) near that of human tissue $(Z_{eff} = 7.6)$ are desirable. The dose equivalent range of interest is from approximately 10 μSv to 1Sv, with a required uncertainty better than approximately 10%. The OSL using aluminium oxide doped with carbon $(Al_2O_3$:C) has been used extensively for personnel radiation dosimetry for many years. It has the advantages of low detection threshold and re-analysis. Table 7.12 gives general characteristics of borate-based available thermoluminescent/optically stimulated luminescence dosimeters relevant for personnel dosimetry.

7.5.2 Environmental Dosimetry

Tissue equivalence is not an issue with dose estimation to the environment, for which the only quantity of interest is the absorbed dose D (in Gy). The primary interest in this field is the impact of 'man-made' radiation on the general public. Sources of such man-made radiations include nuclear waste disposal, emissions from nuclear power, reprocessing plants and the nuclear weapons industry. Major requirements of dosimeters for these applications include high sensitivity (to enable short- or long-term monitoring), stability with respect to adverse weather and changing light levels and temperatures. With natural dose rates of a few mGy (or mSv) per year, the dose range of interest depends on exposure time, but a minimum measurable dose/minimum detectable dose (MMD/MDD) of approximately μGy is desirable for short-term measurements. OSL techniques and materials hold significant promise for these applications due to high sensitivity and ease of use [168,169].

7.5.3 Medical Dosimetry

One of the important applications of OSL dosimetry has been found in the field of medical physics due to their use in radio-diagnosis, nuclear medicine and radiotherapy. OSLDs have become popular in these fields due to their high sensitivity, miniature size, tissue equivalence, high stability to environmental conditions, low OSL fading, reusability, linear dose response, sufficient precision and accuracy. OSL phosphors, such as $Li_2B_4O_7$:Cu-Ag, fluoride-based phosphors have been widely used for clinical dosimetric measurements e.g. central axis depth-dose curves, inphantom measurements, in-vivo dosimetry, surface doses, quality assuarance, etc. of high-energy photon and electron beams. Applications of OSL phosphors in medical dosimetry are described by Akselrod and McKeever [170], Pradhan [171] and Yukihara [172]. OSL dosimeters are popular in many hospitals for external dosimetry during these treatments. OSL has the potential for the development of near-real-time dosimetry in which the measured quantity can be either dose (Sv) or dose rate (Sv/s). Typical doses of interest can be up to 20 Sv [173,174].

7.5.4 Retrospective Dosimetry

'Retrospective dosimetry' is the term used when determining the dose of absorbed radiation to environmental or locally available materials in situations where conventional, synthetic dosimeters were not in place at the time of radiation exposure. Two major categories exist: dating and accident dosimetry. In the luminescence dating, if one wishes to establish the dose delivered to natural materials during exposure to natural radiation over the lifetime of the sample in a certain environment. The other category is accident dosimetry, where one is interested in the determination of absorbed dose due to a radiation accident or other event, over and above the normal background radiation. OSL is proving to be a useful tool in these applications [175].

7.5.5   HIGH DOSE

The use of optically stimulated dosimeters in monitoring high dose radiation (for example, from 10^2 Gy up to 10^6 Gy) is a further example of one of the mainstream uses of the technology. Such high doses may be found, for example inside nuclear reactors or during food sterilization and materials testing.

7.6   LIMITATIONS AND CHALLENGES

The main drawback of these TL materials is that not many are low-Z (tissue equivalent, i.e. $Z_{eff} \approx 7.4$) materials. Low-Z (tissue equivalence) materials are usually preferred in radiation monitoring for due to their energy independent TL response which makes them suitable for the dosimetry even in a mixed field. There are some tissue equivalent phosphors available commercially but they cannot be considered as ideal ones. For example, $CaSO_4$:Dy (TLD 900) is a sensitive but the shape of its glow curves change at high doses and on annealing to high temperatures adding inaccuries in measurements. LiF:Mg,Ti (TLD 100) is considered to be a '*good*' one but suffers from some drawbacks, e.g. it is not a relatively very sensitive material and has also a very complicated glow curve structure, another improved one is LiF:Mg,Cu,P and is very sensitive but reusability is a problem, if heated beyond 523 K and if not the remaining deep traps add inaccuracies in measurements. This problem exists in CaF_2:Mn also. BeO doped with alkali ions is another highly sensitive tissue equivalent TLD phosphor but it is toxic and handling is a problem during synthesis and radiation monitoring.

7.7   FUTURE PROSPECTS AND SCOPE

From literature, the radiation monitoring could be effective only when the radiation measurements are accurate. At present, thermoluminescence (TL) and optically stimulated luminescence (OSL) techniques are widely used for radiation measurements. Also, these techniques are important tools in nuclear and medical research programmes. However, the OSL technique is the most widely used technology for evaluating the personal and environmental dosimetry because of its advantages.

Also from literature, basically various different synthesis methods such as the combustion method, modified solid-state method, Czochralski method, co-precipitation method and sol-gel method have been adopted for synthesis of borate phosphor. Our experience is that the modified solid-state diffusion method gave phosphors with better sensitivity, dose response, reusability and MDD. The only drawback is phosphors showed fast fading. To overcome this drawback, the synthesis method can be optimized for getting better performance of phosphors. Additionally, the efforts need to be made for reducing fading of the OSL signal in phosphors by using suitable co-dopants and activators. For instance, Li, Na and K can be added as co-dopants in the MgB_4O_7:Tb^{3+} phosphor for better OSL performances.

Borate-based phosphors are also reported to be good optically stimulated luminescence dosimeters. An intense thermoluminescence was observed in MgB_4O_7:RE,, $Li_2B_4O_7$:RE, $LiKB_4O_7$: RE, CaB_4O_7:RE, $LiMgBO_3$:RE, $KSr(BO_3)_3$, $LiCaBO_3$:RE, SrB_4O_7: RE, $NaSr_4(BO_3)_3$, $LiBaBO_3$, $MM'_4(BO_3)_3$ (M = Na, M' = Ca; M = K, M' = Ca, Sr) etc. during alpha (α)/Beta (β)/gamma (γ) irradiation. Hence, thermoluminescence measurements and feasibility of using the phosphors as optically stimulated luminescence dosimeters is another direction of future work.

It is well known that the OSL technique is completely dependent on traps in material and energy absorbed by traps from radiation. It is necessary to investigate the performance of materials under different high ionizing radiations such as neutron irradiation, alpha irradiation, etc.

Important requirements for a clinical or personnel dosimeter is to have a dosimeter with a flat energy response at all energies. This mainly depends on the effective atomic number. Normally, tissue equivalent materials have a flat energy response. From literature, $Li_2B_4O_7$:RE and MgB_4O_7: RE are tissue equivalent, whereas $LiKB_4O_7$:RE, CaB_4O_7:RE, $LiMgBO_3$:RE, $KSr(BO_3)_3$, SrB_4O_7: RE, $NaSr_4(BO_3)_3$, $LiBaBO_3$ and $MM'_4(BO_3)_3$ (M = Na, M'= Ca; M = K, M' = Ca, Sr) are not tissue equivalent. Hence, these phosphors may have an over response at lower energies. For using a material as a dosimeter, the response of the phosphor at lower energies has to be carried out and the suitable filters that can be used to flatten the energy response at all energies should be identified.

REFERENCES

[1] Cameron, J. (1991). Radiation dosimetry. *Environ. Health Perspect.*, 91, 45–48.

[2] Li, J., Hao, J. Q., Li, C. Y., Zhang, C. X., Tang, Q., Zhang, Y. L., Sua, Q., and Wang, S. B. (2005). Thermally stimulated luminescence studies for dysprosium doped strontium tetraborate. *Radia. Meas.*, 39, (2005), 229–233.

[3] McKeever, S. W. S., and Moscovitch, M. (2003). On the advantages and disadvantages of optically stimulated luminescence dosimetry and thermoluminescence dosimetry. *Radiat. Prot. Dosim.*, 104, 263270.

[4] McKeever, S. W. S., Moscovitch, M., and Townsend, P. D. (1995). *Thermoluminescence Dosimetry Materials: Properties and Use.* Nuclear Technology Publishing, Ashford, 28, 210.

[5] McKeever, S. W. S. (2002). New millennium frontiers of luminescence dosimetry. *Radiat. Prot. Dosim.*, 100, 27–32.

[6] McKeever, S. W. S., Blair, M. W., Bulur, E., Gaza, R., Gaza, R., Kalchgruber, R., Klein, D. M., and Yukihara, E. G. (2004). Recent advances in dosimetry using the optically stimulated luminescence of Al_2O_3:C. *Radiat. Prot. Dosim.*, 109, 269–274.

[7] Bhatt, B. C., and Kulkarni, M. S. (2014). Thermoluminescent phosphors for radiation dosimetry. *Defect Diffus. Forum*, 347, 179–227.

[8] Markey, B. G., Colyott, L. E., and Mckeever, S. W. S. (1995). Time-resolved optically stimulated luminescence from α-Al_2O_3:C. *Radiat. Meas.*, 24, 457–463.

[9] Mckeever, S. W. S. (2001). Optically stimulated luminescence dosimetry. *Nucl. Instrum. Methods Phys. Res. Sect. B*, 184, 29–54.

[10] Pradhan, A. S., Lee, J. I., and Kim, J. L. (2008). Recent developments of optically stimulated luminescence materials and techniques for radiation dosimetry and clinical applications. *J. Med. Phys.*, 33, 85–99.

[11] Bhatt, B. C. (2011). Thermoluminescence, optically stimulated luminescence and radio-photoluminescence dosimetry: aA perspective, 34, 6–16.

[12] Pradhan, A. S., Lee, J. I., and Kim, J. L. (2008). Recent developments of optically stimulated luminescence materials and techniques for radiation dosimetry and clinical applications. *J. Med. Phys.*, 33, 85–99.

[13] Akselrod, M. S., Bøtter-Jensen, L., and McKeever, S. W. S. (2006). Optically stimulated luminescence and its use in medical dosimetry. *Radiat. Meas.*, 41, S78.

[14] Guide to the Riso TL/OSL reader, RISO, DTU, Denmark (2010).

[15] McKeever, S. W. S., and Moscovitch, M. (2003). On the advantages and disadvantages of optically stimulated luminescence dosimetry and thermoluminescence dosimetry. *Radiat. Prot. Dosim.*, 104, 263–270.

[16] Yukihara, E. G., Milliken, E. D., and Doull, B. A. (2014). Thermally stimulated and recombination processes in MgB_4O_7 investigated by systematic lanthanide doping. *J. Lumin.*, 154, 251–259.

[17] Sahare P. D., Singh M., and Kumar P. (2015). A new high sensitivity Na_2LiPO_4:Eu OSL phosphor. *RSC. Adv.*, 5 3474–3481.

[18] Barve, R., Patil, R. R., Gaikwad, N. P., Kulkarni, M. S., Mishra, D. R., Soni, A., Bhatt, B. C., and Moharil, S. V. (2013). Optically stimulated luminescence and thermoluminescence in some Cu^+ doped alkali fluoro-silicates. *Radiat. Meas.*, 59, (2013), 73–80.

[19] Bulur, E. (1996). An alternative technique for optically stimulated luminescence (OSL) experiment. *Radiat. Meas.*, 26, 701–709.

[20] Kuhns, C. K., Larsen, N. A., and McKeever, S. W. S. (2000). Characteristics of LM-OSL from several different types of quartz. *Radiat. Meas.*, 32, 413–418.

[21] Bulur, E. (2000). A simple transformation for converting CW-OSL curves to LM-OSL curves. *Radiat. Meas.*, 32, 141–145.

[22] McKeever, S. W. S. (2001). Optically stimulated luminescence dosimetry. *Nucl. Instrum. Methods Phys. Res. B*, 184, 29–54.

[23] Yukihara, E. G., Whitley, V. H., McKeever, S. W. S., Akselrod, A. E., and Akselrod, M. S. (2004).

Effect of high-dose irradiation on the optically stimulated luminescence of Al$_2$O$_3$:C. *Radiat. Meas.*, 38, 317–330.

[24] Yukihara, E. G., and McKeever, S. W. S. (2006). Ionisation density dependence of the optically and thermally stimulated luminescence from Al$_2$O$_3$:C. *Radiat. Prot. Dosim.*, 119, 206–217.

[25] Yukihara, E. G., Whitley, V. H., Polf, J. C., Klein, D. M., McKeever, S. W. S., Akselrod, A. E., and Akselrod, M. S. (2003). The effects of deep trap population on the thermoluminescence of Al$_2$O$_3$:C. *Radiat. Meas.*, 37, 627–638.

[26] Yukihara, E. G., Whitley, V. H., McKeever, S. W. S., Akselrod, A. E., and Akselrod, M. S. (2004). Effect of high-dose irradiation on the optically stimulated luminescence of Al$_2$O$_3$:C. *Radiat. Meas.*, 38, 317.

[27] Akselrod, M. S., and Gorelova, E. A. (1993). Deep traps in highly sensitive α-Al$_2$O$_3$:C TLD crystals. *Radiat. Meas.*, 21, 143–146.

[28] Polf, J. C., Yukihara, E. G., Akselrodand, M. S., and McKeever, S. W. S. (2004). Real-time luminescence from Al$_2$O$_3$ fiber dosimeters. *Radiat. Meas.*, 38, 227–240.

[29] Evans, B. D., Pogatshnik, G. J., and Chen, Y. (1994). Optical properties of lattice defects in α-Al$_2$O$_3$. *Nucl. Instrum. Methods Phys. Res. B*, 91, (1994), 258–262.

[30] Akselrod, M. S., and McKeever, S. W. S. (1999). Radiation dosimetry method using pulsed optically stimulated luminescence. *Radiat. Prot. Dosim.*, 81, 167–175.

[31] Piesch, E., Burgkhardt, B., Fischer, M., Rrber, H. G., and Ugi, S. (1986). Properties of radio-photoluminescent glass dosemeter systems using pulsed laser UV excitation. *Radiat. Prot. Dosim.*, 17, 293–297.

[32] Piesch, E., Burgkhardt, B., and Vilgis, M. (1990). Photoluminescence dosimetry: Progress and present state of art. *Radiat. Prot. Dosim.*, 33, 215–226.

[33] Piesch, E., Burgkhardt, B., and Vilgis, M. (1993). Progress in phosphate glass dosimetry: Experiences and routine monitoring with a modern dosimetry system. *Radiat. Prot. Dosim.*, 47, 409–414.

[34] Chen, C. T., Wu, Y. C., and Li, R. K. (1990). The development of new NLO crystals in the borate series. *J. of Cryst. Growth*, 99, 790–798.

[35] Lu, C. H., Godbole, S. V., and Natarajan, V. (2005). Luminescence characteristics of strontium borate phosphate phosphors. *Mater. Chem. Phys.*, 93, 73–77.

[36] Wu, L., Chen, X. L., Li, H., He, M., Dai, L., Li, X. Z., and Xu, Y. P. (2004). Structure determination of a new compound LiCaBO$_3$. *J. Solid State Chem*, 177, 1111–1116.

[37] Bajaj, N. S., and Omanwar, S. K. (2012). Combustion synthesis and luminescence characteristic of rare earth activated LiCaBO$_3$. *J. Rare Earths*, 30, 1005–1008.

[38] Jose, M. T., Anishia, S. R., Annalakshmi, O., and Ramasamy, V. (2011). Determination of thermoluminescence kinetic parameters of thulium doped lithium calcium borate. *Radiat. Meas.*, 46, 1026–1032.

[39] Nagpure, P. A., Bajaj, N. S., Sonekar, R. P., and Omanwar, S. K. (2011). Synthesis and luminescence studies of novel rare earth activated lanthanum pentaborate. *Indian J. Pure Appl. Phys.*, 49, 799–802.

[40] Wu, L., Chen, X. L., Li, H., He, M., Dai, L., Li, X. Z., and Xu, Y. P. (2004). Structure determination of a new compound LiCaBO$_3$. *J. Solid State Chem.*, 177, 1111–1116.

[41] Chen, C. T., Wu, B., Jiang, A., and You, G. (1985). A new-type ultraviolet shg crystal—β-BaB$_2$O$_4$. *Sci. Sin. B*, 18, 235–243.

[42] Chen, C. T., Wu, Y., Jiang, A., Wu, B., You, G., Li, R., and Lin, S. (1989). New nonlinear-optical crystal: LiB$_3$O$_5$. *J. Opt. Soc. Am. B*, 6, 616–621.

[43] Chen, C. T., Wang, Y., Wu, B., Wu, K., Zeng, W., and Yu, L. (1995). Design and synthesis of an ultraviolet-transparent nonlinear optical crystal Sr$_2$Be$_2$B$_2$O$_7$. *Nature*, 373, 322–324.

[44] Takenaga, M., Yamamota, and Yamashita, T. (1980). Preparation and characteristics of Li$_2$B$_4$O$_7$:Cu phosphor. *Nucl. Instrum. Methods B*, 175, 77–78.

[45] Shahare, D. I., Deshmukh, B. T., Moharil, S. V., Dhopte, S. M., Muthal, P. L., and Kondawar, V. K. (1994). Synthesis of Li$_2$B$_4$O$_7$:Cu phosphor. *Phys. Status Solidi (A)*, 141, 329–334.

[46] Palan, C. B., Chauhan, A. O., Sawala, N. S., Bajaj, N. S., and Omanwar, S. K. (2016). Synthesis and preliminary TL/OSL properties of Li$_2$B$_4$O$_7$: Cu-Ag phosphor for radiation dosimetry. *Optik*, 127, 6419–6423.

[47] Shahare, D. I., Dhoble, S. J., and Moharil, S. V. (1993). Preparation and characterization of magnesium-borate phosphor. *J. Mater. Sci. Lett.*, 12, 1873–1874.

[48] Palan, C. B., Chauhan, A. O., Sawala, N. S., Bajaj, N. S., and Omanwar, S. K. (2015). Thermoluminescence and optically stimulated luminescence properties of MgB4O7: Ag phosphor. *Int. J. Lumin. Appl.*, 5, 408–410.

[49] Yamashita, T., Nada, N., Onishi, H., and Kitamura, S. (1971). Proc. 2nd Int. Conf. On Lum. Dosim., Gathilinburg, CONF-680920, 1968, PP 4; Health Phys. 21:295.

[50] Palan, C. B., Koparkar, K. A., Bajaj, N. S., Soni, A., and Omanwar, S. K. (2016). Synthesis and thermoluminescence/optically stimulated luminescence properties of CaB_4O_7:Ce phosphor. *J. Mater. Sci: Mater. Electron*, 27, 5600–5606.

[51] Fukuda Y., Mizuguchi K. M., and Takeuchi N. (1986). Thermoluminescence in Sintered CaB_4O_7:Dy and CaB_4O_7:Eu. *Radiat. Prot. Dosim.*, 19, 397–401.

[52] Fukuda, Y., Tomita, A., and Takeuchi, N. (1990). Electron affinity of CaB_4O_7:PbO ceramic obtained by a simultaneous measurement of TL and TSEE. *J. Mater. Sci. Lett.*, 9, 867–868.

[53] He, M., Chen, X. L., Hu, B. Q., Zhou, T., Xu, Y. P., and Xu, T. (2002). The ternary system $Li_2O–Al_2O_3–B_2O_3$: Compounds and phase relations. *J. Solid State Chem.*, 165, 187–192.

[54] He, M., Chen, X. L., Gramlich, V., Baerlocher, Ch, Zhou, T., and Hu, B. Q. (2002). Synthesis, structure, and thermal stability of $Li_3AlB_2O_6$. *J. Solid State Chem.*, 163, 369–376.

[55] Wu, L., Chen, X. L., Tu, Q. Y., He, M., Zhang, Y., and Xu, Y. P. (2002). Phase relations in the system $Li_2O–MgO–B_2O_3$. *J. Alloys Compd.*, 333, 154–158.

[56] Wu, L., Chen, X. L., Tu, Q. Y., He, M., Zhang, Y., and Xu, Y. P. (2003). Phase relations in the system $Li_2O–CaO–B_2O_3$. *J. Alloys Compd.*, 358, 23–28.

[57] Norrestam, R. (1989). The crystal structure of monoclinic $LiMgBO_3$. *Z. Kristallogr. Cryst. Mater.*, 187, 103–110.

[58] Schla¨eger, M., Hoppe, R., and Anorg., Z. (1993). Darstellung und Aufbau von BaLi [BO3]. *Allg. Chem.*, 619, 976–982.

[59] Wu, L., Chen, X. L., Xu, Y. P., and Sun, Y. P. (2005). *Inorg. Chem.*, 44, 2005, 6409–6414.

[60] Jiang, L. H., Zhang, Y. L., Li, C. Y., Hao, J. Q., and Su, Q. (2007). Phase relations in the system $Li_2O–CaO–B_2O_3$ thermoluminescence properties of Ce^{3+}-doped $LiSr_4(BO_3)_3$ phosphor. *Mater. Lett.*, 61, 5107–5109.

[61] Jiang, L., Zange, Y., Li, C., Pang, R., Shi, L., Zhang, S., Hao, J., and Su, Q. (2009). Thermoluminescence characteristics of $NaSr_4(BO_3)_3$:Ce^{3+} under β-ray irradiation. *J. Rare Earths*, 27, 320–322.

[62] Jiang, L. H., Zhang, Y. L., Li, C. Y., Hao, J. Q., and Su, Q. (2009). Synthesis, photoluminescence, thermoluminescence and dosimetry properties of novel phosphor $KSr_4(BO_3)_3$:Ce. *J. Alloys Compd.*, 482, 313–316.

[63] Annalakshmi, O., Jose, M. T., Madhusoodanan, U., Venkatraman, B., and Amarendra, G. (2013). Kinetic parameters of lithium tetraborate based TL materials. *J. Lumin.*, 141, 60–66.

[64] Rawat, N. S., Kulkarni, M. S., Tyagi, M., Ratna, P., Mishra, D. R., Singh, S. G., Tiwari, B., Soni, A., Gadkari, S. C., and Gupta, S. K. (2012). TL and OSL studies on lithium borate single crystals doped with Cu and Ag. *J. Lumin.*, 132, 1969–1975.

[65] Schulman, J. H., Kirk, R. D., and West, E. J. (1967). Luminescence dosimetry, U.S. Atomic Energy Commission Symposium Series 8, CONF-650637, 113.

[66] Takenaga, M., Yamamato, O., Yamashita, T. (1983). A new phosphor Li2B4O7: Cu for TLD. *Health Phys.*, 44, 1983, 387–393.

[67] Shahare, D. I., Deshmukh, B. T., Moharil, S. V., Dhopte, S. M., Muthal, P. L., and Kondawar, V. K. (1994). *Phys. Status Solidi A*, 141, 329.

[68] Prokic, M. (2002). Dosimetric characteristics of $Li_2B_4O_7$:Cu,Ag,P solid TL detectors. *Radiat. Prot. Dosim.*, 2002, 265–268.

[69] Rawat, N. S., Kulkarni, M. S., Tyagi, M., Ratna, P., Mishra, D. R., Singh, S. G., Tiwari, B., Soni, A., Gadkari, S. C., and Gupta, S. K. (2012). TL and OSL studies on lithium borate single crystals doped with Cu and Ag. *J. Lumin.*, 132, 1969–1975.

[70] Kerikmäe, M., Danilkin, M., Lust, A., Nagirnyi, V., Pung, L., Ratas, A., Romet, I., and Seeman, V. (2013). Hole traps and thermoluminescence in Li2B4O7:Be. *Radiat. Meas.*, 56, 147–149.

[71] Özdemir, A., Z. Yeğingil, N. Nur, K. Kurt, T. Tüken, T. Depçi, G. Tansuğ, V. Altunal, V. Güçkan, G. Sığırcık, Yu, Yan; Karataşlı, Muhammet, and Dolek, Y., 2016. "Thermoluminescence study of Mn doped lithium tetraborate powder and pellet samples synthesised by solution combustion synthesis". *Department of Radiation Oncology Faculty Papers.* Paper 78. https://jdc.jefferson.edu/radoncfp/78

[72] Brant, A. T., Kananan, B. E., Murari, M. K., McClory, J. W., Petrosky, J. C., Adamiv, V. T., Burak, Y. V., Dowben, P. A., and Halliburton, L. E. (2011). Electron and hole traps in Ag-doped lithium tetraborate ($Li_2B_4O_7$) crystals. *J. Appl. Phys*, 110, 093719.

[73] Danilkin, M. I., Koksharov, Yu. A., Romet, I., Seeman, V. O., Vereschagina, N. Yu., Zubova, A. I., and Selyukova, A. S. (2019). Manganese agglomeration and radiation damage in doped $Li_2B_4O_7$. *Radiat. Meas.*, 126, 106134–106137.

[74] Horowitz, Y. S. (1981). The theoretical and microdosimetric basis of thermoluminescence and applications to dosimetry. *Phys. Med. Biol.*, 26, 765–824.

[75] Tiwari, B., Rawat, N. S., Desai, D. G., Singh, S. G., Tyagi, M., Ratna, P., Gadkari, S. C., and Kulkarni, M. S. (2010). Thermoluminescence studies on Cu-doped $Li_2B_4O_7$ single crystals. *J. Lumin.*, 130, 2076–2083.

[76] Drozdowski, W., Brylew, K., Kaczmarek, S. M., Piwowarska, D., Nakai, Y., Tsuboi, T., and Huang, W. (2014). Effect of doping with cobalt on radioluminescence and low temperature thermoluminescence of $Li_2B_4O_7$ crystals. *Radiat. Meas.*, 63 (2014), 26–31.

[77] Holovey, V. M., Sidey, V. I., Lyamayev, V. I., and Puga, P. P. (2007). Influence of reducing annealing on the luminescent properties of $Li_2B_4O_7$:Cu single crystals. *J. Lumin.*, 126, 408–412.

[78] Kar, S., Bairagi, S., Debnath, C., Verma, S., and Bartwal, K. S. (2012). Thermoluminescence studies on γ-irradiated $Mn:Li_2B_4O_7$ single crystals. *Appl. Phys. Lett.*, 101, 071904–071904-4.

[79] Laxamanan, A. R., Chandra, B., and Bhatt, R. C. (1982). Further studies on the radiation dosimetry characteristics of thermoluminescent $Li_2B_4O_7$:Cu phosphor. *Radiat. Meas.*, 2, 231–239.

[80] Patra, G. D., Singh, S. G., Tiwari, B., Singh, A. K., Desai, D. G., Tyagi, M., Sen, S., and Gadkari, S. C. (2016). Optically stimulated luminescence in Ag doped $Li_2B_4O_7$ single crystal and its sensitivity to neutron detection and dosimetry in OSL mode. *Radiat. Meas.*, 88, 14–19.

[81] Ratasa, A., Danilkina, M., Kerikmäea, M., Lusta, A., Mändarb, H., Seemanb, V., and Slavin, G. (2012). *Proceedings of the Estonian Academy of Sciences*, 61, 279–295.

[82] McKeever S. W. S., Moscovitch M., and Townsend P. D. (1995). *Thermoluminescence Dosimetry Materials: Properties and Uses*. Nuclear Technology Publishing, Ashford, 1995.

[83] Prokić, M. (1980). Development of highly sensitive $CaSO_4$: Dy/Tm and MgB_4O_7: Dy/Tm sintered thermoluminescent dosimeters. *Nucl. Instrum. Methods B.*, 175, 83–86.

[84] McKeever S. W. S., Moscovitch M., and Townsend P. D. (1995). *Thermoluminescence Dosimetry Materials: Properties and Uses*. Nuclear Technology Publishing, Ashford.

[85] Yukihara, E. G., Coleman, A. C., and Doull, B. A. (2014). Passive temperature sensing using thermoluminescence: Laboratory tests using Li{sub 2}B{sub 4}O{sub 7}:Cu,Ag, MgB{sub 4}O {sub 7}:Dy,Li and CaSO{sub 4}:Ce,Tb. *J. Lumin.*, 146, 515–526.

[86] Gustafson, T. D., Milliken, E. D., Jacobsohnc, L. G., and Yukiharaa, E. G. (2019). Progress and challenges towards the development of a new optically stimulated luminescence (OSL) material based on MgB_4O_7:Ce,Li. *J. Lumin.*, 212, 242–249.

[87] Yukihara, E. G., Doull, B. A., Gustafson, T., Oliveira, L. C., Kurt, K., and Milliken, E. D. (2017). Optically stimulated luminescence of MgB_4O_7:Ce,Li for gamma and neutron dosimetry. *J. Lumin.*, 183, 525–532.

[88] Takenago, M., Yamanoto, O., and Yamashita, T. (1977). In Proceedings of the 5th International Conference on Lum Dosimetry (P. 148). Sao Paulo, Brazil.

[89] Prokic, M. (1980). Development of highly sensitive $CaSO_4$:Dy and MgB_4O_7:Dy/Tm sintered thermoluminescent dosimeters. *Nucl. Instrum. Methods B*, 175, 83–86.

[90] Prokic, M. (1986). Magnesium borate in TL dosimetry. *Radiat. Prot. Dosim.*, 17, 393–396.

[91] Lochab, S., Pandey, A., and Sahare, P. (2007). Nanocrystalline MgB_4O_7:Dy for high dose measurement of gamma radiation. *Phys. Status Solidi A-Appl. Mater.*, 204, 2416–2425.

[92] Souza, L. F., Vidal, R. M., and Souza, S. O. (2014). Thermoluminescent dosimetric comparison for two different MgB_4O_7:Dy production routes. *Radiat. Phys. Chem.*, 104, 100–103.

[93] Paluch-Ferszt, M., Kozłowska, B., de Souza, S. O., de Souza, L. F., and Nukleonika, D. N. S. (2016). Analysis of dosimetric peaks of MgB4O7:Dy (40% Tefl on) versus LiF:Mg,Ti TL detectors. *Nukleonika*, 61(1), 49–52.

[94] T. L. D. Poland. (2015). *LiF:Mg,Ti thermoluminescent phosphor & pellets*. Retrieved from http://www.tld.com.pl/tld/mts.html. 2001–2005.

[95] Furetta. (2003). *Handbook of Thermoluminescence*. World Scientific Publishing, Italy.

[96] Palan, C. B., and Omanwar, S. K. (2017). Synthesis and preliminary TL/OSL properties of MgB_4O_7:Tb^{3+} phosphor for radiation dosimetry. *IJRITCC*, 5, 53–54.

[97] Souza, L. F., Silva, A. M. B., Antonio, P. L., Caldas, L. V. E., Souza, S. O., d'Errico, F., and Souza, D. N. (2017). Dosimetric properties of MgB_4O_7: Dy, Li and MgB_4O_7: Ce, Li for optically stimulated luminescence applications. *Radiat. Meas.*, 106, 2017, 196–199.

[98] Gustafsona, T. D., Millikenb, E. D., Jacobsohn, L. G., and Yukihara, E. G. (2019). Progress and challenges towards the development of a new optically stimulated luminescence (OSL) material based on MgB_4O_7:Ce,Li. *J. Lumin.*, 212, 242–249.

[99] Kitagawa, Y., Yukihara, E. G., and Tanabe, S. (2021). Development of Ce^{3+} and Li^+ co-doped magnesium borate glass ceramics for optically stimulated luminescence dosimetry. *J. Lumin.*, 232, 117847–117855.

[100] Atuchin, V. V., Kesler, V. G., Zaitsev, A. I., Molokeev, M. S., Aleksandrovsky, A. S., Kuzubov, A. A., and Ignatova, N. Y. (2013). Electronic structure of α-SrB_4O_7: experiment and theory. *J. Phys. Condens. Matter.*, 25, 085503.

[101] Wang L., Wang Y., Wang D., and Zhang J. (2008). Electronic structure calculations of SrB_4O_7 and SrB_4O_7:Eu crystals. *Solid State Commun.*, 148 331–335.

[102] Meijerink, A., Nuyten, J., and Blasse, G. (1989). Luminescence and energy migration in (Sr,Eu) B_4O_7, a system with a $4f^7$-$4f^65d$ crossover in the excited state. *J. Lumin.*, 44, 19–31.

[103] Stefani, R., Maia, A. D., Teotonio, E. E. S., Monteiro, M. A. F., Felinto, M. C. F. C., and Brito, H. F. (2006). Photoluminescent behaviour of SrB_4O_7:RE^{2+} (RE=Sm and Eu) prepared by Pechini, combustion and ceramic methods. *J. Solid State Chem.*, 179, 1086–1092.

[104] Pei, Z., Su, Q., and Zhang, J. (1993). The valence change from RE^{3+} to RE^{2+} (RE=Eu, Sm, Yb) in SrB_4O_7: RE prepared in air and the spectral properties of RE^2. *J. Alloys Compd.*, 198, 51–53.

[105] Jiao, Z., Li, S., Yan, Q., Wang, X., and Shen, D. (2011). *J. Phys. Chem. Solids*, 72, 252–255.

[106] Sakirzanovas, S., Katelnikovas, A., Dutczak, D., Kareiva, A., and Jüstel, T. (2012). Concentration influence on temperature-dependent luminescence properties of samarium substituted strontium tetraborate. *J. Lumin.*, 132, 141–146.

[107] Iwamoto, C., and Fujihara, S. (2009). Fabrication and optical properties of NUV-emitting SiO_2–SrB_4O_7:Eu^{2+} glass–ceramic thin films. *Opt. Mater.*, 31, 1614–1619.

[108] Stefani, R., Maia, A. S., Kodaira, C. A., Teotonio, E. E.S., Felinto, M. C. F. C., and Brito, H. F. (2007). Highly enhanced luminescence of SrB_4O_7:Eu^{2+} phosphor prepared by the combustion method using glycine as fuel. *Opt. Mater.*, 29, 1852–1855.

[109] Rodnyĭ, P. A., Berezovskaya, I. V., Voloshinovskiĭ, A. S., Stryganyuk, G. B., and Potapov, A. S. (2003). Translated from Optika i Spektroskopiya. *Opt. Spectrosc.*, 94, 603–608.

[110] Meijerink, A., Van Hjxk, M. M. E., and Blasse, G. (1993). Luminescence of Ag^+ in crystalline and glassy SrB_4O_7. *J. Phys. Chem. Solids*, 54, 90–906.

[111] Blasé, G., Meijerink, A., Noms, M., and Zuidema, J. (1994). Unusual bismuth luminescence in strontium tetraborate (SrB4O7: Bi). *J. Phys. Chem. Solid*, 55, 171–174.

[112] Sun, Jiayue, Jicheng, Z., Xiaotang, L., and Haiyan, D. U. (2012). Luminescence properties of SrB_4O_7:Sm^{2+} for light conversion agent. *J. Rare Earths*, 30, 1084-1087.

[113] Yavetskiy, R. P., Dolzhenkova, E. F., Tolmachev, A. V., Parkhomenko, S. V., Baumer, V. N., and Prosvirnin, A. L. (2007). Radiation defects in SrB_4O_7:Eu^{2+} crystals. *J. Alloys Compd.*, 441 202–205.

[114] Schipper, W. J., Meijerink, A., and Blasse, G. (1994). The luminescence of Tm^{2+} in strontium tetraborate. *J. Lumin.*, 62, 55–59.

[115] Gou, J., Wang, Y., and Li, F. (2008). The luminescence properties of Dy^{3+}-activated SrB_4O_7 under VUV excitation. *J. Lumin.*, 128, 728–731.

[116] Lia, J., Hao, J. Q., Li, C. Y., Zhang, C. X., Tang, Q., Zhang, Y. L., Su, Q., and Wang, S. B. (2005). *Radiat. Meas.*, 39, 229–233.

[117] Bajaj, N. S., and Omanwar, S. K. 2013. Thermo Luminescence Study of Srb4 O7: Cu Phosphor Prepared by Combustion Synthesis International Journal of Modern Physics: *Conference Series* 22:404–407.

[118] Mishra, G. C., Upadhyay, A. K., Kher, R. S., and Dhoble, S. J. (2011). Thermoluminescence and mechanoluminescence of gamma-ray-irradiated SrB_4O_7:Dy phosphors. *Micro Nano Lett.*, 6, 978–981.

[119] Kadam, R. M., Rajeswari, B., Mohapatra, M., Porwal, N. K., Hon, N. S., Seshagiri, T. K., and Natarajan, V. (2015). Radiation induced centres in irradiated SrB_4O_7 doped europium and their role in thermally stimulated reactions: Thermally stimulated luminescence, fluorescence and electron paramagnetic resonance studies. *J. Lumin.*, 158, 475–483.

[120] Santiago, M., Lavat, A., Caselli, E., Lester, M., Perisinotti, L. J., Perisinotti, L. J., de Figuereido, A. K., Spano, F., and Ortego, F. (1998). Thermoluminescence of strontium tetraborate. *Phys. Status Solids A*, 167, 233–236.

[121] Schulman, J. H., Kirk, R. D., and West, E. J. (1967). Use of lithium borate for thermoluminescence dosimetry. In: Proc. Int. Conf. on Luminescence Dosimetry, US AEC Symposium series CONF-650637, Stanford University, USA: 113–117.

[122] Takenaga, M., Yamamoto, O., and Yamashita, T. (1980). Preparation and characteristics of Li2B4O7:Cu phosphor. *Nucl. Instrum. Methods*, 175, 77–78.

[123] Santiago, M., Graseli, C., Caseli, E., Lester, M., Lavat, A., and Spano, F. (2001). Thermoluminescence of SrB4O7:Dy. *Phys. Status Solidi (A)*, 185, 285–289.

[124] Sabharwal, S. C., and Sangeeta (1998). Effect of sodium doping on thermoluminescence and optical properties of barium borate (BaB_2O_4) single crystals. *J. Crystal Growth*, 187, 253–258.

[125] Bajaj, N. S., and Omanwar, S. K. (2012). Combustion synthesis and luminescence characteristic of rare earth activated $LiCaBO_3$. *J. Rare Earths*, 30, 1005–1008.

[126] Yang, F., Liang, Y., Liu, M., Li, X., Zhang, M., and Wang, N. (2013). Photoluminescence properties

of novel red-emitting NaSrBO$_3$:Eu^{3+} phosphor for near-UV light-emitting diodes. *Opt. Laser Technol.*, 46, 14–19.

[127] Kumar, V. A. K., Bedyal, S. S., Pitale, O. M., Ntwaeaborwa, and Swart, H. C. (2013). Synthesis, spectral and surface investigation of NaSrBO$_3$: Sm^{3+} phosphor for full colour down conversion in LEDs. *J. Alloys Compd.*, 554, 214–220.

[128] Bajaj, N. S., and Omanwar, S. K. (2014). Advances in synthesis and characterization of LiMgBO$_3$:Dy^{3+}. *Int. J. Light Electron Opt.*, 125, 4077–4088.

[129] Bajaj, N. S., and Omanwar, S. K. (2014). Combustion synthesis and luminescence characteristics of NaSr$_4$(BO$_3$)$_3$:Tb^{3+}. *J. Lumin.*, 148, 169–173.

[130] Fukuda, Y., Tomita, A., and Takeuchi, N. (1987). Thermoluminescence and thermally stimulated exoelectron emission of sintered CaB$_4$O$_7$ doped with Pb, Eu, or Dy. *Phys. Status Solidi A*, 99, K135–K138.

[131] Fukuda, Y., Tomita, A., and Takeuchi, N. (1990). Electron affinity of CaB$_4$O$_7$:Pb ceramic obtained by a simultaneous measurement of TL and TSEE. *J. Mater. Sci. Lett.*, 867, 867–868.

[132] Fukuda Y., Mizugchi K., and Takeuchi N. (1986). Thermoluminescence in Sintered CaB$_4$O$_7$:Dy and CaB$_4$O$_7$:Eu. *Radiat. Prot. Dosim.*, 17 397–401.

[133] Tekin, D., Ege, A., Karali, T., Townsend, P. D., and Proki, M. (2010). Thermoluminescence studies of thermally treated CaB$_4$O$_7$:Dy. *Radiat. Meas.*, 45, 764–767.

[134] Haghiri, M. E., Saion, E., Soltani, N., Abdullah, W. S. W., Navasery, M., Saraee, K. R. E., and Deyhimi, N. (2014). Thermoluminescent dosimetry properties of double doped calcium tetraborate (CaB$_4$O$_7$:Cu–Mn) nanophosphor exposed to gamma radiation. *J. Alloys Compd.*, 582, 392–397.

[135] Akın, A., Ekdal, E., Arslanlar, Y. T., Ayvacıklı, M., Karal, T., and Can, N. (2015). Thermally stimulated luminescence glow curve structure of β-irradiated CaB$_4$O$_7$:Dy. *Luminescence*, 30, 830–834.

[136] Palan, C. B., Koparkar, K. A., Bajaj, N. S., Soni, A., and Omanwar, S. K. (2016). Synthesis and thermoluminescence/optically stimulated luminescence properties of CaB$_4$O$_7$:Ce phosphor. *J. Mater. Sci.: Mater. Electron*, 27, 5600–5606.

[137] Aydın, T., Demirtas, H., and Aydın, S. (2013). TL/OSL studies of Li2B4O7:Cu dosimetric phosphors. *Radiat. Meas.*, 58, 24–32.

[138] Rawat, N. S., Kulkarni, M. S., Tyagi, M., Ratna, P., Mishra, D. R., Singh, S. G., Tiwari, B., Soni, A., Gadkari, S. C., and Gupta, S. K. (2012). *J. Lumin.*, 132, (8), 1969–1975.

[139] Palan, C. B., Bajaj, N. S., and Omanwar, S. K. (2016). Luminescence properties of Eu2+ doped SrB4O7 phosphor for radiation dosimetry. *Mater Res. Bull.*, 76, 216-221.

[140] Annalakshmi, O., Jose, M. T., Madhusoodanan, U., Venkatraman, B., and Amarendra, G. (2013). Kinetic parameters of lithium tetraborate based TL materials. *J. Lumin.*, 141, 60–66.

[141] Palan, C. B., Koparkar, K. A., Bajaj, N. S., Soni, A., and Omanwar, S. K. (2016). Synthesis and thermoluminescence/ optically stimulated luminescence properties of CaB$_4$O$_7$:Ce phosphor. *J. Mater. Sci.: Mater. Electron*, 27, 5600–5606.

[142] Ravikumara, N., ArunKumara, R., Panigrahic, B. S., Madhusoodananc, U., Palan, C. B., and Omanwar, S. K. (2018). Spectral and thermoluminescence characteristics of high gamma dose irradiated Dy:LiKB$_4$O$_7$ single crystals. *Nucl. Instrum. Method B*, 436, 203–210.

[143] Ravikumar, N., Kumar, R. A., Madhusoodanan, U., Panigrahi, B. S., Palan, C. B., and Omanwar, S. K. (2018). Luminescence properties of europium doped lithium potassium tetraborate (Eu:LiKB$_4$O$_7$) single crystal for dosimetry applications. *J. Mater. Sci.: Mater. Electron*, 29, 18511–18518.

[144] Palan, C. B., Chauhan, A. O., Sawala, N. S., Bajaj, N. S., and Omanwar, S. K. (2016). Synthesis and preliminary TL/OSL properties of Li$_2$B$_4$O$_7$: Cu-Ag phosphor for radiation dosimetry. *Optik*, 127, 6419–6423.

[145] Palan, C. B., Chauhan, A. O., Sawala, N. S., Bajaj, N. S., and Omanwar, S. K. (2015). Thermoluminescence and optically stimulated luminescence properties of MgB$_4$O$_7$: Ag Phosphor. *Int. J. Lumin. Appl.*, 5, 408–410.

[146] Yukihara, E. G., Doull, B. A., Gustafson, T., Oliveira, L. C., Kurt, K., and Milliken, E. D. (2017). Optically stimulated luminescence of MgB$_4$O$_7$:Ce,Li for gamma and neutron dosimetry. *J. Lumin*, 183, 525–532.

[147] Souza, L. F., Silva, A. M. B., Antonio, P. L., Caldas, L. V. E., Souza, S. O., d'Erricoc, F., and Souza, D. N. (2017). Dosimetric properties of MgB$_4$O$_7$:Dy,Li and MgB$_4$O$_7$:Ce,Li for optically stimulated luminescence applications. *Radiat. Meas.*, 106, 196–199.

[148] Palan, C. B., and Omanwar, S. K. (2017). Synthesis and preliminary TL/OSL properties of MgB$_4$O$_7$:Tb^{3+} phosphorfor radiation dosimetry. *IJRITCC*, 5, 53–54.

[149] González, P. R., Ávila, O., Mendoza-Anaya, D., and Escobar-Alarcón, L. (2021). Effect of sintering temperature on sensitivity of MgB$_4$O$_7$:Tm,Ag obtained by the solution combustion method. *Appl. Radiat. Isot.*, 167, 109459.

[150] Aydın, Talat, Demirtas, H., and Aydın, S. (2013). TL/OSL studies of Li$_2$B$_4$O$_7$:Cu dosimetric phosphors. *Radiat. Meas.*, 58, 24–32.

[151] Li, J., Zhang, C. X., Tang, Q., Zhang, Y. L., Hao, J. Q., Su, Q., and Wang, S. B. (2007). Synthesis, photoluminescence, thermoluminescence and dosimetry properties of novel phosphor Zn(BO$_2$)$_2$:Tb. *J. Phys. Chem.*, 68, 143–147.

[152] Liyan, L., Jingquan, H., Chengyu, Li, Qiang, T., Chunxiang, Z., Su, Q., and Shubin, W. (2006). Thermoluminescence characteristics of SrB$_6$O$_{10}$:Tb. *J. Rare Earths*, 24, 276.

[153] Blasse, G. (1983). *Rev. Inorg. Chem.*, 15, 319–381.

[154] Ternane, R., Adad, M. T. C., Panczer, G., Goutaudier, C., Dujardin, C., Boulon, G., Ariguib, N. K., and Ayedi, M. T. (2002). Structural and luminescent properties of new Ce^{3+} doped calcium borophosphate with apatite structure. *Solid State Sci.*, 4, 53–59.

[155] Blasse, G., and Grabmaier, B. C. (1994). A general introduction to luminescent materials luminescent materials. *Lumin. Mater.*, 1–9. Springer, Berlin, Heidelberg. 10.1007/978-3-642-79017-1_1

[156] Caldino, U., Speghini, A., and Bettinelli, M. (2006). Optical spectroscopy of zinc metaphosphate glasses activated by Ce^{3+} and Tb^{3+} ions. *J. Phys. Condens. Matter*, 18, 3499.

[157] Leskela, M., and Niinitso, L. (1992). Applications of rare earths in full-colour EL displays. *Mater. Chem. Phys.*, 31, 7–11.

[158] Raue, R., Vink, A. T., and Welkar, T. (1998). *Phil. Tech. Rev.*, 44, 335.

[159] Smets, B. M. (1987). Phosphors based on rare-earths, a new era in fluorescent lighting. *J. Mater. Chem. Phys.*, 16, 283–299.

[160] Dhoble, S. J., Moharil, S. V., Dhopte, S. M., Muthal, P. L., and Kondawar, V. K. (1993). Preparation and characterization of the K$_3$Na(SO$_4$)$_2$: Eu phosphor. *Phys. Status Solidi (A)*, 135, 289–297.

[161] Gahane, D. H., Kokode, N. S., Muthal, P. L., Dhopte, S. M., and Moharil, S. V. (2009). Luminescence of Eu^{2+} in some iodides. *Opt. Mater.*, 32, 18–21.

[162] Hoffman, M. V. (1971). Alkaline earth aluminum fluoride compounds with Eu^{+2} activation. *J. Electrochem. Soc.*, 118, 933.

[163] United Nations Sources and Effects of Ionizing Radiation. Volume I: Report to the General Assembly, Scientific Annexes A and B, UNSCEAR Report (2008).

[164] Czarwinski, R., and Crick, M. J. (2011). Occupational exposures worldwide and revision of international standards for protection. *Radiat. Prot. Dosim.*, 144, 2–11.

[165] Bhatt, B. C., and Kulkarni, M. S. (2013). Worldwide status of personnel monitoring using thermoluminescent (TL), optically stimulated luminescent (OSL) and radiophotoluminescent (RPL) dosimeters. *Int. J. Lumin. Appl.*, 3, 6–10.

[166] Bhatt, B. C., and Kulkarni, M. S. (2014). Thermoluminescent phosphors for radiation dosimetry. *Defect Diffus. Forum*, 347, 179–227.

[167] Patra, G. D., Singh, S. G., Tiwari, B., Singh, A. K., Desai, D. G., Tyagi, M., Sen, S., and Gadkari, S. C. (2016). Optically stimulated luminescence in Ag doped Li$_2$B$_4$O$_7$ single crystal and its sensitivity to neutron detection and dosimetry in OSL mode. *Radiat. Meas.*, 88, 14–19.

[168] Jensen, L. B., and Thompson, I. M. G. (1995). An international intercomparison of passive dosemeters, electronic dosemeters and dose rate meters used for environmental radiation measurements. *Radiat. Prot. Dosim.*, 60, 201.

[169] Bajaj, N. S., and Omanwar, S. K. (2013). Combustion synthesis and characterization of phosphor KSr$_4$ (BO$_3$)$_3$:Dy^{3+}. *Opt. Mater.*, 35, 1222–1225.

[170] Akselrod, M. S., Jensen, L. B., and McKeever, S. W. S. (2007). Optically stimulated luminescence and its use in medical dosimetry. *Radiat. Meas.*, 41, S78–S99.

[171] Pradhan, A. S., Lee, J. I., and Kim, J. L. (2008). Recent developments of optically stimulated luminescence materials and techniques for radiation dosimetry and clinical applications. *J. Med. Phys.*, 33, 85–99.

[172] Yukihara E. G., and McKeever S. W. S. (Eds.) (2011). *Optically Stimulated Luminescence: Fundamentals and Applications*. Wiley, Oklahoma, USA.

[173] Huston, A. L., Justus, B. L., Falkenstein, P. L., Miller, R. W., Ning, H., and Altemus, R. (2001). Remote optical fibre dosimetry. *Nucl. Instrum. Methods B*, 184, 55–67.

[174] Prokic, M. (2002). Nuclear technology publishing dosimetric characteristics of Li$_2$B$_4$O$_7$:Cu,Ag,P solid TL detectors. *Radiat. Prot. Dosim.*, 100, 265–268.

[175] Jensen, L. B., Banerjee, D., Jungner, H., and Murray, A. S. (1999). Retrospective assessment of environmental dose rates using optically stimulated luminescence from Al$_2$O$_3$:C and quartz. *Radiat. Prot. Dosim.*, 84, 537–542.

8 Borate
NIR Quantum Cutting Phosphors

N. S. Sawala

CONTENTS

DOI: 10.1201/9781003207757-8

8.1 SOLAR RADIATION

Solar energy can be used in one of form (heat and electricity) as the source of nearly all energy on the earth. Nowadays, wind turbines convert wind power into electricity. Even hydroelectricity is generated from the sun. PV is a simple and smart method of harnessing the energy of sun. PV devices (solar cells) are unique as they directly convert the incident solar radiation into electricity without noise or pollution, making them sustainable, reliable and long-lasting source of energy.

A photon is characterized by either a wavelength (λ) or energy (E). According to Planck's equation:

$$E = \frac{hc}{\lambda}$$

where h is Planck's constant and c is the speed of light.

When dealing with "particles" such as photons or electrons, a commonly used unit of energy is the electron volt (eV) rather than the joule (J). An electron volt is the energy required to raise an electron through 1 volt; thus, a photon with an energy of 1 eV = 1.602×10^{-19} J. By expressing the equation for photon energy in terms of eV and nm, one arrives at a commonly used expression that relates the E and λ of a photon in the following equation:

$$E\,(eV) = \frac{1240}{\lambda\,(nm)}$$

8.1.1 SOLAR RADIATION INCIDENT ON THE EARTH'S ATMOSPHERE

While the solar radiations enter the atmosphere of Earth are relatively constant while the radiations reaching the Earth's surface varies widely because of:

- atmospheric effects, such as absorption and scattering
- variations in the atmosphere, such as water vapour, clouds and pollution
- latitude of the location; and
- season of the year and the time of day.

The above effects have numerous impacts on the solar radiation received on the surface of Earth. These changes affect the overall power received, the spectral distribution of photons of the radiation and the angle from which radiations are incident on the surface. Additionally, an important change is that the unpredictability of the solar radiations at a specific location increases/decreases noticeably. The unpredictability is due to both effects such as clouds and seasonal changes, as well as other effects such as the length of the day at a particular latitude. The amount of energy reaching the surface of the Earth every hour is greater than the amount of energy used by the Earth's population over an entire year [1].

8.1.2 ATMOSPHERIC EFFECTS

Atmospheric effects have several impacts on the solar radiation at the Earth's surface. The major effects for photovoltaic applications are as follows:

- reduction in the solar power due to absorption, scattering and reflections in the atmosphere
- the introduction of a diffuse or indirect component into the solar radiation and
- changes in the atmosphere (such as water vapour, clouds and pollution), which affects the incident solar power, spectrum and directionality.

These effects are summarized in the following Figure 8.1.

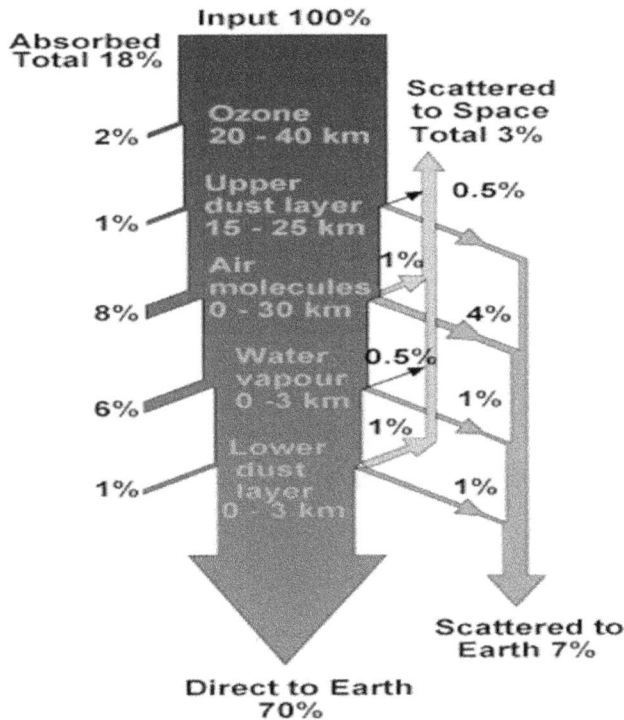

FIGURE 8.1 Atmospheric effects on solar radiation [2].

8.1.3 ABSORPTION IN THE ATMOSPHERE

As solar radiation passes through the atmosphere, gasses, dust and aerosols absorb the incident light. Specific gasses such as ozone (O_3), carbon dioxide (CO_2) and water vapour (H_2O), have very high absorption of photons that have energies close to the bond energies of these atmospheric gases molecules. This absorption causes deep troughs in the spectral radiation curve. For example, most of the UV light less than 300 nm is absorbed by the ozone layer [3].

8.1.4 DIRECT AND DIFFUSE RADIATION DUE TO SCATTERING

Light is absorbed when it passes through the atmosphere and at the same time it undergoes scattering process. One of the mechanisms for light scattering in the atmosphere is known as Rayleigh scattering, which is caused by molecules in the atmosphere. Rayleigh scattering is particularly effective for short wavelength light since it has λ^{-4} dependence. Other than the Rayleigh scattering, aerosols and dust particles contribute to the scattering of incident light known as Mie scattering. Scattered light is undirected and it appears to be coming from any region of the sky. This light is called diffuse light. Since diffuse light is primarily "blue" light, the light that comes from regions of the sky other than where the sun is, appears blue. In the absence of scattering in the atmosphere, the sky would appear black, and the sun would appear as a disk light source. On a clear day, about 10% of the total incident solar radiation is diffused.

8.1.5 EFFECT OF CLOUDS AND OTHER LOCAL VARIATIONS IN THE ATMOSPHERE

The final effect of the atmosphere on incident solar radiation is due to variations in the atmosphere. Depending on the type of cloud cover, the incident power is severely reduced. An example of heavy cloud cover is shown in Figure 8.2.

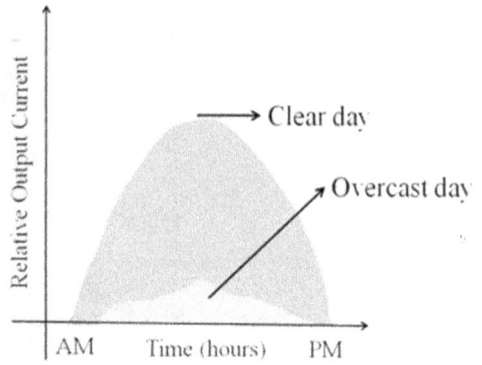

FIGURE 8.2 Relative output current from a PV array on a sunny and a cloudy [4].

8.1.6 AIR MASS (AM)

The AM is the path length which light takes through the atmosphere normalized to the shortest possible path length; that is, when the sun is directly overhead, as shown in Figure 8.3.

The air mass quantifies the drop in the power of solar radiation when it passes through the atmosphere and is absorbed by air and dust. The AM is defined as

$$AM = \frac{1}{\cos\theta}$$

where θ is the angle from the vertical (zenith angle). An easy method to determine the air mass is from the shadow of a vertical pole. AM is the length of the hypotenuse divided by the object height (h):

$$AM = \sqrt{1 + \left(\frac{s}{h}\right)^2}$$

The above calculation for AM assumes that the atmosphere is a level horizontal layer, but due to the curvature of the atmosphere, the AM is not quite equal to the atmospheric path length when the sun is close to the horizon. At sunrise, the angle of the sun from the vertical position is 90° and the

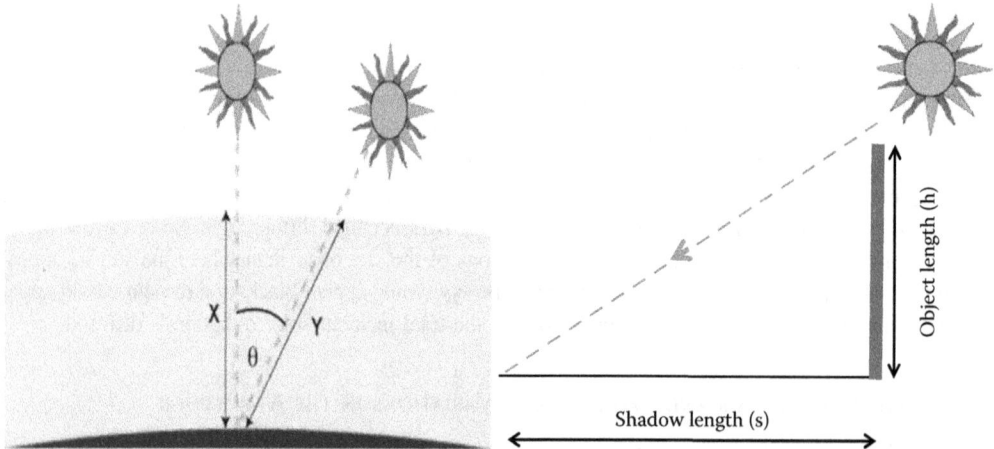

FIGURE 8.3 The AM shows the proportion the light must pass through the atmosphere before striking the Earth relative to its overhead path length, and is equal to Y/X [5].

air mass is infinite, whereas the path length clearly is not. An appropriate equation includes the curvature of the Earth is given by [5]

$$AM = \frac{1}{\cos\theta + 0.50572\cdot(96.07995 - \theta)^{-1.6364}}$$

8.1.7 STANDARDIZED SOLAR SPECTRUM AND SOLAR IRRADIATION

The efficiency of a solar cell is sensitive to changes in both the solar power and the spectral distribution of solar radiation (photons). In order to make the comparison between solar cells measured at different times and locations, a standard spectrum and power density has been defined for both radiations outside the Earth's atmosphere and on the Earth's surface. The standard spectrum at the Earth's surface is called AM1.5G (the G stands for global and includes both direct and diffuse radiation) or AM1.5D (which includes direct radiation only). The spectrum outside the Earth's atmosphere is called AM0 because at no stage does the light pass through the atmosphere, as shown in Figure 8.4.

This spectrum is typically used to predict the expected performance of cells in space. The intensity of AM1.5D radiation can be approximated by reducing the AM0 spectrum by 28% (18% due to absorption and 10% to scattering). The global spectrum is 10% higher than the direct spectrum. These calculations give approximately 970 W/m^2 for AM1.5G. However, the standard AM1.5G spectrum has been normalized to give 1,000 W/m^2 due to the convenience of the round figure and the fact that there are natural changes in incident solar radiation.

The solar spectrum changes throughout the day and with location. The standard spectra were refined in the early 2000s to increase the resolution and to co-ordinate the standards internationally. The previous solar spectrum, ASTMG159, was withdrawn from use in 2005. In most cases, the difference between the spectrums has little effect on device performance and the newer spectra are easier to use [6].

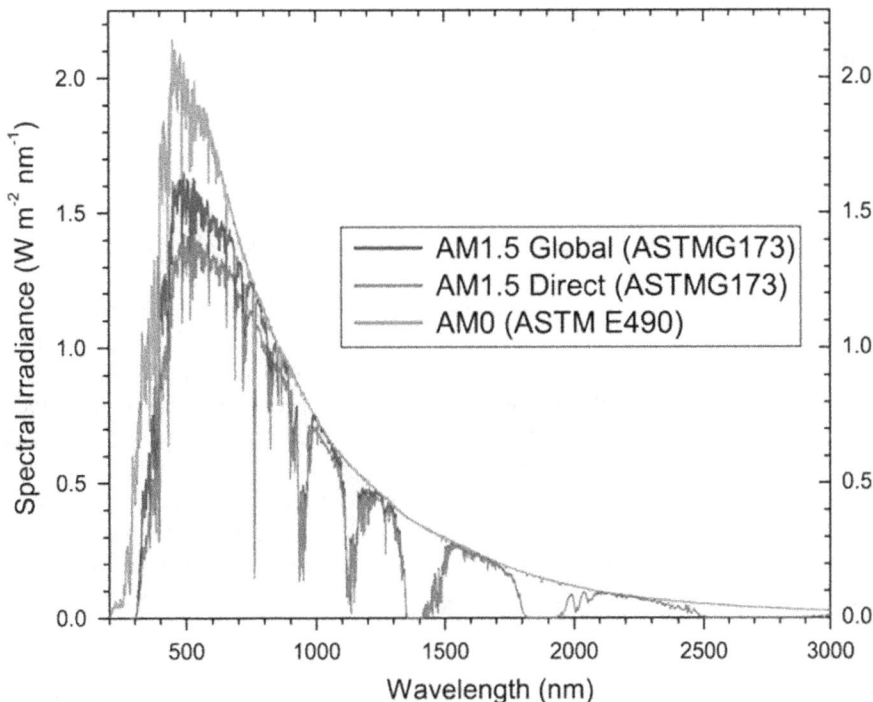

FIGURE 8.4 Standard solar spectra for space and terrestrial applications [6].

8.1.8 Absorption of Radiation: Overview

- When the energy of incident photon is equal to or greater than the energy bandgap of the material, the photon is absorbed by the material and excites an electron into the conduction band.
- Both a minority and majority carrier are generated when a photon is absorbed.
- The generation of charge carriers by photons is the basis of the PV production of energy.

The photons of sunlight incident on the surface of a semiconductor will be either reflected from the upper surface, will be absorbed in the material or, failing either of the above two processes, will be transmitted through the material. For the PV devices, reflection and transmission are the losses since photons that are not absorbed do not contribute for power generation. If the photon is absorbed it has the probability of exciting an electron from the valence band to the conduction band. If the photon has sufficient energy, the electron will be excited from the valence band into the conduction band. The photons having energy (E_p) falling onto a semiconductor material can be divided into three parts based on their energy compared to that of the semiconductor bandgap (E_g).

- $E_p < E_g$, photons interact only weakly with the semiconductor, passing through it as if it were transparent.
- $E_p = E_g$, have just enough energy to create an electron hole pair and are efficiently absorbed.
- $E_p > E_g$, photons are strongly absorbed. But for PV applications, the photon energy greater than the bandgap is wasted as electrons quickly thermalize back down to the edges of conduction band.

The absorption of photons causes the generation of both a majority and minority carriers. In many PV applications, the numbers of light generated carriers are less than the number of majority carriers already present in the solar cell due to doping. Subsequently, the number of majority carriers in an illuminated semiconductor does not alter significantly. However, the opposite is true for the number of minority carriers. The number of photo-generated minority carriers outweigh the number of minority carriers existing in the doped solar cell in the dark (because in doping the minority carrier concentration is so small), and hence the number of minority carriers in an illuminated solar cell can be approximated by the number of light-generated carriers.

8.2 CRYSTALLINE SILICON (C-SI) SOLAR CELLS AND SPECTRAL MISMATCH

When sunlight incidents on a solar cell, the photons activate the electrons in the cell and promote them into conduction band. Those electrons can then be utilized to constitute an electric current. To prepare a solar cell, the semiconductor is doped with either accepter materials (positive charge carriers, p-type) or donor materials (negative charge carriers, n-type). When two differently doped semiconductor layers are combined then p–n junction forms across the boundary of the layers. The PV conversion efficiency of solar cell is a very important factor that governs the performance of a PV device or panels. For any regular single p–n junction solar cell, the characteristic value of energy bandgap of the semiconductor materials from which the solar cell is made plays key role on its conversion efficiency. Currently, solar cells based on crystalline, polycrystalline and amorphous silicon dominates more than 90% of the world production [7]. The use of c-Si enables PV devices to achieve a maximum conversion efficiency of 25% [8]. Figure 8.5 shows normalized spectral response (SR) of a typical c-Si solar cell and solar spectrum available on earth surface i.e. AM1.5 Global. As the c-Si is an indirect bandgap semiconductor there is not a sharp cut-off at the wavelength corresponding to the bandgap ($E_g = 1.12$ eV) [9].

It is showed that the c-Si solar cells work most efficiently and resourcefully in the spectral range 950–1,100 nm and it showed very low SR to the short wavelength radiation from sunlight due to characteristic energy bandgap value of silicon. This is called spectral mismatching. This spectral mismatching along with thermalization of charge carrier due absorption of high energy (more than 1.14 eV) photons are accountable for poor efficiency and low performance of c-Si solar cells.

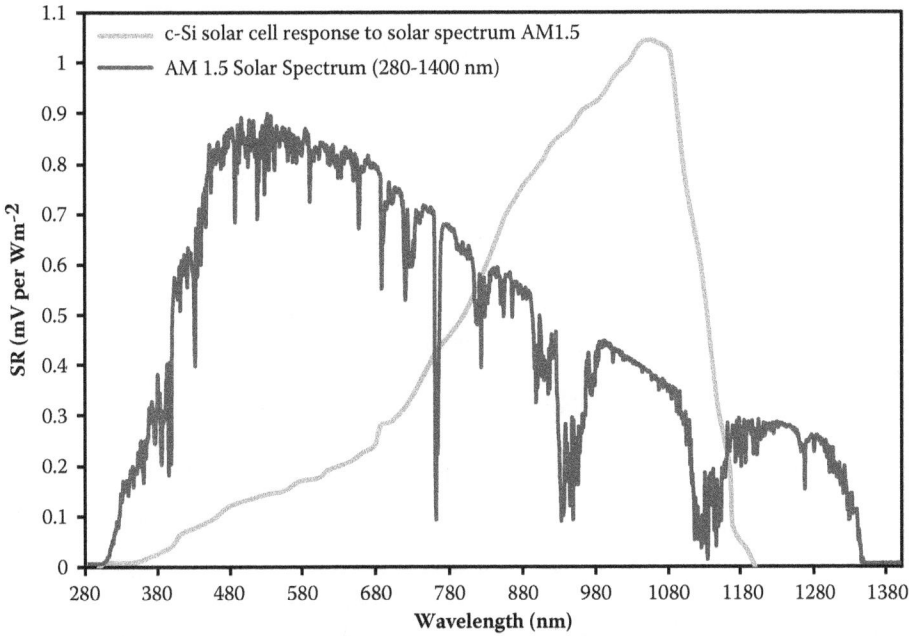

FIGURE 8.5 Normalized spectral response of a typical c-Si solar cell (courtesy of Apogee Instruments, Inc., 2008) along with solar spectrum AM1.5G [9].

8.2.1 Amorphous Silicon Solar Cells

Amorphous Si solar cells have a bandgap of around 1.75 eV and can only absorb NIR light of wavelength less than 700 nm. To enlarge the absorption, UC of sub bandgap NIR light VIS emission from $NaYF_4:Er^{3+}/Yb^{3+}$ phosphors can be combined with the amorphous Si solar cells. In 2010, Zhang et al. reported the invention of an amorphous Si solar cell device composed of $NaYF_4:Er^{3+}/Yb^{3+}$ nanocrystals [10]. It was reported that the introduction of an UC phosphor layer into the solar cell could effectively increase the short circuit current density. The same UC phosphor was also used by de Wild et al. to illustrate increased in conversion efficiency of amorphous Si solar cells [11]. The UC powders were mixed with polymethyl methacrylate and then made into an UC layer of thickness of 200 to 300 mm. A maximum current improvement of 6.2 mA was measured under 980 nm diode lasers source [12].

8.2.2 Solar Cell Parameters

8.2.2.1 Short Circuit Current (I_{SC})

The short-circuit value of current is the current through the solar cell when the voltage across the solar cell is zero (i.e. when the solar cell is short circuited) denoted by I_{SC}. The short circuit current is due to the generation and collection of light-generated charge carriers. The short-circuit current is the largest current which may be drawn from the solar cell. The short-circuit current depends on a number of factors that are discussed in the following:

- the area of the solar cell. To remove the dependence of the solar cell area, it is more common to define the short-circuit current density (J_{sc}) instead of the short-circuit current.
- I_{sc} of the solar cell is directly dependant on the light intensity (number of photons);
- the solar spectrum of the incident sunlight. For most solar cell measurement, the spectrum is standardized to the AM1.5 spectrum;
- the optical properties (absorption and reflection) of the solar cell; and

- the collection probability of the solar cell, which depends chiefly on the surface passivation and the minority carrier lifetime in the base.

When comparing solar cells of the same material type, the most critical material parameter is the diffusion length and surface passivation. In a cell with perfectly passivated surface and uniform generation, the equation for the short-circuit current can be approximated as:

$$J_{SC} = qG\left(L_n + L_p\right)$$

where G is the generation rate, and L_n and L_p are diffusion lengths of electron and hole, respectively. Although this equation makes several assumptions which are not hold good for the conditions encountered in most of the practical solar cells. The above equation indicates that the short-circuit current depends strongly on the generation rate and the diffusion length. Silicon solar cells under an AM1.5 spectrum have a maximum possible current of 46 mA/cm^2, whereas laboratory devices have measured I_{SC} of over 42 mA/cm^2 and commercial solar cell have I_{SC} between about 28 mA/cm^2 and 35 mA/cm^2 [13].

8.2.2.2   Open-Circuit Voltage (V_{OC})

The open-circuit voltage, V_{OC}, is the maximum voltage accessible from a solar cell and this occurs at zero current. The V_{OC} corresponds to the amount of forward bias on the solar cell due to the bias of the solar cell junction with the light-generated current. The short-circuit current and open-circuit voltage are shown on the I-V curve shown in Figure 8.6.

An equation for V_{OC} is derived by setting the net current equal to zero in the solar cell equation and gives

$$V_{OC} = \frac{nkT}{q}\ln\left(\frac{I_L}{I_0} + 1\right)$$

It shows that V_{oc} depends on the saturation current (I_0) of the solar cell and the light-generated current (I_L). While I_{sc} typically has a small discrepancy and the key effect is the saturation current. The I_0 depends on the recombination in the solar cell. Open-circuit voltage is then a measure of the amount of recombination in the device. The V_{OC} can also be determined from the carrier concentration [14].

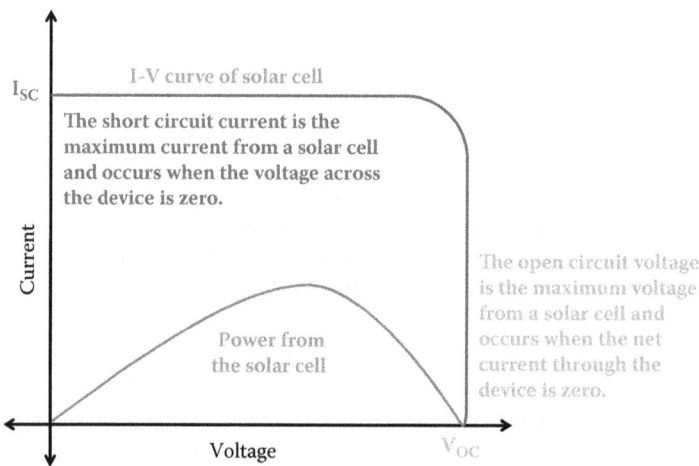

FIGURE 8.6   I-V characteristics of solar cell with V_{OC} and I_{SC} [14].

$$V_{OC} = \frac{kT}{q} \ln\left(\frac{\Delta n \,(N_A + \Delta n)}{n_i^2}\right)$$

where kT/q is the thermal voltage, N_A is the doping concentration, Δn is the excess carrier concentration and n_i is the intrinsic carrier concentration. The determination of V_{OC} from the carrier concentration is also termed 'implied V_{OC}'.

8.2.2.3   Fill Factor (FF)

The I_{SC} and V_{OC} are the maximum current and voltage, respectively, from a solar cell. However, at both of these operating points, the power from the solar cell is zero. The fill factor (FF) is a parameter that is in conjunction with V_{OC} and I_{SC} determines the maximum power from a solar cell. The FF is defined as the ratio of the maximum power from the solar cell to the product of V_{oc} and I_{sc}. Graphically, the FF is a measure of the squareness of the solar cell and is also the area of the largest rectangle which will fit in the IV curve. The FF is illustrated in Figure 8.7.

A solar cell with a higher voltage has a more possible FF since the rounded portion of the I-V curve takes up less area. The maximum theoretical FF from a solar cell can be calculated by differentiating the power from a solar cell with respect to voltage and equating it with zero.

However, large variations in V_{OC} within a given material system are relatively uncommon. For example, at one sun, the difference between the maximum V_{OC} measured for a silicon laboratory cell and a typical commercial solar cell is about 120 mV but maximum FF are 0.85 and 0.83, respectively. However, the variation in maximum FF can be significant for solar cells made from different materials.

The ideality factor or n-factor is a measure of the junction quality and the type of recombination occur in a solar cell. Ideally, the n-factor has a value of 1. A high n-value not only degrades the FF but it gives low open-circuit voltages as it will also usually signal high recombination. The equations described in Figure 8.7 represent a maximum possible FF but in practice the FF will be lower due to the presence of parasitic resistive losses. Hence, the FF is most commonly determined from measurement of the IV curve and is defined as the maximum power divided by the product of $V_{OC}I_{SC}$ [15]:

$$FF = \frac{V_{mp}I_{mp}}{V_{OC}I_{SC}}$$

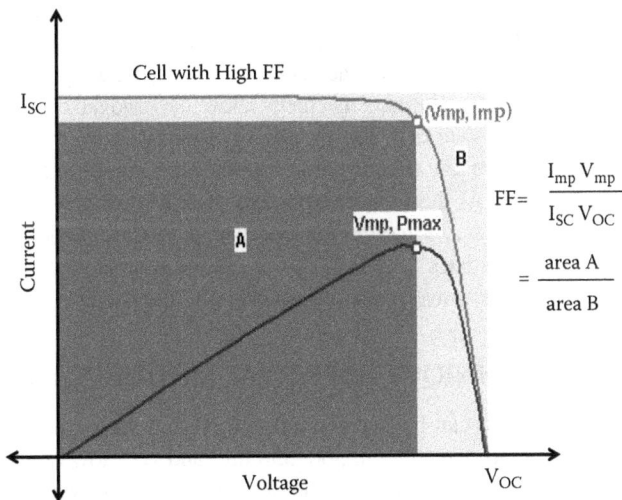

FIGURE 8.7   Plot of cell output current (red line) and power (blue line) as function of voltage [14].

8.2.2.4   Efficiency of Solar Cell

The efficiency is the most generally used term to compare the performance of one solar cell to another one. It is defined as the ratio of energy output from the solar cell to input energy (incident energy) from the sun. The efficiency of a solar cell depends on the spectrum of incident sunlight, sunlight reflecting from the solar cell itself and intensity of the incident sunlight along with the temperature of the solar cell. Therefore, it is necessary to control the conditions under which efficiency is measured so as to compare the performance of solar cells. Terrestrial solar cells are measured under AM1.5 conditions and at a temperature of 25°C, whereas solar cells planned for space use are measured under AM0 conditions. The efficiency of a solar cell is determined as the fraction or degree of incident power (sunlight) which is converted to electricity and is given by:

$$\eta = \frac{V_{OC} I_{SC} FF}{P_{in}} = \frac{P_{max}}{P_{in}}$$

$$P_{max} = V_{OC} I_{SC} FF$$

where V_{OC}, I_{SC} and FF have usual meaning as discussed above η is the efficiency. The input power for efficiency calculations is 1 kW/m^2 or 100 mW/cm^2. Thus, the input power for a 156×156 mm^2 cell is 24.3 W and for a 100×100 mm^2 cell is 10 W.

8.2.2.5   Spectral Response (SR)

The SR is conceptually similar to the quantum efficiency (QE). The QE gives the number of output electrons by the solar cell compared to the number of photons incident on the surface of device. SR is significant since it is the spectral response that is measured from a solar cell, and from this the QE is calculated. The QE can be determined from the SR by replacing the power of the light at a particular wavelength with the photon flux for that wavelength. This gives the equation:

$$SR = \frac{q\lambda}{hc} QE$$

The SR is the ratio of the current generated by the solar cell to the power incident on the solar cell. A spectral response curve is shown in Figure 8.8.

The ideal SR is limited at long wavelengths (beyond 1.2 μm) due to the lack of ability of the semiconductor to absorb photons with energies below the characteristic energy bandgap. The SR decreases at small wavelength region. At these wavelengths, each photon has a large energy (more than bandgap), and hence the ratio of photons to power is reduced. Any energy above the bandgap energy is not utilized by the solar cell and it causes heating the solar cell. The incapability to fully utilize the incident energy at high energies and the incapacity to absorb low energies of light manifests a significant power loss in solar cells consisting of a single p-n junction. At short wavelengths less than 400 nm, the glass absorbs most of the light and the cell response is very low. At intermediate wavelengths, the cell approaches the ideal case. At long wavelengths, the response falls back to zero as shown. Silicon is an indirect bandgap semiconductor so there is not a sharp cut-off at the wavelength corresponding to the bandgap ($E_g = 1.12$ eV) [15].

8.3   LUMINESCENCE TECHNIQUE – SPECTRAL MODIFICATIONS

The problem of spectral mismatch can be overcome by modifying the available spectrum AM1.5G. The photons of high energy (short wavelength, 300–550 nm) and very low energy (high wavelength, 1,400–1,800 nm) successfully transformed to photons of energy (wavelength, 950–1,100 nm). In order to improve the efficiency of single-junction solar cells, luminescence processes such as

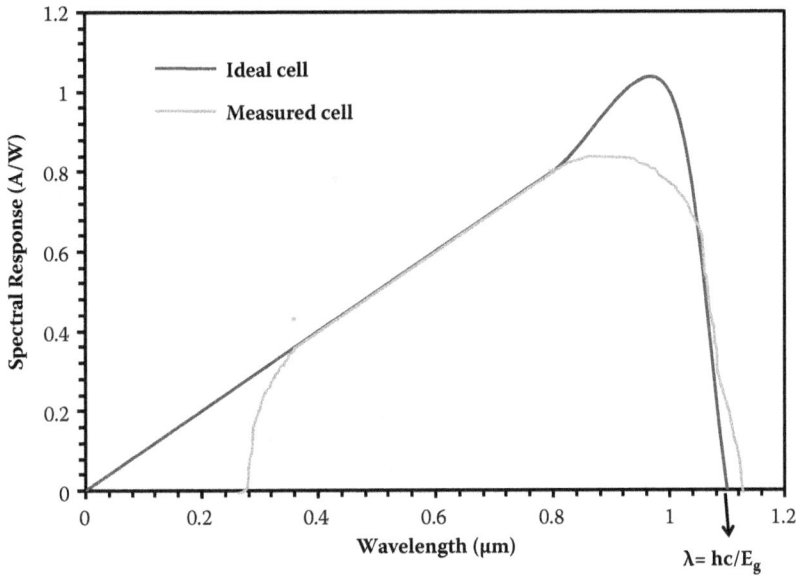

FIGURE 8.8   The spectral response of a silicon solar cell under glass [15].

quantum cutting (QC) via *down-conversion (DC)* and *down shifting (DS)* and *up-conversion (UC)* are presently employed for developing efficient and reliable PV devices.

8.3.1   UP-CONVERSION (UC)

UC is luminescence process where two or more very low energy photons can get transformed into one high-energy photons. It was first discovered by Auzel in the 1960 [16]. by conventional approaches. The use of UC materials may provide a solution to the transmission loss by converting two very low-energy photons into one bandgap photon. Trupke et al. 32 demonstrated that the theoretical efficiency of solar cell can be modified with an up-converter and can reach 63.2% for concentrated sunlight and 47.6% for non-concentrated sunlight [17].

Since the active 4f electrons are well shielded by the 5s and 5p of living in the outer electronic shell, most of the lanthanide ions, apart from of Yb^{3+}, have more than one metastable level, and this is the basic condition for UC. Hence, lanthanide ions are much more competitive for UC with multi-phonon relaxation. Among the lanthanides, UC research for solar cells is focused on single (lanthanides) Ln^{3+} such as Pr^{3+}, Ho^{3+} and Er^{3+} and their dual dopant using Yb^{3+}, as sensitizer in a host like chloride, bromide, fluoride and oxide compounds. And according to the literature [18], the most efficient UC systems are based on fluoride hosts like $NaYF_3$, LaF_3 and YF_3, where the most efficient dopants are Yb^{3+} and Er^{3+} [19].

An extensive constraint lies in the fact that UC processes are possible only in trivalent lanthanide ions with metastable and long-lived intermediate levels. The use of UC materials for developing PV cell was first reported by Gibart's group in 1996 using the Er^{3+}–Yb^{3+} pair in bifacial GaAs solar cells. Since then, many groups have pursued the development of various up-conversion-based PV devices to enhance the conversion efficiency [20].

8.3.2   DOWN-CONVERSION (DC)

DC is a luminescence process that is able to split one absorbed high-energy photon into two (or more) lower-energy photons with quantum efficiency higher than 100% [21]. The suggestion to obtain quantum yields more 100% by creating multiple photons by *'cutting'* a single photon into two lower-energy photons was first proposed by Dexter in 1957 [22]. The mechanism involved the immediate energy transfer from a donor (sensitizer) to two acceptors (activators), each accepting equally half of the energy of the excited donor (sensitizer). In 1974, the experimental evidence for quantum yields above 100% was obtained in YF_3: Pr^{3+} phosphor. However, the mechanism was not the same as proposed by Dexter because it involved two sequential emission steps from the high-energy state $^1S_0$ level of Pr^{3+} ($^1S_0$–$^1I_6$ which is followed by relaxation to the $^3P_0$ state) and emission of a second visible photon from $^3P_0$ state [23,24]. Researcher had concentrated on the combination of two ions, where the energy of the sensitizer ion could be transferred stepwise to two activators via a DC process. A well-known example is the Gd^{3+} – Eu^{3+} pair. The interesting and substantial QE in the visible spectral region has earlier been reported in $LiGdF_4$: Eu (QE = 190%) [25] and BaF_2: Gd, Eu (QE = 194%) phosphors [26].

The recent literature review by Strumpel et al. [27] showed that the examples of DC from UV to VIS in the literature are not practical for improving c-Si solar cell efficiency, as the vacuum ultraviolet (VUV) excitation wavelengths involved are not present in the AM1.5G. But VIS to NIR DC process is still of promising advantage to c-Si solar cells.

8.3.2.1 Narrowband DC Process

The potential application of DC for enhancing the efficiency of solar cells was realized soon. The first experimental demonstration of DC for solar cells involved the Tb^{3+} – Yb^{3+} couple where QC was achieved through co-operative energy transfer (CET) from Tb^{3+} to two Yb^{3+} ions and the mechanism involved was same as suggested by Dexter. This narrowband QC in which absorption by sensitizer (Tb^{3+}) is sharp and narrow. Although this is not supposed to be ideal for c-Si solar cell application which requires broad absorption in the UV-VIS range is still useful to achieve QC. Other than the Tb^{3+}–Yb^{3+} pair, the Tm^{3+}–Yb^{3+} [28,29], Pr^{3+}–Yb^{3+} [30,31] Er^{3+}–Yb^{3+} [32], Ho^{3+}–Yb^{3+} [33] and Dy^{3+}–Yb^{3+} [34] pairs are also the examples of narrowband QC phenomenon.

8.3.2.2 Broadband DC Process

Despite their expediency, the DC materials based on Tb^{3+}, Tm^{3+}, Pr^{3+}, Er^{3+}, Ho^{3+} and Dy^{3+} couple with Yb^{3+} are far away from practical applications due to low excitation efficiency. A key limiting factor for this is the low-absorption cross section ($\sim 10^{-21}$ cm^2) of the lanthanide ions because of the parity forbidden 4f–4f transitions [35]. On the other hand, the dipole allowed 4f–5d transitions have relatively higher absorption cross sections ($\sim 10^{-18}$ cm^2). Recently, much attention has been given to broadband NIR DC by using Ce^{3+}, Eu^{2+} and Bi^{3+} having strong absorption intensity in the UV–blue spectral region as a sensitizer for activator Yb^{3+}. A considerable parallel development for achieving efficient NIR emission of Yb is to make use of host sensitization through direct co operative energy transfer (CET) from the host to Yb^{3+} [36–38]. The broadband DC process in considered to more effective than the narrowband DC process as it provides better spectral modification for c-Si solar cells. The important lanthanide pairs those have been investigated are discussed below.

8.3.3 DOWN-SHIFTING (DS)

DS is a phenomenon that involves transformation of one incident high-energy photon into one lower-energy photon. This process obeys the Stokes law and change in wavelength known as the Stokes shift. The QE of materials never exceed 100%. Still, DS can be useful for improving PV efficiency by changing short-wavelength sunlight (UV-VIS) to the longer-wavelength (NIR) region where the spectral response of the c-Si solar cell is maximum. The ideal phosphors for luminescent DS should have the characteristics like the broadband absorption predominantly in the range where the spectral response of the solar cell is poor; high absorption co-efficient and high PL quantum efficiency in order that all incident light results in emission; high transmittance and narrowband emission mostly in the range where the solar cell performance is high; large Stokes shift to minimize the self-absorption energy losses due to the spectral overlap between the absorption and emission bands and long-term stability. So far, inorganic phosphors and glasses, colloidal materials like quantum dots (QDs) and organolanthanide complexes have been extensively investigated as probable candidates [39].

Up to now, two important types of DS material-based PV devices have been investigated and are 1) luminescent solar concentrator and 2) planar DS layer [40]. The first one is used to concentrate sunlight onto PV cells in order to reduce the amount of cost PV materials required. The concentrator generally composed of polymer sheets doped with luminescent species [41,42]. On the other hand, a planar DS layer is generally placed directly onto the front surface of a solar cell to enhance the performance of a device by overcoming the poor spectral response of the solar cell in the short wavelength region [43].

8.4   ROLE AND SIGNIFICANCE OF SENSITIZERS AND ACTIVATORS

The c-Si solar cell shows considerably different SR in short-wavelength (UV-VIS) and long-wavelength (NIR) region of sunlight makes ideal conditions for exploring DS processes. Among different lanthanide elements, the Nd^{3+} and Yb^{3+} ions have attracted significant attention due their characteristic NIR emitting properties. The NIR emission from these two ions is around 1,000 nm, which is just above the bandgap of c-Si where the solar cell and exhibits the greatest spectral response [44–46]. However, the PL of Nd^{3+} and Yb^{3+} are relatively weak due to 4f–4f transitions that are parity forbidden [46]. The energy levels of activators Nd^{3+} and Yb^{3+} along with the transitions are shown in Figure 8.9.

In order to overcome this problem, some researchers used other luminescent species as sensitizers (such as Ce^{3+}, Bi^{3+}, Cr^{3+} and Eu^{2+}) to sensitize the Yb^{3+} and Nd^{3+} ions [47–52]. Also, host sensitization through ET from an excited host (such as YVO_4) to Nd^{3+} and Yb^{3+} ions also provide an efficient way to enhance the PL properties [53–56].

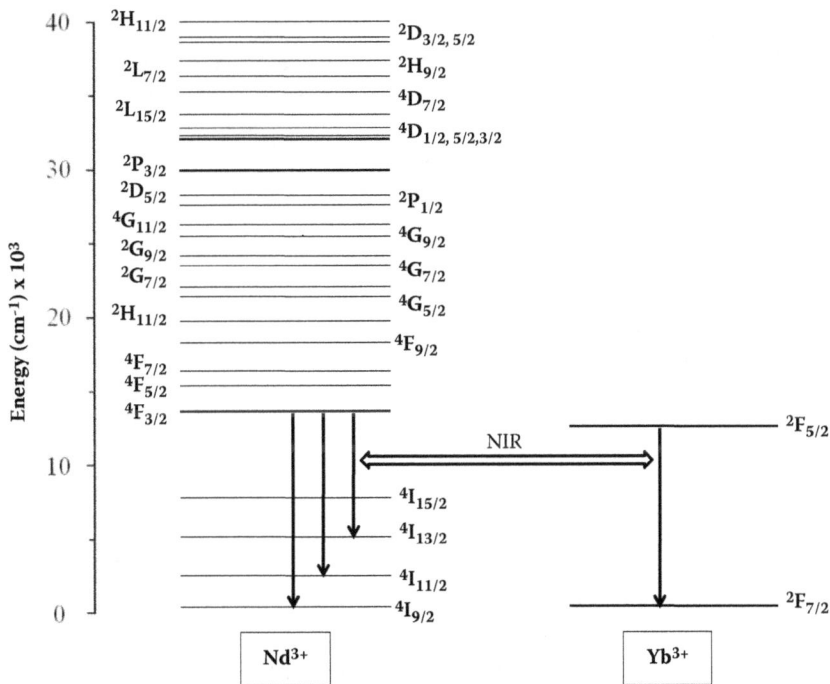

FIGURE 8.9   Energy levels of Nd^{3+} and Yb^{3+} activators [9].

The possible excitation lines and corresponding transitions of Nd^{3+} ions are $^4I_{9/2} \rightarrow ^2P_{1/2}$ (434 nm), $^4I_{9/2} \rightarrow ^4G_{11/2}$ (459 nm), $^4I_{9/2} \rightarrow ^4G_{9/2}$ (480 nm), $^4I_{9/2} \rightarrow ^4G_{9/2}$ (515 nm), $^4I_{9/2} \rightarrow ^4G_{7/2}$ (528 nm), $^4I_{9/2} \rightarrow ^4G_{5/2}$ (588 nm), $^4I_{9/2} \rightarrow ^2H_{11/2}$ (631 nm), $^4I_{9/2} \rightarrow ^4F_{11/2}$ (689 nm), $^4I_{9/2} \rightarrow ^4S_{3/2}$ (743 nm), $^4I_{9/2} \rightarrow ^4F_{7/2}$ (753 nm), $^4I_{9/2} \rightarrow ^2H_{9/2}$ (798 nm), $^4I_{9/2} \rightarrow 4F_{5/2}$ (808 nm) and $^4I_{9/2} \rightarrow ^4F_{3/2}$ (824 nm) [57–59]. The NIR emissions of Nd^{3+} observed at 910 nm, 1,066 nm and 1,342 nm correspond to $^4F_{3/2} \rightarrow ^4I_J$ (J = 9/2, 11/2, 13/2) transitions, respectively. Theses stark levels of Nd^{3+} are split further due to the crystal field effect [60].

In the case of Yb^{3+} activator there are not so many energy levels like in Nd^{3+} activators. The NIR emission of Yb^{3+} observed at 980 nm correspond to $^2F_{5/2} \rightarrow ^2F_{7/2}$ transition. The energy levels of Yb^{3+} are perfectly suited for use in the DC process for c-Si solar cells [60]. The Yb^{3+} ion has a single excited state just about 10,000 cm^{-1} above the $^2F_{7/2}$ ground state, equivalent to an emission about 1,000 nm. The lack of other energy levels allows Yb^{3+} to wholly pick up energy packages of 10,000 cm^{-1} from a sensitizer and emit photons at ~1,000 nm, which can be absorbed by c-Si solar cell. An efficient DC using Yb^{3+} via resonant energy transfer requires donor ions with an energy level situated at about 20,000 cm^{-1} and an intermediate energy level situated at approximately 10,000 cm^{-1} [61]. There is no characteristic excitation of Yb^{3+} in the UV-VIS range but it can be excited by either host or sensitizer to give characteristic NIR emission via ET process.

8.4.1 Ce^{3+} – Yb^{3+} Pair

The Ce^{3+} ion could use as an perfect broadband sensitizer for Yb^{3+}, because its 4f–5d transition covers a broad spectral range and, more importantly, the energy of its 4f–5d transition can be tuned by changing the strength of crystal field of host matrix as well as by the covalency of the host. The NIR QC phenomenon for the Ce^{3+} – Yb^{3+} couple have been reported for many host materials, and the CET ($Ce^{3+} \rightarrow 2Yb^{3+}$) as shown in Figure 8.10 found to be the dominant relaxation way to get the characteristics NIR emission of Yb^{3+} [62,63].

As compared to the regular Ln^{3+} – Yb^{3+} (Ln = Tb, Tm, and Pr) pairs, the Ce^{3+} – Yb^{3+} pair could yields broad solar spectral range to give rise to strong and intense NIR emissions. But few researchers consider that the ET from Ce^{3+} to Yb^{3+} is possibly due to metal to metal charge transfer (CT) through the simple redox reaction of $Ce^{3+} + Yb^{3+} \rightarrow Ce^{4+} + Yb^{2+}$[64,65]. Some researchers and their groups also used the Ce^{3+} ion to sensitize the Tb^{3+} – Yb^{3+} pair for the QC process [66,67]. The NIR emission of Yb^{3+} in the Ce^{3+} – Tb^{3+} – Yb^{3+} system can be increased by about 10 times as compared to that obtained in the Tb^{3+} – Yb^{3+} pair.

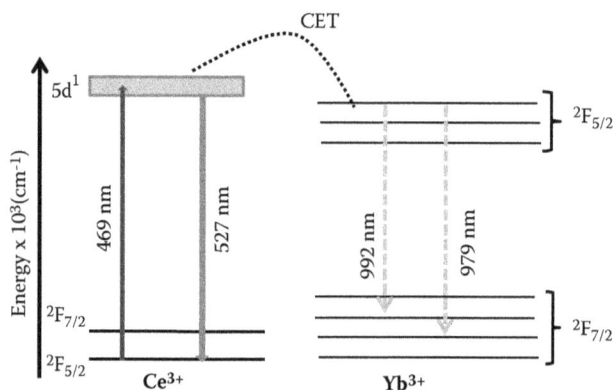

FIGURE 8.10 Proposed CET mechanism in Ce^{3+} – Yb^{3+} pair [9,21].

8.4.2 Bi³⁺–Yb³⁺ Pair

The Bi^{3+} ion with the $6s^2$ electronic configuration has been broadly investigated as both the activator and sensitizer for phosphors [68]. The energy states of Bi^{3+} are mixed by spin orbit coupling, and as a consequence, $^1S_0 - {}^3P_1$ and $^1S_0 - {}^1P_1$ transitions have strong absorption. Two research groups recently demonstrated efficient NIR DC in $Bi^{3+} - Yb^{3+}$ co-doped Ln_2O_3 (Ln = Gd, Y) phosphors and their thin films. In these hosts, Bi exhibits strong absorption in the UV region of 300–400 nm through the allowed $^1S_0 - {}^3P_1$ transition. After the direct excitation of Bi^{3+}, strong NIR emission band at 977 nm from Yb^{3+} was recorded. Additionally, it was shown that the CT between O^{2-} and Yb^{3+} did not have any effect on Bi^{3+} and Yb^{3+} emission. Hence, it can conclude that the nonradiative energy transfer contributes to the energy transfer (ET) process, as shown in Figure 8.11.

The luminescence decay lifetime measurements for Bi^{3+} emission showed that the occurrence of CET from Bi^{3+} to Yb^{3+} ions [69]. The NIR QC of the $Bi^{3+} - Yb^{3+}$ pair was also recently reported in YVO_4 and $YNbO_4$ phosphors [70].

8.4.3 Host-Yb³⁺ Combination

A broadband DC luminescence was also obtained using the Yb^{3+} ions in the host. The NIR DC in YVO_4: Yb^{3+} phosphor was reported by Wei et al. recently [57]. The CET from the excited vanadate group to two neighbouring Yb^{3+} ions resulted in the conversion of UV light (250–350 nm) into NIR emission around 980 nm. The calculated theoretical QE value found to be 185.7% [58]. However, this estimated value must be interpreted in the context of some assumptions. Shestakov et al. $Li^+ - Yb^{3+}$ co-doped nanocrystals of ZnO were tested for NIR QC. ZnO is a direct wide gap semiconductor with large absorption cross section in the 250–400 nm range and high degree of transparency in the VIS and NIR ranges. Upon UV excitation, two broadband emissions cantered at 770 and 1,000 nm were observed [58].

However, Yb^{3+} has an excitation around 940 nm that is not suitable for spectral modification through the DC or DS process but useful for the up-conversion (UC) process. Yb^{3+} ion is typically co-doped as an exceptional UC sensitizer due to its large absorption cross section in the 900–1,100 nm NIR region, corresponding to the $^2F_{5/2} \rightarrow {}^2F_{7/2}$ transition. Also the $Er^{3+} - Yb^{3+}$ pair is most studied upconversion system.

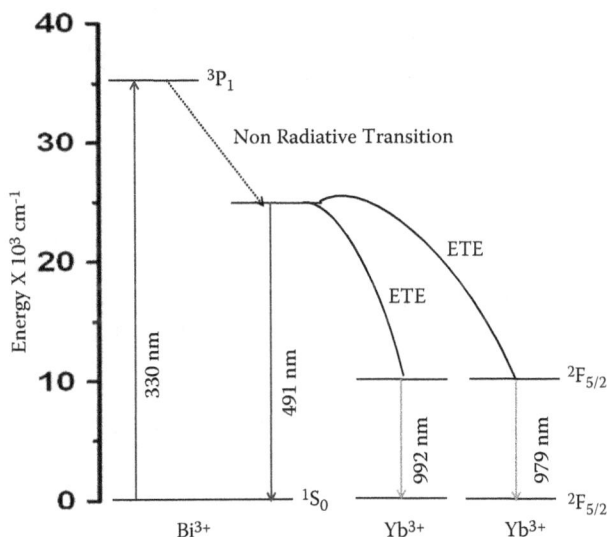

FIGURE 8.11 Proposed CET mechanisms in $Ce^{3+}-Yb^{3+}$ pair [9,21].

8.4.4 HOST-Er^{3+} COMBINATION

The Er-doped UC phosphors are the most promising UC materials for c-Si solar cells due to the ground state absorption of Er^{3+} in the range of 1,480–1,580 nm which corresponds to $^4I_{15/2} \rightarrow ^4I_{13/2}$ transition [59].

8.4.5 HOST-Ho^{3+} COMBINATION

The Ho^{3+} ion has a relatively broad absorption band in the range 1,150–1,225 nm due to $^5I_8 \rightarrow ^5I_6$ transition. Lahoz had reported the use of Ho^{3+} singly doped oxyfluoride glass ceramics as promising UC material for efficiency enhancement in c-Si solar cells [71].

8.5 BORATE-BASED UC, DC AND DS PHOSPHORS

As per the requirement of UC phosphor, there are very few attempts done to explore the borate-based UC phosphor, which are discussed below.

S. Das et al. studied green emission in Er^{3+} – Yb^{3+} co-doped KCaBO$_3$ UC phosphor with varying concentration of Yb^{3+} by conventional solid-state reaction method. It was suggested that this borate-based Er^{3+} – Yb^{3+} co-doped KCaBO$_3$ as potential candidate for green UC phosphor. The increment in the green emission intensities depends on concentration of sensitizer Yb^{3+} ions was explained based on variation of inter ionic distance, two photon absorption and efficient energy transfer between the Er^{3+} and Yb^{3+} ions [72].

Yawalkar et al. have successfully synthesized Li$_6$Y$_{1-x}$Eu$_x$(BO$_3$)$_3$ phosphor was using a modified solid-state diffusion method and proposed that this borate-based red phosphor as good material for LED application, however the phenomenon of UC process had not been exploited in this phosphor [73].

S. P. Hargunani has prepared Er^{3+} – Yb^{3+} co-doped LiLaBO$_3$ UC phosphor first time by employing solution combustion method. The UC luminescence occurred at absorption of NIR light and emission of Natrium yellow and amber in visible range [74].

Borate-based DC and DS phosphors are frequently attempted by researchers and sometimes novel applications or method is reported as follow.

Rare earth (RE)–doped borates and some orthoborates generally have high degree of ultraviolet transparency and excellent optical damage threshold, which causes them to be prime candidates for various practical applications. In recent years, YBO$_3$ doped with different rare earth ions were developed and studied for variety of applications such as fluorescent lighting, VUV absorption for plasma display panel (PDP), etc. And such applications are possible because the host matrix YBO$_3$ possesses excellent efficiency, high colour purity, very good refractory properties, good stability, wide bandgap and high transparency characteristics [75,76]. LaBO$_3$ intrinsically possess characteristics that are advantages for optical properties such as wide transparency range, large electronic bandgap, good thermal and chemical stability, optical stability with non-linear characteristics and remarkably high optical threshold damage [77,78]. Here, the participation of many researchers in this regard is discussed along with their important breakthrough.

Jin Zhao Das et al. developed Ba$_3$Y(BO$_3$)$_3$:Ce^{3+}, Nd^{3+} having NIR emission via solid-state reaction method. It provide wide blue violet light due to Ce^{3+} ions and intense NIR light ascribed to Nd^{3+} ions required for spectral response of silicon solar cell. The NIR emission intensity of Ce^{3+} – Nd^{3+} co-doped phosphor was increased as compared to that of Nd^{3+} single doped phosphor, which reveals that Ce^{3+} ion act as an effective sensitizer with broad absorption cross section. Hence it was concluded that Ba$_3$Y(BO$_3$)$_3$:Ce^{3+}, Nd^{3+} co-doped DS phosphor have promising utility as solar spectral convertor for c-Si solar cells[79].

V. A. Krut'ko et al. synthesized La$_{4-x-y}$Gd$_{10}$Ge$_2$B$_6$O$_{34}$: $_x$Yb^{3+}, $_y$Er^{3+} germanates borates liquid phase homogenization of the starting chemicals with tartaric acid, followed by annealing at temperature 1,000°C and 1,200°C. The structural properties of these phosphors were analyzed.

Their luminescence spectra consists of two groups of bands that corresponds to green emission due to $^2H_{11/2}$, $^4S_{3/2} \rightarrow ^4I^{15/2}$ and red emission due to $^4F_{9/2} \rightarrow ^4I_{15/2}$ transitions of the Er^{3+} ion, with the red band being more intense than the green one. The $La_{2.88}Gd_{10}Yb_{0.7}Er_{0.42}Ge_2B_6O_{34}$ and $La_{2.6}Gd_{10}Yb_{0.98}Er_{0.42}Ge_2B_6O_{34}$ phosphor with Yb:Er ratios of 5:3 and 7:3 were revealed the most intense red PL which may be suitable for amorphous silicon solar cells, however there was no such analysis done [80].

H. Li et al. have suggested a series of NIR range DC in Tb^{3+}–doped $Tb_{1-x}Zn(B_5O_{10}):_xYb^{3+}$ phosphor prepared by a high temperature solid-state method. The co-operative energy transfer mechanism between Tb^{3+} ions and Yb+ ions shown by the overlap between the excitation spectrum, emission spectrum as well as the fluorescence lifetime decay curve. It is concluded that $Tb_{1-x}Zn(B_5O_{10}):_xYb^{3+}$ is a potential DC phosphor for silicon-based solar cells [81].

D. Hou et al. have synthesized Yb^{3+} and Tb^{3+} singly doped and Tb^{3+} –Yb^{3+} co-doped $Ca_5(BO_3)_3F$ phosphors by a solid state reaction process. From the analysis of PL properties and decay curves, it was concluded that energy transfer from Tb^{3+} to Yb^{3+} ion in this phosphor. The co-doped of Tb^{3+} acts as sensitizer ions can broaden the absorption of Yb^{3+} ion in UV-VIS range. This confirmed that such phosphors can be useful to achieve desired DC process required for solar cell applications [82].

A. D. Sontakke et al. have studied both mechanisms in Ce^{3+} – Yb^{3+} co-doped Borate glass namely quantum cutting via DC and electron transfer. In this work, Ce^{3+} sensitized Yb^{3+} luminescence at around 1,000 nm. However, the intensity of sensitized Yb^{3+} PL was found to be very weak as compared to the strong quenching effect occurred in Ce^{3+} PL in Yb^{3+} co-doped glasses. The outcome were accredited to the energetically suitable electron transfer interactions followed by Ce^{3+} – $Yb^{3+} \rightleftharpoons Ce^{4+}$ – Yb^{2+} inter-valence charge transfer in the studied borate glass [83].

Jin Deng Chen et al. studied the near infra-red quantum cutting by Ce^{3+} – Yb^{3+} co-doped $LuBO_3$ DC phosphor having maximum quantum efficiency about 181% at optimum critical concentrations. By this phosphor new platform for borate-based DC phosphor for c-Si applications are open [84].

J. Zhang et al. have reported Tb^{3+}-doped $BaGdB_9O_{16}$ phosphor synthesized by typical solid-state reaction rout. Visible quantum cutting through DC process has been discussed which motivates towards the borate-based DC phosphors for potential use in solar cell application [85].

Z. Hao et al. have developed Ce^{3+} – Yb^{3+} co-activated $GdBO_3$ phosphor for efficient NIR quantum cutting via DC process. The phosphor has been synthesized by modified solid-state reaction method where step-wise heating is done at 500°C, 900°C and 1,200°C for certain period of time. The co-operative energy transfer (CET) has been illustrated and maximum theoretical quantum efficiency 164% has been achieved at critical concentration quenching threshold [86].

P. Song et al. has found the Ce^{3+} – Tb^{3+}/Yb^{3+} triple-doped YBO_3 system as a DC phosphor. An effective theoretical model for simulating Ce^{3+} – Tb^{3+}/Yb^{3+} triple-doped YBO_3 system, accurately calculated conversion efficiency and quantum efficiency of solar spectral DC, and acquired optimal system parameters has been set up using MATLAB [87].

P. Liu et al. has develop highly efficient Cr^{3+} and Yb^{3+}/Nd^{3+} co-doped $YAl_3(BO_3)_4$ (YAB) phosphors as spectral convertors for improving solar cell conversion efficiency. The conventional solid-state reaction method was used and PL studies confirmed that the strong NIR emissions from activators (Yb^{3+} and Nd^{3+}) due to typical DS process. The effect introduction of trivalent ions Gd, Bi and La into $YAl_3(BO_3)_4$:Cr^{3+},Yb^{3+} have been studied for stronger NIR PL intensity. Yb^{3+}, Nd^{3+} co-doped YAB shows much broader NIR emission. Because of the effectual absorption of Cr^{3+} in the VIS region in YAB and the efficient energy transfer to Yb^{3+}, these materials can be useful spectral convertors to improve silicon solar cell performance [88].

Omanwar et al. have reported comparative studies of spectral DS process in Nd^{3+}-doped MBO_3 (M = Y and La) where UV/VIS light transformed into NIR useful for c-Si solar cell. The stearic acid sol-gel technique has been adapted for preparation of this DS phosphor and it is concluded that Nd^{3+}-doped YBO_3 showed better results than that of Nd^{3+}-doped $LaBO_3$ material [89].

Sawala et al. have reported the spectral DS in Ce^{3+} – Yb^{3+} co-doped YBO_3 phosphor synthesized by combustion method and PL propertied has been investigated. It is found that the phosphor successfully prepared at a low temperature and PL spectra indicated that the Ce^{3+} – Yb^{3+} co-doped YBO_3 phosphor is a fine candidate to achieve spectral DS in order to improve solar cell performance [90].

Guerbou et al. has investigated PL properties of Ce^{3+} – Pr^{3+} doped $LnBO_3$ (Ln = Lu, Y and La) where DS process dominated. These phosphors were prepared by conventional solid state diffusion method. The heating temperatures are 900°C and 1450°C for 10h each, for YBO_3 and $LaBO_3$ phosphor, respectively [91].

Chen et al. has studied the DS luminescence and energy transfer in Sb^{3+} – and Gd^{3+}-doped YBO_3 phosphor synthesized by solid-state diffusion method and heated at 800°C and 1,150°C for 2h and 4h, respectively [92].

H. Guo et al. have studied the red-emitting $Ca_3Y(AlO)_3(BO_3)_4$:Eu^{3+} (CYAB:Eu^{3+}) DS phosphors with varying concentrations of Eu^{3+} ion were synthesized by a usual solid-state method and their structural and PL properties along with decay curves and quantum efficiency were calculated. These phosphors can emit red light peaking at 621 nm at narrow 397 nm excitation and the nearly all intense red emission was obtained at the concentration of 50 mol% of Eu^{3+} ion. From the concentration dependence PL studies of the phosphors, the concentration quenching mechanism was dominated by dipole-dipole interaction. The emitted red light is suitable for amorphous silicon solar cell application via DS process [93].

A. Shyichuk et al. have reported the Tb^{3+}-doped $Sr_3RE_2(BO_3)_4$:Tb^{3+} (RE = Y, La or Gd) nano phosphors are the another novel example of borate-based DS phosphor prepared by the Pechini sol-gel method, using citric acid and ethylene glycol as gel-forming agents. It was found that the CIE coordinates of the emission of the obtained phosphor are about to the National Television Standards Committee (NTSC) standard for green-emitting phosphors. Hence, the possible application of this phosphor is in red-green-blue (RGB) tricolour systems. Additionally, due to the occurrence of excitation bands in the range of mercury vapour discharges UV the phosphors can be potentially used in high frame rate 3D plasma display panel (PDP) screens. [94].

A. A. Shyichuk et al. have reported the PL properties of nanosized Eu^{3+}-doped strontium yttrium borate phosphor ($Sr_3Y_2(BO_3)_4$) prepared by the sol-gel Pechini method. Judd-Ofelt study showed that Eu^{3+} ions occupied Y^{3+} sites substitutionally in the crystalline network. This phosphor showed a red emission with the quantum yield of 54–55% and can be potentially utilized as phosphor for PDPs and luminescent tubes but there was no argument about utility of this phosphor as DS candidate for solar cell application [95].

A. Kruopyte et al. have prepared Eu^{3+}-doped polycrystalline GdB_5O_9:Eu^{3+} by an aqueous sol-gel process. About 394 nm excitation wavelengths, quantum efficiency of the phosphors is 100% at 25% doping concentration of Eu^{3+} ion in it. The temperature dependent emission spectra showed emission intensity decrease with increasing temperature. Besides, the exclusive emission spectrum makes this borate DS phosphor very attractive for luminescent security pigments application [96].

8.6 CURRENT TRENDS AND INNOVATIONS

Nowadays, the researchers are focusing to apply the phosphors on the commercial solar cells i.e. c-Si solar cells at a laboratory scale in order to experience the actual effect of UC, DC and DS phenomenon. There is a gap between successful synthesis of these phosphors and their actual implementations over solar panels since almost all prepared phosphors are in crystalline powder form. Few researchers believe to make a film of these phosphors apply on solar cells. Some of them effectively achieve it, but not at considerable limits of conversion efficiency.

A group of researchers thought that fabricating the glass or colloidal solution or gels of as mentioned phosphors can resolve the problem and may produce the desired result instead of making the thin films. Here many such attempts are discussed along with some novel utility of UC, DC and DS phosphors.

The design of the composite thin films was assisted by the Finite difference time domain (FDTD). The conversion efficiency of organic inorganic hybrid perovskite solar cells has increased from 3.81% to 22.1% in less than a decade. Even though such remarkable progress, some challenges put off the commercialization of perovskite solar cells and the lack of UV stability is one of them. Thus, it is highly beneficial to utilize UV photons to improve the solar conversion efficiency of solar PV modules [84].

Very recently, Y. He et al. have reported DS and antireflective graded thin films for solar module power improvement. To utilize UV photons and reduce reflection without prying manufacturing process of c-Si solar cells, thin films having DS and antireflection capabilities were made on the cover glass of c-Si solar cells. The thin films were consisting of graded index layers of Eu^{3+}-doped YVO_4 and hollow silica nano-particles (HSNPs). The thin films can yield 31.8% of UV photons for solar modules. To further advance in the performance, the utilization of different DS materials to produce spectral modification, tailored for solar cells, and the utilization of state-of-the-art UC nano-phosphors [97] can be designed using this novel scheme to attain greater power improvement. Since implementing these kinds of films on cover glass is well appropriate, the current solar technologies can make use of this technology without changing the manufacturing process of solar modules [98].

In recent times, C. Xie et al. have developed UC luminescence in $Ca_3Y(GaO)_3(BO_3)_4$: Yb^{3+}, Mn^{2+} (CYGB) phosphors by regular high temperature solid state method. The PL analysis of phosphor showed that the orange broadband emission under 410 nm excitation due to Mn^{2+} ions occupying two distinct sites of Ca^{2+} having different crystal field environments, while the yellow emission band at 980 nm laser excitation is due to Mn^{2+} ions located only at a definite low coordinated Ca^{2+} site. The suggested UC mechanism of yellow emission was dominated by the ground state absorption and excited state absorption routes of $Yb^{3+} - Mn^{2+}$ ion pairs. The phosphor offers momentous insight into exploring new transition metal ions based UC phosphors for lighting and displays [99].

Recently, D. Zhao et al. has successfully prepared a series of UC phosphors $CsNa_2Yb_{2-x}Er_x(BO_3)$ $(PO_4)_2$: (x = 0.01 − 0.10) by using high temperature flux method with superfluous Cs_2O–NaF–B_2O_3–P_2O_5 constituents as flux and Ln_2O_3 as solute. It is advocated that $CsNa_2Yb_{2-x}Er_x(BO_3)(PO_4)_2$ could be used as a capable temperature sensor with relative sensitivity of 0.011 K^{-1} at 303 K which opens new door of applications of UC phosphors in thermometry [100].

In 2018, A. Maurya et al. succeeded in enhancement UC and DS green photoluminescence in Tb^{3+}, Yb^{3+} co-doped $CaZrO_3$ phosphor in presence of Li^+ ions. The phosphor synthesized by solid-state reaction method and its structural and optical properties has been studied. It shows an increase in the particle size and shrivels in crystal lattice due to Li^+ ion. The phosphor sample emits intense green UC emission due to Tb^{3+} ions on excitation with 980 nm radiation which is further enhanced nearly about 28 times on Li^+ co-doping. The lifetime of $^5D_4$ level of Tb^{3+} ion decreases in the presence of Li^+ ions due to increase in asymmetry in crystal field. The DS emission intensity monitored on 378 and 487 nm excitations is also enhanced in the presence of Li^+ ions. Consequently, this phosphor can be appropriate candidate for UC solid state lighting [101].

R. A. Talewar et al. in 2018 first reported the intense NIR-emitting alkaline-earth chloro-borate (AECB) glasses doped with Nd^{3+} and Yb^{3+}. It synthesized by a usual melt quenching technique and characterized with photoluminescence (PL) and PL decay spectral measurements. For Nd^{3+} ion, intense NIR emission in the range 850–1,450 nm was observed at 824 nm excitation. The intense and typical emission bands of Nd^{3+} found at 883, 1,067 and 1,340 nm and for Yb^{3+} the same at 980 nm were observed with 585 nm excitation [102].

H. T. Kim et al. have recently phosphorescent energy DS for withdrawing surface recombination in silicon nano-wire solar cells. It was demonstrated that a phosphorescent energy DS system employing Ir(III) complexes at short wavelengths to reduce the surface recombination that occurs in c-Si nano-wire solar cells. These developed Ir(III) complexes can be regarded as promising DS converters because they display better properties such as a high quantum yield, a large Stokes shift and a long diffusion length in crystalline film with a reproducible synthesis method.

Using the developed Ir(III) complexes, highly crystalline energy downshifting layers were fabricated by ultrasonic spray deposition to increase the PL efficiency by increasing the radiative decay. Using optimized energy DS layer, c-Si nano-wire solar cells with Ir(III) complexes showed a higher internal quantum efficiency value for short-wavelength light as compared with that of bare c-Si nano-wire solar cells without Ir(III) complexes [103].

A. K. Soni et al. have first developed $Na_2Y_2B_2O_7$:Er^{3+}/Yb^{3+} crystalline phosphors by urea-assisted solution combustion technique. The detail study photon UC emission has been carried at a 980 nm laser diode excitation source. The highest improvement of about 1,433 times for the UC emission intensity of the green band in the co-doped phosphor as compared to that of the individually Er^{3+}-doped phosphor has been reported. An optical temperature sensing analysis performed in this phosphor indicates that the phosphor can be used in a temperature-sensing application based on the fluorescence intensity ratio (FIR) technique in 300–613 K temperature range having maximum sensitivity nearly 7.9×10^{-3} K^{-1} at 300 K. Also, the low-voltage cathode luminescence measurements have been performed for same phosphors as a function of voltage and filament current. The phosphors display highly intense green emission with some weak emission bands in violet, blue and red regions at electron beam excitation. It was accomplished that this phosphors can be used in fabricating the NIR to green UC lighting devices, temperature-sensing probe and also may be a prime candidate to produce green light for flat panel devices based on field emission display [104].

In 2016, Y. Hao et al. have synthesized YBO_3 co-doped with Ce^{3+}–Yb^{3+} viz hydrothermal method and demonstrate the QC by DC mechanism having CET efficiency 41.9%. Additionally, anti-reflection (AR) bi-functional thin film using SiO_2 sol of mentioned phosphor had been prepared and investigated to observe effect on solar cell performance. These comparative studies revealed that photo-conversion efficiency of cell was proved by 0.252% compared to pure SiO_2 AR films and 0.521% compared to glass. In a word, this work provides a simple approach for improving the photoelectric conversion efficiency of silicon-based solar cell [105].

K. Keshavamurthy et al. in 2017 has claimed that the DC process in silver lead borate glasses doped with Eu^{3+} is prepared by the melt quenching method. The PL properties along with thermal and structural properties have been investigated and finally concluded the red light emitting applications in LEDs [106].

Cadmium-Telluride (CdTe) thin-film solar cells have a poor spectral response in the UV and blue range owing to absorption and thermalization losses in the CdS buffer layer. CdS absorbs more than its bandgap energy of 2.4 eV, so that enormous quantity of energy from the incident spectrum AM1.5 is not involved in electron-hole pair generation. To overcome this poor spectral response in the short-wavelength range luminescence DS layer can be applied on the solar cell to convert the high-energy photons into low-energy photons [107].

S. Loos et al. have discovered trivalent rare earth–doped barium borate glass for its potential as DS phosphors cover glass on the CdTe cell. The glass is doped with either Tb^{3+} or Eu^{3+} up to critical limit of 2.5% of concentration giving to enough absorption in the UV/blue wavelength range. It was found that the Tb^{3+} shows intense emission bands in the green range while Eu^{3+} emits in the orange/red range. However, Tb^{3+} is not a capable candidate by reason of its less absorption in the UV and blue wavelength range. Thus, for only CdTe solar cells having a thickness of a CdS buffer layer of 300 nm a slight improvement was found [108].

On the same way, in 2015, F. Steudel et al. have developed RE-doped borate glasses for their prospective as spectral DS cover glass for efficiency enhancement of CdTe solar cells efficiency. The barium borate-based glass is doped with trivalent RE ions such as Eu^{3+}, Sm^{3+} and Tb^{3+} showing an intense red luminescence in the spectral range upon excitation at UV and blue light. Moreover, the glasses are double doped with two RE ions for making absorption broad was prepared. Though the single-doped glasses manifest a slight boost in short-circuit current density, the double-doped glasses allow for higher efficiency gains due to significantly broader spectral range is available for absorption. For a Tb^{3+} double-doped glass with a concentration of 1% each, an efficiency increase of 1.32% was achieved [109].

S. Thomas et al. recently reported spectroscopic properties of Sm^{3+}-doped $Sr_{1.0}Ba_{2.0}B_6O_{12}$ DS phosphor by solid-state reaction method and investigated its structural and luminescence properties along with Judd–Ofelt analysis. Colorimetric analysis using the PL studies showed that the $Sr_{1.0}Ba_{2.0}B_6O_{12:0.5}Sm^{3+}$ phosphor emits a cool white light. The higher values of the spectroscopic quality factor, stimulated-emission cross-section, quantum efficiency and the white-light emission of the phosphor suggest that this phosphor is useful for display and lighting applications. Additionally, it can be proposed that this kind of phosphors may be suitable to improve the performance of amorphous silicon solar cell [110].

S. M. Borisov et al. has observed the thermographic application of chromium-doped yttrium aluminum borate (YAB) single crystal DS phosphor. The phosphor shows no decrease in luminescence intensity at higher temperatures. The phosphor is predominantly promising for applications in temperature-compensated optical chemosensors including those based on NIR-emitting indicators [111].

W. Chang et al. in their work first time illustrated an effective dual mode solar spectral converter Tm^{3+} – Yb^{3+} co-doped oxyfluoride glasses prepared by using high temperature melt quenching approach. The transmittance, PL spectra and decay dynamics have been thoroughly investigated. The experimental results revealed that two distinct luminescence techniques, DC and DS were achieved through the sensitization of Tm^{3+}-excited multiplets and a broadband attributing to Yb^{3+} – O^{2-} charge transfer state, respectively. The synthesized phosphors could convert the UV and blue range wavelength into broad NIR emission that matches well with the efficient spectral response of c-Si solar cells. The resultant Tm^{3+} – Yb^{3+} co-doped oxy-fluoride glasses with dual mode spectral modification indicates its impending application in improving the performance of c-Si based cells [112]. The similar effort can be trying in borate-based glass.

B. C. Jamalaiah et al. has observed fluorescence properties of Sm^{3+} ions in yttrium aluminum borate (YAB) phosphors for various optical applications including DS process for amorphous silicon based solar cell. The $YAl_3(BO_3)_4:Sm^{3+}$ phosphors show intense orange-red emission after 404 nm excitation [113].

8.7 LIMITATIONS AND CHALLENGES

Two of the foremost problems associated with the UC approach that limits the practical application of UC phosphor for solar cells. The first one is the efficiency of current Ln-doped UC phosphor prepared for a solar cell is very low [114]. The UC efficiency may be enhanced by either plasmonic resonance [115] or photonic crystals [116]. Very high incident excitation density, such as lasers, requires getting a high UC efficiency, but it is not possible in the case of sunlight, which is diffused and not concentrated light. In this case, low-power UC in organic molecules is mainly capable for wide bandgap solar cells, provided that sufficient NIR to VIS UC can be achieved.

The second problem is, only a small portion of sub-bandgap sunlight can be up-converted because of the small absorption cross sections of Ln^{3+} ions. The transition metal ions, QDs and organic dyes with large absorption cross sections can be explored as sensitizers to get broadband absorption [117–119].

QC via DC has been established to be a linear process independent of the incident power of excitation source. This permits for the utilization of diffused and non-concentrated sunlight unlike UC phenomenon. But QC with external quantum efficiency larger than 100% has been reported seldom [120]. Hence, there is no direct proof of this concept that has been reported for solar cell applications. Immediate challenges for the distant future in this field are developing new characterization techniques to provide direct experimental verification of the existence of NIR QC and exploring the effect of QC phosphors on the performance of solar cells.

In case of Ds implementation the formation of a extremely effective luminescent solar concentrator remains tricky and the best possible configuration may involve a combination or design layout of different luminescent phosphors [121].

The organic binders can be explore to make thin/thick film of borate based UC, Dc and DS phosphors either on substrate like glass or directly put a layer on solar cells. From literature following few binders are scripted that are useful to achieve the desired objectives.

The polycrystalline borate-based phosphors are prepared in the form of fine powder. The fine powder of phosphor has to apply on surface of available solar cell. For that purpose one need binder to hold the powder of phosphor. The binder must be optically and chemically inert so that it should not affect the PL properties of prepared phosphor and should not react with it. From literature survey, it found the organic binder such as polyvinyl alcohol (PVA) and polydimethylsiloxane (PDMS) are suitable for requirements.

PVA is soluble in water and other solvents and is extensively used in synthetic fibre, paper, contact lens, textile, coating and binder industries, due to its excellent chemical and physical properties, nontoxicity, processability, good chemical resistance, high dielectric strength, good charge storage capacity, wide range of crystallinity, good film formation capacity, complete biodegradability and optical properties [122–125].

Reddy et al. has developed highly transparent and stable Gd and Ho: PVA polymer films which are prepared by solution casting method. The magnetic properties have been studied and summarized their applications in electrochemical display system [126]. Sadhu et al. have studied the mechanical characterizations of Starch PVA blend using potato [127]. Ghoshal et al. studied the formation of PVA films with a brief study of nuclear magnetic resonance (NMR) [128].

Owing to its excellent biocompatibility and less toxicity, PDMS is widely used in personal care, pharmaceutical and medical device applications [129]. The PDMS elastomer is also used in the fabrication of various microfluidic devices and optical waveguides [130,131].

Wang et al. has reported the fabrication of PDMS thin films using a spin coating method by dissolving it in different solvents [132]. Dinh et al. has investigated the mechanical properties of PDMS films for the optimization of polymer-based flexible capacitive pressure microsensors [133]. Schnyder et al. has investigated UV irradiation-induced modifications of PDMS films by using X-ray photoelectron spectroscopy (XPS) and spectroscopic ellipsometry [134].

It is found that for a CdTe solar cell having bandgap about 1.45 eV, the efficiency increases from 30.5% to 36% [135]. One can improve this by considering reflection losses at the interfaces results in an efficiency improvement from 21% for the bare cell with bandgap energy of 1.1 eV to 26% for a cell with perfect DC layer on top [136]. For solar cells having a bandgap energy of 1.45 eV, the efficiency can be increased to 23.5%. The efficiency of the CdTe module with a DS phosphor doped PMMA (polymethyl methacrylat) sheet in front could be increased to 11.2% compared to a module covered with an un-doped PMMA sheet with an efficiency of 9.6%. These calculations make the application of a DS layer attractive for photovoltaic devices.

I-V analysis from few researchers has showed that the efficiency of bare solar cell is not enhanced by the application of DC and DS phosphor on it. The factors responsible for decrease of solar performance are reflection, scattering and opaqueness of DC phosphor-coated films.

Due to high cost of solar cell and less availability, one cannot apply a layer of DC/DS phosphor using binder directly on surface of solar cell because once apply the cell cannot be reused for other observations. But the comparative results of I-V characterizations of bare cell, undoped binder film and DC/DS phosphor doped binder films manifested that the phenomenon of DC/DS is useful for enhancing the efficiency of solar cell via spectral modification. But the other losses due to morphological and structural drawback of prepared phosphors are dominating the small improvement in the conversion efficiency of solar cell. These negative results also might be due to

1. Unequal thickness since films since thickness monitoring has not been done.
2. Non-uniform distribution of particles of selected phosphor since particle size analysis (PSA) has not been investigated.
3. Non-separation of uniform and small-sized particles of selected phosphors.
4. Use of glass plates as substrate instead of direct coating.

This does not mean that the UC, DC and DS luminescent phenomenon are not feasible for solar cell applications.

8.8 FUTURE PROSPECTS AND SCOPE

From literature, basically three different synthesis method, co-precipitation method, sol-gel method and conventional solid-state method, have been adopted. From I-V characteristics, analysis from a few researchers is understood that the particle size of DS/DC phosphors should be of the order of 200 nm so that the scattering reflection loss due to phosphor avoided. Thus, the synthesis method needs to be optimized and modified. The combustion method and sol-gel method can be adapted and modified.

Most of the phosphors reported have the excitation range of 250–400 nm. The phosphors can be found out and explore having much broader excitation in 300–600 nm. The self-activated phosphors, such as vanadates, even tungstates, need to be more explored instead of co-doped DS/DC phosphors. This will help to reduce the cost of UC, DC and DS phosphors. The sensitizers (Bi^{3+}, Eu^{2+} and Ce^{3+}) that are RE are quite costly. Also, the mixed type host, for instance the combination of phosphate, sulphate, oxides with borates, vanadates and tungstates, may provide good results. The quantum efficiency have measured in the discussed work is theoretical measurement with the help of luminescence decay curves of phosphors. It is necessary to instigate quantum efficiency practically or by using specific and well calibrated instrument such as the integrating sphere.

The application of these UC, DC and DS phosphors over the surface of solar cells is a daunting task. The different binding agents need to explore without any base or substrate (glass) in a specific manner so that the noticeable improvement in conversion efficiency of a solar cell is observed.

The comparative analysis of I-V various films showed that the phenomenon of UC, DC and DS is useful for enhancing the efficiency of solar cell via spectral modification. The other losses due to morphological and structural drawback of prepared phosphors are dominating the small improvement in the conversion efficiency of solar cell. These negative results also can be improve by

1. Controlling the thickness of prepared films and optimizing it.
2. Study of particle size distribution of phosphors by PSA.
3. Separation of uniform and nanosized particles of prepared phosphors to avoid the scattering losses.
4. Use of better binding or dispersion medium for phosphors so that it can be directly coated on surface of solar cell. This will remove the necessity of substrate (glass plate).

This introduces the new kind of research work in the field of thin or thick film methods and techniques.

One can choose the glass over crystalline or polycrystalline powder materials because of its number of advantages. Glass has various characters that can make it as an ideal host material for thin/thick film, laser and optical fiber amplifier [137,138]. It can be prepared in various shapes and sizes from fibres of a few microns to 2 metres long or 7.5 cm in diameter [139]. As compared to the crystalline powder host, the absorption and emission cross sections of Yb^{3+} ions are much broader in glasses [140,141]. Due to the more absorption cross sections, light absorption and energy storage of material increases to a better level. Hence, the overlap of emission and absorption of the Nd^{3+} and Yb^{3+} becomes more encouraging in the glass host. This produces better resonance between the two ions than would be the case in crystalline powder phase phosphors. With the plan of investigating the NIR emission through the sensitization of Yb^{3+} or Nd^{3+} ions through various energy levels covering UV-VIS to NIR regions one can adopt various methods to develop borate-based glass for solar cell applications.

As per conclusive remarks, the application of PL phosphors as spectral converters to PV applications could not come at a more appropriate time given the large increase in the universal demand of energy in recent years. Despite the daunting challenges of realizing low cost and highly efficient solar cells, further exploration and attempts to utilize the spectral converting UC, DC and DS phosphors for PV cells, devices and systems will certainly be exciting.

REFERENCES

[1] http://www.PV Lighthouse hosts Altermatt's lectures on the solar spectrum

[2] Honsberg, C.B., and Bowden, S.G. Atmospheric effects on solar radiation. page on www. pveducation.org

[3] Sekuler R., and Blake R. (1985). *Perception*, Alfred A. Knopf Inc, New York.

[4] Honsberg, C.B., and Bowden, S.G. PV array on a sunny and a cloudy. page on www. pveducation.org

[5] Honsberg, C.B., and Bowden, S.G. Photovoltaic education website. www.pveducation.org

[6] Honsberg, C.B., and Bowden, S.G. Standard solar spectra for space and terrestrial applications. page on www.pveducation.org

[7] Powell, D.M., Winkler, M.T., Choi, H.J., Simmons, C.B., Needleman, D.B., and Buonassisi, T. (2012). Crystalline silicon photovoltaic: a cost analysis framework for determining technology pathways to reach base load electricity costs. *Energy Environ. Sci.*, 5, 5874–5883.

[8] Green, M.A., Emery, K., Hishikawa, Y., and Warta, W. (2011). Solar cell efficiency tables (version 37). *Prog. Photovoltaics*, 19, 84–92.

[9] Honsberg, C.B., and Bowden S.G. Normalized spectral response of a C-Si solar cell. page on www. pveducation.org

[10] Zhang, X.D., Jin, X., Wang, D.F., Xiong, S.Z., Geng, X.H., and Zhao, Y. (2010). Synthesis of $NaYF_4$: Yb, Er nanocrystals and its application in silicon thin film solar cells. *Phys. Status Solidi C*, 7, 1128–1131.

[11] de Wild, J., Meijerink, A., Rath, J.K., van Sark, W.G.J.H.M., and Schropp, R.E.I. (2010). Towards upconversion for amorphous silicon solar cells. *Sol. Energy Mater. Sol. Cells*, 94, 1919–1922.

[12] de Wild, J., Rath, J.K., Meijerink, A., van Sark, W.G.J.H.M., and Schropp, R.E.I. (2010). Enhanced near-infrared response of a-Si:H solar cells with β-$NaYF_4$: Yb^{3+} (18%), Er^{3+} (2%) upconversion phosphors. *Sol. Energy Mater. Sol. Cells*, 94, 2395–2398.

[13] Honsberg, C.B., and Bowden, S.G. Solar cell parameter. page on www.pveducation.org

[14] Honsberg, C.B., and Bowden, S.G. I-V characteristics of solar cell. page on www.pveducation.org

[15] Honsberg, C.B., and Bowden, S.G. The spectral response of a silicon solar cell. page on www. pveducation.org

[16] Auzel, F. (2004). Upconversion and anti-stokes processes with f and d ions in solids. *Chem. Rev.*, 104, 139–173.

[17] Trupke, T., Green, M.A., and Wurfel, P. (2002). Improving solar cell efficiencies by up-conversion of sub-band-gap light. *J. Appl. Phys.*, 92, 4117–4122.

[18] Wegh, R.T., Donker, H., Oskam, K.D., and Meijerink, A. (1999). Visible quantum cutting in $LiGdF_4$: Eu^{3+} through downconversion. *Sci.*, 283, 663–666.

[19] Lian, H., Shang, Z.H.M., Geng, D., Zhang, Y., and Lin, J. (2013). Rare earth ions doped phosphors for improving efficiencies of solar cells. *Energy*, 57, 270–283.

[20] Gibart, P., Auzel, F., Guillaume, J.C., and Zahraman, K. (1996). Below band-gap IR response of substrate-free GaAs solar cells using two-photon up-conversion. *Jpn. J. Appl. Phys.*, Part 135, 4401–4402.

[21] Zhang, Q.Y., and Huang, X.Y. (2010). Recent progress in quantum cutting phosphors. *Prog. Mater. Sci.*, 55, 353–427.

[22] Dexter, D.L. (1957). Possibility of luminescent quantum yields greater than unity. *Phys. Rev.*, 108, 630–633.

[23] Piper, W.W., DeLuca, J.A., and Ham, F.S. (1974). Cascade fluorescent decay in Pr^{3+}-doped fluorides: achievement of a quantum yield greater than unity for emission of visible light. *J. Lumin.*, 8, 344–348.

[24] Sommerdijk, J.L., Bril, A., and de Jager, A.W. (1974). Two photon luminescence with ultraviolet excitation of trivalent praseodymium. *J. Lumin.*, 8, 341–343.

[25] Niranjana, V., Janani, K., and Venkata Ravikumar Darba Ramasubramanian, S. (2019). Investigations on the synthesis dependent luminescence property of $LiGdF_4$: Eu^{3+} microcrystals. *Optik.*, 198, 163233, 1–5.

[26] Liu, B., Chen, Y., Shi, C., Tang, H., and Tao, Y. (2003). Visible quantum cutting in BaF_2: Gd, Eu via downconversion. *J. Lumin.*, 101, 155–159.

[27] Strumpel, C., McCann, M., Beaucarne, G., Arkhipov, V., Slaoui, A., Svrcek, V., del Canizo, C., and Tobias, I. (2007). Modifying the solar spectrum to enhance silicon solar cell efficiency—an overview of available materials. *Sol. Energy Mater. Sol. Cells*, 91, 238–249.

[28] Xie, L.C., Wang, Y.H., and Zhang, H.J. (2009). Applications of rare earth Tb^{3+} – Yb^{3+} co-doped down-conversion materials for solar cells. *Appl. Phys. Lett.*, 94, 061905, 1–3.

[29] Zheng, W., Zhu, H., Li, R., Tu, D., Liu, Y., Luo, W., and Chen, X. (2012). Visible-to-infrared quantum cutting by phonon-assisted energy transfer in YPO_4: Tm^{3+}, Yb^{3+} phosphors. *Phys. Chem. Chem. Phys.*, 14, 6974–6980.

[30] van der Ende, B.M., Aarts, L., and Meijerink, A. (2009). Near-infrared quantum cutting for photovoltaics. *Adv. Mater.*, 21, 3073–3077.

[31] van Wijngaarden, J.T., Scheidelaar, S., Vlugt, T.J.H., Reid, M.F., and Meijerink, A. (2010). Energy transfer mechanism for downconversion in the (Pr^{3+}, Yb^{3+}) couple. *Phys. Rev. B: Condens. Matter.*, 81, 155112-1.

[32] Eilers, J.J., Biner, D., van Wijngaarden, J.T., Kraemer, K., Guedel, H.U., and Meijerink, A. (2010). Efficient visible to infrared quantum cutting through downconversion with the Er^{3+}-Yb^{3+} couple in $Cs_3Y_2Br_9$. *Appl. Phys. Lett.*, 96, 151106-1.

[33] Lin, H., Chen, D., Yu, Y., Yang, A., and Wang, Y. (2011). Near-infrared quantum cutting in Ho^{3+}/Yb^{3+} co-doped nanostructured glass ceramic. *Opt. Lett.*, 36, 876-878.

[34] Bai, Z., Fujii, M., Hasegawa, T., Imakita, K., Mizuhata, M., and Hayashi, S. (2011). Efficient ultraviolet-blue to near-infrared downconversion in Bi–Dy–Yb-doped zeolites. *J. Phys. D: Appl. Phys.*, 44, 455301, 1–3.

[35] Chen, D., Wang, Y., Yu, Y., Huang, P., and Weng, F. (2008). Quantum cutting downconversion by cooperative energy transfer from Ce^{3+} to Yb^{3+} in borate glasses. *J. Appl. Phys.*, 104, 116105, 1–3.

[36] Cheng, X., Su, L., Wang, Y., Zhu, X., Wei, X., and Wang, Y. (2012). Near-infrared quantum cutting in YVO_4:Yb^{3+} thin-films via downconversion. *Opt. Mater.*, 34, 1102–1106.

[37] Shestakov, M.V., Tikhomirov, V.K., Kirilenko, D., Kuznetsov, A.S., Chibotaru, L.F., Baranov, A.N., Van Tendeloo, G., and Moshchalkov, V.V. (2011). Quantum cutting in Li (770 nm) and Yb (1000 nm) co-dopant emission bands by energy transfer from the ZnO nano-crystalline host. *Opt. Express*, 19, 15955–15964.

[38] Wei, X., Huang, S., Chen, Y., Guo, C., Yin, M., and Xu, W. (2010). Energy transfer mechanisms in doped near-infrared downconversion phosphor. *J. Appl. Phys.*, 107, 103107, 1–3.

[39] Rowan, B.C., Wilson, L.R., and Richards, B.S. (2008). Advanced material concepts for luminescent solar concentrators. *IEEE J. Sel. Top. Quantum Electron.*, 14, 1312–1322.

[40] Yamada, N., Anh, L.N., and Kambayashi, T. (2010). Escaping losses of diffuse light emitted by luminescent dyes doped in micro/nanostructured solar cell systems. *Sol. Energy Mater. Sol. Cells*, 94, 413–419.

[41] de Boer, D.K.G., Broer, D.J., Debije, M.G., Keur, W., Meijerink, A., Ronda, C.R., and Verbunt, P.P.C. (2012). Progress in phosphors and filters for luminescent solar concentrators. *Opt. Express*, 20, A395–A405.

[42] Debije, M.G., and Verbunt, P.P.C. (2012). Thirty years of luminescent solar concentrator research: solar energy for the built environment. *Adv. Energy Mater.*, 2, 12–35.

[43] Klampaftis, E., Ross, D., McIntosh, K.R., and Richards, B.S. (2009). Enhancing the performance of solar cells via luminescent down-shifting of the incident spectrum: a review. *Sol. Energy Mater. Sol. Cells*, 93, 1182–1194.

[44] Stouwdam, J.W., and van Veggel, F.C.J.M. (2002). Near-infrared emission of redispersible Er^{3+}, Nd^{3+}, and Ho^{3+} doped LaF_3 nanoparticles. *Nano Lett.*, 2, 733–737.

[45] Hebbink, G.A., Stouwdam, J.W., Reinhoudt, D.N., and vanVeggel, F.C.J.M. (2002). Lanthanide(III) - doped nanoparticles that emit in the near-infrared. *Adv. Mater.*, 14, 1147–1150.

[46] Zhang, M.F., Shi, S.G., Meng, J.X., Wang, X.Q., Fan, H., Zhu, Y.C., Wang, X.Y., and Qian, Y.T. (2008). Preparation and characterization of near-infrared luminescent bifunctional core/shell nano-composites. *J. Phys. Chem. C*, 112, 2825–2830.

[47] Meng, J.X., Li, J.Q., Shi, Z.P., and Cheah, K.W. (2008). Efficient energy transfer for Ce to Nd in Nd/Ce codoped yttrium aluminum garnet. *Appl. Phys. Lett.*, 93, 221908, 1–3.

[48] Fu, H., Cui, S., Luo, Q., Qiao, X., Fan, X., and Zhang, X. (2012). Broadband downshifting luminescence of Cr^{3+}/Yb^{3+}-codoped fluorosilicate glass. *J. Non-Cryst. Solids*, 358, 1217–1220.

[49] Ye, S., Jiang, N., Zhou, J., and Wang, D. (2012). Optical property and energy transfer in the ZnO-$LiYbO_2$ hybrid phosphors under the indirect near-UV excitation. *J. Qiu, J. Electrochem. Soc.*, 159, H11–H15.

[50] Tikhomirov, V.K., Vosch, T., Fron, E., Rodriguez, V.D., Velazquez, J.J., Kirilenko, D., van Tendeloo, G., Hofkens, J., Van der Auweraer, M., and Moshchalkov, V.V. (2012). Luminescence of oxyfluoride glasses co-doped with Ag nanoclusters and Yb^{3+} ions. *RSC Adv*, 2, 1496–1501.

[51] Liu, G.X., Zhang, R., Xiao, Q.L., Zou, S.Y., Peng, W.F., Cao, L.W., and Meng, J.X. (2011). Efficient $Bi^{3+} \rightarrow Nd^{3+}$ energy transfer in Gd_2O_3:Bi^{3+}, Nd^{3+}. *Opt. Mater.*, 34, 313–316.

[52] Zhang, G.G., Liu, C.M., Wang, J., Kuang, X.J., and Su, Q. (2011). An intense charge transfer broadband sensitized near-infrared emitting $CaLaGa_3S_6O$:Yb^{3+} phosphor suitable for solar spectral convertor. *Opt. Express*, 19, 24314-24319.

[53] Luo, Q., Qiao, X.S., Fan, X.P., and Zhang, X.H. (2011). Near-infrared emission of Yb^{3+} through energy transfer from ZnO to Yb^{3+} in glass ceramic containing ZnO nanocrystals. *Opt. Lett.*, 36, 2767–2769.

[54] Ye, S., Jiang, N., Zhou, J., Wang, D., and Qiu, J. (2012). Optical property and energy transfer in the ZnO-$LiYbO_2$ hybrid phosphors under the indirect near-UV excitation. *J. Electrochem. Soc.*, 159, H11–H13.

[55] Luo, W.Q., Li, R.F., and Chen, X.Y. (2009). Host-sensitized luminescence of Nd^{3+} and Sm^{3+} ions incorporated in anatase titania nanocrystals. *J. Phys. Chem. C*, 113, 8772-8777.

[56] Stouwdam, J.W., and van Veggel, F.C.J.M. (2004). Sensitized emission in Ln^{3+}-doped TiO_2 semiconductor nanoparticles. *Chem. Phys. Chem.*, 5, 743–746.

[57] Sawala, N.S., Koparkar, K.A., Bajaj, N.S., and Omanwar, S.K. (2016). Near-Infrared down-conversion in $Y_{(1-x)}Yb_xVO_4$ for sensitization of c-Si solar cell. *Optik*, 127, 4375–4378.

[58] Sawala, N.S., Bajaj, N.S., and Omanwar, S.K. (2016). Near-infrared quantum cutting in Yb^{3+} ion doped strontium vanadate . *Infrared Phys. Technol.*, 76, 271–275.

[59] Chen, D.Q., Lei, L., Yang, A.P., Wang, Z.X., and Wang, Y.S. (2012). Ultra-broadband near-infrared excitable upconversion core/shell nanocrystals. *Chem. Commun.*, 48, 5898–5900.

[60] Pathak, A.A., Talewar, R.A., Joshi, C.P., and Moharil, S.V. (2017). Sensitization of Yb^{3+} emission in $CaYAl_3O_7$ host. *Opt. Mater.*, 64, 217–223.

[61] vanderEnde, B.M., Aarts, L., and Meijerink, A. (2009). Lanthanide ions as spectral converters for solar cells. *Phys. Chem. Chem. Phys.*, 11, 11081–11095.

[62] Zhou, J., Teng, Y., Lin, G., and Qiu, J. (2011). Ultraviolet to near-infrared spectral modification in Ce^{3+} and Yb^{3+} co-doped phosphate glasses. *J. Non-Cryst. Solids*, 357, 2336–2340.

[63] Zhou, W., Yang, J., Wang, J., Li, Y., Kuang, X., Tang, J., and Liang, H. (2012). Study on the effects of 5d energy locations of Ce^{3+} ions on NIR quantum cutting process in Y_2SiO_5: Ce^{3+}, Yb^{3+}. *Opt. Express*, 20, A510–A518.

[64] Ueda, J., and Tanabe, S. (2009). Visible to near infrared conversion in Ce^{3+}-Yb^{3+} co-doped YAG ceramics. *J. Appl. Phys.*, 106, 0431011–0431015.

[65] van der Kolk, E., Ten Kate, O.M., Wiegman, J.W., Biner, D., and Kramer, K.W. (2011). Enhanced $^1G_4$ emission in $NaLaF_4$: Pr^{3+}, Yb^{3+} and charge transfer in $NaLaF_4$: Ce^{3+}, Yb^{3+} studied by Fourier transform luminescence spectroscopy. *Opt. Mater.*, 33, 1024–1027.

[66] Zhang, Q., Wang, J., Zhang, G., and Su, Q. (2009). UV photon harvesting and enhanced near-infrared emission in novel quantum cutting Ca_2BO_3Cl: Ce^{3+}, Tb^{3+}, Yb^{3+} phosphor. *J. Mater. Chem.*, 19, 7088–7092.

[67] Zhou, J., Teng, Y., Ye, S., Zhuang, Y., and Qiu, J. (2010). Enhanced downconversion luminescence by co-doping Ce^{3+} in Tb^{3+}-Yb^{3+} doped borate glasses. *Chem. Phys. Lett.*, 486, 116–118.

[68] Blasse, G., and Bril, A. (1968). Investigations on Bi^{3+}-activated phosphors. *J. Chem. Phys.*, 48, 217–222.

[69] Zhou, R., Kou, Y., Wei, X., Duan, C., Chen, Y., and Yin, M. (2012). Broadband downconversion based near-infrared quantum cutting via cooperative energy transfer in $YNbO_4$: Bi^{3+}, Yb^{3+} phosphor. *Appl. Phys. B*, 107, 483–487.

[70] Huang, X.Y., Wang, J.X., Yu, D.C., Ye, S., Zhang, Q.Y., and Sun, X.W. (2011). Spectral conversion for solar cell efficiency enhancement using YVO_4:Bi^{3+}, Ln^{3+} (Ln = Dy, Er, Ho, Eu, Sm, and Yb) phosphors. *J. Appl. Phys.*, 109, 113526, 1–7.

[71] Lahoz, F. (2008). Ho^{3+} doped nanophase glass ceramics for efficiency enhancement in silicon solar cells. *Opt. Lett.*, 33, 2982–2984.

[72] Das, S., Amarnath Reddy, A., and Vijaya Prakash, G. (2011). Strong green upconversion emission from Er^{3+}-Yb^{3+} co-doped $KCaBO_3$ phosphor. *Chem. Phys. Lett.*, 504, 206–210.

[73] Yawalkar, M.M., Zade, G.D., Dabre, K.V., and Dhoble, S.J. (2016). Luminescence study of Eu^{3+} doped $Li_6Y(BO_3)_3$ phosphor for solid-state lighting. *J. Bio. Chem. Sci.*, 31, 1037–1042.

[74] Hargunani, S.P. (2016). Synthesis and upconversion properties of Er^{3+}-Yb^{3+} co-doped $LiBaBO_3$ phosphor engineering and technology. *Int. Adv. Res. J. Sci.*, 3, 216–218.

[75] Palan, C.B., Chauhan, A.O., Sawala, N.S., Bajaj, N.S., and Omanwar, S.K. (2016). Synthesis and preliminary TL/OSL properties of $Li_2B_4O_7$: Cu-Ag phosphor for radiation dosimetry. *Optik Int. J. Light Electron. Opt.*, 127, 6419–6423.

[76] Palaspagar, R.S., Gawande, A.B., Sonekar, R.P., and Omanwar, S.K. (2015). Synthesis and luminescent properties of Tb^{3+} activated novel magnesium alumino-borate phosphor. *J. Chin. Adv. Mater. Soc.*, 3, 170–176.

[77] Chauhan, A.O., Palan, C.B., Sawala, N.S., and Omanwar, S.K. (2017). Combustion synthesis and photoluminescence study of UV emitting $LaBaB_9O_{16}$ phosphors. *J. Mater. Sci. Mater. Electron.*, 28, 7643–7649.

[78] Chauhan, A.O., Koparkar, K.A., Bajaj, N.S., and Omanwar, S.K. (2016). Synthesis and photoluminescence properties of Pb^{2+} doped inorganic borate phosphor $NaSr_4(BO_3)_3$. *AIP Conf. Proc.*, 1728, 0204701–0204704.

[79] Zhao, J., Wang, X., Pang, Q., and Zhang, A. (2021). Enhanced near-infrared luminescence in $Ba_3Y(BO_3)_3$: Nd^{3+} by co-doping with Ce^{3+}. *ECS J. Solid State Sci. Technol.*, 10, 016004, 1–3.

[80] Krut'ko, V.A., Komova, M.G., Pominova, D.V., and Nikiforova, G.E. (2019). Enhanced red up-conversion luminescence in $La_{4-x-y}Gd_{10}Ge_2B_6O_{34}$:$Yb_x$,$Er_y$ synthesized by the liquid-phase homogenization with tartaric acid. *J. Sol-Gel Sci. Tech.*, 92, 442–448.

[81] Lia, H., Lu, Y., Lia, C., Dengb, D., Rong, K., Jing, X., Wang, L., and Xu, S. (2019). Near-infrared down-conversion luminescence in Yb^{3+} doped self-activated $TbZn(B_5O_{10})$ phosphor. *J. Optoelectr. Adv. Mater.*, 21, 373–384.

[82] Hou, D., Li, J., Lin, H., and Zhang, J. (2016). Energy transfer and downconversion near-infrared material of Tb^{3+} and Yb^{3+} doped $Ca_5(BO_3)_3F$. *Physica B: Condensed Matter*, 500, 44–47.

[83] Sontakke, A.D., Ueda, J., Katayama, Y., Zhuang, Y., Dorenbos, P., and Tanabe, S. (2015). Role of electron transfer in Ce^{3+} sensitized Yb^{3+} luminescence in borate glass. *J. Appl. Phys.*, 117, 013105, 1–7.

[84] Chen, J.D., Zhang, H., Li, F., and Guo, H. (2011). High efficient near-infrared quantum cutting in Ce^{3+}-Yb^{3+} co-doped $LuBO_3$ phosphors. *Mater. Chem. Phys.*, 128, 191–194.

[85] Zhang, J., Wang, Y., Chen, G., and Huang, Y. (2014). Investigation on visible quantum cutting of Tb^{3+} in oxide hosts. *J. Appl. Phys.*, 115, 093108, 1–7.

[86] Hao, Z., Jindeng, C., and Hai, G. (2011). Efficient near-infrared quantum cutting by Ce-Yb couple in $GdBO_3$ phosphors. *J. Rare Earths*, 29, 822–825.

[87] Song, P., and Jiang, C. (2013). Broadband solar spectral conversion in near-infrared quantum cutting. *IEEE Photonic J.*, 5, 8400110, 1–10.

[88] Liu, P., Liu, J., Zheng, X., Luo, H., Li, X., Yao, Z., Yu, X., Shi, X., Houa, B., and Xi, Y. (2014). An efficient light converter YAB:Cr^{3+},Yb^{3+}/Nd^{3+} with broadband excitation and strong NIR emission for harvesting c-Si-based solar cells. *J. Mater. Chem. C*, 2, 5769–5777.

[89] Omanwar, S.K., and Sawala, N.S. (2017). Spectral downshifting in MBO_3:Nd^{3+} (M = Y, La) phosphor. *Appl. Phys. A*, 123, 673, 1–5.

[90] Sawala, N.S., Chauhan, A.O., Palan, C.B., and Omanwar, S.K. (2015). Spectral downshifting in Ce^{3+}- Yb^{3+} co-doped YBO_3 phosphor. *Int. J. Lumin. Appl.*, 5, 456–459.

[91] Guerbous, L., Seraiche, M., and Krachni, O. (2013). Photoluminescence and electron-vibrational interaction in 4fn-15d states of Ce^{3+} or Pr^{3+} ions doped $LnBO_3$ (Ln = Lu, Y, La) orthoborates materials. *J. Lumin.*, 134, 165–173.

[92] Chen, L., Luo, A., Deng, X., Xue, S., Zhang, Y., Zhu, J., Yao, Z., Jiang, Y., Chen, S., and Liu, F. (2013). Luminescence and energy transfer in the Sb^{3+} and Gd^{3+} activated YBO_3 phosphor. *J. Lumin.*, 143, 670–673.

[93] Guo, H., Sun, L., Liang, J., Li, B., and Huang, X. (2019). High-efficiency and thermal-stable Eu3+-activated $Ca_3Y(AlO)_3(BO_3)_4$:Eu^{3+} red emitting phosphors for near-UV-excited white LEDs. *J. Lumin*, 205, 115–121.

[94] Shyichuk, A., and Lis, S. (2013). Green-emitting nanoscaled borate phosphors $Sr_3RE_2(BO_3)_4$:Tb^{3+}. *Mater. Chem. Phys.*, 140, 447–452.

[95] Shyichuk, A.A., and Lis, S. (2011). Photoluminescence properties of nanosized strontium-yttrium borate phosphor obtained by the sol-gel Pechini method. *J. Rare Earths*, 29, 1161–1165.

[96] Kruopyte, A., Giraitis, R., Juskenas, R., Enseling, D., Jüstel, T., and Katelnikovas, A. (2017). Luminescence and luminescence quenching of efficient GdB_5O_9:Eu^{3+} red phosphors. *J. Lumin.*, 192, 520–526.

[97] Sun, Y., An, X., Chen, L., Gao, Q., Zhang, X., Duan, L., and Lü, W. (2018). Upconverting nanophosphor incorporated photoanodes for improved photoelectric performances of quantum dot sensitized solar cells. *Mater. Res. Lett.*, 6, 314–320.

[98] He, Y., Liu, J., Sung, S., and Chang, C. (2021). Downshifting and antireflective thin films for solar module power enhancement. *Mater. Design*, 201, 109454, 1–14.

[99]   Xie, C., Xie, S., Yi, R., Cao, R., Yuan, H., and Xiao, F. (2020). Site-selective occupation for broadband upconversion luminescence in $Ca_3Y(GaO)_3(BO_3)_4$:Yb^{3+}, Mn^{2+} phosphors. *J. Phys. Chem. C*, 124, 6845–6852.

[100] Zhao, D., Xue, Y., Fan, Y., Zhang, R., and Zhanga, S. (2021). A new series of rare-earth borate-phosphate family $CsNa_2Ln_2(BO_3)(PO_4)_2$ (Ln = Ho, Er, Tm, Yb): Tunnel structure, upconversion luminescence and optical thermometry properties. *J. Alloys Compd.*, 866, 158801, 1–5.

[101] Maurya, A., Bahadur, A., and Bahadur Rai, S. (2019). Enhanced upconversion and downshifting emissions from Tb^{3+}, Yb^{3+} co-doped $CaZrO_3$ phosphor in presence of Li^+ ions. *Methods Appl. Fluoresc.*, 7, 015002, 1–7.

[102] Talewar, R.A., S.K. Mahamuda, Swapna, K., and Rao, A.S. (2018). Sensitization of Yb^{3+} by Nd^{3+} emission in alkaline-earth chloro borate glasses for laser and fiber amplifier applications. *J. Alloys Compd.*, https://www.sciencedirect.com/science/article/abs/pii/S0925838818331591, 10.1016/j.jallcom.2018.08.270

[103] Kim, H.T., Lee, K., Jin, W., Um, H.D., Lee, M., Hwang, E., Kwon, T.H., and Seo, K. (2018). Phosphorescent energy downshifting for diminishing surface recombination in silicon nanowire solar cells. *Sci. Rep.*, 8, 16974.

[104] Soni, A., Rai, V., and Mahata, M. (2017). $Yb^{3}+$ sensitized $Na2Y2B2O7$:Er^{3+} phosphors in enhanced frequency upconversion, temperature sensing and field emission display. *Mater. Res. Bull.*, 89, 116–124.

[105] Hao, Y., Wang, Y., Hu, X., Liua, X., Liu, E., Fan, J., Sun, Q., and Miao, H. (2016). YBO_3:Ce^{3+},Yb^{3+} based near-infrared quantum cutting phosphors: synthesis and application to solar cells. *Ceramics Int.*, 42, 9396–9401.

[106] Keshavamurthy, K., and Eraiah, B. (2017). Silver lead borate glasses doped with europium ions for phosphors applications. *Bull. Mater. Sci.*, 40, 859–863.

[107] Hovel, H.J., Hodgson, R.T., and Woodall, J.M. (1979). The effect of fluorescent wavelength shifting on solar cell spectral response. *Solar Energy Mater.*, 2, 19–29.

[108] Loos, S., Steudel, F., Ahrens, B., and Schweizer, S. (2014). Optical properties of down-shifting barium borate glass for CdTe solar cells. *Opt. Mater.*, 41, 143–145.

[109] Steudel, F., Loos, S., Ahrens, B., and Schweizer, S. (2015). Luminescent borate glass for efficiency enhancement of CdTe solar cells. *J. Lumin.*, 164, 76–80.

[110] Thomas, S., George, R., Qamhieha, N., Gopchandran, K.G., Mahmoud, S.T., and Quatela, A. (2021). Sm^{3+} doped strontium barium borate phosphor for white light emission: spectroscopic properties and Judd–Ofelt analysis. *Spectrochim. Acta Part A: Mol. Biomol. Spectrosc*, 248, 119187.

[111] Borisov, S.M., Gatterer, K., Bitschnau, B., and Klimant, I. (2010). Preparation and characterization of chromium(III)-activated yttrium aluminum borate: a new thermographic phosphor for optical sensing and imaging at ambient temperatures. *J. Phys. Chem. C*, 114, 9118–9124.

[112] Chang, W., Li, L., Doub, M., Yana, Y., Jiang, S., Pan, Y., Wu, Z., Zhou, X., and Cui, M. (2019). Dual-mode downconversion luminescence with broad near-ultraviolet and blue light excitation in Tm^{3+}/Yb^{3+} co-doped oxy-fluoride glasses for c-Si solar cells. *Mater. Res. Bull.*, 112, 109–114.

[113] Chinna, B., Jamalaiah, and Rasool, S.N. (2015). Fluorescence properties of Sm^{3+} ions in yttrium aluminum borate phosphors for optical applications. *J. Mol. Struct.*, 1097, 161–165.

[114] Boyer, J.C., and van veggel, F.C.J.M. (2010). Absolute quantum yield measurements of colloidal $NaYF_4$: Er^{3+}, Yb^{3+} upconverting nanoparticles. *Nanoscale RSC*, 2, 1417–1419.

[115] Zhang, W., Ding, F., and Chou, S.Y. (2012). Large enhancement of upconversion luminescence of $NaYF_4$:Yb^{3+}/Er^{3+} nanocrystal by 3D plasmonic nano-antennas. *Adv. Mater.*, 24, OP236–OP241.

[116] Johnson, C.M., Reece, P.J., and Conibeer, G.J. (2011). Slow-light-enhanced upconversion for photovoltaic applications in one-dimensional photonic crystals. *Opt. Lett.*, 36, 3990–3992.

[117] Suyver, J.F., Aebischer, A., Biner, D., Gerner, P., Grimm, J., Heer, S., Kramer, K.W., Reinhard, C., and Gudel, H.U. (2005). Novel materials doped with trivalent lanthanides and transition metal ions showing near-infrared to visible photon upconversion. *Opt. Mater.*, 27, 1111–1130.

[118] Pan, A.C., del Canizo, C., Canovas, E., Santos, N.M., Leitao, J.P., and Luque, A. (2010). Enhancement of up-conversion efficiency by combining rare earth-doped phosphors with PbS quantum dots. *Sol. Energy Mater. Sol. Cells*, 94, 1923–1926.

[119] Zou, W., Visser, C., Maduro, J.A., Pshenichnikov, M.S., and Hummelen, J.C. (2012). Broadband dye-sensitized upconversion of near-infrared light. *Nat. Photonics*, 6, 560–564.

[120] Fan, B., Chlique, C., Merdrignac-Conanec, O., Zhang, X., and Fan, X. (2012). Near-infrared quantum cutting material Er^{3+}/Yb^{3+} doped La_2O_2S with an external quantum yield higher than 100%. *J. Phys. Chem. C*, 116, 11652–11657.

[121] Hernandez-Noyola, H., Potterveld, D.H., Holt, R.J., and Darling, S.B. (2012). Optimizing luminescent solar concentrator design. *Energy Environ. Sci.*, 5, 5798–5802.

[122] Billmeyer F. (1984). *Text Book of Polymer Science*, Wiley, Singapore.

[123] Jeong, S.K., Jo, Y.K., and Jo, N.J. (2006). Decoupled ion conduction mechanism of poly (vinyl alcohol) based Mg-conducting solid polymer electrolyte. *Electrochim. Acta*, 52, 1549–1555.

[124] Zhang, H., and Wang, J. (2009). Vibrational spectroscopic study of ionic association in poly (ethylene oxide)–NH$_4$SCN polymer electrolytes. *Spectrochim. Acta Part A: Mol. Biomol. Spectrosc.*, 71, 1927–1931.

[125] Hirankumar, G., Selvasekarapandian, S., Kuwata, N., Kawamura, J., and Hattori, T. (2005). Thermal, electrical and optical studies on the poly (vinyl alcohol) based polymer electrolytes. *J. Power Sources*, 144, 262–267.

[126] Obula Reddy, M., and Chandra Babu, B. (2015). Structural, optical, electrical, and magnetic properties of PVA:Gd^{3+} and PVA:Ho^{3+} polymer film. *Indian J. Mater. Sci.*, 927364, 1–8.

[127] Sadhu, S.D., Soni, A., Garg, M., and Varmani, S.G. (2014). Preparation of starch-poly vinyl alcohol (PVA) blend using potato and study of its mechanical properties. *Int. J. Pharma Sci. Invention*, 3, 33–37.

[128] Ghoshal, S., Denner, P., and Stapf, C.S. (2012). Study of the formation of poly (vinyl alcohol) films. *Macromolecule*, 45, 1913–1923.

[129] Ratner B.D., Hoffman A.S., Schoen F.J., and Lemons J.E. (2004). *Biomaterial Science: An Introduction to Materials in Medicine*, 2nd Ed. Elsevier Academic Press.

[130] Werber, A., and Zappe, H. (2005). Tunable microfluidic microlenses. *Appl. Opt.*, 44, 3238–3245.

[131] Yen, D.A.C., Eich, R.K., and Gale, B.K. (2005). A monolithic PDMS waveguide system fabricated using soft-lithography techniques. *J. Light wave Technol.*, 23, 2088–2093.

[132] Wang, S., Kallur, A., and Goshu, A. (2011). Fabrication and characterization of PDMS thin film. *Proc. of SPIE*, 7935, 79350M, 1–6.

[133] Bilent, S., Hong, T., Dinh, N., Martincic, E., and Joubert, P.Y. (2019). Influence of the porosity of polymer foams on the performances of capacitive flexible pressure sensors. *Sensors*, 1968, 1–12.

[134] Lippert, B.S.T., Graubner, V.M., Kotz, U., Nuyken, O., and Wokaun, A. (2003). UV irradiation induced modification of PDMS films investigated by XPS and spectroscopic ellipsometry. *Surf. Sci.*, 532–535, 1067–1071.

[135] Trupke, T., Green, M.A., and Würfel, P. (2002). Improving solar cell efficiencies by down-conversion of high-energy photons. *J. Appl. Phys.*, 92, 1668–1674.

[136] Badescu, V., De Vos, A., Badescu, A.M., and Szymanska, A. (2007). Improved model for solar cells with down-conversion and down-shifting of high-energy photons. *J. Phys D: Appl. Phys.*, 40, 341–352.

[137] Pugliese, D., Boetti, N.G., Lousteau, J., Ceci-Ginistrelli, E., Bertone, E., Geobaldo, F., and Milanese, D. (2016). Concentration quenching in an Er-doped phosphate glass for compact optical lasers and amplifiers. *J. Alloys Compd.*, 657, 678–683.

[138] Dvoyrin, V.V., Mashinsky, V.M., Bulatov, L.I., Bufetov, I.A., Shubin, A.V., Melkumov, M.A., Kustov, E.F., Dianov, E.M., Umnikov, A.A., Khopin, V.F., Yashkov, M.V., and Guryanov, A.N. (2006). Bismuth-doped-glass optical fibers - a new active medium for lasers and amplifiers. *Opt. Lett.*, 31, 2966–2968.

[139] Shaw, R., and Uhlmann, D. (1973). Neodymium glass laser having room temperature output at wavelengths shorter than 1060 nm U.S. Patent US3717583 A.

[140] Boulon, G. (2008). Why so deep research on Yb-doped optical inorganic materials? *J. Alloys Compd.*, 451, 1–11.

[141] Brenier, A. (2001). A new evaluation of Yb^{3+} -doped crystals for laser applications. *J. Lumin.*, 92, 199–204.

9 Borate Phosphor
Mechanoluminescence and Lyoluminescence Phosphors

V. R. Raikwar

CONTENTS

DOI: 10.1201/9781003207757-9

9.1   HISTORICAL BACKGROUND

9.1.1   MECHANOLUMINESCENCE (ML)

Luminescence is a phenomenon of light emission by certain materials caused by some external parameters such as chemical reaction, incident light, applied pressure, force, temperature or electric current. The type of luminescence induced by any mechanical action on solids is known as mechanoluminescence. This mechanical action can be elastic deformation, plastic deformation and fracture of solids. The light emissions induced by these actions are respectively called elastico ML (EML), plastico ML (PML) and fracto ML (FML) [1]. Nearly 50% of all inorganic salts and organic molecular solids show FML; however, only a few solids exhibit EML and PML. Although the phenomenon of ML is known for a long time, up to the end of 20th century no remarkable practical application of mechanoluminescent materials could be made because of the low ML intensity and lack of reproducibility. In earlier days, ML was known as triboluminescence [2]; nowadays it corresponds to luminescence due to the contact of two dissimilar materials. Mechanoluminescence was a known phenomenon as light was generated from striking stone or quartz. But, the first discovery of ML was reported by Francis Bacon (1605) [3]. It was mentioned in his work on "The advancement of learning, divine and human", the scraping of the hard sugar with a knife; "it is not the property of fire alone to give light...loaf-sugar in scraping or breaking". In 1664, Robert Boyle reported that a particular diamond, when pressed upon with a steel bodkin, produced a short-lived bright glow [4]. For several centuries, the phenomenon of light emission associated with rock breaking in mines and during earthquakes has been reported. The sugarcane production was taken by the people worldwide from hundreds of years. In earlier days, a sugar product was very hard. The process to separate it from the container might have produced a tinkling spark, which is nothing but ML. This was a period that was before the invention of photomultiplier tube (PMT), where most observations were carried out with the observer's naked eyes who might be efficiently trained for taking observations in the dark. The observations and conclusions which were in 17th-century and 18th-century research articles mostly included a number of minerals and pottery materials, sugar, rock salt, agate and jasper. The workers in this field concluded that these materials were capable of holding light that they receive and release it when fractured. By the end of 1920, there were nearly 500 known ML materials. Longchambon recorded the ML spectra of sugar and other crystals for the first time [5]. In 1925, Longchambon in his pioneer work on organic compounds, reported ML in 70% of 305 aromatic compounds and in 20% of 90 alkaloids. Longchambon suggested that ML emission arises from the dielectric breakdown of surrounding gases (although the spectra may contain additional lines if the materials can be excited to photoluminescence by ultraviolet in the gas discharge) [5]. Bhatnagar et al. [6] reported the ML of saccharin crystals. Inoue et al. [7] recorded the ML spectra of sugar and tartaric acid crystals. Carriere, in 1946 [8], studied the ML produced during the stretching of rubbers. Hoff and Boord [9] found that cis-4-octane shows ML, while trans-4-octane does not when they are both subjected to rapid cooling by immersing a test tube containing these compounds in liquid air. The ML of II–VI semiconductors, particularly ZnS:Mn, coloured alkali halide crystals and Si were also reported during the first generation of Jenny reported triboluminescence in semiconductors [10]. The leading workers during the first generation of ML were: Tschugaeff [11], Trautz [12], Garnez [13], Longchambon [5] and Wick [14]. The PMTs were used for ML studies after 1950. As per the literature survey, very few reported ML research in the 17th, 18th and 19th centuries. About this important phenomenon of ML, the workers showed interest in 20th century and with the advancement of technology and possibility of using ML in applications in various fields of life, published around 372 research papers. In the 21st century, more than 500 research papers have been published to date. That shows the importance of ML in myriad applications from sensing to displays [15]. Many materials were discovered, which depicted ML and they were tried and tested in various applications. ML materials can be conductors, semiconductors and insulators. Several materials have been discovered or synthesised, after the development of photomultiplier tube in the mid-20th century, with one-half of inorganic crystals and almost one-third of organic compounds showed ML [16]. In the last three decades (mostly in 1990s), a range of materials have been discovered that emit an intense and repeatable EML during their elastic deformation without any destruction [17–33].

Major applications of ML are listed here: as a stress sensor, "the real-time visualization of the stress distribution in solids", for the visualization of internal defects in a pipe, in the analysis of artificial legs, for the "real-time visualization of quasidynamic crack-propagation" in solids, to determine crack-growth resistance and other parameters of crack propagation, as a novel "ML-driven solar cell" system, for the "real-time visualization of the stress field near the tip of a crack", as a source of ML light, in the determination of laser and ultrasonic powers, in writing secret messages, as a "EML-based safety-management monitoring system", in the non-destructive testing of materials, in radiation dosimetry, in mechanoluminescence damage sensors, in fracture sensors, in impact sensors, in pigments and paints for security printing and fracture studies, in a "fuse-system for army warheads", in the evaluation of the design of a milling machine, in "online monitoring of grinding processes", as a potential earthquake indicator, in the determination of several parameters of solids, and a triboluminescence X-ray unit has been designed in which X-ray imaging can be performed [34]. A new era of application includes artificial tooth, artificial skin, encrypted communication in a defence system, information storage, patient care and health monitoring [35].

The progress in mechanoluminescence has been relatively slow because of multiple reasons, such as low emission of light by the crystals in the aggregate state and absence of proper explanation of an intrinsic mechanism of this special type of light emission. One of the possible reasons of light emission in ML can be the electrons (along with ions and neutral species) were emitted from the surface of crystals when the ML crystals were fractured by external force stimulus. Subsequently, the luminescent centres could be excited by electron bombardment to yield mechanoluminescence.

9.1.2 LYOLUMINESCENCE (LL)

Lyoluminescence is a form of chemiluminescence in which the emission of light occurs when a pre-irradiated solid is dissolved into a specific liquid solvent. In general, lyoluminescence effect is observed, when solid samples are irradiated heavily by ionizing radiation and later they are dissolved in water. The total amount of light emitted by the material increases proportionally with the total radiation dose received by the material up to a certain level called the saturation value. "Many gamma-irradiated substances are known to be lyoluminescent; these include spices, powdered milk, soups, cotton and paper". Several organic substances and halide crystals produce a glow when lengthy exposure to high-energy radiation like X- rays, γ-rays, etc. is followed by immersion in a liquid solvent. This glow is called the "lyoluminescent glow". The first reported phenomenon of lyoluminescence was by Wiedemann and Schmidt, who observed lyoluminescence in glucose [36]. The glucose after dissolution and after exposure to ionizing radiation produced a glow that was initially enhanced with time and then saturated after about a minute. Ahnstrom and Ehrenstein investigated that unirradiated glucose did not show any glow on dissolution [37]. Investigating further, Ahnstrom and Ehrenberg subsequently used the glow to determine the density of F-centres in an inorganic sample as a function of the incident gamma dose [38,39]. Lelievre and Adloff worked on halide crystals and studied same characteristics. In their investigation, it was observed that when the pH of the solution was increased, LL of glucose was enhanced [40,41].

The LL intensity is affected by various factors. It is proportional to the amount of solute dissolved and also to the radiation dose to which the sample was exposed. The theoretical and quantitative analysis on effect of these parameters on LL intensity was studied by Burns and Williams [42] and Mittal [43]. Realizing its potential as a dosimeter, various work on lyoluminescence was noted in the 1970s (Atari et al. [44]; Atari and Ettinger [45]; Bartlett [46]; Bartlett et al. [47]; Burkill [48]; Takavar et al. [49]; Hanig [50]). Most of the experiments, however, were carried out as experimental investigations on this front. A theoretical effort towards the quantitative estimation of LL from organic substances was due to Chatterjee et al. [51]. Their approach was centred around the development of "a rate equation from phenomenological considerations" using mechanisms already suggested by Ettinger and Puite [52,53] and Russel and Vassil'ev [54,55]. Kundu and his co-workers have studied the LL of certain organic solids [56–58]. In the past, extensive measurements on the LL of γ-irradiated alkali halide crystals have been made. Most of the results can be understood on the basis of the theory explored by Russel and Vassil'ev. Arnikar et al. have found that when γ-irradiated alkali halide crystal of a given mass is dissolved in a fixed volume of distilled water, the LL intensity increases linearly with time, reaches an optimum value, shows gradual decrease and finally disappears. It has been found that the slope of log (I) versus t plot decreases with the increasing time. This may primarily be due to the fact that the rate of dissolution decreases with increasing concentration of the solute in the solvent [59]. The study of LL investigation of inorganic phosphors resulted in very fruitful results of which investigations begin in early 1990s. Before that, alkali halides were investigated a lot. LL during dissolution of calcite irradiated with gamma rays in HCl was studied by Copty Wergles et al. [60]. Terbium-doped potassium peroxodisulphate dissolved in aqueous solution and its LL after irradiation by UV was reported by Kulmala and Haapaka [61]. The mechanism of analytical application of luminol-specific extrinsic LL of UV irradiated potassium peroxidisulphate was reported by Matachescu et al. [62]. Vickery et al. investigated the LL phenomenon in the super molecular aggregation of antinuclear gold complex [63]. Chattopadhyay et al. developed "a nonlinear differential equation and analysed it by a dynamic system analysis as well as exact numerical integration" [64]. Raman et al. in 2001 studied the LL of trehalose dihydrate and reported that LL shows linear gamma sensitivity in the dose range 0.1×10^2 to 5×10^2 Gy [65]. Lyoluminescence property of gamma irradiated Eu activated $KNaSO_4$ and $K_3Na(SO_4)_2$ phosphors were reported by S. J. Dhoble in 2002. He studied the fading of LL intensity and found that this phosphor is suitable for LL dosimetry [66]. Gour et al. studied dependence of LL of divalent impurity doped NaCl on temperature [67]. Banerjee et al. studied particle size dependence of lyoluminescence of gamma irradiated sodium bromide [68]. Hakansson et al. studied the lyoluminescence of aluminium induced by lanthanides chelates dissolved in alkaline aqueous solution [69]. LL of sulphate phosphors doped with Dy, Tb and Eu were investigated for possible applications in dosimetry for ionizing radiations [70–72]. Sahu et al. investigated LL intensity of gamma irradiated KCl:Sr. The particle size dependent LL was reported [73]. The temperature and mass dependence of LL intensity of gamma irradiated KCl was studied by Nayar et al. in 2010 and reported that peak lyoluminescence intensity increases with increase in amount of solute [74]. The study of phosphate phosphors was reported [75]. Borate phosphors are not studied much for lyoluminescence. A few reports on borate phosphors are published to date. Mishra et al. studied thermoluminescence and LL in gamma irradiated $YCa_4O(BO_3)_3$: Ce^{3+} phosphors [76]. Mishra et al. studied LL of gamma irradiated $GdCa_4O(BO_3)_3$ doped with rare earth Ce. They studied the effect of a mass of solute and pH of a solvent on the LL intensity [77].

9.2 FUNDAMENTALS OF MECHANOLUMINESCENCE

9.2.1 MECHANOLUMINESCENCE THEORY

The luminescence caused by any mechanical action on solids is called mechanoluminescence (ML). The actions comprise scratching, rubbing, cutting, crushing cleaving, shaking, compressing or grinding of solids. It can also be generated by sudden change in temperature viz. cooling or heating of materials or ML can be created by the shock waves produced during exposure of samples to powerful laser pulses. The deformation caused by the phase transition or during separation of two dissimilar materials in contact may be the possible reasons for ML. "The phenomenon of ML has been known by many other names, such as trennugslicht, triboluminescence, piezo luminescence, deformation luminescence, and stress-activated luminescence"[1]. Initially it was thought that only rubbing and scratching a sample could have produced luminescence. But if the sample undergoes any kind of mechanical stress or deformation it produces luminescence. As the word "mechano" is generally correlated with different mechanical concepts, such as "tribo, stress, deformation, piezo, fracto, plastic and elastico, the various processes like cutting, cleaving, grinding, rubbing, compressing, and crushing" etc.; hence, this type of luminescence is better known as "mechanoluminescence" [1]. Since the first scientific report of ML by Francis Bacon in 1605, the workers in this field have reported ML from more than 50% of all inorganic and one-third of organic molecular solids; however, most of them are fracture induced and only few of them show ML accompanied by elastic deformation. The use of the fracture-involved process has a limitation in terms of reproducibility. On the other hand, piezo or elastico-luminescence has a better applicability in reproducible sensing systems because it operates in the elastic range without failure of the material, owing to the rechargeable trap-releasing mechanism. In addition to reproducibility, intense emission is an important requirement for ML materials while considering their application in sensors [78].

Although several mechanisms of ML excitation have been proposed [1,21,79–82], the two basic mechanisms which are applicable to a large number of materials are: "(i) electron capture or electron trapping (ET) mechanism, and (ii) electron bombardment (EB) mechanism". Whereas the ET mechanism can be explained by the motion of dislocations and vacancies, the EB mechanism becomes possible by the electric field generated by "piezoelectricity, charged dislocations and other processes of polarization". As explained by Chandra et al. (2013), in the mechanism of plastico-ML [83], "the plastic deformation in any material causes dislocations movement. Capturing of electrons takes place by the moving dislocations from the interacting F-centres lying in the expansion region of dislocations. The captured electrons from F-centres move with the dislocations and they also drift along the axes of dislocations. When these electrons recombine with the holes lying in the dislocation donor band, the light emission characteristic of the hole centres take place". In case of X or γ-irradiated alkali halide crystals, some of the atoms from the impurity after their recombination with dislocation electrons can induce ML emission called impurity emission.

The physical processes involved in inducing ML in solids indicate that basically there should be two types: "(1) deformation-ML (DML) and (2) tribo-ML (TML)" as shown in Figure 9.1. The first type of ML generates light due to the physical processes induced during deformation of solids, whereas, the latter is due to contact phenomena such as triboelectricity, tribochemical reactions and tribothermal generation induced because of the contact or separation of two dissimilar materials. DML depends only on the material under deformation and is independent of both; the contact phenomenon and the material used to produce the deformation. It is further subdivided into three types: "fracto-ML(FML), elastico-ML(EML) and plastico-ML (PML)". The luminescence generated due to the creation of new surfaces during the fracture of solids is called fracto-ML. The

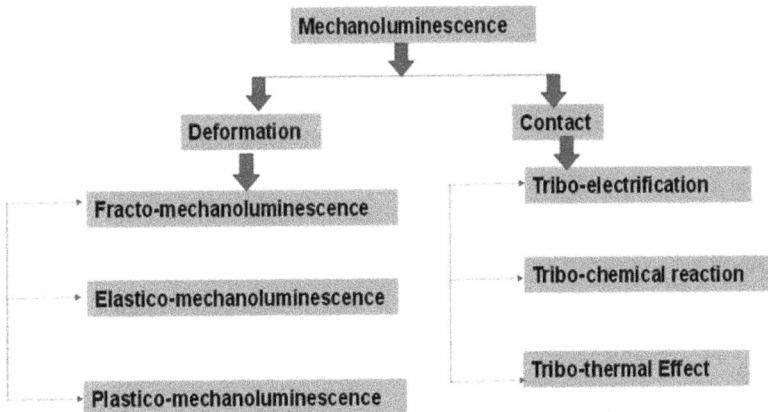

FIGURE 9.1 Mechanoluminescence classification.

luminescence produced due to the plastic deformation of solids, where fracture is not required is called plastico-ML and the luminescence produced during elastic deformation of solids, where neither plastic deformation nor fracture is required, is called elastico-ML.

9.2.1.1   Fractomechanoluminescence (FML)

If a crystal undergoes fracture of any kind, several effects arise in addition to light emission. It includes production of highly localized heat, intense electric fields, emission of electrons and positive ions or acoustic wave production. The mechanism of FML has been explained in detail by Chandra et al. [83,84]; accordingly, "the after effect of fracture are the surfaces which are highly charged, generating electric fields up to $10^6 - 10^8$ V·cm^{-1}". This produces piezoelectricity, which, in turn, excites the activator. The ML intensity increases and then shows an exponential decrease. When more compression is applied to the sample, more activator centres are excited and hence the intensity depends on the impact strength. In a classic FML experiment, a piston, or a metal ball, moving along the z-axis, is projected onto the sample that is fractured upon impact. The measure of the deformation of the sample is given by the strain σ according to the following equation:

$$\sigma = \frac{x}{l} = \frac{v_0}{l.\,\xi}[1 - e^{-\tau\xi}] \tag{9.1}$$

where l is the thickness of sample, x is the contraction sustained by the sample, v_0 is the velocity of metal ball or piston at the time of impact and $\xi = 1/\tau_r$ is the time constant for the relaxation of the impactor.

9.2.1.2   Elastico-Mechanoluminescence (EML)

When a film or pellet, formed by mixing luminescent materials in epoxy resin is deformed, under a threshold pressure, elastico-mechanoluminescence (EML) emission takes place. EML spectra is generally observed by exerting pressure on the sample with the help of a press machine, or by means of the impact of a hard ball onto the sample which results in a reversible deformation of the sample. Two main mechanisms have been proposed and investigated for ML induced by elastic deformation [75]. In the first one, an electric field generated due to deformation of crystalline structures reduces the trap depth of the carriers or to move electrons into the conduction bands. These electrons may transfer their energy to the activator ion triggering its luminescence, which is termed 'intrinsic luminescence'. The fantastic features of EML phosphors like linearity, reusability, durability and non-destructive light emission are especially attractive for applications like displays, sensors and light sources [85,86]

9.2.1.3   Plastico-Mechanoluminescence (PML)

This luminescence arises during irreversible deformation of the sample upon applying pressure, without fracturing it. The effect of pressure materialises in the formation of dislocations in the sample. If K is the inverse of the strength coefficient of the material, the induced plastic strain s can be written as

$$\sigma = \left(\frac{P}{K}\right)^n = \left(\frac{P_0}{K}\right)^n[1 - e^{-\xi t}]^n \tag{9.2}$$

where P_0 is the final pressure, $\xi = 1/\tau_r$ is the time constant for the relaxation of the impactor.

9.2.1.4   Tribo-Electrification

In this type of ML, light emission occurs due to an electric field produced during rubbing of two dissimilar materials. A familiar example is when silky clothes are rubbed with our skin and electric charges are generated, and thereby produce an electric field. This phenomenon was studied by Meyer, during low-energy impact of sapphire needle on the surface of alkali halides and observed by Keszthelyi and Bard during shaking of mercury in a glass container [87,88].

9.2.1.5 Tribo-Chemical Reaction

Here, the light emission occurs during the chemical reaction that takes place when two dissimilar materials are rubbed. It's a well-known phenomenon in chemistry. It was reported that when mercury was shaken in the reduced pressure of CO_2, hydrocarbon, H_2–N_2 mixture and acetone etc., a chemical reaction takes place, producing visible luminescence [89].

9.2.1.6 Tribo-Thermal Effect

The blackbody emission produced during the cutting of certain glasses with a diamond-impregnated saw blade produces heat due to the friction between the blade and the samples. Hence, it is a thermally driven tribo effect. The luminescence occurs at different temperatures for different materials [90].

9.2.2 Experimental Setups for Measurement of ML

The behaviour of a system under pressure can be studied by luminescence measurements of solids at high hydrostatic pressures. The pressure can gradually be increased so that the surrounding molecular environment can be changed in a controlled manner. This way, the effect of pressure on energy levels can be studied [91]. Mechanoluminescence is an interesting phenomenon, which is a light emission caused by friction, rubbing, striking, grinding, cutting, etc. The emission of light intensity depends on how the material is deformed. The apparatus used to measure ML should have a device for a deforming sample and another for the spectral measurements [1]. The deformation of a sample can be done by various techniques such as compression, loading, piston impact, needle impact, bending, stretching, scratching, air blast, rubbing, grinding, etc. Figure 9.2 shows the device used for this purpose by Chandra [1]. Figure 9.2(a) is a desktop model Instron testing machine. In this technique, ML intensity is measured with a load cell and strain with a linear variable differential transducer. Figure 9.2(b) is a schematic diagram of a piston impact technique

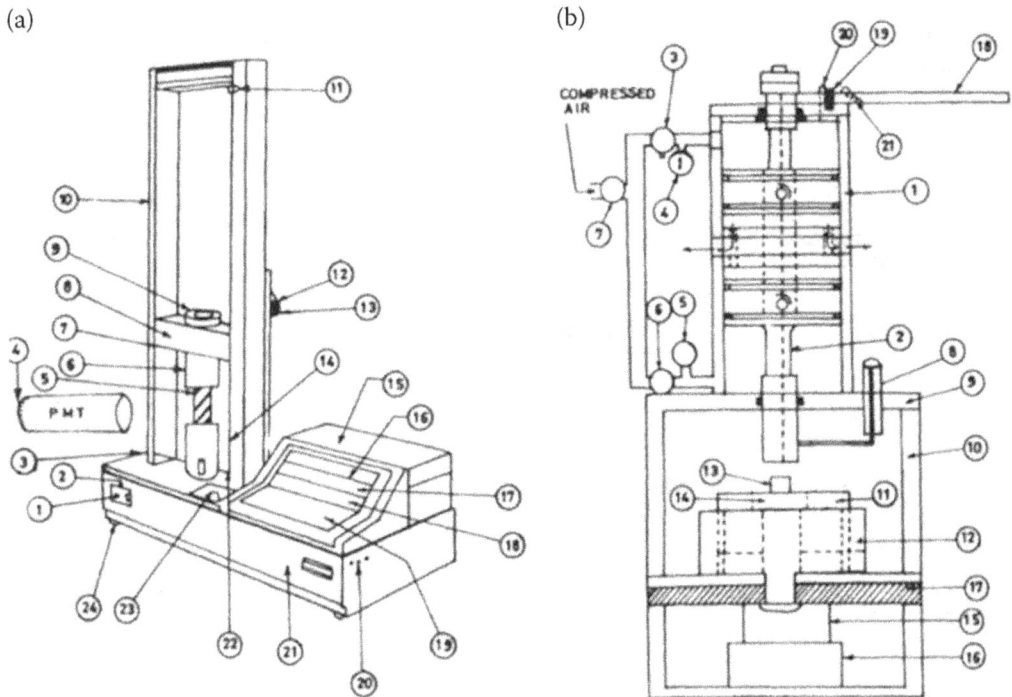

FIGURE 9.2 (a) A stress-strain measurement Instron testing machine, (b) schematic showing piston impact technique {Adapted from Chandra B P (1983) [1], for parts description readers are referred to [1]}.

used by Chandra [1]. The ML is excited by the impact of a moving piston on the crystal with the velocity of the moving piston being measured by a transducer.

The compression [92,93] and the piston impact or impulsive techniques [94] are the highly likely choice for ML measurements.

ML detection using high-speed camera experimental setup used by L. Belyaev et al. shows promising method of ML intensity detection [95,96]. As shown in Figure 9.3, they used interference filters and a monochromator of great light power. The samples were put in a special chamber and a uniaxial pressure was applied. The light stream passes through the quartz window of the chamber and gets divided into two beams by quartz plates. The two photomultipliers are used for recording the monochromatic and integral light streams. The signals from the two photomultipliers are displayed on an oscillograph [96].

Rai et al. [97] used a setup to measure ML, as shown in Figure 9.4, by applying continuously varying pressure on the sample. It had a photomultiplier tube RCA 931A connected with a digital

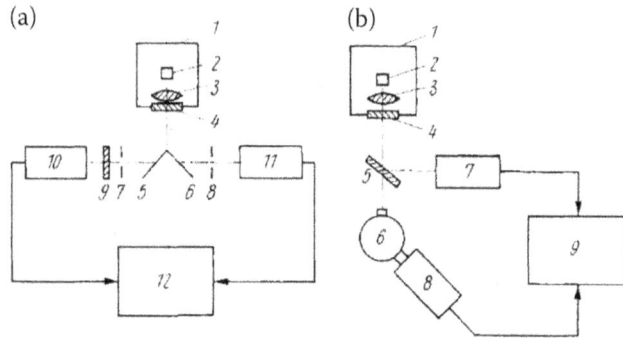

FIGURE 9.3 (a) Experimental arrangement for the registration of triboluminescence spectra. 1-Chamber, 2-Specimen, 3-Quartz lens, 4-Quartz window, 5,6-Mirrors, 7, 8-Diaphragms, 9-Interference filter, 10,11-Photomultipliers, 12-Oscillograph, (b) experimental arrangement for the registration of triboluminescence spectra 1-Chamber, 2-Specimen, 3-Quartz lens, 4-Quartz window, 5-Quartz plate, 6- Monochromator, 7, 8-Photomultipliers, 9-Oscillograph {Adapted from Belyaev L (1969) [96]}.

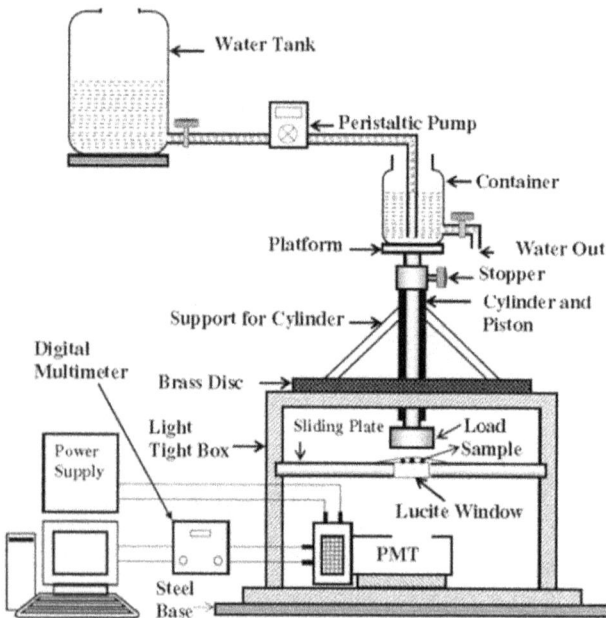

FIGURE 9.4 ML measurement system by applying continuously varying pressure on the sample {Adapted from Rai et al. (2012) [97]}.

multimeter. The piston is provided with a cylindrical load head and a platform. The load head is rested on the sample and pressure is applied by pumping water in the container that is placed on the platform through peristaltic pump. The rate of pressure applied on the sample can be varied by the flow rate of water from the container.

The impact-based method, also called a weight drop tower, was developed by Xu et al. in 1998 [98]. It was modified by a researcher group, Fontenot et al. [99], as shown in Figure 9.5(a). There is a polyvinyl chloride (PVC) pipe, in which a steel ball is dropped from a certain height that impacts the sample. The stress resulting on the powder sample is related to the weight of the steel ball and the drop distance. The ML signal thus generated will be detected by a photodiode and oscilloscope arrangement. This weight drop tower method provides a low cost, facile setup for investigating ML materials in both bulk and powder form.

Sakai and co-workers constructed a combined apparatus comprising "a photon counting system with an atomic force microscopy (AFM)" (Figure 9.5(b)) [100]. The AFM probe applies pressure directly on a EML crystal and the light emission of the particle is recorded by the integrated photon counting system. This measurement gives the relationship between the strain and ML intensity of a single particle.

For measurement of ML in thin film and pellet samples, several prototypes were designed. One of them was a universal testing machine developed by Yamada et al., as shown in Figure 9.5(c). The sample powder is generally mixed with resin and converted into pellet form and pressed under the universal testing machine and stress-strain curve is recorded. A high-speed camera and pho-tomultiplier tube record the EML images and light current, respectively [101].

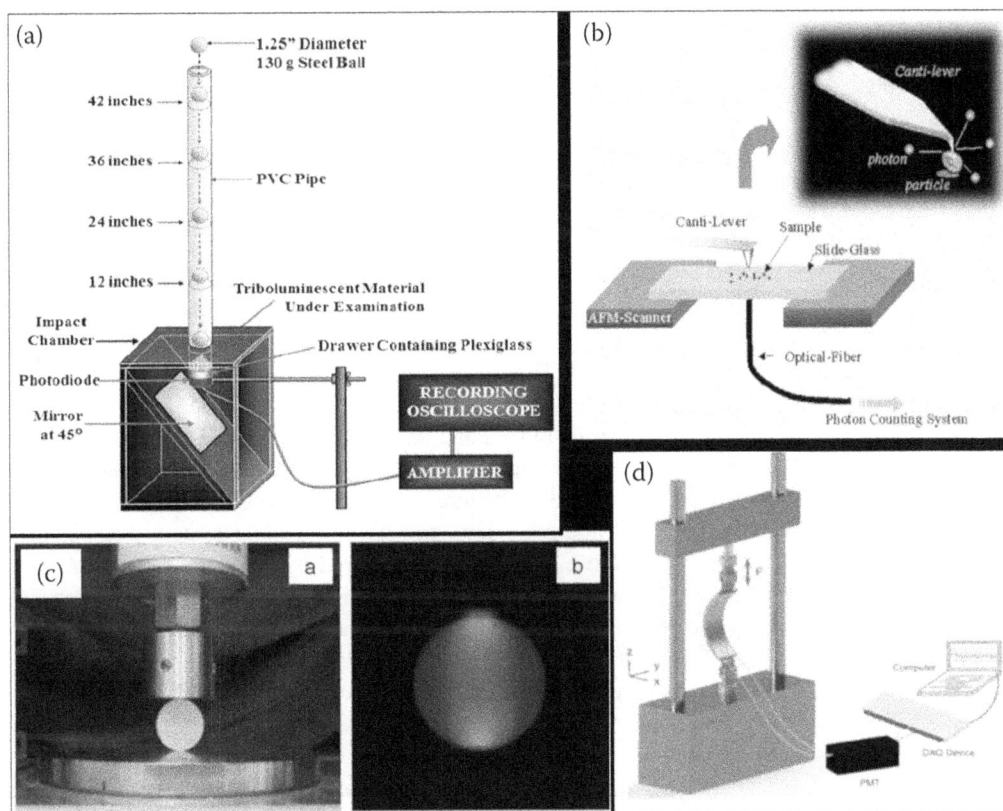

FIGURE 9.5 Schematics of typical setups for characterizing EML phosphors: (a) weight drop tower [99], (b) AFM coupled with photon counting system [100], (c) universal testing machine for pellet sample [101], (d) universal testing machine for thin films [102]. {Adapted from [99–102]}.

A film can be created using the ML material/polymer composite and the measurement can also be implemented in the universal testing system (Figure 9.5(d)) [102,103]. Here, both tensile and compressive stress can be applied on the sample film.

9.2.3 APPLICATIONS OF ML

ML provides detection methods of processes occurring due to deformation and fractures of solids. The ML materials which can be applied as a thin film may be possibly used in myriad fields of life. The applications of ML are mainly in a dosimeter. The other applications include illumination source and display, stress detection and multifunctional sensing [104–107]. As the ML intensity can be recorded at time intervals of fraction of seconds, it can provide sensitive optical tool for studying time resolved fracture dynamics. ML can be used to study kinetics of emergence of mobile dislocations onto the metal surfaces [108]. It was shown that FML can be used for the investigation of rock fractures in mines and during earthquakes [109]. The FML materials can also be used for a wireless fuse-system for army warheads [110,111]. The triboluminescence-based triboluminometer has potential applications in radiology, toxicology, cardiology and pediatrics [112]. The TML makes it possible to write a message that can be seen in the dark [113]. Mainly EML phosphors can convert mechanical energy to light energy. Terasaki et al. have demonstrated the ML material acted as a light source for solar cell, stress historical-log recording system and fluorescent molecular probe for bioimaging [114–116]. They have investigated ultrasound-induced ML (USML), and successfully detected the ML that depends on the input intensity of ultrasonic wave irradiation. They demonstrated the application of USML as the light source using TiO_2 photocatalyst for the first time [117]. Stress sensors can be created using the EML materials and visualization of stress distribution in solids [31], impact sensors [118], damage sensors [119], stress field near the tip of crack [120] and internal defects of a pipe [121]. It was shown that EML materials can be used for the stress imaging of the orthopaedic devices and hence they have potential for analysing the design of the artificial organs [122]. A crack that occurs in the concrete of bridges and buildings can be visualised using the safety-management monitoring system that has been developed using EML [123]. Freund and his co-workers have reported that the ML material have potential for their use in earthquake sensors [124–126]. Some other applications include wind-driven nighttime solar energy converters [127], intelligent solid lubricant [128], aerospace impact sensors [129], foreign planet ice detection [130], environment friendly billboards [131], micro-plasma surface engineering [132], fraud protection [133], etc. It was found that the excitation of small hair-like fibres made of silicone and containing EML material (ZnS), cause emission of light. The elastic stress on such an arrangement will create a piezo-electric field that is directly proportional to applied stress. When charge is in sufficient amount, the trapped electrons in ML materials get detrapped and cause luminescence. The areas where the stress is more are brighter due to proportional relation between applied stress and generated charge. When the elastic stress is removed, the fibre-like structure returns to their original shape and position [127]. Quickenden et al. explained that there are icy surfaces in our solar system that could be emitting light that is detectable when those surfaces are shadowed by the sun [130]. The newly discovered area of applications are multi-touched pressure flexible sensors made by using ML phosphors that are mixed anion compounds doped with rare earth and transition metal materials like $SrZnSO:Pr^{3+}$, $SrZn_2S_2O:Mn^{2+}$, β-SiAlON:Eu^{2+} and $SrSi_2O_2N_2:Yb^{3+}$ [134]. They can be widely extended to other applications like "stress intelligent sensors, electronic skins and human-machine interfaces". X. Wang et al. used flexible pressure sensor matrix (PSM) based on the ML process for personalised signature [133]. They realised the need of pressure sensitised dynamic mapping in a very short time span, which is the requirement of security signature. Keeping this in mind, they devised a PSM system that uses ZnS:Mn as ML material sandwiched by two polymeric layers at the top and bottom. The transparent polymeric films are favourable to the transmission of yellow light emitted by the ML material under pressure. Such systems can find applications in "real time pressure mappings, smart sensor networks, high level security systems and human machine interfaces" [133].

9.2.4 Mechanoluminescence as a Dosimeter

The need of the hour is to search for materials that exhibit phenomena that would be applicable to fast-neutron dosimetry. Although there are various types of neutron dosimeters and they have their own advantages and disadvantages, all neutron dosimeters must have the following characteristics or requirements:

i. A dosimeter should respond over a range of nine decades in energy for fast neutron dosimetry and, over this energy range, its response must be parallel to the fluence-to-dose-equivalent conversion factor.
ii. One important requirement for a fast-neutron dosimeter is that it should be capable of detecting a neutron-absorbed dose in mixed radiation fields and it should have gamma ray-to-neutron-dose ratios greater than ten.
iii. The dosimeter should also be able to detect a fast-neutron dose equivalent as small as ^10 mrem and have good signal stability.
iv. It should also be non-toxic, which is an important health aspect for the people working in this field and cost effective too, for the nuclear industry.
v. It should be relatively insensitive to other types of radiations. It should have sufficient accuracy and precision.
vi. It should have a good signal stability with negligible fading due to ambient conditions.
vii. It should be rugged and easy to handle for a large scale of monitoring purposes.

ML and LL dosimeters may fulfil many of these characteristics. For this important application, the research fraternity must accelerate their efforts as in the future nuclear power may be the only source of energy that can fulfil an increased need of energy. As the use of nuclear energy will increase, it requires the safety from radiations and for the quantitative analysis; dosimeters are the ultimate choice.

$MgAl_2O_4$: Dy phosphor was synthesised and studied for the possible application as a dosimeter. It was shown that ML intensity increases, attains an optimum value at a particular time and then decreases again. For higher gamma dose, ML intensity is maximum. The fading was observed for around 5 to 6% when recorded after 15 days of irradiation. When the mass of the sample increases, the number of crystallites in the sample increases and thereby the ML intensity and total ML intensity increases to a maximum for a higher dose. The light emission in $MgAl_2O_4$:Dy corresponds to the movement of dislocations and the recombination of activated electrons and holes. The carriers from the traps are excited by movement of dislocations. The light emission is thus caused by a recombination of activated electrons and holes in defect centres. If there is an increase in gamma dose, the density of defect centres increases. It was reported that a large number of lattice defects can be served as trapping centres and will be provided by the cation disorder and nonstoichiometry of aluminates like $MgAl_2O_4$ [135].

9.3 FUNDAMENTALS OF LYOLUMINESCENCE

9.3.1 Lyoluminescence Theory

When high-energy radiation like X- rays, gamma rays, etc. are allowed to fall on several organic substances and halide crystals, followed by immersion in a liquid solvent, they produce a glow. This glow is called the lyoluminescent glow. The first known instance of observation of lyoluminescence (LL) was credited to Wiedemann and Schmidt [36], who observed lyoluminescence in glucose. They exposed glucose to ionizing radiation and later dissolved glucose in water at room temperature; this gets a glow that gives a steady illumination after about a minute. The LL emission intensity depends upon the amount of solute dissolved in the solvent and the radiation exposure dose as discussed by several groups working in LL field. There were several papers published in determining the quantitative dependence of LL intensity on these parameters [43,44].

Several mechanisms were proposed to explain the LL in various materials. An accepted scheme proposed by Russel [54] and Vasilev [55] predicts that the peroxy radicals are formed during irradiation that react with the oxygen present in the sample. These peroxy radicals undergo recombination reaction during dissolution and the deexcitation of reaction products gives rise to emission in the visible range. Using this scheme, postulates of Ettinger and Puite [52,53] were proposed. According to these postulates, alkyl radicals are produced in the sample during irradiation that are then oxidized to convert into peroxy radicals on dissolution. On combination, excited carbonyls in their triplet state are formed which then transit to ground state and emit LL in visible region [52].

For alkali halides or other inorganic materials, LL is believed to be caused by electrons freed from F-centres [44]. An electron trapped in a halide ion is called an F-centre and a V-centre is a hole analogue of an F-centre. It has been proposed that dissolution of these materials in water results in the hydration of the electron-hole pairs created by radiation. These pairs recombine, which results in the emission of light. This phenomenon is shown in the following reaction:

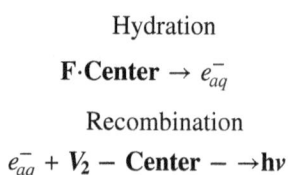

Hydration

$$\textbf{F·Center} \rightarrow e_{aq}^{-}$$

Recombination

$$e_{aq}^{-} + V_2 - \textbf{Center} - \rightarrow \textbf{h}\nu$$

Although most of the studies on LL dosimetry have involved the use of gamma radiation (usually ^{60}Co or ^{137}Cs sources), it would appear that the most important use of LL would be in neutron personnel dosimetry. The character of the observed light may depend on several competing mechanisms and on conditions of dissolution, type and nature of radiation dose and on the type of phosphor. Other mechanisms for LL involve the generation of singlet oxygen or perhaps decomposition of hydroperoxide.

There are several parameters that affect the LL response. One of them includes the nature of solvent. Atari et al. [136] used distilled water and TlCl solution. They found that the TlCl solution when used as solvent increased emission intensity by 200 times when irradiated with a dose of 55 Krad. Also, the conditions during dissolution and yield in organic lyoluminescence have a great effect on LL. The pH of the solution affects the light emission intensity [137]. The temperature of the solution is another factor on which emission intensity depends. The LL yield in pure water drops with an increase of temperature. The oxygen dissolved in the solvent also plays an important role in LL. It oxidises alkyl radicals to peroxyls that later become involved in the light emitting process. It also involves quenching excited triplets that are also responsible for the LL process. It was found that the content of oxygen determines the shape of the dose response curve. The LL yield also depends on phosphor grain size. Becker studied some other factors like the insensitive surface layers that may play an important role in response to phosphor powders to gamma radiation [138].

It is a well-known fact that the incorporation of bivalent cations in the crystal network of alkali halides gives rise to interstitial aggregation processes [139–141] and that new types of centres are formed. Bivalent cations act as traps for interstitials. This leads to the formation of D_3 centres. These newly formed more stable centres will promote growth of F-centres and enhance the emission of light. Ramani [142] and Kalkar [143] have shown that dissolving irradiated NaCl in a solution containing transition metal ions (Co^{2+}, Cu^{2+}, Ni^{2+}) enhances the light emission. The well explained reaction between the hydrated electron and the bivalent cation produces the reduced metal cation in an excited state. The enhancement of light emission would then arise from deexcitation of this excited cation. Chattopadhyay et al. explained theoretically the LL mechanism by developing a non-linear differential equation and analysing it by dynamic system analysis and exact numerical integration and showed that LL emission intensity depends on radiation dose, rate of radiative recombination, rate of dissolution in the solvent, etc. [64]. When the gamma-irradiated halide crystal containing N molecules at any time, t is dissolved in a solvent, the rate of decrease in the number of solute atoms may be written as

$$-\frac{dN}{dt} = \alpha N \tag{9.3}$$

where α is the rate of dissolution in the solvent. The rate of LL also depends on radiation exposure. As the dose is increased, first the LL intensity is increased and then attains a saturation value. In this saturation zone, if the term related to radiation dose is neglected, and taking N = N0 at t = 0, integration of equation (9.3) gives

$$N = N_0 e^{-\alpha t} \tag{9.4}$$

The rate of generation of the hydrated electrons R is directly proportional to $-\frac{dN}{dt}$ and also it has a linear relation with the density of F-centres n_F; thus R is given by the equation

$$R = \Gamma n_F \left(-\frac{dN}{dt}\right) \tag{9.5}$$

From equations (9.3), (9.4) and (9.5),

$$R = \Gamma n_F \; \alpha N_0 e^{-\alpha t} \tag{9.6}$$

where Γ is a factor defining the correlation of the number of hydrated electrons with the number of dissolved color centres of the crystals. As the crystals get dissolved into solvent, there will be an increase in the number of hydrated electrons; at the same time, there will be a decrease in the number of electrons as these electrons will recombine with holes on the crystal surface. The rate equation is now given by

$$\frac{dn}{dt} = \Gamma n_F \; \alpha N_0 e^{-\alpha t} - \beta n \tag{9.7}$$

where $\beta = \sigma_r N_r v$ = rate constant for the recombination of hydrated electrons with holes.
  Now, the lifetime of hydrated electrons is given by

$$\tau = \frac{1}{\beta} = \frac{1}{\sigma_r N_r v} \tag{9.8}$$

Integrating equation (9.7), for n = 0 at t = 0

$$n = \frac{\Gamma n_F \; \alpha N_0}{\beta - \alpha} (e^{-\alpha t} - e^{-\beta t}) \tag{9.9}$$

If η is a radiative recombination probability, then LL intensity is given by

$$I = \frac{\eta \beta \Gamma n_F \; \alpha N_0}{\beta - \alpha} (e^{-\alpha t} - e^{-\beta t}) \tag{9.10}$$

It can be shown from equation (9.10) that the LL intensity should be maximum at a particular value of t, say tm. The LL intensity is zero at t = 0 and at t = ∞ [144]. These theoretical models were well coordinated with experimental results, as shown by many researchers. In the theoretical discussion by Chattopadhyay et al., the temperature dependence and overall time dependence of the LL intensity was well explained [64].

9.3.2 Experimental Setups for Measurement of LL

Galand et al. [145] used an apparatus (Figure 9.6) for measurement of LL, in which there were two compartments separated by a diaphragm that would be closed when solvent and sample material were being introduced inside the apparatus. After putting on the cover, diaphragm is opened. There is an N_2 gas inlet and outlet. Continuous stirring of the mixture of solvent and solid was carried out under a controlled nitrogen flow. The light emission was measured using a photon counter system.

C. Hunter et al. used a transient emission spectrometer (TES), as shown in Figure 9.7. The samples which were irradiated up to a saturation dose were placed in the dissolution cell of the

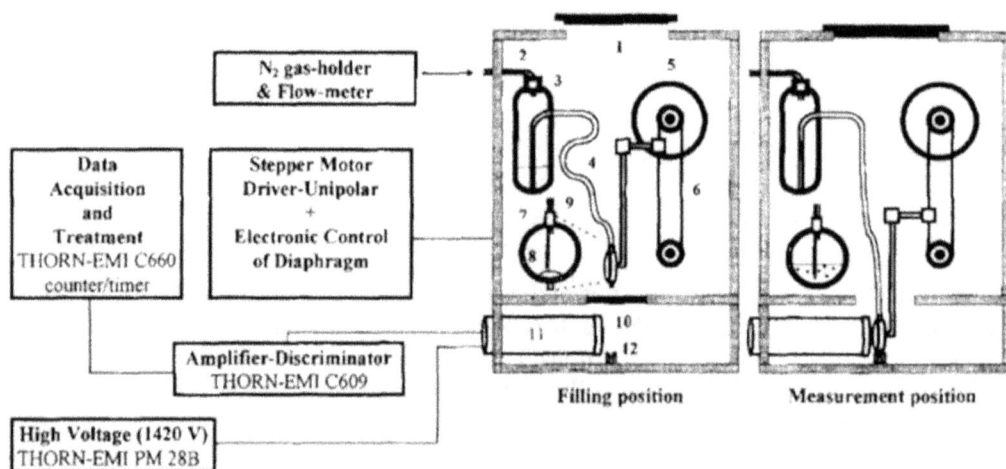

FIGURE 9.6 Lyoluminescence device. (1) Cover, (2) N_2 inlet, (3) solvent cell, (4) silicon rubber tubing, (5) stepper motor, (6) belt, (7) cylindrical cell, (8) solvent, (9) N_2 outlet, (10) diaphragm, (11) photomultiplier bialkali photocathode, (12) thrust block {Adapted from Galand et al. [145]}.

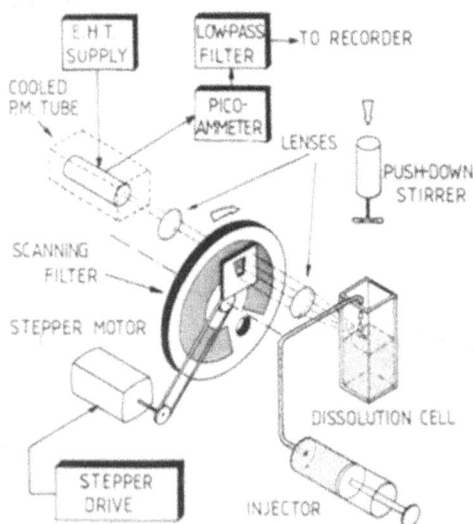

FIGURE 9.7 Transient emission spectrometer {Adapted from Hunter et al. [146]}.

TES. The solvent was then introduced into the cell. The emitted light was analysed by a rotating graded interference filter that covered the visible light range. The stepper motor drives the filter disc, which has an aperture, the position of which acts as a reference point in the spectral scan. The light from the dissolution cell was focused onto the photomultiplier tube using a set of lenses [146].

Figure 9.8 shows the experimental arrangement that has been used to measure the LL emission [147]. A sample was placed in the vessel and solvent was injected through a hole at room temperature. Amplification of the photomultiplier tube output was connected to a voltage-to-frequency converter which electronically integrated the total amount of light. The LL yield was expressed as the number of counts per mg weight of sample dissolved.

Figure 9.9 shows the LL reader design used by Atari et al. The output current from the photomultiplier tube was fed to a varactor bridge electrometric amplifier operating as a current-to-

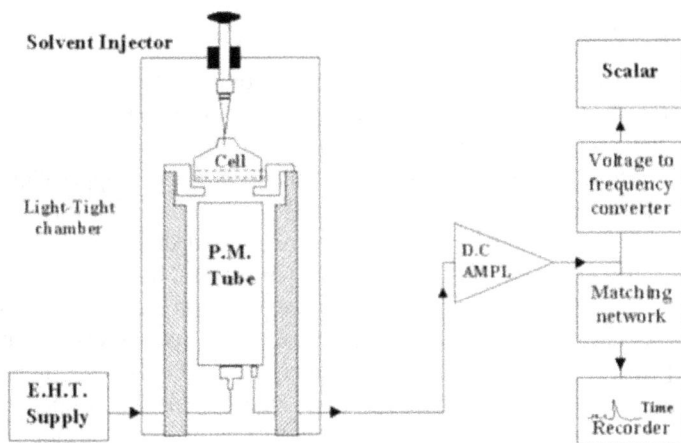

FIGURE 9.8   The lyoluminescent reader {Adapted from Shindl A et al. [147]}.

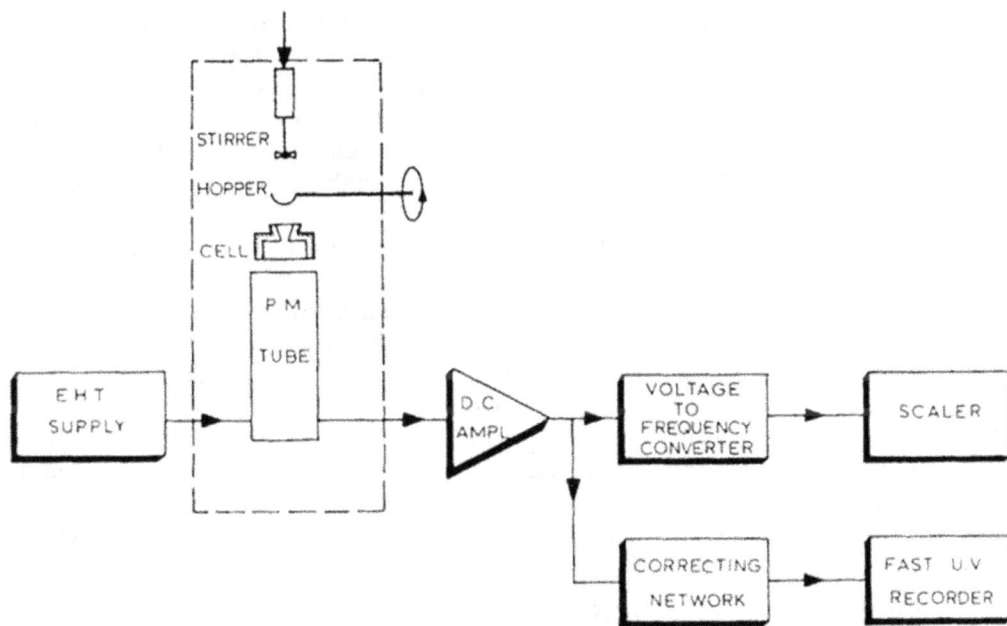

FIGURE 9.9   LL reader design {Adapted from Atari et al. [45]}.

voltage converter. The amplified signal was recorded by a u.v. recorder in order to provide the dissolution glow curve, but for most of the quantitative work, the glow curve waveform was electronically integrated by a voltage-to-frequency converter and scalar [45].

9.3.3 APPLICATIONS OF LL

The LL materials have applications as a system, especially for tissue equivalent radiation dosimetry, detection of irradiated food materials, radiation damage studies in solids, accident dosimetry and in the detection of irradiated food materials. Undoubtedly, the device called a dosimeter plays a significant role in the field of radiation protection and therapeutic treatment. The radiation dosimeter measures the exposure of dose amounts, absorbed dose, kinetic energy released in matter and equivalent dose.

9.3.4 LYOLUMINESCENCE AS A DOSIMETER

There is growing awareness about the risk of accidental exposure due to ageing nuclear power installations or dumping of nuclear waste in nearby areas. The residents are aware of the health risks and consequences and also social and economic disturbances irrespective of any nation. The facile and cheap measurement techniques like LL dosimeters thus find attractive application in the radiation field. For several organic compounds, the "light yield" is proportional to dose over a wide range; hence, lyoluminescence provides the basis of a dosimetric method suitable for measuring therapeutic doses and accidental doses of irradiation. The light yield is obtained by integrating the area under the glow curve which is the graph of LL intensity versus time. Different types of materials are discovered for the fabrication of a dosimeter. Based on their physical and chemical properties, they are specified and selected for a particular application to detect different range of energies. Currently various types of radiation detectors are available for many medical and environmental applications. There are several advantages of the LL dosimetry. It does not need "a programmed heating and good contact" with the heater as in case of TL or a high-power source of monochromatic light and continuous/pulsed driver as in case of optically stimulated luminescence. Just a small transparent container with an optical detector enclosed in a black-box for getting the irradiated sample dissolved and for recording the glow curves is sufficient. A storage oscilloscope or a computer is used to save the output data. Therefore, lyoluminescence dosimetry (LLD) is preferred, especially, in personnel, medical dosimetry during radiotherapy and accidental radiation monitoring as well as in food processing, where high doses need to be estimated. The advantage of this method is that some organic, edible and commonly available materials like common salt (NaCl) and sucrose could also be used as LLD phosphors.

As per Ettinger [148], in LLD with organic materials, free radicals are formed when irradiated and these radicals can be used as quantitative indicators of irradiation. These indicators are called dosimeters. If radicals get suitable solvent, then on dissolution these radicals react with themselves and with the solvent through various reactions and rearrangements. The energies of these radicals being high, they are capable of forming excited molecules. The de-excitation of which gives emission of visible or near-UV light emission. The emission spectra are as broad as they are at molecular origin.

LL dosimetry has been used to monitor skin doses in radiotherapy in the clinical range of 0.5 to 5.0 Gy. They are accurate and reproducible as comparison to TLD. LLD also can be used during radiotherapy and to measure oral doses in dentistry. Mannose dosimeters with plastic spacers inside a flexible, sealed plastic tube can be used for dose measurements in the lower part of the gastrointestinal tract. LL can also be used for the measurement of the radical scavenging activity of chemical compounds, especially those which have potential for use as radioprotective agents [149].

9.4 PHOSPHORS FOR MECHANOLUMINESCENCE

Strain electric fields are produced by inorganic crystalline materials upon deformation. Many inorganic compounds, like simple oxides, fluorides, chalcogenides, phosphates, silicates, oxysulfides, oxynitrides, tungstates, titanates, stannates, aluminates and borates [29,150–154] satisfy this prerequisite and show ML emissions. However, ML intensities are observed to vary significantly from one compound to another. For the applications in biological imaging and bright-field stress sensing, high-brightness ML materials with much longer NIR emissions are highly required and need to be developed urgently. C. Chen et al. developed lanthanide and transition metal doped mixed-anion compounds $SrZn_2S_2O$ and $SrZnSO$ showing visible to near IR (470–1,600 nm) ML for the possible applications in stress intelligent sensors, electronic skins and human-machine interfaces, damage diagnosis and dynamic force detection [155]. Apart from various novel phosphors like ZrO_2:Ti [156], the most studied materials can be broadly divided into silicate based and aluminate based. The borate-based phosphors are the new emerging class of materials that are reported for ML. As the focus is on borate materials, we will restrict the description of other inorganic groups like silicates, aluminates and alkali halides.

9.4.1 SILICATE-BASED PHOSPHORS

For silicate phosphors, Eu^{2+} has been the most used dopant, and these materials have shown strong water resistance and good thermal stability [157,158]. A series of Eu-doped silicate phosphors have been discovered, which exhibited composition-dependent emission light, such as blue for $Sr_2MgSi_2O_7$:Eu, blue greenish for $SrCaMgSi_2O_7$:Eu and green for $Ca_2MgSi_2O_7$:Eu [159,160]. Besides emission wavelength, their EML intensity can be tuned by introducing co-dopants. For example, the green ML intensity of $Ca_2MgSi_2O_7$: Eu, Dy was found to be approximately two orders of magnitude greater than that of $Ca_2MgSi_2O_7$: Eu.

9.4.2 ALUMINATE-BASED PHOSPHORS

Doped aluminate phosphors with EML character have attracted much attention since the early report of $Sr_3Al_2O_6$: Eu by Akiyama et al. [161]. Xu et al. later systematically studied a series of Eu-doped SrAlxOy [24], including $SrAl_{12}O_{19}$, $Sr_4Al_{14}O_{25}$, $SrAl_4O_7$, α-$SrAl_2O_4$, β-$SrAl_2O_4$, Sr_3AlO_6 and their mixed phases. They found that the Eu^{2+}-doped α-$SrAl_2O_4$ with a monoclinic structure produced the strongest EML among all investigated phosphors. A green phosphor, $SrAl_2O_4$: Eu,Dy (SAO), has been considered one of the most potential ML materials since Xu et al. discovered the ML trait of this traditional long phosphorescent material [162]. Other lanthanides like Ce, Ho, Pr and Er have also been used to dope aluminates to realize tunable emission ranging from UV to the near-infrared (NIR) range [163–165]. Similar to some of the aforementioned alkali halide crystals, the EML of SrAl2O4: Eu is non-repetitive unless it is reactivated by light irradiation [166]. $CaYAl_3O_7$:Eu ML phosphor was reported by Zhang H et al. They revealed that $CaYAl_3O_7$:Eu emits blue light under the application of mechanical stress. The effect of temperature on ML intensity was studied. Also, it was found that the ML intensity increases proportionally with the increase of mechanical load.

9.4.3 ALKALI HALIDE–BASED PHOSPHORS

Light emission by alkali halides irradiated by gamma radiation undergoing deformation or other mechanical working was reported by several workers. Metz et al. investigated tiboluminescence in LiF, KBr, and NaCl [167]. Molotskii and Shmurak [168] varied the amplitude of the "stress pulse (50 µs pulse width)" from less than 10% of the elastic limit into the plastic regime for γ-irradiated impurity doped alkali halides such as KCl:Cu. KCl:Ag, KCl:Tl, KBr:Cu. NaCl:Cu, KI:Tl and LiF:Tl. For the lowest stress, the ML emission was very faint and it developed slowly. They suggested that, in this case the bending rate is much less as compared to the velocity of dislocation captured electrons along the dislocation axes, and therefore, the ML emission does not involve movement of dislocations, but it involved migration of electrons along the dislocations with assistance from the stress.

They assigned the fast ML emission that tracked the stress pulse in time at higher stress to the dislocation motion, with the dislocations still pinned in the elastic regime. The elastico ML in X- or γ-irradiated alkali halide crystals can occur due to the electron trapping mechanism.

9.4.4 BORATE-BASED PHOSPHORS

There are a limited number of ML materials, but for most compounds, the emission centres of ML are found to be the same as those of thermoluminescence, photoluminescence or electro-luminescence, if present. To analyse the relation between the defect centres produced by the gamma-irradiation and dislocation in the processes of deformation, Kher et al. [169] compared mechanoluminescence (ML) and thermoluminescence (TL) of gamma-irradiated $CaSO_4$:Dy. They found that the same transition states are responsible for ML and TL emission. Brahme et al. studied ML and TL in Dy-, Ce-, Er- and Gd-doped CaF_2 crystals, which were gamma irradiated. Both the ML and TL intensities of the reported phosphors were increased with doping of rare earth impurities [170]. "The order of ML and TL intensity for dopants were found similar and their order for decreasing intensity is CaF_2: Dy> CaF_2: Ce> CaF_2: Er> CaF_2:Gd"[170]. The recorded ML spectra showed similarity with that of TL spectra. Hence, they suggested that though different excitation processes were applied for TL and ML, the luminescence centres in the crystal structure which are responsible for TL and ML emission may be the same.

The ML phenomenon occurs in a particular material because of "a specific crystal structure, point defects and their agglomeration, and microstructures, such as domains or domain walls. Holes or electrons traps are created by point defects in phosphors. Due to effect of strain, there is a change of their geometric configuration that brings about the change in binding energy for trapped charges, which may facilitate the escape of trapped charges. Centrosymmetry of the material structure can be disturbed due to point defects and piezoelectricity will therefore be observed. Thus, for the development of ML phosphors, "compounds with anisotropy in elasticity or piezoelectricity are favourable because electric fields in certain crystallographic directions can be large enough to induce observable ML" [171].

The borate crystals play a very important role in the field of nonlinear optics, especially in ultraviolet (UV) applications due to their high UV transmittance and high damage threshold [172]. Many borates were chosen as host lattices as they exhibit large bandgap, high luminescence efficiency, high thermal and chemical stability and low cost during past years and widely used for optical applications [173,174]. Borate materials were a focus of attention for various applications ranging from UV emitting lamps [175], as back illuminators for PLLCDs (photoluminescent liquid crystal displays) [176], for phototherapy [177,178], for reading secret marks and photocopying [179], tricolour lamps [180], thermoluminescent radiation dosimeter [181] and solid-state lighting [182], to list a few.

Much less literature has been published describing ML in borate-based materials. Undoped and rare earth (RE = Ce, Dy, Eu) doped MB_4O_7 (M = Ba, Ca, Sr) phosphors were synthesised using solid-state diffusion method by Mishra G. In this method, requisite mass of carbonates ($BaCO_3$, $CaCO_3$, $SrCO_3$), boric acid and rare earth oxides (Ce_2O_3, Dy_2O_3, Eu_2O_3) were mixed and grounded thoroughly in a mortar and pestel. The prepared mixture was then heated in a furnace at 725°C temperature for 12 hours. The sample was then cooled to room temperature gradually. After that, the sample was grounded again and heated at 750°C for another 12 hours. The as-prepared phosphor powder was cooled slowly up to room temperature. Similarly, Undoped and rare earth (RE = Ce, Dy, Eu) doped $LnCa_4O(BO_3)_3$ (Ln = Gd, La, Y) phosphors were synthesised by the same technique. The stoichiometric ratios of oxides (Gd_2O_3, La_2O_3, Y_2O_3), calcium carbonate, boric acid, and rare earth oxides (Ce_2O_3, Dy_2O_3, Eu_2O_3) were grounded thoroughly in a pestel and mortar. The mixture was then heated at 725°C for 12 hours and then cooled slowly. Obtained sample was grounded again and then fired at 750°C for another 12 hours, and then cooled slowly up to room temperature. Different concentrations of dopants (0.05, 0.1, 0.2, 0.5, 1, 2, 3 mol%) were used and the effect of variation of dopants and their concentrations on ML intensity were tested.

In this investigation, he found that the ML phosphors BaB_4O_7:Dy and $GdCa_4O(BO_3)_3$:Ce have remarkable intensity, they show linear dose response, have low fading effect and exhibit simple glow curve. Moreover, these phosphors show emission in that spectral region which makes these phosphors a potential candidate for ML and LL dosimeter, respectively [183].

The other investigations are based on the fact that the total ML intensity increases when the mass of the phosphors is increased. The increase in the number of crystallites (volume) of the phosphor may be the reason. The total ML intensity also increases due to an increase in strain rate and surface charge density of deformed sample. In conclusion, the peak ML intensity and total ML intensity of gamma-irradiated borate-based phosphors that are doped with rare earth cations show a proportionate increase in intensity with increasing gamma dose given to the sample. The intensity also increases with dopant concentration.

It is well-known fact that the spectroscopic properties of rare earth cations can be divided into two categories [184]. The first category is the band emission from the 5d→4f transition (e.g. Eu^{2+}, Ce^{3+}). These transitions can strongly depend on the host lattice as the empty 5d states are not shielded from the crystal field [185]. The second category is the charge transfer band emission (e.g. Pr^{3+}, Sm^{3+}, Eu^{3+}), which can be due to the transfer of an electron from 4f states of the rare earth cation to atoms of the matrix. Since, 4f electrons are shielded by 5s and 5p electrons, the host lattice atmosphere has little effect on the charge transfer transitions. The luminescence properties of rare earth trivalent and bivalent cations in a particular host can be related to each other according to the energy level diagram originally developed by Dorenbos [186,187]. Accordingly, the energy of the 4f and 5d states of the trivalent and bivalent cation, with respect to the valence (VB) and conduction bands (CB), use luminescence and optical spectroscopy data.

To analyse the emission at different wavelengths at a particular excitation wavelength in rare earth–activated phosphors, one must study the transitions in excitation spectra and emission spectra of rare earth ions. It is well understood that Ce^{3+} ions can be easily doped into different host materials and shows efficient broad band luminescence due to its 4f-5d parity allowed electric dipole transition. Ce^{3+} ion has a larger Stokes shift than the other rare earth ions and also greater oscillator strength. In the case of Ce^{3+}-doped compounds when optical absorption occurs, there is a transition from the $^2F_{5/2}$ ground state to the $^5d_1$ excited states, whereas emission occurs whenever there are transitions from the lowest $^5d_1$ excited state to the $^2F_{5/2}$ and $^2F_{7/2}$ ground states. A large electron phonon coupling for the $^2F_{5/2}$ state produces a large Stokes shift energy between the absorption and emission spectra [188].

As far as Dy^{3+} dopant is considered, the transitions $^4F_{9/2} \rightarrow {}^6H_J$ ((J = 7/2, 9/2, 11/2, 13/2 and 15/2) are responsible for emission in the visible regions. $^4F_{9/2} \rightarrow {}^6H_{13/2}$ transition at 575 nm causes yellow emission, it is hypersensitive and its emission intensity highly depends on the host matrix (inorganic phosphor); while $^4F_{9/2} \rightarrow {}^6H_{15/2}$ transition at 485 nm causes blue colour emission and it is less sensitive to the host matrix. The proper ratio of yellow to blue intensity depends on the environment of dopant cations viz. Dy^{3+} ions. Dy^{3+} cations, when doped in a proper host matrix environment with an optimum ratio of yellow and blue emission, will emit nearly white colour. Hence, such luminescent materials doped with Dy^{3+} cations may be used as potential phosphors for various optoelectronic applications [189].

Eu^{3+} ion is another important rare earth ion and often emits red light corresponding to the transitions $^5D_0 \rightarrow {}^7F_J$ (J = 1–6). According to the Judd–Ofelt theory [190,191], the intensity of the magnetic dipole transition ($^5D_0 \rightarrow {}^7F_1$) does not depend upon the local environment, whereas the electric dipole transition ($^5D_0 \rightarrow {}^7F_2$) is very sensitive to the symmetry of the cation surroundings. When the Eu^{3+} ions act as an activator, occupy sites with noninversion symmetry, $^5D_0 \rightarrow {}^7F_2$ transitions will be dominant, but if the Eu^{3+} ions occupy sites with inversion symmetry, the emission corresponding to $^5D_0 \rightarrow {}^7F_{21}$ is prominent.

Further, to continue with discussion of rare earth cations as luminescent centres, borate materials are suitable hosts and provide ambient atmosphere to create charge transfer between oxygen anions from host to activator cations, which are nothing but rare earth cations. Many borates were chosen as host lattices as they exhibit a large bandgap, high luminescence efficiency, high thermal and chemical stability and low cost during the past years [192], and are known due to their chemical and physical stability and widely used for optical applications [181]. Considering the fact that borate materials are not studied for ML and LL, exploring the possibility of plausible photoluminescence, mechanoluminescence and lyoluminescence in some borate phosphors, G. Mishra investigated photoluminescence (PL) of borates and found that under 247 nm excitation, PL emission spectra showed two peaks at 384 and 388 nm for BaB_4O_7:Ce (0.1 mol%) phosphor, for CaB_4O_7:Ce(0.1 mol%) phosphor two peaks at 380 and 391 nm were observed; however, SrB_4O_7:Ce (0.5 mol%) phosphor showed two peaks at 362 nm and 377 nm in its PL spectrum. The PL emission spectra of Dy-doped MB_4O_7 (M = Ba, Ca, Sr) and $LnCa_4O(BO_3)_3$ (Ln = Gd, La, Y) phosphors when recorded under excitation wavelength of 354 nm, showed two peaks. The first was in the range 470–485 nm and the second was in the range of 567–582 nm for varying concentration of dopants. PL emission spectra of Europium-doped MB_4O_7 (M= Ba, Ca, Sr) and $LnCa_4O(BO_3)_3$ (Ln = Gd, La, Y) phosphors, when recorded under excitation wavelength of 354 nm, showed two peaks at 590 nm and 615 nm for all the concentrations of dopants. Thus, it was concluded that the rare earth impurity Ce, Dy and Eu when doped in reported borates, they entered into host lattice and showed their characteristic photoluminescence emission [183].

As ML emission corresponds to stress, fracture and damage of solids, maximum ML materials have been studied for possible applications as mechanoluminescent stress, fracture and damage sensors. After PL emission in the reported phosphors was confirmed by PL excitation and emission spectra, the borates were investigated for mechanoluminescence (ML) and lyoluminescence (LL) emission by Mishra G. For studying these types of luminescence, the samples have to undergo radiation exposure. He used γ-rays exposure obtained from ^{60}Co source. The device that uses slow deformation rate is not suitable for studying ML in borate phosphors. Hence, the device which uses impulsive excitation technique was used to investigate ML properties and ML kinetics at different impact velocities of these phosphors [183]. The device consists of transparent lucite plate on which a sample can be placed. Below it, a photomultiplier tube is kept, which acts as a light detector. The load of different masses can be dropped from the variable heights (h). By changing the distance between the load to be dropped and the phosphor sample on the Lucite plate, the impact velocity of the load could be changed up to 4 m/s. A digital storage oscilloscope displays the spectral response curve. The effect of change in the parameters like mass of load "m" and the height "h" on ML intensity was studied.

Mishra G. observed that after the impact of mass m, the ML intensity increases, attains a maximum value, decreases with time and finally disappears. Among three dopants, ML intensity of the Dy-doped BaB_4O_7 sample is greater than the Ce- and Eu-doped sample. However, in $LnCa_4O(BO_3)_3$, the Ce-doped sample shows a maximum ML [183]. Some of the important observations are as follows:

1. The total ML intensity increases with an increase in impact velocity and saturates at higher value of impact velocity. But, the time corresponding to ML peak (t_m) shifts towards shorter time value with increase in impact velocity for all the samples.
2. It was also observed that for all the borate samples, ML intensity increases when there is an increase in the mass of the sample, without any considerable change in peak positions. Same results are observed in aluminates and silicate phosphors also. This shows that borate phosphors can be used for ML applications.
3. The ML intensity of the rare earth–doped borate-based phosphors that are irradiated with gamma rays depends upon the concentration of the impurity ions also. It was observed in the reported borate phosphors that ML intensity first increases with concentration of dopant; the reason being the increment in defect centres and luminescence centres, attains an optimum value for a particular concentration for a particular dopant, decreases with further increase in dopant concentration. Undoped samples show a very weak ML intensity.

4. ML intensity showed increment almost linearly with gamma dose given to the samples and seems to be saturated for higher gamma doses. That means that the ML materials don't show luminescence due to dielectric breakdown occurred during fracture of crystallites. It's a function of gamma dose.

5. The defect centres like cation vacancies and borate radicals are created, if borate-based phosphors are exposed to ionizing radiation [193]. When the gamma dose is increased, the density of defect centres (Nc) increases. According to kinetics, when a sample of a particular mass is deformed at a given impact velocity, ML intensity and total intensity should increase with the density of defect centres. Experimental results showed the corelation with the theory [183].

6. The time for ML peak intensity tm is an important parameter. It should be independent of the density of defect centre. It doesn't depend on dopant concentration. With an increase in impact velocity, tm shifts towards shorter time values.

7. When compared with PL measurement, the ML spectra showed a spectral shift. The reason for this shift was given by the researcher to the fact that PL and ML spectra were recorded by different equipment with different spectral response [183].

9.5   PHOSPHORS FOR LYOLUMINESCENCE

According to Ettinger et al., the ideal lyoluminescent material would (a) give a high light yield; (b) have a linear dose–light yield relationship over a wide dose range; (c) give reproducible light yields after storage under a variety of conditions (humidity, temperature, aerobic and anaerobic atmospheres) [194]. While developing LL-based applications, all these points should be taken into consideration.

9.5.1   ALKALI HALIDE–BASED PHOSPHORS

After the first report of light emission on dissolution of irradiated alkali chlorides in water by Wiedemann and Schmidt in 1895 [36], most alkali halides, saccharides and some organic compounds were studied for the similar effect. According to Ahnstrom [39], the primary species formed on dissolution of irradiated NaCl in water are the hydrated electron (e_{aq}^-) and the hydrated chlorine atom (Cl_{aq}^-). On the combination of these two species, an excited chloride ion $(Cl^-)*$ is formed and light is emitted. The light emission can be either in the solvent or at the solid-solvent interphase. Atari et al. [45] exposed alkali halides like NaCl, NaBr, KCl and KBr single crystals and CsCl and CsBr powders to ionizing radiations producing both trapped electron colour centres (F-centres) and trapped hole colour centres (V-centres). But their colouration by the addition of stoichiometric excess of alkali metal introduces only F-centres. These additively coloured crystals are relatively stable as only one type of colour centres is formed. They measured afterglow and lyoluminescence of the irradiated NaCl crystals [44].

9.5.2   LYOLUMINESCENCE IN ORGANIC SOLIDS

Lyoluminescence has been observed for virtually all organic compounds; but till now the most useful at this stage appear to be amino acids and sugars. These are also said to have "tissue equivalence" and hence work on such substrates has relevance to radiation damage in biological systems. In the saccharides group, mainly mannose, xylose, glucose, lactose, fructose, raffinose, ribose, arabinose, sucrose and galactose show LL and amino acid groups contains L-glutamine, L-threonine, L-glutamic acid, L-valine, L-phenylalanine, L-isoleucine, L-serine, L-arginine.HCl, L-leucine, L-tyrosine, L-asparagine. H_2O, L-lysine, L-proline, glycine, L-aspatic acid and L-methionine show lyoluminescence. According to the Russel-Vassilev scheme, primary or secondary alkyl radicals are generated during gamma irradiation react with oxygen to give the corresponding peroxyl radicals which are disproportionate via a tetroxide. According to the Wignar conservation rule, the oxygen is generated in the singlet state with ground-state singlet carbonyl or the carbonyl is generated in the triplet state with ground-state triplet state.

9.5.3 LYOLUMINESCENCE IN INORGANIC SALTS

Though not much work has been reported on LL in inorganic phosphors, a few studies showed that there are inorganic groups like sulphates. Kulama and Haapaka [61] reported LL induced by the dissolution of UV-irradiated Tb-doped potassium peroxodisulphate in aqueous solution. Matachescu et al. [62] have reported the mechanism of analytical application of luminol-specific extrinsic LL of UV-irradiated potassium peroxidisulphate. LL phenomenon in the super molecular aggregation of antinuclear gold complex was studied by Vickery et al. [63]. Chattopadhyay et al. [64] developed "a nonlinear differential equation and analyzed it by a dynamic system analysis as well as exact numerical integration". Raman et al. [65] reported the LL of trehalose dihydrate and showed that LL had linear gamma sensitivity in the dose range 0.1×10^2 to 5×10^2 Gy.

LL property of gamma-irradiated Eu-activated $KNaSO_4$ and $K_3Na(SO_4)_2$ phosphors were reported by Dhoble [66]. He observed linear response of absorbed gamma dose from 0.06 to 10 Ckg^{-1}. He also studied the fading of LL intensity and found that this phosphor is suitable for LL dosimetry. Dependence of the lyoluminescence of divalent impurity doped NaCl on temperature was studied by Gour et al. [67]. Particle size dependence of lyoluminescence of gamma-irradiated sodium bromide was reported by Banerjee et al. [68]. Hakansson et al. studied the lyoluminescence of aluminium induced by lanthanides chelates in an alkaline aqueous solution [69].

9.5.4 BORATE-BASED LL PHOSPHORS

Borate phosphors are generally studied for photoluminescence (PL). They are chemically and physically stable materials and found applications in myriad fields such as phototherapy, lamp phosphors, plasma display panels and solid-state lighting [177,195–199]. Borate materials have attracted researchers as dosimeter materials since their effective atomic numbers are very close to that of tissue. These materials as a bulk have negligible luminescence. But when they are doped with impurities, the phosphor material shows excellent photoluminescent properties [200–204]. Mishra et al (2011) studied lyoluminescence in borate phosphors doped with rare earth elements Dy, Ce, etc. When the rare earth–doped borate-based phosphors undergo exposure of gamma rays, many trap centres like BO_3^{2-}, O_2^- are formed in which holes are trapped and thermoluminescence (TL) emission may be caused by the recombination of holes from hole-trapped radicals with Ce^{3+} ions. The electron–hole recombination process releases energy which is used for the excitation of Ce^{3+} that may result in TL emission. When the $YCa_4O(BO_3)_3$:Ce phosphor was dissolved in dilute HCl, the LL phenomena was observed. The reason is the release of trapped energy during dissolution. The mechanism is the same as proposed by Ahnstrom [39]. If X is taken as radicals generated when the sample is exposed to gamma radiation, then the mechanism of LL emission can be explained as follows:

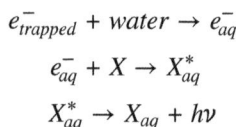

$$e^-_{trapped} + water \rightarrow e^-_{aq}$$

$$e^-_{aq} + X \rightarrow X^*_{aq}$$

$$X^*_{aq} \rightarrow X_{aq} + h\nu$$

The LL glow curve of these borate phosphors which is the graph of LL intensity versus time, showed that the LL intensity increases at first, attains a maximum value and decreases again. The LL intensity also depends on the gamma dose to which the sample is exposed. As dopant concentration is increased, the LL intensity has been increased. The possible cause of increased emission is the creation of a greater number of active luminescent centres due to increased gamma dose. During dissolution interconversion of Ce^{2+} to Ce^{3+} cations take place and that may be the cause of LL emission. They also showed that the LL intensity increases almost linearly with increased gamma dose.

The borate phosphor $GdCa_4O(BO_3)_3$ doped with Ce was synthesised and studied. The effect of mass of the solute and pH value of the solvent on the LL intensity was studied. It was found that the LL intensity of the sample increases with increase in mass of solute. They found that when Ce-, Dy- and Eu-doped $LnCa_4O(BO_3)_3$ (Ln = Gd, La, Y) samples are dissolved in a solvent like dilute HCl, the LL intensity increases linearly with time, it attains an optimum value. It shows gradual decrease and finally LL intensity vanishes. The reason of increment of LL intensity is the formation of more hydrated electrons in a large quantity of the solute added to the solvent. The LL intensity also shows the time dependence characteristics. When gamma-irradiated Ce-doped $GdCa_4O(BO_3)_3$ phosphor with different quantities of solute is dissolved in dilute hydrochloric acid, the LL intensity increases initially, attains optimum value and then decreases. The decay time increases with the increasing amount of solute added to the solvent [183].

Sahare et al. (2015) studied $MgB_4O_7:Mn^{2+}$ for the application of radiation dosimetry of high-energy radiations. They synthesised borate phosphor doped with transition element Mn^{2+} and irradiated it with gamma rays from the ^{60}Co radioactive source. They studied the effect of particle size on LL intensity and concluded that particles with an average size of 85 μm were more suitable for dosimetry. The pH effect of the solvent was also studied and it was observed that the NaOH solution with pH 12.1 gives the maximum LL yield. The reduction of the LL intensity with time after irradiation was also studied. It was observed that a reduction of LL takes place by approximately 25% after storing the material for 21 days in the dark at room temperature [205].

Thus, the borate-based phosphors doped with impurity like rare earth materials after exposure to gamma rays can create various defect centres. When such borates are dissolved in solvents (like dilute HCl), the trapped energy is released during dissolution and LL is observed. The rare earth ions cause the LL emission. The released energy may be transferred to rare earth ions non-radiatively which on de-excitation emits characteristic LL. This can lead to good reason for use of borate-based phosphors for LL applications.

9.6 CURRENT TRENDS AND INNOVATIONS

The research work in the ML and LL fields was very slow due to weak emission intensity and lack of systematic study in this field. In 2015, the combination of the ML process and the well-known aggregation induced emission (AIE) offered a new hope for "the molecular design of bright ML luminogens, providing new ways to explore the mysterious ML excitation process. C. Wang [180] reported the first example of an AIEgen with fluorescence–phosphorescence dual mechan-oluminescence and purely organic ML luminogens with room temperature phosphorescence in 2016 and 2017, respectively, to understand the inherent mechanisms deeply. Coupled with encouraging work of other scientists, the ML process is highly related to the molecular packing in the solid state, and generally occurs accompanying the fracture of crystals upon mechanical stimuli. Accordingly, some more bright ML luminogens have been developed recently.

The application of ML materials in responsive flexible electronics can be a good example of current trends in ML. In recent years, wearable devices and intelligent medicine have shown rapid development. For human beings, it is now essential to monitor the health condition of different weak forces produced by the human body, such as heartbeat, muscular movement and breath. A flexible ML device has the ability to convert force into light signals. When such a device is worn at the vulnerable area of certain patients or elderlies, it can be used to monitor the external impact force through a visual signal, which will help to monitor the vital parameters of the body. Besides external forces, the designed ML device can also be used in monitoring internal weak forces produced by the human body. For instance, the ML device discussed by Wu et al. [206] shows that the intangible heartbeats can be converted to visible light signal, which is very convenient in monitoring. Through studying the material packing mode in crystalline state and the corresponding theoretical model calculation, Li and co-workers have recently provided some new ideas for the design of efficient ML materials: "(1) AIE (Aggregation induced emission) characteristics are valuable to enhancing the brightness of ML materials because ML materials are usually used in solid state, and (2) the introduction of self-assembling groups, e.g. thiophene rings in their work, is beneficial to re-establishing the ordered crystal structure, leading to recyclable and even quantifiable ML properties" [207,208]. Encrypted communication is one of the fundamental needs of today's information-based era. By setting a threshold, number codes comprising 0 and 1 could be generated at various terminals and translated into different letters, words or sentences [209].

For professional sports, an effective sport-analytic method is an important part. That can be proven to improve and sustain performance. Athletic analytics play an important role in all sports games and competitions as well as physical education and training. EML materials have been developed with the performance of battery-free motion-driven light emission that can be coated on the surface of a moving object to sense the intensity of the applied force. The object will emit light in response to the stimuli and this emitted light could last for several seconds. If paved into the table, court or racket, it can record the athletic contact position and force distribution. Most of the inorganic EML materials show stable ML with linear response to the applied force over a wide range. Strong weather resistance, chemical stability and operation repeatability are the other features. On this basis, the hitting or falling positions are judged through the luminous points, while the actual stress is estimated by the emission intensity. These features can be used in athletic analytics by monitoring force position and making force magnitude visible, thereby, the trainees would accurately judge and control games and improve their athletic performance. It is particularly suitable in sports on campus, gymnasium, indoor and outdoor, as well as some low lighting circumstances [210].

9.7   LIMITATIONS AND CHALLENGES

As compared to other luminescence types like photoluminescence and electroluminescence, ML and LL have been studied minimally. Few research groups worldwide are currently doing work on ML materials to keep hope for this type of luminescence. The systematic approach along with new age applications is the need of the hour. The characterization facilities are available at a very few places. Mechanoluminescence and lyoluminescence has potential applications, still the research path in this field is being hurdled with various problems in understanding the mechanism of ML and in developing promising ML and LL materials. There is not clarity in guidelines for discovering ML and LL materials for novel applications. The literature study of proposed mechanisms and reported materials for various applications could not make all the processes clear. The light emission wavelength and its tunability with applied pressure is also very limited. ML phosphors that emit near infra-red light, which can be potential candidates in biomedical imaging using ultrasound radiation are not studied much. With the change in stress level how the energy levels of the dopant change are not understood properly. Despite numerous mechanisms proposed by Chandra et al., the study of the mechanism behind ML has been at the level of phenomenological modelling. Thus, in-depth studies on the change in defect levels under various strains at the atomic level is the immediate issue. Density functional theory is a precision method that can come to rescue which with utmost accuracy can predict the position of impurity levels in the bandgap with the help of hybrid functionals, or GW method. Hysteresis is one of the several phenomena that is based on the shifting of domain states. As the ML phenomenon is irreversible under the application of strains, the existence of "domain states" in ML phenomenon are not studied yet. Thus, this can be a novel application area for future research. Unexpected outcomes during the discovery of novel ML phosphors can lead to the explanation of the mechanism behind ML. The change in temperature or chemical substitution can cause phase transitions. Thus, these strategies can be used to tune the properties of ML because phase transitions may cause "lattice distortion, cation or anion orderings or domain states" which are key parameters for ML. As most of the ML phosphors show a particular kind of phase transitions, a novel and promising research lead would be the effect of phase transition on ML, elastic and electronic properties of these compounds. If there would be a standardised procedure to characterize ML and LL phosphors, it would make comparison of different ML and LL phosphors possible and thus lower the difficulties of discovering novel materials. Despite the challenges ahead, the development of ML and LL can be enhanced by deeper understanding of the mechanisms, by the systematic efforts in tuning ML and LL properties. It can also be improved by carefully analysing and replacing the trial-and-error synthesis attempts by well-designed and simulated synthesis of ML and LL materials. In this sense, borate materials, which are a minimally touched area of inorganic phosphors, can be a pioneer for use in designing novel applications. The fascinating mechanisms and promising applications deserve an intense research in this particular field of luminescence, which can be boosted by combined efforts from chemists, physicists and engineers.

9.8   FUTURE PROSPECTS AND SCOPE

Though a good amount of work has been done to study ML and LL in inorganic salts like alkali halides, some sulphides like ZnS, borates like $MCa_4O(BO_3)_3$ (M = Gd, La, Y), aluminates like $SrAl_2O_4$, organic materials like saccharides and amino acids and silicates like $Ca_2MgSi_2O_7$: Eu, Dy, these aren't sufficient to develop any device using these materials for continuous use in practical fields. It is the need of the hour to intensify the research both in the manufacturing and application areas that will use ML and LL phenomena as a tool. The advancement in information technology, nuclear technology, space technology and health sector and food industry have opened new application areas for using luminescence as a detector. The need of sustainable devices in various atmospheres with long-lasting luminescence have created new areas of research. Most of the applications in the past are directly or indirectly related to detection of radiation dose of gamma or X-rays. Thus, the focus was limited: to synthesise a material whose luminescence intensity depends on various parameters related to radiation dose and the material properties. The luminescence output is also less. That makes these materials restrictive in many fields. Efforts should be taken to increase the luminescence output without fading effect.

During the past two decades, efforts are slowly increasing to find a new field of applications using ML and LL materials. The use of a simulation technique to explain any mechanism comes as a boon from a computer science advancement technique. With the help of simulations, one can predict the light output from a particular system. Efforts are going on to find new material groups; doped or undoped and new application areas in sports, health monitoring, imaging and food technology fields. That is the positive direction that can be speeded further in novel application areas. There are so many untouched groups of materials that can be taken up as ML and LL materials and check their luminescence for various applications adopting a systematic way. The research and development window has been opened for these materials with numerous probabilities and possibilities.

REFERENCES

[1] Chandra, B. P. (1998). *Mechanoluminescence in Luminescence of Solids*, 1st ed., Vij, O.-R., Ed., Springer Science Business Media, LLC, New York, NY, USA, pp. 361–387, ISBN 978-1-4613-7446-6.
[2] Harvey, K. N. (1957). Triboluminescence, Piezoluminescence, Crystalloluminescence and Lyoluminescence. In *A History of Luminescence: From Earliest Times until 1900*, 1st ed., Harvey, K.-N., Ed., The American Philosophical Society, Philadelphia, PA, USA, pp. 378–379. 10.5962/bhl.title.14249
[3] Bacon, F. (1901). *The Advancement of Learning*, 1st ed., Devey, J., Ed., Press of P. F Collier & Son, New York, NY, USA, pp. 208–209. Fourth Book. https://oll.libertyfund.org/title/bacon-the-advancement-of-learning

[4] Boyle, R. (1664). *Experiments and Consideration Touching Colours (London)*. 413. eBook #14504.

[5] Longchambon, H. (1922). Study of the spectrum of the light emitted in the triboluminescence of sugar. *J. Franklin Inst.*, 195 (2), 269–270. 10.1016/S0016-0032(23)90249-8

[6] Bhatnagar, S. S., Mathur, K. G., and Budhiraja, K. L. (1933). Untersuchungen ueber Triboluminiszenz. *Z. Phys. Chem. A*, 163, 8–16.

[7] Inoue, T., Kunitomi, M., and Shibata, E. (1939). Triboluminescence. *Hiroshima J. Sci.*, 9, 129–136.

[8] Carriere, Z. (1964). Triboluminescence of rubbers. *Cah Phys.*, 46, 63. (in French) .

[9] Hoff, M. C., and Boord, C. E. (1950). Triboluminescence in cis-4-octane. *J. Am. Chem. Soc.*, 72, 2770–2771. 10.1021/ja01162a511

[10] Jenny, D. A. (1957). Triboluminescence in semiconductors. *J. Appl. Phys.*, 28 (12), 1515. 10.1063/1.1722694

[11] Tschugaeff, L. (1901). Ueber Triboluminiscenz. *Eur. J. Inorg. Chem.*, 34, 1820–1825.

[12] Trautz, M. (1910). Triboluminescence. *Ion*, 2, 77–89.

[13] Gernez, D. (1905). Triboluminescence des composes metaliques. *C. R. hebd. Seanc. Acad. Sci*, 140, 1337–1339.

[14] Wick, F. G. (1940). Triboluminescence of sugar. *J. Opt. Soc. Am.*, 30 (7), 302–306. 10.1364/JOSA.30.000302

[15] Jha, P., and Chandra, B. P. (2014). Survey of the literature on mechanoluminescence from 1605 to 2013. *Luminescence*, 29, 977–993. 10.1002/bio.2647

[16] Chandra, B. P. (2008). Mechanoluminescence induced by elastic deformation of coloured alkali halide crystals using pressure steps. *J. Lumin.*, 128, 1217–1224. 10.1016/j.jlumin.2007.12.001

[17] Chandra, B. P., and Bisen, D. P. (1992). Electronic excitation during elastic deformation of γ-irradiated LiF single crystals. *Phys. Status Solidi (A)*, 132(2), K101-K104. https://doi.org/10.1002/pssa.2211320237

[18] Akiyama, M., Xu, C. N., Liu, Y., Nonaka, K., and Watanabe, T. (2002). Influence of Eu, Dy co-doped strontium aluminate composition on mechanoluminescence intensity. *J. Lumin.*, 97, 13–18. 10.1016/S0022-2313(01)00419-7

[19] Wang, X., Xu, C. N., Yamada, H., Nishikubo, K., and Zheng, X. G. (2005). Electro-mechano-optical conversions in Pr^{3+}-doped $BaTiO_3$-$CaTiO_3$ ceramics. *Adv. Mater.*, 17, 1254–1258. 10.1002/adma.200401406

[20] Akiyama, M., Nishikubo, K., and Nonaka, K. (2003). Intense visible light emission from stress-activated $SrMgAl_6O_{11}$:Eu. *Appl. Phys. Lett.*, 83, 650–652. 10.1063/1.1594828

[21] Jia, Y., Yei, M., and Jia, W. (2006). Stress induced mechanoluminescence in $SrAl_2O_4$:Eu^{2+}, Dy^{3+}. *Opt. Mater.*, 28, 974–979. 10.1016/j.optmat.2005.05.014

[22] Akiyama, M., Xu, C. N., Matsui, H., and Nonaka, K. (1999). Recovery phenomenon of mechanoluminescence from $Ca_2Al_2SiO_7$:Ce by irradiation with ultraviolet light. *Appl. Phys. Lett.*, 75, 2548–2550. 10.1063/1.125073

[23] Xu, C. N. (2002). Coatings. In: *Encyclopedia of Smart Materials*, 1, Schwartz, M. , Ed., John Wiley and Sons, Inc., pp. 190–201.

[24] Xu, C. N., Yamada, H., Wang, X., and Zheng, X. G. (2004). Strong elasticoluminescence from monoclinic-structure $SrAl_2O_4$. *Appl. Phys. Lett.*, 84, 3040–3042. 10.1063/1.1705716

[25] Fu, X., Yamada, H., and Xu, C. N. (2009). Property of highly oriented $SrAl_2O_4$:Eu film on quartz glass substrates and its potential application in stress sensor. *J. Electrochem. Soc.*, 156, J249–J253. 10.1149/1.3156652

[26] Sohn, K. S., Seo, S. Y., Kwon, Y. N., and Park, H. D. (2002). Direct observation of crack tip stress field using the mechanoluminescence of $SrAl_2O_4$:(Eu,Dy,Nd). *J. Am. Ceram. Soc.*, 85, 712–714. 10.1111/j.1151-2916.2002.tb00158.x

[27] Xu, C. N., Watanabe, T., Akiyama, M., and Zheng, X. G. (1999). Artificial skin to sense mechanical stress by visible light emission. *Appl. Phys. Lett.*, 74, 1236–1238. 10.1063/1.123510

[28] Reynold, G. T., and Ausin, R. H. (2001). Mechanoluminescence of plastic scintillation counters. *J. Lumin.*, 92, 79–82. 10.1016/S0022-2313(00)00230-1

[29] Chandra, B. P. (2010). Persistent mechanoluminescence induced by elastic deformation of ZrO_2:Ti phosphors. *J. Lumin.*, 130 (11), 2218–2222. 10.1016/j.jlumin.2010.06.023

[30] Chandra, B. P., Goutam, R. K., Chandra, V. K., Raghuwanshi, D. S., Luka, A. K., and Baghel, R. N. (2010). Mechanoluminescence induced by elastic and plastic deformation of coloured alkali halide crystals at fixed strain rates. *Radiat. Eff. Defects Solids*, 165, 907–919. 10.1080/10420150.2010.487903

[31] Xu, C. N., Watanabe, T., Akiyama, M., and Zheng, X. G. (1999). Direct view of stress distribution in solid by mechanoluminescence. *Appl. Phys. Lett.*, 74, 2414–2416. 10.1063/1.123865

[32] Gerlach, R. L., Houston, J. E., and Park, R. L. (1970). Ionization spectroscopy of surfaces. *Appl. Phys. Lett.*, 16, 179. 10.1063/1.1653152

[33] Kim J. S., Kibble K., Kwon Y. N., and Sohn K. S. (2009). Rate-equation model for the loading-rate-dependent mechanoluminescence of $SrAl_2O_4:Eu^{2+}$, Dy^{3+}. *Opt. Lett.*, 34, 1915–1917. 10.1364/OL.34.001915

[34] Jha, P., and Chandra, B. P. (2014). Survey of literature on mechanoluminescence from 1605 to 2013. *Luminescence*, 29 (8), 977–993. 10.1002/bio.2647

[35] Wenbo, W., Yukun, D., and Bin, L. (2020). Mechanoluminescence: Quantitative pressure-brightness relationship enables new applications. *Matter*, 2, 291–296. 10.1016/j.matt.2020.01.007

[36] Wiedemann, E., and Schmidt, G. C. (1895). *Ann. Phys. Leipzig*, 56, 210.

[37] Ahnstrom, G., and Ehrenstein, G. V. (1959). Luminescence of aqueous solutions of substances irradiated with ionizing radiationin the solid state. *Acta Chem. Scand.*, 13, 855–856. 10.3891/acta.chem.scand.13-0855

[38] Ahnstrom, G., and Ehrenberg, L. (1960). Dosimetry of mixed neutron-gamma radiations in the Stockholm reactor. *RISO Rep.*, 16, 15.

[39] Ahnstrom, G. (1965). Physical and chemical processes accompanying the dissolution of irradiated substances studies by means of luminescence measurements. *Acta Chem. Scand.*, 19 (300–312), 1965. 10.3891/acta.chem.scand.19-0300

[40] Lelievre, B., and Adloff, J. P. (1964). Étude de la luminescence accompagnant la dissolution du chlorure de sodium irradié par le rayonnement γ du cobalt 60. *J. Phys.*, 25 (8–9), 789–796. 10.1051/jphys:01964002508-9078900ff.jpa-00205871

[41] Reynolds, G. T. (1992). Lyoluminescence. *J. Lumin.*, 54, 43–69. 10.1016/0022-2313(92)90048-E

[42] Burns, W., and Williams, T. (1955). Chemical effects associated with 'Colour Centres' in alkali halides. *Nature*, 175, 1043–1044. 10.1038/1751043a0

[43] Mittal, J. P. (1970). Proc. Symp. on Radiat. Chem. Trombay, BARC. Bombay; Rept. BARC-489, 40.

[44] Atari, N. A. (1980). Lyoluminescence mechanism of gamma and additively coloured alkali halides in pure water. *J. Lumin.*, 21 (3), 305–316. 10.1016/0022-2313(80)90009-5

[45] Atari, N. A., Ettinger, K. V., and Fremlin, J. H. (1973). Lyoluminescence as a possible basis of radiation dosimetry. *Radiat. Eff.*, 17, 45–48. 10.1080/00337577308232596

[46] Bartlett, D. T. (1973). A comparison of responses to irradiation by photons, electrons, protons and neutrons of some lyoluminescent materials. M.Sc. Thesis, University of Birmingham, U.K.

[47] Bartlett, D. T., Brown, J. K., and Durrani, S. A. (1982). Lyoluminescence and electron spin measurements on D-mannose. Effect of storage time and annealing. *Radiat. Effects*, 66, 233. 10.1080/00337578208222483

[48] Burkill, J. A. (1977). A study of high sensitivity dosimetry with lyoluminescence materials. M Sc. Thesis, University of Aberdeen, U. K.

[49] Takavar, A., Ettinger, K. V., Mallard, J. R., Atari, N. A., and Srirath, S. (1977) Progress in lyoluminescence dosimetry using saccharides. In Proc. 5th In:. Conf on Luminescence dosimetry Isot. 33, 1139. Dosimetry, Sao Paulo, pp. 169.

[50] Hanig, R. E. (1984). Enhancement of lyoluminescence by radiation sensitization and chemical dopants. *Int. J. Appl. Radiat. Isot.*, 35 (10), 987–989. 10.1016/0020-708X(84)90220-5

[51] Chatterjee, B. K., Sur, T., and Roy, S. C. (1993). Theory of light emission from γ-ray-irradiated organic substances when dissolved. *Phys. Rev. B*, 47, 555. (R) 10.1103/PhysRevB.47.555

[52] Ettinger, K. V., and Puite, K. J. (1982). Lyoluminescence dosimetry Part I. Principle. *Int. J. Appl. Radiat. Isot.*, 33, 1115–1138. 10.1016/0020-708X(82)90239-3

[53] Puite, K. J., and Ettinger, K. V. (1982). Lyoluminescence dosimetry Part II. State- of- the-Art. *Int. J. Appl. Radiat. Isot.*, 33, 1139–1157. 10.1016/0020-708X(82)90240-X

[54] Russel, G. A. (1957). Deuterium isotope effects in the autoxidation of aralkyl hydrocarbons. Mechanism of the interaction of peroxy radicals. *J. Am. Chem. Soc.*, 79, 3871–3877. 10.1021/ja01571a068

[55] Vassil'ev, R. F. (1967). Chemiluminescence in liquid phase reactions. *Prog. React. Kinet.*, 4, 305–352.

[56] Kundu, H. K. (1992). In: Murthy, K. V. R., Prasad, L. H. H., Joshi, T. R., Eds., *Thermoluminescence and its Applications*, Tata McGraw Hill, New Delhi, p. 3X. ISBN:0074622757.

[57] Kundu, H., and Mitra, B. (1984). A study of self-quenching of lyoluminescence yield from gamma-irradiated saccharides and amino acides. *Radiat. Effects Lett.*, 85, 123.

[58] Kundu, H. K., Banerjee, D., and Mitra, B. (1992). Mass-effect in the lyoluminescence yield from the tissue-equivalent dosimetric materials. *Ind. J. Pure Appl. Phys.*, 30, 340.

[59] Arnikar, H. J., Rao, B. S. M., Gijare, M. A., and Sardesai, S. S. (1975). Variation of aquolumi-nescence during the decay and regeneration of thermoluminescence under F-light. *J. Chim. Phys. (Paris)*, 72, 654–658. 10.1051/jcp/1975720654

[60] Copty Wergles, K., Nowotny, R., and Hille, P. (1990). Lyoluminescence of calcium carbonate and possible applications in the dating of loess and soils. *Rad. Prot. Dos.*, 34, 79–82. 10.1093/oxfordjournals.rpd.a080852

[61] Kulmala, S. M., and Haapaka (1994). Terbium (III) lyoluminescence induced by the dissolution of UV-irradiated potassium paroxodisulfate in aqueous solutions. *Ann. Chem. Acta*, 294, 13-25SSDI 0003-2670(94)00106-V .

[62] Matachescu, C., Kulama, S., Loine, E., and Racrinne, P. (1997). Luminol-specific extrinsic lyolu-minescence of X-ray irradiated sodium chloride. *Anal. Chem. Acta*, 349, 1–10. 10.1016/S0003-2670(97)00233-X

[63] Vickery, J. C., Olmstead, M. M., Fung, E. Y., and Balch, A. L. (1997). Solvent-stimulated lumines-cence from the supramolecular aggregation of a trinuclear gold(I) complex that displays extensive intermolecular AuċAu interactions. *Angew. Chem. Int. Ed. Engl.*, 36, 1179–1181. 10.1002/anie.1 99711791

[64] Chattopadhyay, A. K., Mahapatra, G. S., and Chaudhury, P. (2000). Lyoluminescence: A theoretical approach. *Phys. Rev. B*, 62, 906–908. 10.1103/PhysRevB.62.906

[65] Raman, A., Oommen, I. K., and Sharma, D. N. (2001). Lyoluminescence characteristics of trehalose dihydrate. *Appl. Radiat. Isot.*, 54 (3), 387–391.

[66] Dhoble, S. J. (2002). Preparation of (K: Eu) NaSO4 phosphor for lyoluminescence dosimetry of ionising radiation. *Radiat. Prot. Dosim.*, 100 (1-4), 285–287.

[67] Gour, A. S., Choudhary, V., Chourasia, B., and Chandra, B. P. (2004). *Proc. Int. Conf. Lumin. Appl.*, 2004, 377–379.

[68] Banerjee, M., Mehta, M., Gulahre, V., Kher, R. S., and Khokhar, M. S. K. (2005). Lyoluminescence of gamma-irradiated sodium bromide samples. *Proc. Natl. Conf. Lumin. Appl.*, 2005, 430–431.

[69] Håkansson, M., Jiang, Q., Spehar, A. M., Suomi, J., and Kulmala, S. (2006). Extrinsic lyolumi-nescence of aluminum induced by lanthanide chelates in alkaline aqueous solution. *J. Lumin.*, 118 (2), 272–282. 10.1016/j.jlumin.2005.09.012

[70] Upadhyay, A., Dhoble, S. J., Rai, R., and Kher, R. S. (2008). Synthesis of KNaSO$_4$:Tb^{3+} and MgSO$_4$:Dy^{3+} phosphors for lyoluminescence dosimetry. *Nucl. Instrum. Methods Phys. Res. B*, 266, 2594–2598. 10.1016/j.nimb.2008.03.236

[71] Kher, R. S., Upadhyay, A. K., Dhoble, S. J., and Khokhar, M. S. K. (2008). Luminescence studies of MgSO$_4$:Dy phosphors. *Indian J. Pure Appl. Phys.*, 46 (9), 607–610.

[72] Sahare, P. D., and Saran, M. (2016). Effect of pH on Lyoluminescence of K$_3$Na(SO$_4$)$_2$:Eu^{3+} phosphor for its application in dosimetry of high energy radiations. *J. Lumin.*, 179, 254–259. 10.1016/j.jlumin.2016.06.040

[73] Sahu, V., Brahme, M., Bisen, D. P., and Sharma, R. (2009). Effect of lyoluminescence decay in impurity doped KCl microcrystalline powder in lyoluminescence dosimetry of ionization radiations. *J. Optoelectron. Biomed. Mater.*, 1 (3), 297–302.

[74] Nayar, V., Chowdhary, P. S., Bhujbal, P. M., and Dhoble, S. J. (2011). Studies on the mass and temperature dependence of lyoluminescence intensity of microcrystalline powder of KCl. *Luminescence*, 26 (5), 324–330. 10.1002/bio1232

[75] Puppalwar, S. P., Dhoble, S. J., and Kumar, A. (2011). Rare earth Ce^{3+}, Dy^{3+} activated Li$_2$BPO$_5$ phosphors for lyoluminescence dosimetry. *Indian J. Pure Appl. Phys.*, 49 (4), 239–244, http://nopr.niscair.res.in/handle/123456789/11320.

[76] Mishra, G. C., Upadhyay, A. K., Kher, R. S., and Dhoble, S. J. (2011). Thermoluminescence and lyoluminescence in γ-ray irradiated and Ce^{3+}-doped YCa$_4$O(BO$_3$)$_3$ phosphors. *J. Mater. Sci.*, 46, 7275–7278. 10.1007/s10853-011-5687-1

[77] Mishra, G. C., Dhoble, S. J., Srivastava, A., Dwivedi, S., and Kher, R. S. (2018). Study of lyolu-minescence of gamma irradiated GdCa$_4$O(BO$_3$)$_3$:Ce phosphors: Mass of solute and pH of solvent as a key factor. *Optik*, 158, 826–830. 10.1016/j.ijleo.2017.12.189

[78] Timilsina, S., Kim, J. S., Kim, J., *et al.* (2016). Review of state-of-the-art sensor applications using mechanoluminescence microparticles. *Int. J. Precis. Eng. Manuf.*, 17, 1237–1247. 10.1007/s12541-016-0149-y

[79] Camara, C. G., Escobar, J. V., Hird, J. R., and Putterman, S. J. (2008). Correlation between na-nosecond X-ray flashes and stick–slip friction in peeling tape. *Nature*, 455 (7216), 1089–1092. 10.1038/nature07378

[80] Jia Y., Yei M., and Jia W. (2006). Stress-induced mechanoluminescence in $SrAl_2O_4:Eu^{2+}$, Dy^{3+}. *Opt. Mater.*, 28, 974–979. 10.1016/j.optmat.2005.05.014

[81] Xu, C. N., Li, C., Imai, Y., Yamada, H., Adachi, Y., and Nishikubo, K. (2006). Development of elastico-luminescent nanoparticles and their applications. *Adv. Sci. Technol.*, 45, 939–944. 10.4028/www.scientific.net/AST.45.939

[82] Chandra, B. P., and Rathore, A. S. (1995). Classification of mechanoluminescence. *Crystal Res. Technol.*, 30, 885–896. 10.1002/crat.2170300702

[83] Chandra, B. P., Chandra, V. K., and Jha, P. (2013). Models for intrinsic and extrinsic fracto-mechanoluminescence of solids. *J. Lumin.*, 135, 139–153. 10.1016/j.jlumin.2012.10.009

[84] Chandra, B. P., Chandra, V. K., Jha, P., Patel, R., Shende, S. K., Thaker, S., and Baghel, R. N. (2012). Fracto-mechanoluminescence and mechanics of fracture of solids. *J. Lumin.*, 132, 2012–2022. 10.1016/j.jlumin.2012.03.001

[85] Moon, S. J., Song, S., Lee, S. K., and Choi, B. (2013). Mechanically driven light-generator with high durability. *Appl. Phys. Lett.*, 102, 051110-1-5. 10.1063/1.4791689

[86] Jha, P., Khare, A., Singh, P. K., Chandra, V. K., and Sonwane, V. D. (2018). Ball impact induced elastico-mechanoluminescence for impact sensor. *J. Lumin*, 195, 40–43. 10.1016/j.jlumin.2017.10.063

[87] Meyer, K., Obrikat, D., and Rossberg, M. (1970a). Progress in triboluminescence of alkali halides and doped zinc sulphides (II). *Kristall u. Technik*, 5 (1970a), 181–205. 10.1002/crat.19700050202

[88] Keszthelyi, C. P., and Bard, A. J. (1973). Triboluminescence and triboelectrification by the motion of mercury over glass coated with scintillator dyes. *J. Electrochem. Soc.*, 120 (1973), 1726–1729. 10.1149/1.2403353

[89] De Paoli, S., and Strausz, O. P. (1970). Gaseous reactions and luminescence initiated by triboelectricity. *Can. J. Chem.*, 48, 3756–3757. 10.1139/v70-630

[90] Chapman, G. N., and Walton, A. J. (1983). Triboluminescence of glasses and quantz. *J. Appl. Phys.*, 54, 5961–5965. 10.1063/1.331773

[91] Chandra, B. P. (1985). Squeezing light out of crystals: Triboluminescence. *Nucl. Tracks Radiat. Meas.*, 10 (1-2), 225-241. 10.1016/0735-245X(85)90028-6

[92] Chandra, B. P., and Zink, J. I. (1980). Triboluminescence and the dynamics of crystal fracture. *Phys. Rev. B*, 21 (2), 816–826. 10.1103/PhysRevB.21.816

[93] Chandra, B. P., and Zink, J. I. (1980). Mechanical characteristics and mechanism of triboluminescence of fluorescent molecular crystals. *J. Chem. Phys.*, 73, 5933–5941. 10.1063/1.440151

[94] Chandra, B. P. (1983). Mechanoluminescence and high pressure photoluminescence of (Zn, Cd)S phosphors. *Pramana*, 19, 455–465. https://www.ias.ac.in/article/fulltext/pram/019/05/0455-0465.

[95] Kobakhidze, L., Guidry, C. J., Hollerman, W. A., and Fontenot, R. S. (2013). Detecting mechanoluminescence from ZnS: Mn powder using a high speed camera. *IEEE Sensors J.*, 13 (8), 3053–3059. 10.1109/JSEN.2013.2261489

[96] Belyaev, L. M., and Martyshev, Yu. N. (1969). Triboluminescence of some alkali halide crystals. *Phys. Status Solidi*, 34 a (1), 57–62. 10.1002/pssb.19690340105

[97] Rai, R. K., Kher, R. S., Khan, S. A., Dhoble, S. J., and Upadhyay, A. K. (2012). Device for mechanoluminescence excitation by applying gradually varying pressure on the sample. *Indian J. Pure Appl. Phys.*, 50, 534–537.

[98] Xu, C. N., Akiyama, M., Nonaka, K., and Watanabe, T. (1998). Electrical power generation characteristics of PZT piezoelectric ceramics. *IEEE Trans. Ultrason. Ferroelect. Freq. Control*, 45, 1065–1070. 10.1109/58.710589

[99] Fontenot, R. S., Hollerman, W. A., Aggarwal, M. D., Bhat, K. N., and Goedeke, S. M. (2012). A versatile low-cost laboratory apparatus for testing triboluminescent materials. *Measurement*, 45, 431–436. 10.1016/j.measurement.2011.10.031

[100] Sakai, K., Koga, T., Imai, Y., Maehara, S., and Xu, C. N. (2006). Observation of mechanically induced luminescence from microparticles. *Phys. Chem. Chem. Phys.*, 8, 2819–2822. 10.1039/B604656H

[101] Kamimura, S., Yamada, H., and Xu, C. N. (2012). Development of new elasticoluminescent material $SrMg_2(PO_4)_2:Eu$. *J. Lumin.*, 132, 526–530. 10.1016/j.jlumin.2011.09.033

[102] Shohag, M. A. S., Tran, S. A., Ndebele, T., Adhikari, N., and Okoli, O. I. (2018). Designing and implementation of triboluminescent materials for real-time load monitoring. *Mater. Des.*, 153, 86–93. 10.1016/j.matdes.2018.05.006

[103] Persits, N., Aharoni, A., and Tur, M. (2017). Quantitative characterization of ZnS:Mn embedded polyurethane optical emission in three mechanoluminescent regimes. *J. Lumin.*, 181, 467–476. 10.1016/j.jlumin.2016.09.015

[104] Peng, D. F., Chen, B., and Wang, F. (2015). Recent advances in doped mechanoluminescent phosphors. *Chem. Plus Chem.*, 80, 1209–1215. 10.1002/cplu.201500185

[105] Jeong, S. M., Song, S., Lee, S. K., and Ha, N. Y. (2013). Color manipulation of mechanoluminescence from stress-activated composite films. *Adv. Mater.*, 25, 6194–6200. 10.1002/adma.201301679

[106] Shin, S. W., Oh, J. P., Chang, W. H., Kim, E. M., Woo, J. J., Heo, G. S., and Jin, H. K. (2016). Origin of mechanoluminescence from Cu-doped ZnS particles embedded in an elastomer film and its application in flexible electro-mechanoluminescent lighting devices. *ACS Appl. Mater. Interfaces*, 8, 1098–1103. 10.1021/acsami.5b07594

[107] Chen, L., Wong, M. C., Bai, G. X., Jie, W. J., and Hao, J. H. (2015). White and green light emissions of flexible polymer composites under electric field and multiple strains. *Nano Energy*, 14 (2015), 372–381. 10.1016/j.nanoen.2014.11.039

[108] Molotskii, M. I. (1989). *Sov. Sci. Rev. B: Chem.*, 13, 1.

[109] Brady, B. T., and Rowell, G. A. (1986). Laboratory investigation of the electrodynamics of rock fracture. *Nature*, 321, 488–492. 10.1038/321488a0

[110] Glass, C. (1977). A Light Activated Fuse System, BRL MR 2726 (US Army Report).

[111] Glass, C., Dante, J., CjaIella, C., and Golaski, S. (1975). An Electro-Optical Fuse System, BRL MR 2552 (US Army Report).

[112] Orel, V. E., Opov, Va. Z., Goraiskii, E. K., Leshchinskii, I. V., and Khazanovish, D. M.(1989). The TRA-2 triboluminometer, a rapid analyzer of lipid peroxidation of the blood. *Meditsinskaia tekhnika*, 4, 34-37.

[113] Dickinson, J. T., and Donaldson, E. E. (1987). Autographs from peeling pressure sensitive adhesives: Direct recording of fracture-induced photon emission. *J. Adhes.*, 24 (2-4), 199–220. 10.1080/00218468708075427

[114] Terasaki, N., Xu, C. N., Imai, Y., and Yamada, H. (2007). Photocell system driven by mechanoluminescence. *Jpn. J. Appl. Phys.*, 46, 2385–2388. 10.1143/JJAP.46.2385

[115] Terasaki, N., Zhang, H., Imai, Y., Yamada, H., and Xu, C. N. (2009). Hybrid material consisting of mechanoluminescent material and TiO$_2$ photocatalyst. *Thin Solid Films*, 518, 473–476. 10.1016/j.tsf.2009.07.026

[116] Terasaki, N., Yamada, H., and Xu, C. N. (2011). Mechanoluminescent light source for a fluorescent probe molecule. *Chem. Commun.*, 47, 8034–8036. 10.1039/C1CC11411E

[117] Terasaki, N., Yamada, H., and Xu, C. N. (2013). Ultrasonic wave induced mechanoluminescence and its application for photocatalysis as ubiquitous light source. *Catal. Today*, 201, 203–208. 10.1016/j.cattod.2012.04.040

[118] Chandra, B. P. (2011). Development of mechanoluminescence technique for impact studies. *J. Lumin.*, 131, 1203–1210. 10.1016/j.jlumin.2011.02.027

[119] Wang, W. X., Imai, Y., Xu, C. N., Matsubara, T., and Takao, Y. (2011). A new smart damage sensor using mechanoluminescence material. *Mater. Sci. Forum*, 675– 677, 1081–1084. 10.4028/www.scientific.net/MSF.675-677.1081

[120] Sohn, K. S., Seo, S. Y., Kwon, Y. N., and Park, H. D. (2002). Direct observation of crack tip stress field using the mechanoluminescence of SrAl2O4:(Eu, Dy, Nd). *Am. Ceram. Soc.*, 85 (3), 712–714. 10.1111/j.1151-2916.2002.tb00158.x

[121] Zhang, H., Xu, C. N., Terasaki, N., and Yamada, H. (2010). Detection of stress distribution using Ca$_2$MgSi$_2$O$_7$:Eu,Dy microparticles. *Physica E*, 42, 2872–2875. 10.1016/j.physe.2010.02.013

[122] Hyodo, K., Xu, C., Mishima, H., and Miyakawa, S. (2010). In: C. T. Lin, J. C. H. Goh (Eds.), Proceedings of the sixth World Congress of Biomechanics (WCB 2010), August 1–6, 2010, Singapore, IFMBE, vol. 31, Part 2, pp. 545–548.

[123] Xu, C. N. (2010). Monitoring system for safety management of structures using elastico-luminescent materials. *AIST Today*, 10 (7), 18.

[124] Freund, F. (2003). Rocks that crackle and sparkle and glow: Strange pre-earthquake phenomena. *J. Sci. Explor.*, 17 (1), 37–71. http://www.isfep.com/FF_EQ_SSE_2003.pdf.

[125] Freund, F. (2002). Charge generation and propagation in igneous rocks. *J. Geodyn.*, 33, 543–570. 10.1016/S0264-3707(02)00015-7

[126] Takeuchi, A., Lau, B. W. S., and Freund, F. T. (2006). Current and surface potential induced by stress activated positive holes in igneous rocks. *Phys. Chem. Earth*, 31, 240–247. 10.1016/j.pce.2006.02.022

[127] Patel, D. K., Bat-El, C., Etgar, L., and Magdassi, S. (2018). Fully 2D and 3D printed anisotropic mechanoluminescent objects and their application for energy harvesting in the dark. *Mater. Horiz.*, 5 (4), 708–714. 10.1039/c8mh00296g

[128] Xu, H., Wang, F., Wang, Z. F., Zhou, H., Zhang, G., Zhang, J., Wang, J., and Yang, S. (2019). Intelligent solid lubricant materials with failure early-warning based on triboluminescence. *Tribol. Lett.*, 67, 13. 10.1007/s11249-018-1120-0

[129] Sahu, M. K. (2013). Theoretical approach to the laser-stimulated luminescence in II-VI semiconductor. *J. Pure Appl. Indus. Phys.*, 3 (1), 1–67.

[130] Quickenden, T. I., Selby, B. J., and Freeman, C. G. (1998). Ice triboluminescence. *J. Phys. Chem. A*, 102 (34), 6713–6715. 10.1021/jp981657y

[131] Jeong, S. M., Song, S., Joo, K. I., Kim, J., Hwang, S. H., Jeong, J., and Kim, H. (2014). Bright, wind driven white mechanoluminescence from zinc sulphide microparticles embedded in a polydimethylsiloxane elastomer. *Energy Environ. Sci.*, 7, 3338–3346. 10.1039/c4ee01776e

[132] Cheng, J., Ding, W. B., Zi, Y., Lu, Y., Ji, L., Liu, F., Wu, C., and Wang, Z. L. (2018). Triboelectric microplasma powered by mechanical stimuli. *Nat. Commun.*, 9, 11. 10.1038/s41467-018-06198 -x

[133] Wang, X. D., Zhang, H. L., Yu, R. M., Dong, L., Peng, D. F., Zhang, A., Zhang, Y., Liu, H., Pan, C., and Wang, Z. L. (2015). Dynamic pressure mapping of personalized handwriting by a flexible sensor matrix based on the mechanoluminescence process. *Adv. Mater.*, 27 (14), 2324–2331. 10.1002/adma.201405826

[134] Chen, C., Zhuang, Y., Tu, D., Wang, X., Pan, C., and Xie, R. J. (2020). Creating visible-to-near-infrared mechanoluminescence in mixed-anion compounds SrZn2S2O and SrZnSO. *Nano Energy*, 68, 104329. 10.1016/j.nanoen.2019.104329

[135] Satapathy, K., and Mishra, G. (2013). Dose dependence of Mechanoluminescence properties in $MgAl_2O_4$: Dy phosphor. *J. Chem.*, 2013, Article ID 930818, 4. 10.1155/2013/930818

[136] Atari, N. A., and Ettinger, K. V. (1974). Lyoluminescent tissue equivalent radiation dosimeter. *Nature*, 247 (1974), 193–194. 10.1038/247193a0

[137] Takavar, A., Ettinger, K. V., Mallard, J. R., Atari, N. A., and Srirath, S. (1977). In Proc. 5th Int. Cant on Luminescence Dosimetry Sag Paulo, 1977 (Ed. Scharmann A.) pp. 169–174 (Phys. Inst. Giessen, BRD, 1978).

[138] Becker K. (1973). *Solid State Dosimetry*, CRC Press, Cleveland, 1973 ISBN: 0878190465 9780878190461.

[139] McKeever S. W. S. (1985). *Thermoluminescence of Solids*, Cambridge University Press, Cambridge. Online ISBN: 9780511564994 10.1017/CBO9780511564994

[140] Sonder, E., and Sibley, W. A. (1965). Influence of lead impurity on the low temperature color centre production in KCl. *Phys. Rev.*, 140, A539–A546. 10.1103/PhysRev.140.A539

[141] Alvarez-Rivas, J. L. (1980). PLENARY SESSION Thermoluminescence and lattice defects in alkali halides. *Journal de Physique Colloques (J. Phys.)*, 41 (C6), C6-353–C6-358. 10.1051/jphyscol:1980691

[142] Ramani, R. (1984). Effect of inorganic metal ion impurities on lyoluminescence. *Radiat. Eff. Lett.*, 86, 47–56. 10.1080/01422448408205213

[143] Kalkar, C. D. (1984). Aqualuminescence from γ-irradiated NaCl in the presence of metal ions. *J. Radioanal. Nucl. Chem. Lett.*, 86, 65–78. 10.1007/BF02164902

[144] Chandra, B. P., Tiwari, R. K., Mor, R., and Bisen, D. P. (1997). Theoretical approach to the lyoluminescence of alkali halides. *J. Lumin.*, 75 (2), 127–133. 10.1016/S0022-2313(97)00108-7

[145] Galand, E., Pagnoulle, C., Niezette, J., Vanderschueren, J., and Garsou, J. (1997). Lyoluminescence of inorganic salts. *J. Lumin.*, 75, 27–33. 10.1016/S0022-2313(97)00100-2

[146] Hunter, C., Temperton, D. H., Ettinger, K. V., and Forrester, A. R. (1982). Lyoluminescence spectra of typical dosimeter materials. *Int. J. Appl. Radiat. Isot.*, 33 (11), 1291–1297. 10.1016/0020-708X(82)90249-6

[147] Shindl, A., Shindy, A. B., Al-Hashimi, Ali, H. M., AlThaher, S., and Ghania, T. (2004). Tissue equivalent personal X-ray dosimeter. *Bas. J. Surg.*, 10, 2004, https://www.researchgate.net/publication/315099990.

[148] Ettinger, K. V. (1989). Free radical dosimetry techniques and their suitability for precise and accurate measurements of radiation. *Appl. Radiat. Hot.*, 40, 865–870. 10.1016/0883-2889(89)90008-7

[149] Ettinger K. V., Forrester A. R., and Mallard J. R. (1980). *Proc. Int. Symp. Bio-Medical Dosimetry Paris*, IAEA, Vienna, 533-551. ISBN 92-0-010281-6.

[150] Sailaja, S., Dhoble, S. J., Brahme, N., and Reddy, B. S. (2011). Synthesis, photoluminescence and mechanoluminescence properties of Eu^{3+} ions activated $Ca_2Gd_2W_3O_{14}$ phosphors. *J. Mater. Sci.*, 46, 7793–7798. 10.1007/s10853-011-5759-2

[151] Zhang, J. C., Wan, Y., Xin, X., Han, W. P., Zhang, H. D., Sun, B., Long, Y. Z., and Wang, X. (2014). Elastico-mechanoluminescent enhancement with Gd^{3+} codoping in diphase (Ba, Ca)TiO$_3$:Pr^{3+}. *Opt. Mater. Express*, 4 (11), 2300–2309. 10.1364/OME.4.002300

[152] Hollerman, W. A., Fontenot, R. S., Bhat, K. N., Aggarwal, M. D., Guidry, C. J., and Nguyen, K. M. (2012). Comparison of triboluminescent emission yields for 27 luminescent materials. *Opt. Mater.*, 34 (9), 1517–1521. 10.1016/j.optmat.2012.03.011

[153] Ronfard-Haret, J. C., Valat, P., Wintgens, V., and Kossanyi, J. (2000). Triboluminescence of trivalent rare earth ions inserted in polycrystalline zinc oxide. *J. Lumin.*, 91, 71–77. 10.1016/S0022-2313(00)00204-0

[154] Chandra, B. P., Chandra, V. K., and Jha, P. (2015). Mechanoluminescence of coloured alkali halide crystals. *Defect Diffus. Forum*, 361, 121–176. 10.4028/www.scientific.net/DDF.361.121

[155] Chen, C., Zhuang, Y., Tu, D., Wang, X., Pan, C., and Xie, R. (2020). Creating visible-to-near-infrared mechanoluminescence in mixed-anion compounds $SrZn_2S_2O$ and SrZnSO. *Nano Energy*, 68, 104329. 10.1016/j.nanoen.2019.104329

[156] Jha, P., Khare, A., Singh, P., and Chandra, V. K. (2020). Fracto-mechanoluminescence of ZrO2:Ti nanophosphor excited by impulsive excitation. *Opt. Mater.*, 99, 109477. 10.1016/j.optmat.2019.109477

[157] Zhang, H. W., Terasaki, N., Yamada, H., and Xu, C. N. (2009). Mechanoluminescence of europium-doped $SrAMgSi_2O_7$(A=Ca, Sr, Ba). *Jpn. J. Appl. Phys.*, 48, 04C109-1-4 10.1143/JJAP.48.04C109

[158] Zhang, H. W., Yamada, H., Terasaki, N., and Xu, C. N. (2007). Stress-induced mechanoluminescence in $SrCaMgSi_2O_7$:Eu. *Electrochem. Solid State Lett.*, 10, J129–J131. 10.1149/1.2762205

[159] Zhang, H. W., Yamada, H., Terasaki, N., and Xu, C. N. (2008). Green mechanoluminescence of $Ca_2MgSi_2O_7$:Eu and $Ca_2MgSi_2O_7$:Eu,Dy. *J. Electrochem. Soc.*, 155, J55–J57. 10.1149/1.2816215

[160] Sun, H., Wu, C., Dai, K., Chang, J., and Tang, T. (2006). Proliferation and osteoblastic differentiation of human bone marrow-derived stromal cells on akermanite-bioactive ceramics. *Biomaterials*, 27, 5651–5657. 10.1016/j.biomaterials.2006.07.027

[161] Akiyama M., Xu C. N., Nonaka K., and Watanabe T. (1998). Intense visible light emission from $Sr_3Al_2O_6$:Eu, Dy. *Appl. Phys. Lett.*, 73, 3046–3048. 10.1063/1.1494463

[162] Xu, C. N., Watanabe, T., Akiyama, M., and Zheng, X. G. (1999). Direct view of stress distribution in solid by mechanoluminescence. *Appl. Phys. Lett.*, 74 (17), 2414–2416. 10.1063/1.123865

[163] Zhang, H. W., Yamada, H., Terasaki, N., and Xu, C. N. (2007). Ultraviolet mechanoluminescence from $SrAl_2O_4$:Ce and $SrAl_2O_4$:Ce,Ho. *Appl. Phys. Lett.*, 91, 081905-1-3. 10.1063/1.2772768

[164] Jia, D. (2006). Relocalization of Ce^{3+} 5d electrons from host conduction band. *J. Lumin.*, 117, 170–178. 10.1016/j.jlumin.2005.05.008

[165] Terasawa, Y., Xu, C. N., Yamada, H., and Kubo, M. (2011). Near infra-red mechanoluminescence from strontium aluminate doped with rare-earth ions. *IOP Conf. Ser.: Mater. Sci. Eng.*, 18, 212013-1-4. 10.1088/1757-899X/18/21/212013

[166] Jha P., and Chandra B. P. (2017). Mechanoluminescence of $SrAl_2O_4$:Eu,Dy nanophosphors induced by low impact velocity. *Luminescence*. 32, 171–176. 10.1002/bio.3162

[167] Metz, F. I., Schweiger, R. N., Leidler, H. R., and Girifalco, L. A. (1957). Stress activated luminescence in X-irradiated alkali halide crystals. *J. Phys. Chem. B*, 61 (1), 86–89. 10.1021/j150547a016

[168] Molotskii, M. I., and Shmurak, S. Z. (1992). Elementary acts of deformation luminescence. *Phys. Lett. A*, 166 (1992), 286. 10.1016/0375-9601(92)90378-Y

[169] Kher, R. S., Pandey, R. K., Dhoble, S. J., and Khokhar, M. S. K. (2002). Mechano and thermo-luminescence of Gamma irradiated $CaSO_4$:Dy phosphor. *Radiat. Prot. Dosim.*, 100 (1-4), 281–284. 10.1093/oxfordjournals.rpd.a005868

[170] Brahme, N., Bisen, D. P., Kher, R. S., and Khokhar, M. S. K. (2009). Mechanoluminescence and thermoluminescence in γ-irradiated rare earth doped CaF2 crystals. *Phys. Procedia*, 2, 431–440. 10.1016/j.phpro.2009.07.028

[171] Feng, A., and Smet, P. F. (2018). A review of mechanoluminescence in solids: Compounds, mechanism and applications. *Materials*, 11, 484. 10.3390/ma11040484

[172] Becker, P. (1998). Borate materials in nonlinear optics. *Adv. Mater.*, 10, 979–992. 10.1002/(SICI)1521-4095(199809)10:13<979::AID-ADMA979>3.0.CO;2-N

[173] Cavalcante, L. S., Sczancoski, J. C., Albarici, V. C., Matos, J. M. E., Varela, J. A., and Longo, E. (2008). Synthesis, characterization, structural refinement and optical absorption behaviour of $PbWO_4$ powders. *Mater. Sci. Eng.*, 150, 18. 10.1016/j.mseb.2008.02.003

[174] Liu, J., Wang, X. D., Wu, Z. C., and Kuang, S. P. (2011). Preparation, characterization and photoluminescencs properties of BaB_2O_4:Eu^{3+} red phosphor. *Spectrochim. Acta Part A*, 79(2011), 1520. 10.1016/j.saa.2011.05.009.

[175] Gawande, A. B., Sonekar, R. P., and Omanwar, S. K. (2013). Synthesis and Photoluminescence study of $Ca_3B_2O_6$:Pb^{2+}. *AIP Conf. Proc.*, 1536, 601–602. 10.1063/1.4810370

[176] Vecht, A., Newport, A. C., Bayley, P. A., and Crossland, W. A. (1998). Narrow band 390 nm emitting phosphors for photoluminescent liquid crystal displays. *J. Appl. Phys.*, 84, 3827–3829. 10.1 063/1.368561

[177] Sonekar, R. P., Omanwar, S. K., Moharil, S. V., Dhopte, S. M., Muthal, P. L., and Kondawar, V. K. (2007). Combustion synthesis of narrow UV emitting rare earth borate phosphors. *Opt. Mater.*, 30 (4), 622–625. 10.1016/j.optmat.2007.02.016

[178] Thakre, D. S., Omanwar, S. K., Muthal, P. L., Dhopte, S. M., Kondawar, V. K., and Moharil, S. V. (2004). UV-emitting phosphors: Synthesis, photoluminescence and applications. *Phys. Status Solidi*, 201 (3), 574–581. 10.1002/pssa.200306720

[179] Ropp R. C. (2004). *Luminescence and the Solid State*, 2nd ed., Elsevier, Amsterdam, Volume 21, eBook ISBN: 9780080473239.

[180] Zheng, Y., Qu, Y., Tian, Y., Rong, C., Wang, Z., Li, S., Chen, X., and Ma, Y. (2009). Effect of Eu^{3+}-doped on the luminescence properties of zinc borate nanoparticles. *Colloid Surf. A: Physicohem. Eng. Asp.*, 349 (19), 10.1016/j.colsurfa.2009.07.039

[181] Nagpure, P. A., and Omanwar, S. K. (2012). Synthesis and luminescence characteristics of terbium (III) activated $NaSrBO_3$. *J. Rare Earth*, 30 (9), 856–859. 10.1016/S1002-0721(12)60145-8

[182] Chikte (Awade) D., Omanwar S. K., and Moharil S. V. (2013). Luminescence properties of red phosphor $NaSrBO_3$:Eu^{3+}prepared with novel combustion synthesis method. *J. Lumin.*, 142(2013), 180–183, 10.1016/j.jlumin.2013.03.045

[183] Mishra, G. C. (2012). Development of borate based phosphors for mechanoluminescence and lyo-luminescence radiation dosimetry. http://hdl.handle.net/10603/18211)

[184] Ballato, J., Lewis, III, J. S., and Holloway, P. (1999). Display applications of rare-earth-doped materials. *MRS Bull.*, 24, 51, WOS:000082817900020 .

[185] Zhang, Z., Ten Kate, O. M., Delsing, A., van der Kolk, E., Notten, P. H. L., Dorenbos, P., Zhao, J., and Hintzen, H. T. (2012). Photoluminescence properties and energy level locations of RE^{3+} (RE = Pr, Sm, Tb, Tb/Ce) in $CaAlSiN_3$ phosphors. *J. Mater. Chem.*, 22, 9813. 10.1039/c2jm30220a

[186] Dorenbos, P. (2003). f → d transition energies of divalent lanthanides in inorganic compounds. *J. Phys.: Condens. Matter*, 15 (2003), 575. 10.1088/0953-8984/15/3/322

[187] Dorenbos, P. (2009). Lanthanide charge transfer energies and related luminescence, charge carrier trapping, and redox phenomena. *J. Alloys Compd.*, 488 (2009), 568. 10.1016/j.jallcom.2008.09.059

[188] Henderson B., and Imbusch F. C. (1989). *Optical Spectroscopy of Inorganic Solids* (Oxford: Clarendon), Chapter 5. Clarendon Press, Oxford University Press, New York.

[189] Su, Q., Pei, Z., Chi, L., Zhang, H., Zhang, Z., and Zou, F. (1993). The yellow-to-blue intensity ratio (Y/B) of Dy^{3+} emission. *J. Alloys Compd.*, 192 (1–2), 25–27. 10.1016/0925-8388(93)90174-L

[190] Judd, B. R. (1962). Optical absorption intensities of rare earth ions. *Phys. Rev.*, 127, 750–761. 10. 1103/PhysRev.127.750

[191] Ofelt, G. S. (1962). Intensities of crystal spectra of rare earth ions. *J. Chem. Phys.*, 37, 511–520. 10. 1063/1.1701366

[192] Liu, J., Wang, X. D., Wu, Z. C., and Kuang, S. P. (2011). Preparation, characterization and pho-toluminescence properties of BaB_2O_4: Eu^{3+} red phosphor. *Spectrochim. Acta Part A*, 79 (5), 1520–1523. 10.1016/j.saa.2011.05.009

[193] Porwal, N. K., Kadam, R. M., Seshagiri, T. K., Natarajan, V., Dhobale, A. R., and Page, A. G. (2005). EPR and TSL studies on MgB_4O_7 doped with Tm: Role of in TSL glow peak at 470K. *Radiat. Meas.*, 40 (1), 69–75. 10.1016/j.radmeas.2005.04.007

[194] Ettinger, K. V., Forrester, A. R., and Hunter, C. H. (1982). Lyoluminescence and spin trapping. *Can. J. Chem.*, 60, 1549–1559. 10.1139/v82-502

[195] Xie, R., Hirosaki, N., and Takeda, T. (2009). Wide color Gamut backlight for liquid crystal displays using three-band phosphor-converted white light-emitting diodes. *Appl. Phys. Express*, 2 (2), 022401. 10.1143/APEX.2.022401

[196] Peng, M., and Wondraczek, L. (2010). Orange-to-red emission from Bi^{2+} and alkaline earth codoped strontium borate phosphors for white light emitting diodes. *J. Am. Ceram. Soc.*, 93, 1437–1442. 10.1111/j.1551-2916.2009.03590.x

[197] Wang, Y., Yiguang, Wang, Y., Chi, N., Yu, J., and Shang, H. (2013). Demonstration of 575-Mb/s downlink and 225-Mb/s uplink bi-directional SCM-WDM visible light communication using RGB LED and phosphor-based LED. *Opt Express*, 21, 1203-1208. 10.1364/OE.21.001203

[198] Do, Y. R., and Bae, J. W. (2000). Application of photoluminescence phosphors to a phosphor-liquid crystal display. *J. Appl. Phys.*, 88, 4660–4665. 10.1063/1.1311825

[199] Gracia, C. R., Diaz-Torres, L. A., Oliva, J., and Hirata, G. A. (2014). Green EuAlO3:Eu2+ nano-phosphor for applications in WLEDs. *Opt. Mater.*, 37, 520–524. 10.1016/j.optmat.2014.07.016

[200] Raikwar, V., Bhatkar, V., and Omanwar, S. (2018). Morphological and photoluminescence study of NaSrB$_5$O$_9$: Tb^{3+} nanocrystalline phosphor. *J. Asian Ceram. Soc.*, 6 (4), 359–367. 10.1080/21870764 .2018.1529014

[201] Hedaoo, V. P., Bhatkar, V. B., and Omanwar, S. K. (2015). PbCaB$_2$O$_5$ doped with Eu^{3+}: A novel red emitting phosphor. *Opt. Mater.*, 45, 91–96. 10.1016/j.optmat.2015.02.037

[202] Hedaoo (Raikwar), V. P., Bhatkar, V. B., and Omanwar, S. K. (2015). Synthesis, characterization and photoluminescence in novel lead calcium diborate doped with Mn^{2+}. *Optik*, 126 (24), 4813–4816. 10.1016/j.ijleo.2015.09.192

[203] Gawande, A. B., Sonekar, R. P., and Omanwar, S. K. (2014). Synthesis and enhancement of lu-minescence intensity by codoping of M$^+$ (M=Li, Na, K) in Ce^{3+} doped strontium haloborate. *Opt. Mater.*, 36 (7), 1143–1145. 10.1016/j.optmat.2014.02.017

[204] Ingle, J. T., Sonekar, R. P., Omanwar, S. K., Wang, Y., and Zao, L. (2014). Combustion synthesis and photoluminescence study of novel red phosphor (Y1−x−y, Gdx)BaB$_9$O$_{16}$:Eu^{3+y} for display and lighting. *J. Alloys Compd.*, 608, 235–240. 10.1016/j.jallcom.2014.04.079

[205] Sahare, P. D., and Srivastava, S. K. (2016). Lyoluminescence dosimetry of high-energy gamma radiation using MgB$_4$O$_7$:Mn^{2+}. *J. Radioanal. Nucl. Chem.*, 307, 31–36. 10.1007/s10967-015-4117-2

[206] Wu, W., Duan, Y., and Liu, B. (2020). Mechanoluminescence: Quantitative pressure-brightness relationship enables new applications. *Matter*, 2 (2), 291–293. 10.1016/j.matt.2020.01.007

[207] Wang, C., Yu, Y., Yuan, Y., Ren, C., Liao, Q., Wang, J., Chai, Z., Li, Q., and Li, Z. (2020). Heartbeat-sensing mechanoluminescent device based on a quantitative relationship between pressure and emissive intensity. *Matter*, 2 (1), 181–193. 10.1016/j.matt.2019.10.002

[208] Wang, C., Yu, Y., Chai, Z., He, F., Wu, C., Gong, Y., Han, M., Li, Q., and Li, Z. (2019). Recyclable mechanoluminescent luminogen: Different polymorphs, different self-assembly effects of the thio-phene moiety and recovered molecular packing via simple thermal-treatment. *Mater. Chem. Front.*, 3, 32–38. 10.1039/C8QM00411K

[209] Wu, W., Duan, Y., and Liu, B. (2020). Mechanoluminescence: Quantitative pressure-brightness relationship enables new applications. *Matter*, 2, 284–296. 10.1016/j.matt.2020.01.007

[210] Peng, D., Wang, C., Ma, R., Mao, S., Qu, S., Ren, Z., Golovynskyi, S., and Pan, C. (2021). Mechnoluminescent materials for athletic analytics in sports science. *Sci. Bull.*, 66, 206–209. 10.1016/ j.scib.2020.09.029

10 Borate Phosphors for Neutron Radiography

P. K. Tawalare

CONTENTS

10.1 NEUTRON RADIOGRAPHY

Radiography is a non-destructive testing method that produces an image using penetrating radiation, such as X-rays or neutrons, instead of visible light. The attenuation of a radiation beam passing through an object reveals clues about the internal structure of that object. Among various applications of neutrons, neutron radiography (NR) holds a special place. It was developed much later following several other applications like neutron activation analysis (NAA), isotope production, crystallography using neutron diffraction, etc. One of the reasons is that NR requires a special type of equipment and trained personnel, the requirement arising out of safety measures to be taken against the possible hazards. Radiography with gamma rays and X-rays has been long established as a very important technique of non-destructive testing, but the use of neutrons is only just becoming prominent.

DOI: 10.1201/9781003207757-10

The definition of the term given in American Society for Testing and Materials (ASTM) is as follows: neutron radiography – the process of producing a radiograph using neutrons as the penetrating radiation. The term 'radiograph' was already in use, but normally it means an image obtained using X-rays that were used for this purpose almost immediately after their discovery in 1895 by Roentgen. It has received the attention of scientists from various fields. Neutron radiography (NR) has been demonstrated by several researchers to be one of the non-destructive testing methods for over 50 years. Most common NR uses thermal neutrons. Several types of technologies are available for the same. These technologies differ in use of detectors such as "photographic film, imaging plates, amorphous silicon, CCD cameras and CMOS pixel detectors", and also in the neutron sources employed. Without exception, the various techniques exploit the efficient nuclear reaction induced by thermal neutrons of a handful of specific elements like ^{10}B, ^{6}Li and Gd. The secondary radiation resulting after the neutron capture is detected.

In neutron radiography, an object under examination is placed in a neutron beam; after passing through the object, the beam that remains enters a detector that registers the fraction of the initial intensity that has been transmitted by each point in the object [1]. Any inhomogeneity in the object or internal defect (e.g. void, crack, porosity or inclusion) will show up as a change in radiation intensity reaching the detector. The pattern of the penetrating radiation reveals clues about the internal structure of an object [2]. Figure 10.1 shows a schematic of a representative neutron imaging system.

A neutron radiography system typically consists of a neutron source, a collimator (sometimes including filter devices), a shutter system, an imaging system and a radiation beam stop. A neutron source, such as a nuclear reactor, usually produces neutrons traveling in all directions. Fast neutrons collide with a moderator and slow down to thermal energies. Because the thermal neutrons are dispersed over the source and moderator volumes, a collimator is necessary between the source and the object to provide a properly directed neutron beam. An aperture/collimator absorbs neutrons that are not travelling in the desired direction, producing a shaped neutron beam. Filters within the collimator assembly can remove different energy neutrons and gamma rays from the beam. A shutter mechanism opens and closes the beam, and may be a simple slab of absorber material which can be moved into and out of the beamline. A sketch showing elements of a NR system is shown in Figure 10.2. A reactor source of neutrons shown with a divergent collimator images an NR object on a film cassette. The collimator performance is determined by L/D.

An imaging station consists of a method to hold the object in the beam and an imager or detector at the image plane. Potential imaging technologies include activation foils, scintillating screens and multi-channel plates [4–6]. Finally, a beam stop placed behind the imaging station absorbs the radiation beam, and prevents excessive exposure from the beam. A typical neutron imaging apparatus is shown in Figure 10.3.

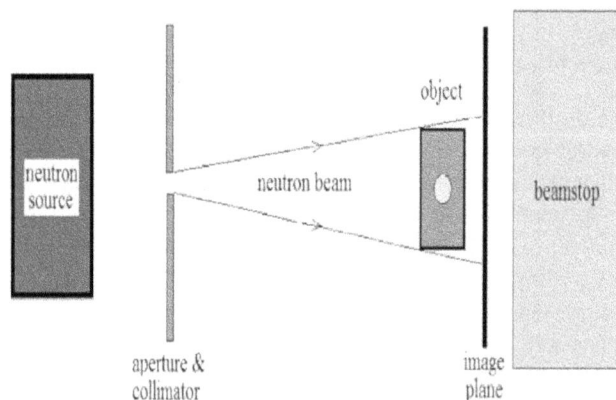

FIGURE 10.1 Schematic of a representative neutron imaging system [1].

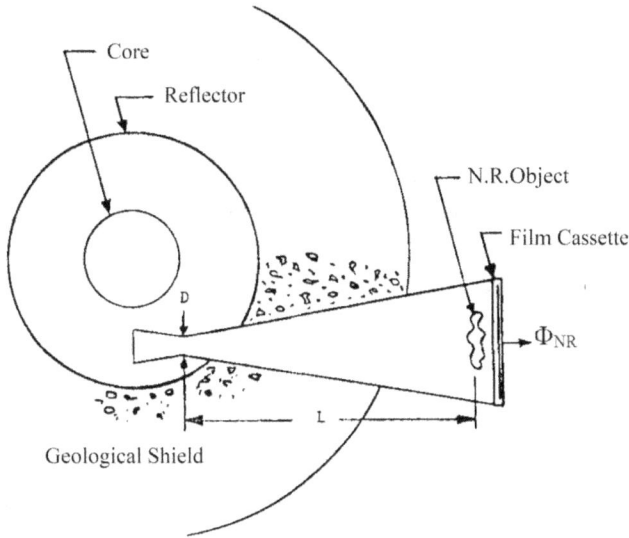

FIGURE 10.2 Sketch showing elements of a NR system [3].

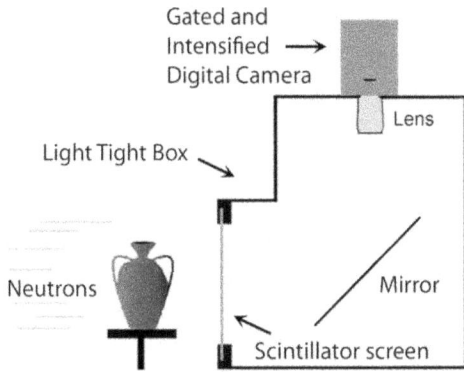

FIGURE 10.3 A typical neutron imaging apparatus [7].

The three parameters on which the merit of the neutron imaging system is decided are: "spatial resolution, time resolution and image quality [8]".

10.1.1 SPATIAL RESOLUTION

The "spatial resolution" is decided by the detector as well as the beam characteristics. If the beam is divergent and poorly collimated, the images will be obviously blurred due to the "geometrical convolution". The beam quality can be quantified by a parameter "beam collimation ratio (L/D) – where L is the collimator length and D is the aperture diameter". If this ratio is 500 or more, the image quality will be influenced by the detector performance alone. Under these conditions, the distance between the object and the detector plane can be minimized.

10.1.2 TIME RESOLUTION

The "time resolution" is mainly governed by the detector efficiency and response time, i.e. it gives an idea about the speed with which an image can be generated in the area of interest. This is the time for which the object to be imaged is exposed to neutrons, t_{exp}, and the signal required for producing the image is generated. Typically, the film radiography using a source with 10^7 neutrons per cm^2 s will need about 10 min while the digital imaging systems need an do the job in only few seconds or less.

10.1.3 Image Quality

To a first approximation, the "image quality" can be described in terms of the "signal-to-noise ratio of the image (S/N)". This will improve with the exposure time. There are three factors that contribute to the noise level; i) the stochastic processes going on during the detection itself (shot noise), ii) that from the readout process (readout noise) and iii) from background effects as dark current or random effects from background radiation (dark noise). These factors will differ from system to system. Which factor becomes decisive also likewise depends on the specific system.

10.1.4 Early History

The pioneering work on NR dates back to 1935. Credit goes to Kallmann and Kuhn who brought it in a public domain in 1948 [9]. Early developments are summarized by Thewlis [10,11] Berger [12] narrated the principles of "neutron radiography" and described some applications reviewed the work carried out on this topic up to 1964. However, due to non-availability of intense neutron sources, it did not catch the attention of the contemporary scientists. After a rather sedate beginning, the work gathered momentum when neutrons from reactors could be used. The advent of adequate neutron sources and sensitive detectors paved the way for accelerating the developments in the field of radiographic applications. Rather poor quality of the radiographs generated during the early period failed to generate interest. Reasonable quality that could compete with other radiology techniques could be achieved only after the availability of neutron beams from the reactor. However, the reactors are not portable. With improvement in the detector technology, the radioactive (isotope) sources linked to small accelerator sources could compete with the reactor beams. Most of the early history is summed up by Whittemore [3] and Hawkesworth and Walker [13].

Capability of neutrons to generate radiographic images of large objects with a spatial resolution comparable to that of conventional X-ray techniques is remarkable [7]. The digital detection methods enable recording images from several angles. A 3-D image can then be obtained using image re-construction software. This is called tomography. It is essentially imaging a cross section of the object from a specific angle. Joining all such cross sections leads to a 3-D image. Moreover, the absorption cross section for neutron does not vary systematically with atomic number. For this reason, the tomography offers a method for obtaining information about a specific area of an inhomogeneous sample. In general, neutrons from commonly used sources will show a wide distribution of velocities. The images obtained with neutrons are usually energy averaged. The image contrast can be improved by making the incident beam as monoenergetic as possible. Such a procedure can give the information about the object which is otherwise lost due to energy averaging. A neutron beam can be made monoenergetic by means of a "Fermi chopper" or with help of a monochromator crystal.

Both fast and thermal neutrons can be used for radiography. Thermal neutrons have short range in the matter they travel. But they can interact with the matter through secondary radiations (α, β, Υ rays) they produce. Neutron imaging using fast neutrons provides a nondestructive testing method for very large size objects of any atomic composition which distinguishes it from methods like X-ray imaging, which has the limitations arising from these factors [14]. An object made up of both low and high Z atoms can be conveniently scanned by a fast neutron beam as the absorption co-efficient does not depend on atomic number. This is a specific feature of fast neutron radiography which is not observable for radiographies using photon beams or even thermal neutrons. This feature arises from the fact that in the fast energy range (at few MeV) the macroscopic neutron interaction cross sections are generally low and almost independent of energy and atomic number. This permits high penetration depth irrespective of the sample composition. In many practical situations, fast neutron radiography offers a viable method of non-destructive testing. The factors that limit the application of fast neutron imaging are the relatively poor spatial resolution and low detection efficiency of the detector arising primarily from the high penetrating power of fast neutrons. As a remedy, the interaction of fast neutron with hydrogen is exploited for the conversion and detection of fast neutrons in the MeV range. The recoil protons created during this interaction can produce scintillation light that can be detected by photodetectors. Therefore, the most commonly used fast neutron imaging screens are either made of hydrogen-rich organic (plastic) scintillators or a plastic converter matrix mixed with an inorganic scintillator, typically ZnS. There are pros and cons for using both types of screens. Plastic scintillators with high hydrogen contents, are also machinable and can be made into various desired shapes. These are also low cost. They have good transmission for the scintillation light and hence thick screens required for high conversion efficiency can be employed.

10.1.5 Neutron vs X-rays

Basic feature of X-radiography and NR is the same, viz. transmission of radiation through an object, and obtain an image with help of the radiation detecting device. Yet the technical details of NR are vastly different from those of X-radiography. NR is not simply additional method for radiography or as a competitor to X- or gamma rays. It is rather complimentary and often can give new information. X- or Υ ray sources used for radiography are usually of small size giving out uncollimated beams. Desired resolution is ensured by keeping the distance between object and source sufficiently large and that between the object and the recording device small. Usually, strong neutron beams are obtained from large size sources like reactors or accelerators. The neutron beam has to be collimated to obtain satisfactory resolution. This is not an obstacle as collimation of neutron beam is straightforward. Thermal neutrons are highly suitable for probing solids as the wavelengths are comparable to the interatomic separations and the energy with that of phonons. As compared to X- or Υ rays, neutron is a massive particle. Though chargeless, it has magnetic moment. For radiography purpose, however, the magnetic moment of neutron, does not play active role. The interaction of neutron with matter is through nuclear forces is most relevant for radiography. Scattering of X-rays is from the electrons which surround the atomic nucleus, in comparison, scattering of slow neutrons is characteristics of the nucleus of an atom and not by the electron cloud. The interaction between neutron and nucleus can involve elastic (n, n) or inelastic (n, n') scattering, Another process is absorption due to radiative capture with charged particle emission e.g. (n, Υ) or (n, α). The type and the magnitude of the interaction does not depend on the atomic charge but on the energy of the neutron and the structure of the nuclide. This is in sharp contrast to the case of photons (X- or Υ rays) where only the atomic number is important. Neutrons can distinguish between isotopes as the interaction cross section can vary markedly from nuclide to nuclide. Even with the same element it is possible to distinguish not only between materials of similar atomic number but also between isotopes. Another important difference is that for X-rays the attenuation coefficient increases with atomic number, which is not the case for thermal neutrons. A striking example is that of lead. Being a high Z element, it can effectively stop X- or Υ rays. However, its attenuation coefficient for neutrons is small. It is hence possible to obtain radiographs of thick sheet of lead using neutrons. Not only this, but using neutrons radiographs can be obtained of an object constituted of very light elements even if it is shielded by a thick sheet of heavy metal. A remarkable example is that of hydrogen and lead. Not only light elements like B or Li, but even hydrogenous material like plastics have high neutron scattering cross section by virtue of free protons. This characteristic makes NR highly effective in the non-destructive examination of corrosion. Corrosion can be detected through O and H. X-ray cross sections for these atoms are much smaller than for heavier elements and X-ray radiography becomes difficult for this application. Boron, Cd and Gd are very suitable for radiography because of their high neutron absorption. Most importantly, neutron radiography can be used when the samples are highly radioactive and themselves emit X- or Υ rays.

10.1.6 Neutron Sources

When neutron radiography is done for inspecting mechanical faults or physical appearance thermal neutrons, i.e. neutrons in the energy range 0.005–0.5 eV are preferred. Attenuation coefficients for thermal neutrons show large variation as compared to those for X-rays. The large differences in attenuation coefficients make neutron radiography useful, usually providing almost a reversal of typical X-ray attenuation. The information about the object that is obtainable from X-ray and neutron radiographics is of complementary nature in many cases. Cold neutrons have also been used as a probing beam for NR. Cold neutrons have energies <0.005 eV. Radiographs with better contrast are often obtainable with such beams. In context of MeV neutrons (fast neutrons), the term 'slow neutrons' are used to cover both the cold and thermal neutrons and energies beyond that up to several keV. Other terms that describe the energy of neutrons are epithermal neutrons (energy ranging from that slightly higher than thermal to about 10 keV) and fast neutrons, which have energies above 10 keV, up to the MeV. Neutrons with still higher energies (>10 MeV) are often called relativistic because of the velocities becoming comparable to that of light. Although radiography is possible with neutrons of all these energy ranges, thermal neutrons are most commonly used.

Some commonly used sources along with their typical features are tabulated below (Table 10.1).

10.1.7 Neutron Detection

Helium isotope and boron trifluoride gases are used in the "proportional counters" are the commonly used detectors for thermal neutrons. However, ^{3}He is an very rare helium isotope. Its abundance in nature is only about 1.37 ppm. Due to use in nuclear sciences, ^{3}He source is depleting. This has resulted in several fold escalation of the cost of these thermal neutron detectors. As regards the use of BF_3 for neutron detection, the corrosive and toxic nature of BF_3 poses several hazards. Considering these factors, alternative methods had to be developed for neutron detection and radiography. Experience in the field of X-radiography came in handy in these developments. By and large, all the techniques available for X-ray imaging could be suitably modified and used for NR imaging. Mention must be made of X-ray films and real-time techniques using electronic recording. Neutron beams do not directly affect photographic emulsions. However, placing a thin foil of metal such as gadolinium, acts as a converter for neutrons producing β rays, which can be more efficiently detected by a photographic film.

For several years, a combination of film and converter metallic layers (gadolinium, dysprosium or indium) constituted the major detection system for neutron radiography. Such detectors could provide adequate "spatial resolution" which was limited only by properties of the probe beam. Introduction of the digital detection systems changed this situation. Such systems enabled much higher sensitivities, larger dynamic ranges and the opportunity to use the digital image information for more quantitative evaluations.

TABLE 10.1

Sources along with their typical features such as typical flux and typical beam

Source	Typical flux (neutrons/cm^2 sec)	Typical beam dimension (cm)	Ref.
Reactor-core moderator	10^7–10^8	20	[13]
Reactor-thermal column	10^3–10^6	30	
Reactor-monochromator	10^4–10^5	30	
Reactor-be filter	10^4–10^6	20	
Accelerator/moderator assembly	10^3–10^5	10	
Am/Bc assembly	10–10^2	10	

10.1.8   Film Radiography

The photographic film is the most versatile detector in NR, similar to its predecessor X-radiography. It offers the best "spatial resolution", indelible record, the possibility of integrating over long times in weak neutron fields and low cost. However, neutrons are inefficient for directly affecting the photographic emulsions; a converter (intensifier) screen is required to facilitate detection. The intensifier absorbs neutron and emits α or Υ rays by (n,α) or (n,Υ) reaction which in turn are recorded by the photographic emulsion. (n,α) reactions are preferred when discrimination against Υ rays is desired. Some converting materials like gadolinium emit β rays. Thermal neutron absorption cross section of gadolinium is amongst the highest, in conjunction with the spontaneous emission of low-energy electrons from "internal conversion of gamma-rays", it makes film radiography viable. The highest number of electrons emitted in a (n,Υ) reaction have energy 70 keV implying that the range is short. Therefore, the combination gadolinium converter with photographic film has a high resolution.

Another method is brighter images for obtaining use of luminescent materials. X-rays or Υ rays do not efficiently affect the photographic film. Hence, a phosphor that emits light upon X-rays or Υ rays incidence is coated on the front side. One of the earliest screens developed is described by Stedman [15].

10.1.9   Radiography Based on Scintillators

Film radiography is rather cumbersome in that the sensitivity of the photographic emulsion is rather low. The photographic process involves exposure, developing and fixing. The reproducibility gets affected due to such steps. Moreover, this is not online, but the results are known only after processing the film.

A fast, online method for neutron radiography is based on the phenomenon of scintillation. When ionizing radiations strike a solid, a flash of light appears. This is known as scintillation. Neutrons, which are chargeless, cannot produce scintillation by themselves, as they just pass through the matter. Thus, the scintillator to be used for neutron detection, should be inclusive of ^{6}Li, ^{10}B, ^{155}Gd, ^{157}Gd or other isotopes that have good interaction cross section for neutrons. The interaction should result in emission of charge particles like α particles or gamma rays. By virtue of for the ^{6}Li (n, α)^{3}H, with the energy of 4.8 MeV and cross section of 942 barn, the most suitable nuclide is ^{6}Li for detection of thermal neutrons. It can also provide good discrimination against gamma rays. The scintillation light can be converted to electrical signal using photodiodes. Scintillating material can be directly deposited on photodiodes. Figure 10.4 shows typical neutron sensitive scintillator screens for neutron imaging. When a beam of neutrons falls on the scintillator, a flash of light is emitted that is converted to electrical signal by the photodiode.

Using an array of photodiodes, a large area can be scanned. This process is in real time as the scintillation decay time is of the order of several nanoseconds. Photodiode signals can be converted to a digital signal and image processing software can be used. The resolution is lower than that of a photographic plate, being limited by the size of the photodiode. However, image quality can be improved using digital techniques.

Scintillators for a thermal neutron "should possess following properties [16],

1. a high light yield (>20,000 photons per neutron)
2. a fast response time (10–100 ns) for time resolution,
3. a high density, and high atomic number Z for efficient gamma ray detection
4. emission wavelength matching with the sensitivity curve of the light sensor (silicon diode or photomultiplier tube)
5. amenability for growing large crystals (>10 × 10 × 10 cm)
6. for thermal neutron detection, isotopes like ^{6}Li, ^{10}B, ^{155}Gd, ^{157}Gd should be constituent".

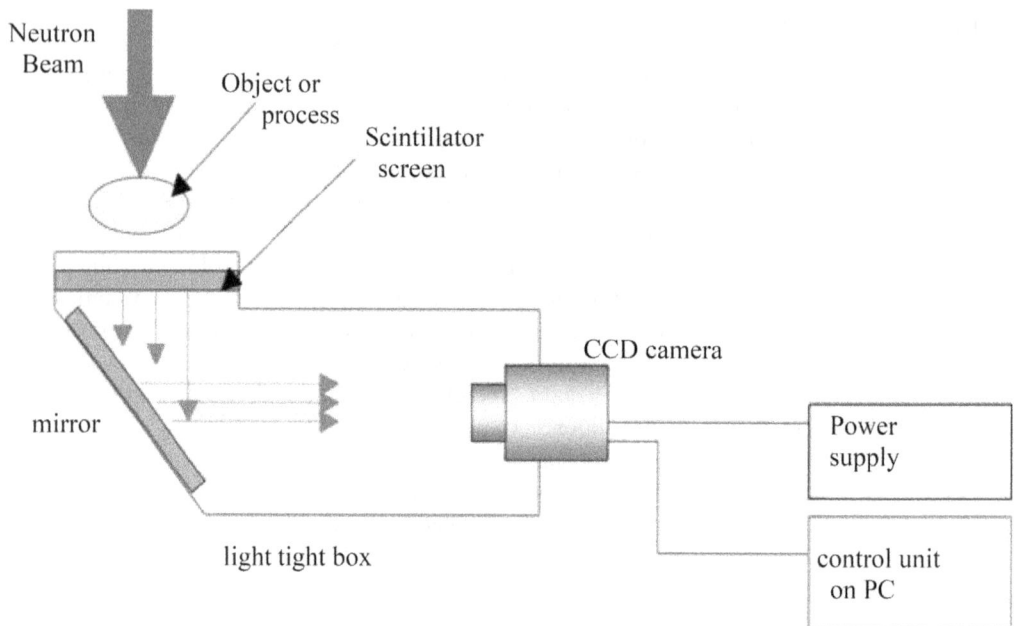

FIGURE 10.4 Typical neutron sensitive scintillator screens for neutron imaging [8].

Thermal neutrons are detected rather indirectly by alpha, beta or gamma rays produced in reactions:

$n + {}^6Li \rightarrow {}^3H$ (2:75 MeV) $+ \alpha$ (2:05 MeV)

$\sigma = 520b$

$n + {}^{10}B \rightarrow {}^7Li$ (1:0 MeV) $+ \alpha$ (1:8 MeV) 7%

$= {}^7Li$(0:83MeV $+ \alpha$ (1:47 MeV) $+ \Upsilon$ (0.48 MeV) (93 %)

$\sigma = 2100$ b

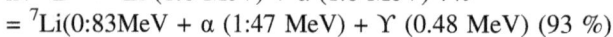

$n + {}^{157}Gd \rightarrow {}^{158}Gd + \Upsilon$ + conversion elements + X-rays (29 keV–182 keV);BR 85%

$\sigma = 70000$ b

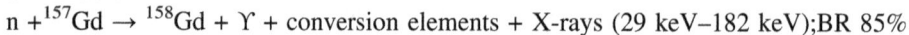

$n + {}^{155}Gd \rightarrow {}^{156}Gd + \Upsilon$ + conversion elements + X-rays (39 keV–199 keV); BR 80%

$\sigma = 17,000$ b

Early scintillators were based on the combination $ZnS/{}^6LiF$ (4:1 wt. ratio) [17]. ${}^6LiF$ produces α particles by a (n,α) reaction. The ensuing α particles produce scintillation in ZnS. Characteristics of various scintillators developed subsequently and used for NR using thermal neutrons have been reviewed by van Eijk et al. [18,19].

10.1.10 RADIOGRAPHY BASED ON STORAGE PHOSPHORS

Real-time methods produce instantaneous radiographs. However, the detector sensitivity has to be high and neutrons sources strong. Another method for NR makes use of storage phosphors that allows integration of the information. Like a scintillator, a storage phosphor also uses a phenomenon of luminescence for detection of radiation. The difference is that the emission of light is not immediate but a significant part of the energy deposited in the interaction with radiation is stored in metastable traps. The trap is the loss mechanism of a scintillator and leads to undesirable afterglow. It acts like a "memory bit" for a storage phosphor. The condition is that the trap should be emptied only after stimulation. Upon stimulation, an electron is released from the trap into the conduction band from where it can recombine with hole and transfer energy to the luminescence centre, or trapped again. This process has been termed 35 (PSL). Figure 10.5 shows the experimental setup for PSL measurement [20].

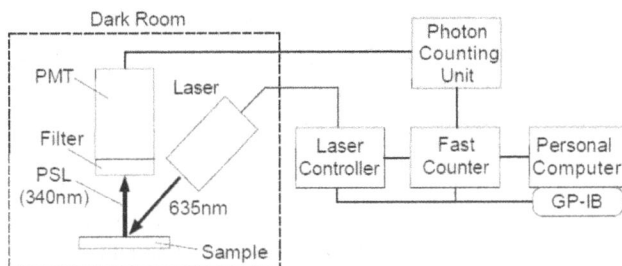

FIGURE 10.5   The experimental setup for PSL measurement [20].

Phenomenon of PSL has been exploited for preparing "image plates" for X-ray imaging and computed tomography since the 1980s. The commercial plates use BaFBr:Eu (Fuji, Agfa, Kodak). The images recorded on plates can be scanned by a He/Ne laser beam. "Imaging plates" based on storage-phosphor image plates are prepared from a mixture of the phosphor powder and a suitable binder. The mixture is coated on a base plate. NR using PSL has also been suggested using mostly borate phosphors [21,22].

10.2   ROLE OF BORATES

Scintillators and storage phosphors for neutrons should be capable of interacting with them. Thermal neutrons have large cross section with elements like ^{6}Li, ^{10}B, ^{155}Gd, ^{157}Gd, Cd. In particular, the ^{10}B isotope has a large thermal neutron cross section, and high-energy release per absorbed neutron (^{10}B(n, α)^{7}Li reaction). Wide bandgap insulating compounds containing these elements as constituents can be potential candidates. Several lithium compounds are hygroscopic, Cd is toxic and Gd is high cost. Boron, on the other hand, especially in the form of borate, leads to large number of insulating compounds with wide optical transmission window. Most borates are chemically and thermally stable with long shelf life. Borates are amenable to crystal growth owing to moderate melting points; e.g. $Li_6Gd(BO_3)_3$ as crystal can be grown at 835 C [23]. A large number of borates have already been studied and found applications as non-linear optical materials (NLO) [24].

During the last few eras, borates have more attention due to their interesting structures and ensuring applications in non-linear optical materials (NLO), birefringent materials, ferroelectric and piezoelectric materials, and luminophores [25,26]. Particularly, the β-BaB_2O_4 (β -BBO), [27] LiB_3O_5 (LBO), [28] $CsLiB_6O_{10}$ (CLBO) [29] and $KBe_2BO_3F_2$ (KBBF) [30] are outstanding NLO materials for efficient second harmonic generation (SHG) of Nd:YAG lasers significantly used in UV or deep-UV wavelength. Though the crystal structures directly affect the properties of materials, many strategies were used to synthesize promising optical materials [31–33]. Some innovative aluminum borate crystals with outstanding NLO properties have been reported, such as $K_2Al_2B_2O_7$ [34], $BaAl_2B_2O_7$ [35] and $BaAlBO_3F_2$ [36]. With speedy progress of science and technology, the researchers silent attempt to plan innovative NLO materials. According to the anionic group theory [37], coplanar and aligned NLO-active BO_3 groups in the structure will give rise to a relatively large SHG response and a sufficient birefringence for the generation of deep-UV coherent light. Based on the research of Becker [38], isolated BO_3 groups could exist. It is feasible to obtain borates with isolated BO_3 groups by increasing the proportion of cations in experiments. It is remarkable that the introduction of cations with strong bonding such as Be^{2+}, Al^{3+}, Mg^{2+}, Zn^{2+}, Si^{4+} [39] etc. can inhibit the polymerization of B–O groups. Hence, the UV-favourable system, Al–B–O, was chosen and anticipated to find a new borate with excellent NLO property. Wide hard work in the Al_2O_3–B_2O_3 system directed to a new NLO crystal of Al_5BO_9. The structure of Al_5BO_9 was first reported by Sokolova et al. [40]. Recently, Gatta et al. [41]

researched behaviour and phase stability of Al_5BO_9 in high-pressure condition, and Ok et al. [42] analysed the SHG properties of $Al_{5-x}Ga_xBO9$ (0.0 < × < 0.5) solid solutions.

Most common coordination for boron is 3 and 4. In combination with oxygen, it forms (BO_3) and (BO_4) polyhedra. Through these polyhedral, a variety of compounds having diverse structures can be formed. By and large, "a borate is made up of clusters of corner sharing (BO_3) and (BO_4) polyhedra, which occur as discrete polyanions to form larger clusters, chains, sheets or frameworks". (BO_3) and (BO_4) polyhedra assume geometrical shapes of trigonal capped prisms (tcp) or regular tetrahedrons. Intercalation of these polyhedral with other cationic groups leads to numerous structural types. Apart from (BO_3) and (BO_4), there are more than 50 types of boron-oxygen anions and polyanions that make up different borates. Depending on the arrangement of the anionic groups borates have been classified as pyroborates, metaborate and polyborates. By virtue of these building blocks, borates possess characteristics that can be exploited for developing optical materials. The notable characteristics are transparency over broad range, large electronic bandgap, good thermal and chemical stability, synthesis at moderate temperatures, good non-linear optical (NLO) properties and exceptionally high optical damage threshold. These properties caught up attention of materials scientists. Several optical materials used in various applications are based on borate compounds. Apart from the crystalline compounds, a variety of borate and borosilicate glasses are known which are good optical materials.

In the next section, we discuss borate-based phosphors for NR.

10.3   PHOSPHORS FOR RADIOGRAPHY

Over the years, several borate phosphors have been considered for use in NR. First report on Ce^{3+}-activated $Li_6Gd(BO_3)_3$ as a scintillator for NR was published in 1999 [43], highlighting the presence of three nuclei with high neutron capture cross section. Neutron detection properties of another member of double borate family $Li_6Y(BO_3)_3$ [44] and $Li_6Lu(BO_3)_3$ [45] were reported much later.

Characteristics of some borate scintillators considered for NR are summarised in Table 10.2.

Other scintillators for NR that did not contain lanthanide ions in the formula were also explored. Single crystals of $Ca_3(BO_3)_2$ belonging to the trigonal structure with R-3c space group [50] had been grown. Spectroscopic properties of Nd^{3+} [51,52], Dy^{3+} [53] and Er^{3+} [54] doped $Ca_3(BO_3)_2$ had been reported for laser application. $Ca_3(BO_3)_2$ doped with cerium activator exhibits desired properties such as fast decay, high light yield, low sensitivity for gamma-ray backgrounds, which are useful for NR

TABLE 10.2
Borate scintillators for neutron radiography

Sr. no.	Phosphor	Emn (nm)	Decay (ns)	Sensitivity photons/n	R.I	Ref.	Remark
1	$Ca_3(BO_3)_2$:Ce^{3+}	400	36	700		[46]	Low cost
2	$Ca_3(BO_3)_2$:Pr^{3+}	265,310	17	1,000		[47]	Low cost
3	$Sr_3Y(BO_3)_3$:Pr^{3+}	608	17, 28,000		1.7–1.8	[48]	
4	$Li_6Y(BO_3)_3$:Ce^{3+}		38			[44]	Includes two boron capturing nuclei
5	$Li_6Gd(BO_3)_3$:Ce^{3+6} $Li_6{}^{160}Gd({}^{10}BO_3)_3$	395	200	52057,000	1.65	[43]	Includes three boron capturing nuclei
6	$Li_6Lu(BO_3)_3$:Ce^{3+}	390	18			[45]	Includes two boron capturing nuclei
7	$Li_2B_4O_7$			160		[49]	Low Z

[46]. Later, adequate properties for NR were also reported in Pr^{3+}-doped [47] CBO. A large family of double borates with hexagonal structure, $M_3Ln(BO_3)_3$ (M = Ba, Sr and Ln = La Lu, Y, Sc) were studied as potential laser host. Among these, $Sr_3Y(BO_3)_3$ exhibited radioluminescence under α rays [48]. $YCa_4O(BO_3)_3$ is a double oxoborate that had been explored as a neutron scintillator [55]. Initially it was studied as laser host [56] and piezoelectric [57] and non-linear optical material [58]. Subsequently, YCOB activated with rare earth ions Ce^{3+}, Tb^{3+} and Eu^{3+} and ns^2 ions Pb^{2+} have been explored for scintillator applications [59,60]. YCOB has low density (3.269 g/cm^3) and hence low absorption coefficient for background gamma rays. This is an advantageous feature for NR applications.

Some borate phosphors that were actively considered for NR are discussed in details in the following sections.

10.3.1 DOUBLE ORTHOBORATES

Double orthoborates of alkali and lanthanides have been studied as far back as 1977 [61,62]. Kbala et al. studied ionic conductivity in double borates involving Li [63]. Chaminade et al. have described the synthesis in details [64]. Luminescence of rare earth activators like Nd^{3+} [65], Gd^{3+} [66], Ce^{3+} [67,68], Tb^{3+} [69,70], Eu^{3+} [71], Er^{3+} [72], Yb^{3+} [73], Tm^{3+} [74], ns^2 Pb^{2+} [75] and Bi^{3+} [76] were studied in the subsequent years. Double borates of Na have also been synthesized and luminescence of lanthanide activators studied [77]. However, neutron scintillation activity has been studied only in lithium compounds.

Among all the double orthoborates studied, $Li_6Gd(BO_3)_3$ is the most promising. In the ternary system Gd_2O_3–Li_2O–B_2O_3 there are two ternary compounds: $LiGd_6O_5(BO_3)_3$ and $Li_6Gd(BO_3)_3$ [78]. The former has not received much attention. $Li_6Gd(BO_3)_3$ belongs to space group $P2_1/c$ with crystal lattice parameters a = 7.2277 Å, b = 16.5057 Å, c = 6.6933 Å, β = 105.37. The three-dimensional network of the compound $Li_6Gd(BO_3)_3$ is built up from isolated BO triangle planes, distorted GdO_8 polyhedra and distorted LiO_4, LiO_5 polyhedra. The gadolinium atoms occupy the centers of distorted GdO_8 triangulated dodecahedra. The lithium atoms Li(1), Li(3), Li(2) and Li(4), Li(5), Li(6) are fourfold and fivefold coordinated, respectively. The tri-angulated dodecahedra connect each other by common edges to form infinite zig-zag chains parallel to the c axis giving rise to a chain with Gd–Gd distance of 3.84 Å. The shortest distance between two Gd ions of parallel chains is equal to 7.15 Å [64]. So the Gd–Gd interactions along the chains should be predominant over the interchain ones, leading to a monodimensional character. These chains form infinite layers along the direction parallel to the (100) plane, which are separated by lithium ions along the axis (Figure 10.6a and b). Borate groups share edges or vertices with the gadolinium and lithium polyhedra.

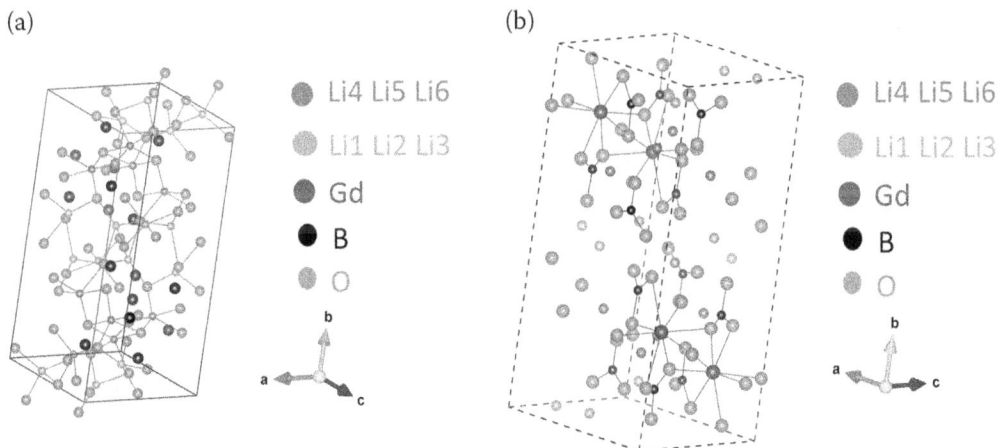

FIGURE 10.6   (a) Unit cell of $Li_6Gd(BO_3)_3$ showing coordination of Lithium at various sites. (b) Unit cell of $Li_6Gd(BO_3)_3$ showing coordination of Gd and B.

Like a typical borate compound, $Li_6Gd(BO_3)_3$ enjoys several features such as high transparency in the UV and high radiation resistance. The presence of three thermal neutron absorbing nuclei offer various channels for neutron detection. For ^{10}B [n,α] ^7Li reaction with the energy of 2.3 MeV and cross section of 3,840 barn, light yield is high (six times higher than ^6Li glass). Crystal growth conditions are easy to achieve. Congruent melting behaviour with very moderate melting point (range from 835 to 860 C) and a high supercooling (200 C), facilitates Czochralski growth of $Li_6Gd(BO_3)_3$ single crystals [79]. The zone melting method can also be applied to obtain $Li_6Gd(BO_3)_3$ single crystals [66].

Czirr et al. [43] were among the earliest workers who calculated and measured neutron detection properties of $Li_6Gd(BO_3)_3$. A great potential of the host for neutron imaging was recognized. The isotopic compositions of all three neutron absorbing nuclei can be adjusted to optimize for specific neutron energy range. For thermal neutrons, Gd that is depleted in the 155 and 157 isotopes, B depleted in ^{10}B and Li enriched in ^6Li can yield a scintillator with high efficiency and a high sensitivity (photons/neutron absorbed). Early results on imaging done with crystals grown from natural isotopic composition in an organic binder, in combination with X-ray film, indicated "that an optimized scintillator will be 10–20 times faster than the direct method".

By way of understanding the mechanism of scintillation in this material, the experiments on the creation and dynamics of short-lived radiation-induced defects on the cation sublattice were conducted by Ogorodnikov et al. [80]. They used time-resolved luminescence and optical spectroscopy under a nanosecond time-range electron beam excitation. Analysis of these results could identify the underlying mechanism of the creation and relaxation of short-lived Frenkel defect pairs on the LGBO cation sublattice. On the basis of these experiments, a conclusion could be reached "that the observed transient optical absorption in LGBO originates in photo induced hole transfer between 2p states of neighboring anions surrounding the lithium vacancy and transient optical absorption decay kinetics is mediated primarily by tunneling electron transfer between the intrinsic lattice defects, namely, the electronic Li^0 and the hole O^- centers".

Apart from $Li_6Gd(BO_3)_3$, $Li_6Y(BO_3)_3$ [44] and $Li_6Lu(BO_3)_3$ have also been evaluated for the neutron imaging properties. Compared to $Li_6Gd(BO_3)_3$ crystal, the light output of $Li_6Y(BO_3)_3$ crystal is lower. While growing $Li_6Gd(BO_3)_3$ crystal using Czochralski technique, there are some difficulties arising from "the instability of growth interface, in which the solid-liquid interface inverses during the growth process stopping the growth". There no such problems with $Li_6Y(BO_3)_3$, but it has lower sensitivity. To overcome these problems solid solutions $Li_6Gd_{1-z}Y_z(BO_3)_3$ have been studied. $Li_6Gd_{0.1}Y_{0.9}(BO_3)_3$:Ce crystal had the best light output, which is 21.8% of BGO crystal and it may be the most promising crystals for thermal neutron detection. $Li_6Lu(BO_3)_3$ has the advantages of short decay time (18 ns) as compared to $Li_6Gd(BO_3)_3$ crystal (200 ns, 800 ns) [81]. Again, its sensitivity is lower. For optimizing $Li_6Gd_{1-z}Lu_z(BO_3)_3$ system had been explored [82]. However, neutron sensitivities were found to be too low.

10.3.2 Borate Imaging Plates Using PSL

In 1983, the imaging plate has been introduced as a two-dimensional position-sensitive detector for X-rays in the field of medical diagnostics [83]. "The great advantages of the IP compared to conventional position sensitive detector are a) the wide dynamic range, b) the high spatial resolution, c) the large detection area, d) the high detective efficiency, and e) reusability." Consequently, the imaging plate for neutron detection has been developed. By modifying the design of X-ray imaging PSL plate, prototype imaging plates for thermal neutron detectors in neutron diffraction [84] and neutron radiography experiments [85] were successfully employed. Niimura et al. developed one of the earliest neutron imaging plate that utilized photostimulated luminescence [86]. The elementary processes occurring in the NIP are as follows:

1. Neutrons are captured by a converter element (like Gd or Li) and the secondary charged particles are emitted.

2. The secondary particles absorbed by PSL materials create colour centers. The spatial variation of these trapped charge carriers (F centers in the case of electrons) represents a latent image of the object of interest.

3. The image is read out by stimulating electron-hole recombination with a scanned red or green laser where the recombination energy is transferred to a Eu^{2+} ion, and appears as photostimulated luminescence. This process includes many complicated processes such as scattering and/or absorption of the incident light, secondary excitation of colour centers, Eu^{2+} emission and so on. The main part of the transmission efficiency is the absorption of the incident laser light and emitted PSL light.

PSL material in the prototype was BaFBr:Eu storage phosphor used for X–ray imaging plate and the converter material was Gd_2O_3 powder in a binder.

A detailed quantitative analysis of these steps had been provided subsequently [87].

Barium is a high Z material, and as such is sensitive to gamma rays as well. In situations where thermal neutrons are to be detected against gamma background, low Z materials are needed. Moreover, if the phosphor itself contains Li and/or B, there will be no need for the converter. Hence, some borates were considered as phosphors for PSL-based NIP.

Crystal structure of haloborates is orthorhombic with space group P_{nn2} or $P4_22_12$. In this structure two different sites are available for the divalent cation (Figure 10.7). On both sites, the cation is coordinated by six oxygen ions at an average distance of about 2.7 Å and by two halide ions at about 3.0 Å. The oxygen ions belong to borate groups [88]. The site symmetry for the divalent cation is C_1. Knittel et al. [89] considered haloborates $M_2B_5O_9X:Eu^{2+}$ (M = Ca, Sr, Ba; X = Cl, Br) for this purpose and compared the performance with the commercial plate based on the BaFBr:Eu^{2+} phosphor.

Their analysis showed that when highly enriched ^{10}B is used, the values for the effective absorption of both plate types do not differ much. This is partly due to the fact that more than 30% of the neutrons absorbed by ^{155}Gd or ^{157}Gd do not yield any energy deposition in the BaFBr:Eu^{2+} powder grains. If a neutron does induce energy deposition, it typically produces 45 keV whereas the energy deposited in a haloborate material is about 2.1 MeV. The essential difference between a IP-Gd and the IP-B is the variance in the number of photoelectrons vN_{phe}. N_{phe} Being the average number of photoelectrons created per incident neutron. They observed [90] highest neutron sensitivity for $Sr_2B_5O_9X:Eu^{2+}$. Fading of PSL signal for this phosphor is much less compared to that of BaFBr:Eu^{2+}. These results were obtained on materials with natural abundance of boron isotopes. Haloborates with enriched ^{10}B are expected to show much improved performance.

Sakasai et al. developed $SrBPO_5:Eu^{2+}$ phosphate borate material for this purpose [91]. X-ray imaging properties of this phosphor were already known [92]. This phosphor also shows the PSL emission at 390 nm like BaFBr:Eu^{2+}.

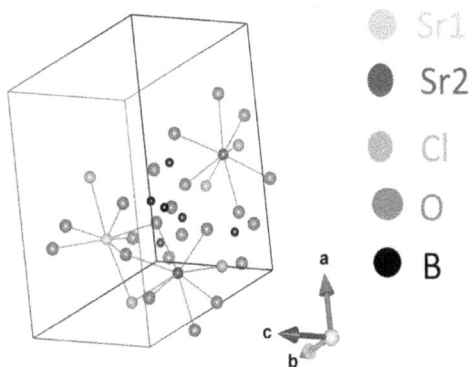

FIGURE 10.7 Part of a unit cell of $Sr_2B_5O_9Cl$ showing coordination of Sr at various sites.

In the $SrBPO_5$ lattice, "Eu^{2+} occupies an irregular nine coordinated large polyhedron occurring in tortuous vertical columns formed by $(BO_4)^{3-}$ and $(PO_4)^{3-}$ tetrahedra with columns interconnected by a 3_1 screw axis" The unit cell of $SrBPO_5$ is shown in Figure 10.8.

This was itself sensitive to neutrons for PSL production without adding any neutron sensitive materials such as Gd. The PL lifetime is 670 ns [93]. Though the neutron sensitivity was less compared to the commercial NIP, gamma discrimination was 10 times better. Sakasai et al. later developed $SrBPO_5:Ce^{3+}$ which had a faster response [20]. The PSL emission of this phosphor is at 340 nm and PL lifetime 23.5 ns. Sakasai et al. also studied $CaBPO_5:Ce^{3+}$, which has a lower effective atomic number and thus expected to provide better gamma discrimination [94]. In this lattice, the PSL emission is at 340 nm. However, the performance of this phosphor was inferior to that of $SrBPO_5:Ce^{3+}$.

All the borate phosphors considered have better gamma discrimination compared to NIP based on $BaFBr:Eu^{2+}$. However, NIP based on lithium borate can be expected to have ultimate properties as it is low Z ($Z_{eff} = 7.4$). Scintillation properties of lithium metaborate Li_2BO_2 and tetraborate $Li_2B_4O_7$ (LBO) were already known [95]. These materials were first used as boron-containing additives in polystyrene luminescent composition for thermal neutron detection [96]. The scintillation compositions, however, were heterogeneous and showed a low transparency for the intrinsic emission in the UV region because those compounds were low soluble both in the initial monomer and in polystyrene. The transparency region of LBO crystals is within the range from 160 to 3,200 nm. These crystals exhibit photoluminescence (PL) with a maximum at 330–340 nm when being excited by vacuum UV at 133 nm. The radioluminescence (RL) spectrum of LBO crystals is identical to the PL one and is characterized by three decay components with $\tau_1 < 2ns$, $\tau_2 \sim 5$ ns and $\tau_3 = 50\sim100$ µs; it is just the process with τ_3 that defines the main part of the light yield. The light yield for undoped lithium tetraborate single crystals is about 100 photons/MeV. The doping with copper (0.08 mol%) results in appearance of a maximum about 370 nm and a longer decay component (perhaps in the millisecond range). The light yield of doped LBO crystals increases by a factor of about 5.

Lithium tetraborate ($Li_2B_4O_7$) is a congruently melting compound with low melting point and small density (melting point: 916 C, $\rho = 2{:}45$ g/cm^3) the single crystals of which are usually grown in bulk form by Czochralski and Bridgman methods. It has tetragonal structure with space group $I4_1 c d$. Li has coordination 5 and linked to BO_4 and BO_3 groups (Figure 10.9).

With the presence of ^6Li and ^{10}B in the crystal, efficient neutron capture is expected, on the contrary, low efficiency for (unwanted) gamma ray background detection can be expected due to low material density and low effective atomic number (Zeff. = 7.3) [97].

$Li_2B_4O_7$ had been studied as a thermoluminescence dosimetry material for a long time. In fact, it is one of the earliest TLD materials [98]. Scintillation in $Li_2B_4O_7$ had also been known. Both Cu-doped $Li_2B_4O_7$ [99,100] and non-doped $Li_2B_4O_7$ [101] were reported as the scintillator. However,

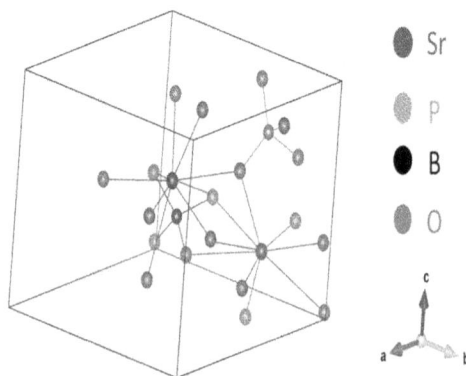

FIGURE 10.8 Part of a unit cell of $SrBPO_5$ showing coordination of Sr.

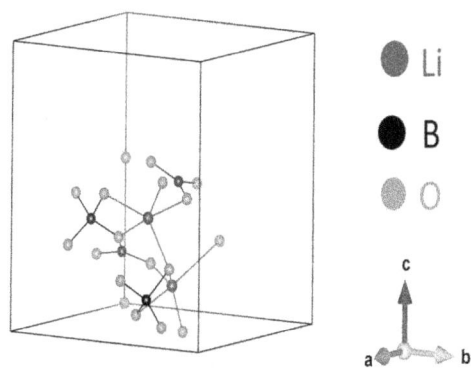

FIGURE 10.9 Unit cell of $Li_2B_4O_7$ showing coordination of Li and B.

OSL in this material was reported as late as 2012 [102]. In recent years, OSL has been reported for Ag- [103] and Cu [104]-doped $Li_2B_4O_7$ single crystal. Much higher OSL signal was obtained for Ag^+ doping. However, Ag^+ emission is in the deep UV region at about 267 nm. The mechanism of OSL in these phosphors has been discussed in detail by Kananen et al. [105]. It is also opined that the ^6Li and ^{10}B nuclei have large cross sections for thermal neutron absorption. Enrichment with either or both of these isotopes may allow $Li_2B_4O_7$ to be used as a neutron detector. Neutron-sensitive calcium aluminate glasses were reported by Kawamura et al. [106]; however, they did not investigate PSL properties of these glasses.

MgB_4O_7:Ce, Li is a low Z material with $Z_{eff} = 8.1$. It can be prepared as a crystal or glass. In crystalline form, MgB_4O_7 crystallizes in the orthorhombic Pbca space group. The structure is three dimensional. Mg^{2+} is bonded in a five-coordinate geometry to five O atoms. There is a spread of Mg–O bond distances ranging from 2.00 to 2.32 Å (Figure 10.10).

There are four inequivalent B^{3+} sites. In the first B^{3+} site, B^{3+} is bonded to four O atoms to form corner-sharing BO_4 tetrahedra. There is a spread of B–O bond distances ranging from 1.45–1.51 Å. In the second B^{3+} site, B^{3+} is bonded in a trigonal planar geometry to three O atoms. There is a spread of B–O bond distances ranging from 1.35–1.40 Å. In the third B^{3+} site, B^{3+} is bonded to four O atoms to form corner-sharing BO_4 tetrahedra. There is a spread of B–O bond distances ranging from 1.44–1.56 Å. In the fourth B^{3+} site, B^{3+} is bonded in a trigonal planar geometry to three O2- atoms. There is one shorter (1.36 Å) and two longer (1.39 Å) B–O bond lengths.

Luminescence properties of this phosphor relevant to OSL were studied by Souza et al. [107]. The reasons for choosing this material were explained. The phosphors present some particularities, such as a light sensitive trapped charge population and a luminescence emission at ultraviolet (UV) spectral range for a better discrimination between stimulation light (generally blue or green) and

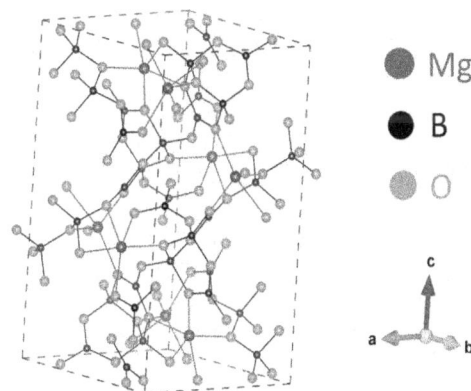

FIGURE 10.10 Unit cell of $Li_2B_4O_7$ showing coordination of Mg and B.

OSL signal. Among all different rare earth elements, cerium is the most promising dopant, because Ce^{3+} is known to have a short luminescence lifetime (10^{-7}–10^{-8} s) and an emission band around 370 nm. Moreover, the use of lithium as a co-dopant effectively increases the OSL sensitivity of MgB_4O_7:Ce, acting as a charge compensator in the substitution of divalent ions (e.g. Mg^{2+}) by trivalent lanthanide (Ce^{3+}). Yukihara et al. calculated the n/Υ ratio for this phosphor as 10 as compared to 4 for OSLD phosphor Al_2O_3:C [108]. Souza et al. optimized the synthesis procedure, and described it as a potential phosphor in OSL applications [109–111]. The PSL lifetime was 2 µs.

10.3.3 LITHIUM BORATE-BARIUM CHLORIDE GLASS-CERAMIC

The commercial NIP is based on Gd_2O_3 as neutron converter and BaFBr:Eu^{2+} as a phosphor. The presence of high Z materials results in gamma sensitivity and poor n/gamma discrimination. To overcome this problem, Gd_2O_3 as was replaced by borosilicate glass, 56.7 B_2O_3–25.8Li_2O–2.5 LiF –15 BaCl_2:Eu^{2+}. BaCl_2:Eu^{2+} acts as a phosphor, while the borosilicate glass as the neutron converter [112]. The PSL arises from orthorhombic $BaCl_2$ nanocrystallites in the glass ceramics, which also contain a PSL-inactive phase of cubic $BaCl_2$. With the addition of LiF, PSL intensity increases more than tenfold [113]. The mechanism for thermal neutron detection is that when irradiated with thermal neutrons, these materials will make use of the high reaction cross sections for the $^{10}B(n,\alpha)^7Li$ or $^6Li(n,\alpha)^3H$ reactions, and the secondary a-particle energy will penetrate to the storage phosphor nanocrystallites and result in electron–hole pair generation in the same way as for X-ray imaging. A glass ceramic IP is expected to possess all the advantages that the Fuji IP has over the other technologies, with the added advantages of transparency for improved spatial resolution, a lower effective atomic number for reduced gamma sensitivity and the ability to be molded into arbitrary shapes, such as optical fibre sensors. ^6Li, ^{10}B-enriched glass increased the neutron sensitivity further [114].

A problem associated with using such plates is related to the Eu content in the glass ceramics. It results in neutron-induced radioactivity with a half-life of 550 min. Thus, the glass ceramics remain significantly radioactive for 2–3 days following neutron irradiation. A possible way of overcoming this problem is to replace Eu^{2+} with Ce^{3+}. Ce^{3+} is also known to exhibit PSL in BaCl_2:Ce^{3+} crystals. It is noteworthy that the neutron capture cross section of Ce ($\sigma = 0.63$ b) is orders of magnitude smaller than that of Eu ($\sigma = 4,600$ b).

10.3.4 OTHER BORATE GLASSES

"Borate glasses are indeed characterized by low melting temperature, extensive glass formation range, high transparency, high thermal and radiation stability. Their drawbacks are the poor chemical stability due to the hygroscopic properties of boron oxide, and high phonon energies" [115].

PSL of Ce^{3+} in alkali borate glasses of the composition $25R_2O.75B_2O_3.0.5CeO_2$ (R = Li, Na or K) had been studied earlier[116]. Maximum PSL was observed for K glasses. However, neutron imaging using these phosphors was not attempted. Valenca et al. [117] observed OSL in lithium-potassium mixed borosilcate glasses, as well as alkali borosilicate-MgO/CaO glasses [118], apparently even without using activator and found these glasses useful for dosimetry. On the other hand, Sroda et al. [119] found that Ce doping improved OSL sensitivity of barium borate glasses. The enhancement was attributed to the traps provided by Ce^{4+} which reduced to trivalent state by electron capture. However, no efforts to make NIP were made in these experiments.

A glass with composition $xSnO-(25-x)SrO-75B_2O_3$ exhibited PSL emission at 390 nm with stimulation in green region. Imaging using ionizing radiations like β and Υ was also done [120]. However, no neutron imaging was reported for this phosphor.

Souza et al. [121] used the borate glass composition ($80MgB_2O_4$–$20MgB_4O_7$). On the other hand, the recipe by Kitagawa et al. [122] uses $25MgO$–$72B_2O_3$–$3Li_2O$–$0.3Ce^{3+}$. Higher B_2O_3 is for making up the evaporation losses. They also proposed the mechanism of OSL in this phosphor.

Barrera et al. [123] studied OSL in compositions $4.5PbO-1.9H_3BO_3$ and lead aluminum borate glass ($11.1\ PbO-2.9H_3BO_3-0.2Al_2O_3$). β radiations were used for creating OSL traps. However, detailed comparison with standard OSL phosphors is lacking. No efforts have been made to prepare an imaging plate.

Thus, we find that ease of production, relatively low cost, suitability to host luminescent activators, tissue equivalence that can lead to gamma discrimination and the possibility to use ^{10}B and thus detect slow neutrons, are some of the advantages of borate glasses for neutron imaging purposes. However, these features are not yet fully exploited.

10.4 NEUTRON RADIOGRAPHY APPLICATIONS

Neutron radiography is a non-destructive testing (NDT) technique similar to those using X-ray and gamma ray. The neutron radiography technique is being used extensively for the design and the inspection of reactor fuels for reactors that are used for research and power generation purposes. It is also being used for non-destructive examination of nuclear fuel and materials, aerospace components, explosives and other industrial products. Apart from applications in industrial sectors, the technique is widely used for research and development activities in biological systems. Most of the neutron radiographic work published is based on thermal neutron radiography however, few applications of high energy neutron radiography have also been reported [124].

10.4.1 NUCLEAR FUEL INSPECTION

The most sought-after application of NR is for inspection of nuclear fuels, mostly uranium oxide pellets. Correct positioning of the fuel within the reactor is critical for its functioning. This can be verified using NR. Hydriding of UO_2 fuel can be highly detrimental for the reactor functioning. This can be conveniently detected by NR. Hydrogen concentrations as low as 150 ppm can be detected by neutron radiography.

A rogue pellet of higher plutonium content in a fast reactor fuel pin would result in local overheating and possible clad failure during irradiation. This can be easily detected by NR.

Presence of elements like Gd, Cd, B, Li in a nuclear fuel can interrupt fission by absorbing neutron. NR can detect such elements within a specific pellet which can then be removed.

10.4.2 BIOIMAGING

In 1985, the usefulness of the imaging plate was demonstrated in a study of muscle contraction. The high DQE and wide dynamic range of the imaging plate resulted in reduction in exposure time to make possible the imaging in <10 sec. NIP enables to obtained accurate data sets for protein crystallography. This is not possible with X-rays as prolonged exposures damage proteins. NIP is also useful for computed radiography. Neutron radiography has also shown particular benefit over X-ray images for the delineation of pathologic intraosseous lesions. Fast neutron radiographic technique using cellulose nitrate foils can be used to see a malignant tumour of the columna vertebralis that is otherwise not possible with other roentgenological examination [125]. Gallstones were subjected to thermal neutron and X-ray radiographic examination and useful information were obtained concerning the formation of gallstones. The use of NR in medical radiography has been frequently reviewed [126–128].

10.4.3 DENTISTRY AND ORAL SURGERY

Neutron radiography has the potential for detecting the pathological changes within bone such as replacement of bone marrow and soft tissue formation inside bone. Tooth is a sort of bone, and hence the application of NT in dentistry and oral surgery. NR can be used for the inspection of materials used in dentistry.

In vivo study of the pulp is difficult and unreliable [129]. The pulp lies within a highly calcified dentinal vault and is not accessible to direct or indirect observation unless some portion of the tooth is destroyed. The pulp cannot be biopsied without risking its vitality. Electrical stimulation and other indirect tests are of limited value. Only after extraction can meaningful studies be made, and these require involved decalcification and sectioning methods. Conventional x-rays does not, permit direct visualization of the pulp substance unless there is a calcified material, such as a pulp stone, within the inter-radicular space. Thus, when pulpal degeneration occurs and is confined wholly within the tooth, the conventional radiograph is of little value. Substances such as calcium, steel, silver and gold do not attenuate the neutron beam significantly and are relatively radiolucent, while hydrogenous substances are radio-opaque. Hence, the use of NR for this purpose.

In dental tissues, the same effect would be anticipated. The hydrogen-rich pulp would be expected to appear as a radiopaque structure through the relatively radiolucent enamel, dentin and bone.

10.4.4 INSPECTION OF MECHANICAL PARTS

Prof. Hartmut Kallman was the first to propagate the idea of employing neutrons for NDT. His published work in 1948, along with patents coauthored by Kuhn, provides the fundamental algorithms of neutron radiography [130].

Using neutrons of various energies for non-destructive testing of materials or process has been followed for considerable time. NR is superior to more prevalent "X-ray radiography" especially when the probing beam has to pass through considerable thickness of heavy elements and scan for the light elements (H, C, Li, B). In many cases, these two imaging techniques are complementary to each other.

The fact that a neutron beam can move through thick steel has been utilised to find out the details of moving lubricant in bearings of operating engines of automobile as well as jet aircrafts. There is no alternative technique available for this task. The most frequent use of NR has been for inspecting jet engine turbine blades. During the production of these blades, a major fault can occur when a part of the casting liquid escapes the leaching process. NR comes in handy for inspecting thick metal parts in other big machines. NR thus is a useful tool for the aircraft, aerospace and automotive industries, for checking "(i) precision in manufacture of turbine engine blades, (ii) corrosion, (iii) flaws in adhesive bindings, (iv) internal flaws, (v) missing or misplaced O-rings or other components, and (vi) cracks, inclusions, and voids or other types of internal defects in materials, etc".

NR has also been used to reveal the interior of geological [131], archeological [132] or paleontological samples. Schillinger et al. carried out pioneering work in 1996 on the application of neutron tomography for the purpose of non-destructive testing of archaeological specimens [133]. Following this, neutron tomography has been successfully applied to various other fields. The progress was made possible due to vast improvements in digital image recording and processing techniques. Advances in the computational methods for reconstructing 3-D images from high-resolution tomographs in relatively short time, also had made an equal contribution to these developments. "Furthermore, new generation detectors with better signal-to-noise characteristics and faster read-out electronics allowed to overcome some of the limits of conventional neutron radiography and tomography concerning spatial and time resolution" [134].

Other fields where NR has contributed are petrology, archeology, geology, etc. It is important to investigate flowing fluids in porous media. NR is specifically suitable for such studies owing to the high neutron cross section of hydrogen. This enables obtaining images with good contrast for water or organic liquids flowing inside a thick covering.

10.5 LIMITATIONS AND CHALLENGES

Using NR techniques have radiation hazards associated with it. Though neutrons themselves do not interact much and pass through the objects, they can "activate" the matter they pass through, making them radioactive. The period of the induced activities can vary from seconds to days. Rigorous "radiation safety measures" must be taken with NR inspection similar to those followed for other radiation testing. Personnel have to be trained for taking preventive measures and tackle possible emergency situations. These procedures are much more rigorous compared to routine X- and gamma radiography. Radiation hazards of neutron exposures are much more severe and varied as compared to those ensuing from X-rays or gamma rays. Working area and personnel are to be monitored for exposures. The equipment and dosimeters used must be responsive to neutrons. In addition, care must be taken with all components that have been exposed in the neutron beam, including the beam stop, cassettes and the inspection object, all of which may exhibit some level of radioactivity caused by neutron activation. Though the radiation levels encountered during NR are comparatively lower than those experienced during other situations, the users have to be cautioned and educated about these hazards.

There are no phosphors that are directly sensitive to fast neutrons. Embedding phosphors in hydrogenous media like plastics, or using layers of hydrogenous material is the only method for using the existing phosphors for detecting fast neutrons.

10.6   FUTURE PROSPECTS AND SCOPE

It is seen that of the various lithium borates, other metal borates and borate glasses, only a few have been explored as neutron storage phosphor/scintillator, and fewer for NR. On the other hand, NR is applied to more and more areas. There is great scope for developing luminescent materials based on borates for these applications. PSL phosphors containing Li, B e.g. Li_2O, $LiAlO_2$, $LiBF_4$, Sr/CaB_4O_7 Eu/Ce, borate glasses, need to be explored. For scintillation, borate glasses could be very attractive as there is no need to grow single crystals for getting transparent media. Ce^{3+}-activated borates can be prospective fast scintillators. It will also prove fruitful to search for cross-luminescence (CRL) in borate hosts. CRL is characterized with very fast, subnano second decay. PSL depends on creation of traps by neutron irradiation. Halides are most susceptible for such processes. To combine advantages of borates and halides, a lattice that incorporates both B and halogen atoms will have to be explored. Hydrogeneous Li B compounds for fast neutron scintillator/PSL, hydrides like LiH, $LiBH_4$; ammonium compounds like ammonium borate; PSL in Ca/SrB_4O_7 Eu/Ce. Thus, haloborates could be more suitable than borates for obtaining storage phosphors for neutrons.

For developing detectors for fast neutrons, compounds incorporating hydrogen and Li or/and B may be investigated. Some such compounds are LiH, $LiBH_4$, $NaBH_4$, ammonium compounds like ammonium borate, etc.

REFERENCES

[1] Craft, A. E. (2013). Design, construction, and demonstration of a neutron beamline and a neutron imaging facility at a Mark-I TRIGA reactor. http://hdl.handle.net/11124/78115

[2] Bossi, R. H., Iddings F. A., and Wheeler, G. C. (2002). *Nondestructive Testing Handbook, Vol. 4, Radiographic Testing*, American Society for Nondestructive Testing, Inc., USA.

[3] Whittemore, W. L. (1990). Neutron radiography. *Neutron News*, 1(3), 24–29. 10.1080/1044863 9008202041

[4] Berger, H. (1976). Detection systems for neutron radiography: Practical applications of neutron radiography and gaging. *Am. Soc. Testing Mater. STP*, 586, 35–57.

[5] Lehmann, E. H., Frei, G., Kuhne, G., and Boillat, P. (2007). The micro-setup for neutron imaging: A major step forward to improve the spatial resolution. *Nucl. Instrum. Methods Phys. Res. A*, 576, 389–396. 10.1016/j.nima.2007.03.017

[6] Tremsin, A. S., McPhate, J. B., Vallerga, J. V., Siegmund, O. H. W., Feller, W. B., Lehmann, E., Butler, L. G., and Dawson, M. (2011). High-resolution neutron microtomography with noiseless neutron counting detector. *Nucl. Instrum. Methods Phys. Res. A*, 652(1), 400–403. 10.1016/j.nima. 2010.08.009

[7] Salvato, G., Aliotta, F., Finocchiaro, V., Tresoldi, D., Vasi, C. S., and Ponterio, R. C. (2010). An apparatus for measuring the timing properties of scintillators for neutron imaging. *Nucl. Instrum. Methods Phys. Res. A*, 621(1–3), 489–492. 10.1016/j.nima.2010.05.054

[8] Lehmann, E. H., Vontobel, P., Frei,G., and Bronimann, C. (2004). Neutron imaging-detector options and practical results. *Nucl. Instrum. Methods Phys. Res. A*, 531(1), 228–237. 10.1016/j.nima.2004.06.010

[9] Kallmann, H. (1948) Neutron radiography. *Research*, 1(6), 254–260. PMID: 18910022.H.

[10] Thewlis, J. (1956). Neutron radiography. *Br. J. Appl. Phys.*, 7(10), 345–350. 10.1088/0508-3443/7/10/301

[11] Thewlis, J. (1958). *Progress in Nondestructive Testing*: edited by E. G. Stanford and J. H. Fearon, Heywood, London, 1, 113–119. 10.1007/BF00551079

[12] Berger, H. (1965). *Neutron Radiography*, Elsevier Publishing Company, Amsterdam, London, New York

[13] Hawkesworth, M. R., and Walker, J. (1969). Review: Radiography with neutrons. *J. Mater. Sci.*, 4, 817–835. 10.1007/BF00551079

[14] Zboray, R., Adams, R., Morgano, M., and Kis, Z. (2019). Qualification and development of fast neutron imaging scintillator screens. *Nucl. Instrum. Methods Phys. Res. A*, 930, 142–150. 10.1016/j.nima.2019.03.078

[15] Stedman, R. (1960). Scintillator for thermal neutrons using Li_6F and ZnS (Ag). *Rev. Sci. Instrum.*, 31, 1156. 10.1063/1.1716833

[16] van Eijk, C. W. E. (2001). Inorganic-scintillator development. *Nucl. Instrum. Methods Phys. Res. A*, 460, 1–14. 10.1088/0031-9155/47/8/201

[17] Spowart, A. R. (1969). Measurement of the absolute scintillation efficiency of granular and glass neutron scintillators, *Nucl. Instrum. Methods Phys. Res. A*, 75(1), 35–42. 10.1016/0029-554X(69)90644-2

[18] van Eijk, C. W. E., Bessiere, A., and Dorenbos, P. (2004). Inorganic thermal-neutron scintillators. *Nucl. Instrum. Methods Phys. Res. A*, 529(1–3), 260–267. 10.1016/j.nima.2004.04.163

[19] Van Eijk, C. W. E. (2012). Inorganic scintillators for thermal neutron detection. *IEEE Trans. Nuclear Sci.*, 59(5), 2242–2247. 10.1109/TNS.2012.2186154

[20] Sakasai, K., Katagiri, M., Toh, K., Takahashi, H., Nakazawa, M., and Kondo, Y. (2004). Characteristics of $SrBPO_5:Ce^{3+}$ based materials as a neutron storage phosphor. *Nucl. Instrum. Methods Phys. Res. A*, 529(1–3), 378–383. 10.1016/j.nima.2004.05.015

[21] Karasawa, Y., Niimura, N., Tanaka, I., Miyahara, J., Takahashi, K., Saito, H., Tsuruno, A., and Matsubayashi, M. (1995). An imaging plate neutron detector. *Physica B*, 213–214, 978–981. 10.1016/0921-4526(95)00341-6

[22] Thoms, M., Lehmann, M. S., and Wilkinson, C. (1997). The optimization of the neutron sensitivity of image plates. *Nucl. Instrum. Methods Phys. Res. A*, 384(2–3), 457–462. 10.1016/S0168-9002(96)00865-0

[23] Chen, J., Li, Y., Song, G., Yao, D., Yuan, L., and Wang, S. (2006). Growth and characterization of pure $Li_6Gd(BO_3)_3$ single crystals by the modified Bridgman method. *J. Crystal Growth*, 294(2), 411–415. 10.1016/j.jcrysgro.2006.06.011

[24] An, D., Zhang, M., Li, D., Pan, S., Chen, H., Yang, Z., Zhu, Y., Sun, Y., Zhang, H., and Li, Y. (2015). Linear and non-linear optical properties of aluminum borate crystal Al_5BO_9: Experiment and calculation, *J. Mater. Res.*, 30(15), 2319–2326. 10.1557/jmr.2015.204

[25] Sasaki, T., Mori, Y., Yoshimura, M., Yap, Y. K., and Kamimura, T. (2000). Recent development of non-linear optical borate crystals: Key materials for generation of visible and UV light, *Mater. Sci. Eng. R*, 30(1–2), 1–54. 10.1016/S0927-796X(00)00025-5

[26] Katsumata, T., Yoshimura, T., Kanazawa, K., and Aizawa, H. (1994). Growth of lithium borate crystals from the vitreous state. *J. Mater. Res.*, 9(8), 2051–2056. 10.1557/JMR.1994.2051

[27] Chen, C. T., Wu, B. C., Jiang, A. D., and You, G. M. (1985). A new-type ultraviolet SHG crystal beta- BaB_2O_4. *Sci. Sin. Ser. B*, 28(3), 235–243. 10.1360/yb1985-28-3-235.

[28] Chen, C. T., Wu, B. C., Jiang, A. D., You, G. M. Li, R. K., and Lin, S. J. (1989). New non-linear optical crystal: LiB3O5. *J. Opt. Soc. Am. B*, 6(4), 616–621. 10.1364/JOSAB.6.000616

[29] Mori, Y., Kuroda, I., Nakajima, S., Sasaki, T., and Nakai, S. (1995). New non-linear optical crystal: Cesium lithium borate. *Appl. Phys. Lett.*, 67, 1818–1820. 10.1063/1.115413

[30] Wu, B. C., Tang, D. Y., Ye, N., and Chen, C. T. (1996). Linear and non-linear optical properties of the $KBe_2BO_3F_2$ (KBBF) crystal, *Opt. Mater.*, 5(1), 105–109. 10.1016/0925-3467(95)00050-X

[31] Wu, H., Yu, H., Yang, Z., Hou, X., Su, X., Pan, S., Poeppelmeier, K. R., and Rondinelli, J. M. (2013). Designing a deep-ultraviolet non-linear optical material with a large second harmonic generation response, *J. Am. Chem. Soc.* 135(11), 4215–4218. 10.1021/ja400500m

[32] Wu, H., Yu, H., Yang, Z., Hou, X., Su, X., Pan, S., Poeppelmeier, K. R., and Rondinelli, J. M. (2014). $Cs_3Zn_6B_9O_{21}$: A chemically benign member of the KBBF family exhibiting the largest second harmonic generation response, *J. Am. Chem. Soc.*, 136(4), 1264–1267. 10.1021/ja4117389

[33] Zhang, M., Su, X., Pan, S., Zheng, W., Zhang, H., Yang, Z., Zhang, B., Dong, L., Wang, Y., Zhang, F., and Yang, Y. (2014). Linear and non-linear optical properties of $K_3B_6O_{10}Br$ single crystal: Experiment and calculation. *J. Phys. Chem. C*, 118(22), 11849–11856. 10.1021/jp500858q

[34] Hu, Z. G., Higashiyama, T., Yoshimura, M., Yap, Y. K., Mori, Y., and Sasaki, T. (1998). A new non-linear optical borate crystal $K_2Al_2B_2O_7$(KAB). *Jpn. J. Appl. Phys.*, 37, L1093. 10.1143/JJAP. 37.L1093

[35] Liu, L. Liu, C. Wang, X., Hu, Z. G., Li, R. K., and Chen, C. T. (2009). Impact of Fe^{3+} on UV absorption of $K_2Al_2B_2O_7$ crystals. *Solid State Sci.*, 11(4), 841–844. 10.1016/j.solidstatesciences. 2009.01.003

[36] Zhou, Y., Yue, Y., Wang, J., Yang, F., Cheng, X., Cui, D., Peng, Q., Hu, Z., and Xu, Z. (2009). Nonlinear optical properties of $BaAlBO_3F_2$ crystal. *Opt. Express*, 17(22), 20033–20038. 10.1364/ OE.17.020033

[37] Chen, C., Lin, Z. and Wang, Z. (2005). The development of new borate-based UV non-linear optical crystals. *Appl. Phys. B*, 80, 1–25. 10.1007/s00340-004-1645-9

[38] Becker, P. (1998). Borate materials in non-linear optics. *Adv. Mater.*, 10(13), 979–992. 10.1002/ (SICI)1521-4095(199809)10:13<979::AID-ADMA979>3.0.CO;2-N

[39] Zhou, J., Fang, W. H., Rong, C., and Yang, G. Y. (2010). A series of open framework alumino borates templated by transition-metal complexes. *Chem. Eur. J.*, 16(16), 4852–4863. 10.1002/chem.2 00902664

[40] Sokolova, Y. V., Azizov, A. V., Simonov, M. A., Leonyuk, N. I., and Belov, N. V. (1978). The crystal structure of the synthetic ortho-tri-borate $Al_5(BO_3)O_6$. *Dokl. Akad. Nauk SSSR*, 243(3), 655–658.

[41] Gatta, G. D., Lotti, P., Merlini, M., Liermann, H. P., Fisch, M., Rotiroti, N., and Armbruster, T. (2013). High-pressure behaviour and phase stability of Al_5BO_9, a mullite-type ceramic material. *J. Am. Ceram. Soc.*, 96(8), 2583–2592. 10.1111/jace.12411

[42] Shin, Y., Lee, D. W., Hong, J., Kawk, K., and Ok, K. M. (2012). Second harmonic generating properties of polar noncentrosymmetric aluminoborate solid solutions Al5-xGaxBO9 (0.0 < x < 0.5). *Dalton Trans.*, 41, 3233–3238. 10.1039/C2DT11971D

[43] Czirr, J. B., MacGillivray, G. M., MacGillivray, R. R., and Seddon, P. J. (1999). Performance and characteristics of a new scintillator. *Nucl. Instrum. Methods Phys. Res. A*, 424 15–19. 10.1016/ S0168-9002(98)01295-9

[44] Fu, Z., Pan, S., Yang, F., Gu, S., Lei, X., Heng, Y., Ren, G., and Qi, M. (2015). Neutron detection properties of $Li_6Y(BO_3)_3$:Ce crystal. *Radiat. Meas.*, 72, 39–43. 10.1016/j.radmeas.2014.11.010

[45] Singh, A. K., Tyagi, M., Desai, D. G., Singh, S. G., Sen, S., and Gadkari, S. C. (2018). A comparative study of $Li_6R(BO_3)_3$; R = Gd, Lu, & Y), single crystals for thermal neutron detection. *Phys. Status Solidi A*, 215, 1800224. 10.1002/pssa.201800224

[46] Fujimoto, Y., Yanagida, T., Tanaka, H., Yokota, Y., Kawaguti, N., Fukuda, K., Totsuka, D., Watanabe, K., Yamazaki, A., and Yoshikawa, A. (2011). Growth and characterization of Ce-doped $Ca_3(BO_3)_2$ crystals for neutron scintillator. *J. Crystal Growth*, 318, 784–787. 10.1016/j.jcrysgro.2010.10.189

[47] Fujimoto, Y., Yanagida, T., Tanaka, H., Yokota, Y., Kawaguti, N., Fukuda, K., Totsuka, D., Watanabe, K., Yamazaki, A., and Yoshikawa, A. (2011). Optical and scintillation properties of Pr^{3+} doped $Ca_3(BO_3)_2$ single crystals. *Phys. Status Solidi B*, 248(2), 444–447. 10.1002/pssb.201000635

[48] Fujimoto, Y., Yokota, Y., Yanagida, T., Pejchal, J., Konno, H., Sugiyama, K., Nikl, M., and Yoshikawa, A. (2010). Crystal growth and characterization of $Sr_3Y(BO_3)_3$. *IEEE Trans. Nucl. Sci.*, 57(3), 1264–1267. 10.1109/TNS.2009.2035695

[49] Kawaguchi, N., Okada, G., Futami, Y., Nakauchi, D., Kato, T., and Yanagida, T. (2020). Scintillation and dosimetric properties of monocrystalline and polycrystalline $Li_2B_4O_7$. *Sens. Mater.*, 32(4), 1419–1426. 10.18494/SAM.2020.2752

[50] Kusachi, I., Henmi, C., & Kobayashi, S. (1995). Takedaite, a new mineral from Fuka, Okayama Prefecture, Japan. *Mineral. Mag.*, 59(396), 549–552. 10.1180/minmag.1995.059.396.1

[51] Lu, X., You, Z., Li, J., Zhu, Z., Jia, G., Wu, B., Tu, C. (2006). Optical transition properties of the Nd^{3+} ions in $Ca_3(BO_3)_2$ crystal. *Phys. Status Solidi A*, 203(3), 551–557. 10.1002/pssa.200521453

[52] Lu, X., You, Z., Li, J., Zhu, Z., Jia, G., Wu, B., and Tu, C. (2005). Growth and properties of pure and rare earth-doped $Ca_3(BO_3)_2$ single crystal. *J. Cryst. Growth*, 281(2-4), 416–425. 10.1016/ j.jcrysgro.2005.04.028

[53] Lu, X., You, Z., Li, J., Zhu, Z., Jia, G., Wu, B., and Tu, C. (2005). Spectroscopy of $Ca_3(BO_3)_2$: Dy^{3+} crystal. *J. Phys. Chem. Solids*, 66, 1801–1805. 10.1016/j.jpcs.2005.08.088

[54] Lu, X., You, Z., Li, J., Zhu, Z.., Jia, G., Wu, B., and Tu, C. (2006). Optical properties of Er^{3+} doped $Ca_3(BO_3)_2$ crystal. *J. Appl. Phys.*, 100,033103. 10.1063/1.2227264

[55] Fujimoto, Y., Yanagida, T., Yokota, Y., Kawaguchi, N., Fukuda, K., Totsuka, D., Watanabe, K., Yamazaki, A., Chani, V., and Yoshikawa, A. (2011). Scintillation and optical properties of Pb-doped YCa_4O $(BO_3)_3$ crystals. *Nucl. Instrum. Methods Phys. Res. A*, 652, 238–241. 10.1016/j.nima.2010.10.046

[56] Mercer, C. J., Tsang, Y. H., Binks, D. J., Zhang, H., and Wang, J. (2009). Q-switched operation of a Nd:YCOB laser. *Opt. Commun.*, 282(1), 97–100. 10.1016/j.optcom.2008.09.076

[57] Shimizu, H., Nishida, T., Takeda, H., and Shiosaki, T. (2009). Dielectric, elastic and piezoelectric properties of $RCa_4O(BO_3)_3$ (R=rare-earth elements) crystals with monoclinic structure of point group m. *J. Cryst. Growth*, 311(3), 916 -920. 10.1016/j.jcrysgro.2008.09.144

[58] ArunKumar, R., and Dhanasekaran, R. (2009). Flux growth of yttrium calcium oxyborate (YCOB) single crystals for non-linear optical applications. *J. Cryst. Growth*, 311(3), 541–543. 10.1016/j.jcrysgro.2008.09.195

[59] Knitel, M. J., Dorenbos, P., van Eijk, C. W. E., Plasteig, B., Viana, B., Kahn-Harari, and Vivien, A. (2000). Photoluminescence, and scintillation/thermoluminescence yields of several Ce^{3+} and Eu^{2+} activated borates. *Nucl. Instrum. Methods Phys. Res. A*, 443(2–3), 364–374. 10.1016/S0168-9002(99)01154-7

[60] Yang, H. C., Li, C. Y., He, H., Tao, Y. Xu, J. H., and Su, Q. (2006). VUV–UV excited luminescent properties of $LnCa_4O(BO_3)_3$:RE3+ (Ln = Y, La, Gd; Re = Eu, Tb, Dy, Ce). *J. Lumin.*, 118(1), 61–69. 10.1016/j.jlumin.2005.06.007

[61] Abdullaev, G. K., and Mamedov, K. S. (1977). Crystal structure of the double lithium ytterbium orthoborate $Li_6Yb(BO_3)_3$. *Sov. Phys. Crystallogr.*, 22(2), 220–222.

[62] Abdullaev, G. K., Mamedov, K. S., Rza-Zade, P. F., Guseinova, S. A., and Dzhafarov, G. G. (1977). Synthesis and structural study of crystals of the double orthoborate of lithium and holmiun. *Russian J. Inorg. Chem.*, 22, 1765–1766.

[63] Kbala, M., Levasseur, A., Fouassier, C., and Hagenmuller, P. (1982). Etude De La Conductivite Ionique De Nouveaux Borates Doubles De Type: $Li_{6-x}Ln_{1-x}T_x(BO_3)_3$. *Solid State Ionic*, 6(2), 191–194. 10.1016/0167-2738(82)90087-X

[64] Chaminade, J. P., Viraphong, O., Guillen, F., Fouassier, C., and Czirr, B. (2001). Crystal growth and optical properties of new neutron detectors Ce^{3+}: $Li_6R(BO_3)_3$ (R = Gd,Y). *IEEE Trans. Nucl. Sci.*, 48, 1158–1161. 10.1109/23.958742

[65] Mascetti, J., Fouassier, C., and Hagenmuller, P. (1983). Concentration quenching of the Nd emission in alkali rare earth borates. *J. Solid State Chem.*, 50(2), 204–212. 10.1016/0022-4596(83)90189-5

[66] Garapon, C. T., Jacquier, B., Chaminade, J. P., and Fouassier, C. (1985). Energy transfer in Li_6Gd $(BO_3)_3$. *J. Lumin.*, 34(4), 211–222. 10.1016/0022-2313(85)90104-8

[67] Pana, S., Zhanga, J., Pana, J., Ren, G., Lic, N., Wuc, Z., and Heng, Y. (2018). Optimized crystal growth and luminescence properties of Ce^{3+} ions doped $Li_6Gd(BO_3)_3$, $Li_6Y(BO_3)_3$ and their mixed crystals. *J. Alloys Compd.*, 751, 129–137. 10.1016/j.jallcom.2018.04.099

[68] Ogorodnikov, I. N., and Pustovarov, V. A. (2013). Luminescence of $Li_6Gd(BO_3)_3$ crystals upon ultraviolet and inner-shell excitations. *J. Lumin.*, 134, 113–125. 10.1016/j.jlumin.2012.09.002

[69] Fawad, U., Kim, H. J., Khan, A., Park, H., and Kim, S. (2015). X-ray and photoluminescence study of $Li_6Gd(BO_3)_3$:Tb^{3+}, Dy^{3+} phosphors. *Sci. Adv. Mater.*, 7 (12), 2536–2544. 10.1166/sam.2015.2436

[70] Zhang, F., Wang, Y., and Tao, Y., (2012). Photoluminescence properties of $Li_6Gd(BO_3)_3$:Tb^{3+} Under VUV/UV excitation. *Phys. Procedia*, 29, 55–61. 10.1016/j.phpro.2012.03.692

[71] Yavetskiy, R., Tolmachev, A., Dubovik, M., Korshikova, T., and Parkhomenko, S. (2007). Growth of $Li_6Gd_{1-x}Y_x(BO_3)_3$:Eu^{3+} crystals for thermoluminescent dosimetry. *Opt. Mater.*, 30(1), 119–121. 10.1016/j.optmat.2006.11.008

[72] Zhao, Y. W., Gong, X. H., Chen, Y. J., Huang, L. X., Lin, Y. F., Zhang, G., Tian, Q. G., Luo, Z. D., and Huang, Y. D. (2007). Spectroscopic properties of Er^{3+} ions in $Li_6Y(BO_3)_3$ crystal. *Appl. Phys. B*, 88, 51–55. 10.1007/s00340-007-2655-1

[73] Delaigue, M., Jubera, V., Sablayrolles, J., Chaminade, J. P., Garcia, A., and Manek Honninger, I. (2007). Mode-locked and Q-switched laser operation of the Yb doped $Li_6Y(BO_3)_3$ crystal. *Appl. Phys. B*, 87, 693–696. 10.1007/s00340-007-2641-7

[74] Ma, X., Li, J., Zhu, Z., You, Z., Wang, Y. and Tu, C. (2008). Thermal and optical properties of Tm^{3+}: $Li_6Gd(BO_3)_3$ crystal: A potential candidate for 1.83 mm lasers. *J. Lumin.*, 128, 1660– 1664. 10.1016/j.jlumin.2008.03.015

[75] Erdogmus, E., and Yildiz, E. (2020). Structural and luminescence properties of Pb^{2+} doped Li_6Gd $(BO_3)_3$ phosphor. *J. Appl. Spectrosc.*, 87(4), 615–620. 10.1007/s10812-020-01044-9

[76] Chen, P., Mo, F., Guan, A., Wang, R., Wang, G., Xia, S., and Zhou, L. (2016). Luminescence and energy transfer of the colour-tunable phosphor $Li_6Gd(BO_3)_3$:Tb^{3+}/Bi^{3+}, Eu^{3+}. *Appl. Radiat. Isotop.*, 108, 148–153. 10.1016/j.apradiso.2015.12.042

[77] Rambabu, U., Annapurna, K., Balaji, T., Satyanarayana, J. V., Rajamohan Reddy, K., and Buddhudu., S. (1996). Fluorescence Spectra of Eu^{3+} and Tb^{3+} doped Na_6 Ln $(BO_3)_3$ (Ln = La, Gd, Y) phosphors. *Spectrosc. Lett.*, 29(5), 833–839. 10.1080/00387019608001614

[78] Zhang, Y., Chen, X. L., Liang, J. K., and Xu, T. (2003). Phase relations of the system $Li_2OGd_2O_3$-B_2O_3 and the structure of a new ternary compound. *J. Alloys Compd.*, 348 (1–2), 314318. 10.1016/S0925-8388(02)00843-5

[79] Shekhovtsov, A. N., Tolmachev, A. V., Dubovik, M. F., Dolzhenkova, E. F., Korshikova, T. L., Grinyov, B. V., Baumer,V. N., and Zelenskaya, O. V. (2002). Structure and growth of pure and Ce^{3+} doped $Li_6Gd(BO_3)_3$ single crystals. *J. Cryst. Growth*, 242(1–2), 167–171. 10.1016/S0022-0248(02) 01137-5

[80] Ogorodnikov, I. N., Poryvay, N. E., Pustovarov, V. A., Tolmachev, A. V., Yavetskiy, R. P., and Yakovlev, V.Yu. (2010). Short-living defects and recombination processes in $Li_6Gd(BO_3)_3$ crystals. *Radiat. Meas.*, 45(3–6), 336–339. 10.1016/j.radmeas.2010.01.039

[81] Yang, F. Pan, S. K., Ding, D. Z., Chen, X. F., Feng, H., and Ren, G. H. (2010). Crystal growth and luminescent properties of the Ce-doped $Li_6Lu(BO_3)_3$. *J. Cryst. Growth*, 312(16–17), 2411–2414. 10.1016/j.jcrysgro.2010.05.007

[82] Fawad, U., Rooh, G., Kim, H. J., Park, H., Kim, S., and Jiang, H. (2014). Crystal growth and scintillation properties of $Li_6Lu_xGd_{1-x}(BO_3)_3$:Ce^{3+} single crystals. *Nucl. Instrum. Methods Phys. Res. A*, 764, 268–272. 10.1016/j.nima.2014.07.040

[83] Sonoda, M., Takano, M., Miyahara, J., and Kato, H. (1983). Computed radiography utilizing scanning laser stimulated luminescence. *RADLAX*, 148(3), 833. 10.1148/radiology.148.3.6878707

[84] Wilkinson, C., Gabriel, A., Lehmann, M., Zemb, T. and Ne, F. (1992). Image plate neutron detector: Proceedings Volume 1737, Neutrons, X Rays, and Gamma Rays. Imaging Detectors, Material Characterization Techniques, and Applications, 1737, 324–329. 10.1117/12.138680

[85] Rausch, C., Bucherl, T., Gahler, R., von Seggern, H., and Winnacker, A. (1992). Proceedings Volume 1737, Neutrons, X Rays, and Gamma Rays: Imaging Detectors, Material Characterization Techniques, and Applications, 1737, 255. 10.1117/12.138666

[86] Niimura, N., Karasawa, Y., Tanaka, I., Miyahara, J., Takahashi, K., Saito, H., Koizumi, S., and Hidaka, M. (1994). An imaging plate neutron detector. *Nucl. Instrum. Methods Phys. Res. A*, 349(2–3), 521–525. 10.1016/0168-9002(94)91220-3

[87] Haga, Y. K., Neriishi, K., Takahashi, K., and Niimura, N., (2002). Optimization of neutron imaging plate. *Nucl. Instrum. Methods Phys. Res. A*, 487(3), 504–510. 10.1016/S0168-9002(01)02204-5

[88] Meijerink, A., and Blasse, G. (1989). Luminescence properties of Eu2+ activated alkaline earth haloborates. *J. Lumin.*, 43(5), 283–289. 10.1016/0022-2313(89)90098-7

[89] Knitel, M. J., Bom,V. R., Dorenbos, P., van Eijk, C. W. E., Berezovskaya, I., and Dotsenko, V. (2000). The feasibility of boron containing phosphors in thermal neutron image plates, in particular the systems $M_2B_5O_9X$:Eu^{2+} (M = Ca, Sr, Ba; X = Cl, Br). Part I: simulation of the energy deposition process. *Nucl. Instrum. Methods Phys. Res. A*, 449(3), 578–594. 10.1016/S0168-9002(99)01474-6

[90] Knitel, M. J., Bom,V. R., Dorenbos, P., van Eijk, C. W. E., Berezovskaya, I., and Dotsenko, V. (2000). Feasibility of boron containing phosphors in thermal neutron image plates, in particular the systems $M_2B_5O_9X$: Eu^{2+} (M = Ca, Sr, Ba; X = Cl, Br). Part II: experimental results. *Nucl. Instrum. Methods Phys. Res. A*, 449(3), 595–601. 10.1016/S0168-9002(99)01475-8

[91] Sakasai, K., Katagiri, M., Toh, K., Takahashi, H., Nakazawa, M., and Kondo, Y. (2003). Characteristics of a $SrBPO_5$:Eu^{2+} material as a neutron storage phosphor. *IEEE Trans. Nucl. Sci.*, 50(4), 788–791. 10.1109/TNS.2003.815309

[92] Karthikeyani, A., and Jagannathan, R. (2000). Eu^{2+} luminescence in stillwellite-type $SrBPO_5$ -a new potential x-ray storage phosphor. *J. Lumin.*, 86(1), 79–85. 10.1016/S0022-2313(99)00171-4

[93] Sakasai, K., Katagiri, M., Toh, K., Takahashi, H., Nakazawa, M., and Kondo, Y. (2001). A $SrBPO_5$:Eu^{2+} storage phosphor for neutron imaging. *Appl. Phys. A*, 74, S1589–S1591. 10.1007/ s003390101255

[94] Sakasai, K., Katagiri, M., Toh, K., Takahashi, H., Nakazawa, M., and Kondo, Y. (2004). Neutron storage characteristics of $CaBPO_5$:Ce^{3+} based phosphors. *IEEE Trans. Nucl. Sci.*, 51(4), 1712–1716. 10.1109/TNS.2004.832287

[95] Chernikov, V. V., Dubovik, M. F., Gavrylyuk, V. P., Grinyov, B. V., Gri, L. A., Korshikova, T. I., Shekhovtsov, A. N., Sysoeva, E. P., Tolmachev, A. V., and Zelenskaya, O. V. (2003). Peculiarities of scintillation parameters of some complex composition borate single crystals. *Nucl. Instrum. Methods Phys. Res. A*, 498(1-3), 424–429. 10.1016/S0168-9002(02)02051-X

[96] Baroni, E. E., Viktorov, D. V., Rozman, I. M., and Shoniya, V. M. (1962). *Nucl. Electron*, 1, 131.

[97] Ishii, M., Kuwano, Y., Asaba, S., Asai, T., Kawamura, M., Senguttuvan, N., Hayashi, T., Koboyashi, M., Nikl, M., Hosoya, S., Sakai, K., Adachi, T., Oku, T., and Shimizu, H. M. (2004). Luminescence of doped lithium tetraborate single crystals and glass. *Radiat. Meas.*, 38(4–6), 571–574. 10.1016/j.radmeas.2004.03.017

[98] Schulman, J. H., Kirk R. D., and West, E. J. (1967). Luminescence dosimetry, U.S. Atomic Energy Commission Symposium Series 8, CONF-650637, 113.

[99] Kobayashi, M., Ishii, M., and Senguttuvan, N. (2015). Scintillation characteristics of undoped and Cu^{2+} doped $Li_2B_4O_7$ single crystals. *Problems of Spectroscopy and Spectrometry*, 34, 71–89. arXiv:1503.03759

[100] Guarneros-Aguilar, C., Cruz-Zaragoza, E., Marcazzo, J., Palomino-Merino, R., and Espinosa, J. E. (2013). Synthesis and TL characterization of $Li_2B_4O_7$ doped with copper and manganese. *AIP Conf. Proc.*, 1544(1), 70–77. 10.1063/1.4813462

[101] Singh, L., Chopra, V., and Lochab, S. P. (2011). Synthesis and characterization of thermo-luminescent $Li_2B_4O_7$ nanophosphor. *J. Lumin.* 131(6), 1177–1183. 10.1016/j.jlumin.2011.02.035

[102] Rawat, N. S., Kulkarni, M. S., Tyagi, M., Ratna, P., Mishra, D. R., Singh, S. G., Tiwari, B., Soni, A., Gadkari, S. C., and Gupta, S. K. (2012). TL and OSL studies on lithium borate single crystals doped with Cu and Ag. *J. Lumin.*, 132(8), 1969–1975. 10.1016/j.jlumin.2012.03.008

[103] Patra, G. D., Singh, S. G., Tiwari, B., Singh, A. K., Desai, D. G., Tyagi, M., Sen, S., and Gadkari. (2016). Optically stimulated luminescence in Ag doped $Li_2B_4O_7$ single crystal and its sensitivity to neutron detection and dosimetry in OSL mode. *Radiat. Meas.*, 88, 14–19. 10.1016/j.radmeas.2016.03.002

[104] Hemam, R., Singh, L. R., Prasad, A. I., Gogoi, P., Kumar, M., Chougaonkar, M. P., Singh, S. D., and, Sharan, R. N. (2016). Critical view on TL/OSL properties of $Li_2B_4O_7$ nanoparticles doped with Cu, Ag and co-doping Cu, Ag: Dose response study: *Radiat. Meas.*, 95,44–54. 10.1016/j.radmeas.2016.11.003

[105] Kananen, B. E., Maniego, E. S., Golden, E. M., Giles, N. C., McClory, J. W., Adamiv, V. T., Burak, Ya.V., and Halliburton, L. E. (2016). Optically stimulated luminescence (OSL) from Agdoped $Li_2B_4O_7$ crystals. *J. Lumin.*, 177, 190–196. 10.1016/j.jlumin.2016.04.032

[106] Kawamura, I., Kawamoto, H., Fujimoto, Y., Koshimizu, M., Okada, G., Koba, Y., Ogawara, R., Suda, M., Takayuki Y., and Asai, K. (2020). Thermoluminescence properties of Dy^{3+} doped $CaO–Al_2O_3–B_2O_3$ glasses for neutron detection. *Nucl. Instrum. Methods Phys. Res. B*, 468, 18–22. 10.1016/j.nimb.2020.02.015

[107] Souza, L. F., Novais, A. L. F., Antonio, P. L., Caldas, L. V. E., and Souza, D. N. (2019). Luminescent properties of MgB_4O_7:Ce,Li to be applied in radiation dosimetry. *Radiat. Phys. Chem.*, 164, 108353. 10.1016/j.radphyschem.2019.108353

[108] Yukihara, E. G., Doull, B. A., Gustafson, T., Oliveira, L. C., Kurt, K., and Milliken, E. D. (2017). Optically stimulated luminescence of MgB_4O_7:Ce,Li for gamma and neutron dosimetry. *J. Lumin.*, 183, 525–532. 10.1016/j.jlumin.2016.12.001

[109] Gustafson, T. D., Milliken, E. D., Jacobsohn, L. G., and Yukihara, E. G. (2019). Progress and challenges towards the development of a new optically stimulated luminescence (OSL) material based on MgB_4O_7:Ce,Li. *J. Lumin.*, 212, 242–249. 10.1016/j.jlumin.2019.04.028

[110] Souza, L. F., de Souza, D. N., Rivera, G. B., Vidal, R. M., and Caldas, L. V. E. (2019). Dosimetric characterization of MgB_4O_7:Ce,Li as an optically stimulated dosimeter for photon beam radio-therapy. *Perspect. Sci.* 12, 100397. 10.1016/j.pisc.2019.100397

[111] Souza, L. F., Silva, A. M. B., Antonio, P. L., Caldas, L. V. E., Souza, S. O., d'Errico, F., and Souza, D. N. (2017). Dosimetric properties of MgB_4O_7:Dy,Li and MgB_4O_7:Ce,Li for optically stimulated luminescence applications. *Radiat. Meas.* 106, 196–199. 10.1016/j.radmeas.2017.02.009

[112] Appleby, G. A., Edgar, A., Williams, G. V. M., and Bos, A. J. J. (2006). Photo-stimulated lumi-nescence from $BaCl_2$:Eu^{2+} nanocrystals in lithium borate glasses following neutron irradiation. *Appl. Phys. Lett.*, 89, 101902. 10.1063/1.2335807

[113] Appleby, G. A., Bartle, C. M., Williams, G. V. M., and Edgar, A. (2006). Lithium borate glass ceramics as thermal neutron imaging plates. *Curr. Appl. Phys.*, 6(3), 389–392. 10.1016/j.cap.2005.11.025

[114] Appleby, G. A., and Vontobel, P. (2008). Optimisation of lithium borate–barium chloride glass ceramic thermal neutron imaging plates. *Nucl. Instrum. Methods Phys. Res. A*, 594, 253–256. 10.101 6/j.nima.2008.06.036

[115] Majerus, O., Tregouet, H., Caurant, D., and Pytalev, D. (2015). Comparative study of the rare earth environment in rare earth metaborate glass (REB_3O_6, RE = La, Nd) and in sodium borate glasses. *J. Non-Cryst. Solids*, 425, 91–102. 10.1016/j.jnoncrysol.2015.05.031

[116] Qiu, J., Shimizugawa, Y., Iwabuchi, Y., and Hirao, K. (1997). Photostimulated luminescence of Ce^{3+} doped alkali borate glasses. *Appl. Phys. Lett.*, 71(1), 43–45. 10.1063/1.119463

[117] Valenca, J. V. B., Silveira, I. S., Silva, A. C. A., Dantas, N. O., Antonio, P. L., Caldas, L. V. E., dErrico, F., and Souza, S. O. (2018). Optically stimulated luminescence of the [20% Li_2CO_3 +x% K_2CO_3+(80-x)% B_2O_3] glass system. *J. Lumin.*, 200, 248–253. 10.1016/ j.jlumin.2018.03.060

[118] Valenca, J. V. B., Silveira, I. S., Silva, A. C. A., Dantas, N. O., Antonio, P. L., Caldas, L. V. E., d'Errico, F., and Souza, S. O. (2017). Optically stimulated luminescence of borate glasses containing magnesia, quicklime, lithium and potassium carbonates. *Radiat. Phys. Chem.*, 140, 83–86. 10.1016/ j.radphyschem.2016.12.017

[119] Sroda, M., Swiontek, S., Wojciech, G., and Bilski, P. (2019). The effect of CeO_2 on the thermal stability, structure and thermoluminescence and optically stimulated luminescence properties of barium borate glass. *J. Non-Cryst. Solids*, 517, 61–69. 10.1016/j.jnoncrysol.2019.03.026

[120] Nanto, H., Nakagawa, R., Takei, Y., Hirasawa, K., Taniguchi, S., Miyamoto, Y., Masai, H., Kurobori, T., and Yanagida, T. (2014). Ionizing radiation sensors utilizing optically stimulated luminescence in SnO-doped $SrO-B_2O_3$ and $ZnO-P_2O_5$ glass. *IEEE SENSORS 2014 Proceedings*, 416–419. 10.1109/ICSENS.2014.6985023

[121] Souza, L. F., Silva, C. R. E., Souza, D. N., and Nogueira, M. S. (2021). Application of optically stimulated luminescence in tandem systems for diagnostic radiology. *Radiat. Phys. Chem.*, 182, 109354. 10.1016/j.radphyschem.2021.109354

[122] Kitagawa, Y., Yukihara, E. G., and Tanabe, S. (2021). Development of Ce^{3+} and Li^+ co-doped magnesium borate glass ceramics for optically stimulated luminescence dosimetry. *J. Lumin.*, 232, 117847. 10.1016/j.jlumin.2020.117847

[123] Barrera, G. R., Souza, L. F., Novais, A. L. F., Caldas, L. V. E., Abreu, C. M., Machado, R., Sussuchi, E. M., and Souza, D. N. (2019). Thermoluminescence and optically stimulated lumines-cence of $PbO–H_3BO_3$ and $PbO–H_3BO_3–Al_2O_3$ glasses. *Radiat. Phys. Chem.*, 155, 150–157. 10.101 6/j.radphyschem.2018.02.005

[124] Richardson, A. E. (1977). Improved images in 14.5 MeV neutron radiography. *Mater. Eval.*, 35(4), 52–58. RN:9374108.

[125] Duehmke, E., and Greim, L. (1978). Fast Neutron Radiography of Extended Biological and Medical Objects. Proceedings of the International Congress on Radiology, Rio de Janeiro, Brazil,23-29 October, 1977, Atomkernenergie (1977). 30, 175–181.

[126] Atkins, H. L. (1965). Biological application of neutron radiography, *Mater. Eval.*, 23, 453–458.

[127] Barton, J. P. (1964). Some possibilities of neutron radiography. *Phys. Med. Biol.*, 9(1), 33.

[128] Tochilin, E., (1965). Photographic detection of fast neutrons: Application to neutron radiography. *Phys. Med. Biol.*, 10(4), 477.

[129] Weisman, M. I., and Brown, M. (1971). Neutron radiography of the dental pulp: A preliminary report. *Oral Surg. Oral Med. Oral Pathol.*, 32(3), 487–492. 10.1016/0030-4220(71)90211-8

[130] Spowart, A. R. (1972). Neutron radiography. *J. Phys. E: Sci. Instrum.*, 5(6), 497–510.

[131] Vontobel, P., Lehmann, E., and Carlson, W. D. (2005). Comparison of X-ray and neutron tomo-graphy investigations of geological materials. *IEEE Trans. Nucl Sci*, 52(1), 338–341. 10.1109/ TNS.2005.843672

[132] Fioria, F., Giunta, G., Hilger, A., Kardjilov N., and Rustichelli, F. (2006). Non-destructive char-acterization of archaeological glasses by neutron tomography, *Physica B*, 385–86, 1206–1208. 10. 1016/j.physb.2006.05.410

[133] Schillinger, B., Gebhard, R., Haas, B., Ludwig, W., Rausch, C., and Wagner, U. (1996). 3D neutron tomography in material testing and archaeology. In: Proceedings of the 5th world conference on neutron radiography, Berlin, 688–693.

[134] Dierick, M., Vlassenbroeck, J., Masschaele, B., Cnudde, V., Hoorebeke, L. Van., and Hillenbach (2005). High-speed neutron tomography of dynamic processes. *Nucl. Instrum. Methods Phys. Res. A*, 542(1–3), 296–301. 10.1016/j.nima.2005.01.152

Index

For Product Safety Concerns and Information please contact our EU
representative GPSR@taylorandfrancis.com
Taylor & Francis Verlag GmbH, Kaufingerstraße 24, 80331 München, Germany

www.ingramcontent.com/pod-product-compliance
Lightning Source LLC
Chambersburg PA
CBHW080929220326
41598CB00034B/5725